Probability and Mathematical Statistics (Continued)

WILKS · Mathematical Statistics
ZACKS · The Theory of Statistical Inference

Applied Probability and Statistics

BAILEY · The Elements of Stochastic Processes with Applications to the Natural Sciences
BAILEY · Mathematics, Statistics and Systems for Health
BARTHOLOMEW · Stochastic Models for Social Processes, *Second Edition*
BECK and ARNOLD · Parameter Estimation in Engineering and Science
BENNETT and FRANKLIN · Statistical Analysis in Chemistry and the Chemical Industry
BHAT · Elements of Applied Stochastic Processes
BLOOMFIELD · Fourier Analysis of Time Series: An Introduction
BOX and DRAPER · Evolutionary Operation: A Statistical Method for Process Improvement
BROWN and HOLLANDER · Statistics: A Biomedical Introduction
BROWNLEE · Statistical Theory and Methodology in Science and Engineering, *Second Edition*
BURY · Statistical Models in Applied Science
CHAMBERS · Computational Methods for Data Analysis
CHATTERJEE and PRICE · Regression Analysis by Example
CHERNOFF and MOSES · Elementary Decision Theory
CHOW · Analysis and Control of Dynamic Economic Systems
CLELLAND, deCANI, BROWN, BURSK, and MURRAY · Basic Statistics with Business Applications, *Second Edition*
COCHRAN · Sampling Techniques, *Third Edition*
COCHRAN and COX · Experimental Designs, *Second Edition*
COX · Planning of Experiments
COX and MILLER · The Theory of Stochastic Processes, *Second Edition*
DANIEL · Application of Statistics to Industrial Experimentation
DANIEL and WOOD · Fitting Equations to Data
DAVID · Order Statistics
DEMING · Sample Design in Business Research
DODGE and ROMIG · Sampling Inspection Tables. *Second Edition*
DRAPER and SMITH · Applied Regression Analysis
DUNN · Basic Statistics: A Primer for the Biomedical Sciences, *Second Edition*
DUNN and CLARK · Applied Statistics: Analysis of Variance and Regression
ELANDT-JOHNSON · Probability Models and Statistical Methods in Genetics
FLEISS · Statistical Methods for Rates and Proportions
GIBBONS, OLKIN and SOBEL · Selecting and Ordering Populations. A New Statistical Methodology
GNANADESIKAN · Methods for Statistical I
tivariate Observations
GOLDBERGER · Econometric Theory
GROSS and CLARK · Survival Distributions
GROSS and HARRIS · Fundamentals of Queu

PARAMETER ESTIMATION
IN ENGINEERING AND SCIENCE

PARAMETER ESTIMATION
IN ENGINEERING AND SCIENCE

JAMES V. BECK

Department of Mechanical Engineering and
Division of Engineering Research
Michigan State University

and

KENNETH J. ARNOLD

Department of Statistics and Probability
Michigan State University

JOHN WILEY & SONS

New York • Chichester • Brisbane • Toronto

Library of Congress Cataloging in Publication Data

Beck, James V 1930-
Parameter estimation in engineering and science.

(Wiley series in probability and mathematical statistics)
Includes bibliographical references and index.
1. Engineering—Statistical methods.
2. Estimation theory. I. Arnold, Kenneth J., 1914- joint author. II. Title.
TA340.B39 620'.001'51954 77-635
ISBN 0-471-06118-2

Printed in the United States of American

10 9 8 7 6 5 4 3

to BARBARA
and PAULINE

PREFACE

Parameter estimation is a modern and exciting discipline that is growing rapidly. One of the indications that it is a young field of study is that the name of the discipline and many of its terms have not yet been agreed upon: other names for parameter estimation are nonlinear estimation, nonlinear regression, optimization of parameters, calibration, model building, identification, and identification of systems.

The objectives of this book are to provide (1) methods for estimating constants (i.e., parameters) appearing in mathematical models, (2) estimates of the accuracy of the estimated parameters, and (3) tools and insights for developing improved mathematical models. The book presents methods that can use all the available statistical information. We would say that the three objectives cover the main problems in parameter estimation. Some of the names mentioned above emphasize certain of them; for example, model building and identification emphasize the third objective.

The book has been used in printed note form for five years and has undergone continual revision and, we hope, improvement. The first author initially wrote all eight chapters, Chapters 2 and 3 later being replaced by those written by the second author. Both authors, however, contributed to all the chapters.

Several different courses can be taught from the book. A beginning course at the undergraduate level could be taken from the first five chapters with the emphasis on Chapter 5, which gives an algebraic treatment of linear parameter estimation. Chapters 2 and 3 provide a concise treatment of the required statistical background so that the student would find the course easier if he had previously taken a probability or statistics course. A graduate course would include Chapters 1, 6, 7, and 8 with the matrix approach being used freely. Ideally, students would have a background in elementary probability, statistics, and matrices for such a course. Because of the first author's background in heat transfer, many of the examples are taken from this field but knowledge of this field or any other particular field in engineering or science is not required.

This book has a number of unique aspects. One of these relates to the care in treatment of statistical assumptions. Eight standard assumptions are identified, and a numbering system is given to provide a convenient means of designating them. These assumptions are listed in Sections 5.1.3 and 6.1.5, with an abbreviated form inside the back cover.

There are many people who have helped in the preparation of this text to whom we express our appreciation. The notes in various revisions were typed by Mrs. Noralee Burkhardt, Ms. Roberta Smith, Miss Gloria Mannino, Mrs. Kathy Winnie, Mrs. Barbara Coolen, Miss Jan Furtaw, Mrs. Constance Geraci, Miss Patti Truax, Mrs. Pat Holewinski, Mrs. Brenda Stott, and Ms. Mary Beth Mac Donell. The encouragement of the editor, Ms. Beatrice Shube, is particularly appreciated. Some of the results given in this book are related to research done by the first author and supported by National Science Foundation Grants GK-16526, GK-240, GK-2075, and GK-41495.

JAMES V. BECK
KENNETH J. ARNOLD

December 1976
East Lansing, Michigan

Contents

CHAPTER 1

INTRODUCTION TO AND SURVEY OF PARAMETER ESTIMATION

1.1 INTRODUCTION

One of the fundamental tasks of engineering and science, and indeed of mankind in general, is the extraction of information from data. Parameter estimation is a discipline that provides tools for the efficient use of data in the estimation of constants appearing in mathematical models and for aiding in modeling of phenomena.

The models may be in the form of algebraic, differential, or integral equations and their associated initial and boundary conditions. An estimated parameter may or may not have a direct physical significance.

Parameter estimation can also be visualized as a study of inverse problems. In the solution of partial differential equations one classically seeks a solution in a domain knowing the boundary and initial conditions and any constants. In the inverse problem not all these constants would be known. Instead discrete measurements of the dependent variable inside the domain must be used to estimate values for these constants, also called parameters.

Parameter estimation is needed in the modern world for the solution of the many diverse problems related to the space program, investigation of the atom, and modeling of the economy. Examples and applications in this book, however, are directed to estimation problems occurring in engineering and science in which partial differential equations as well as ordinary differential and algebraic equations are used to model the phenomena.

Fortunately, simultaneous with the development of increased need of parameter estimation, computers have been built that make parameter estimation practicable for a great array of applications. It should be noted that both digital computational and data acquisition facilities are practical necessities in parameter estimation. Both these facilities have been readily available only since the late 1950s or early 1960s, whereas estimation was first extensively discussed by Legendre in 1806 [1] and Gauss in 1809 [2]. In Gauss's classic paper he claimed usage of the *method of least squares* (still used in parameter estimation) as early as 1795 in connection with the orbit determination of minor planets. For this reason Gauss is recognized as being the first to use this important tool of parameter estimation.

The name "parameter estimation" is not universally used. Other terms are nonlinear least squares, nonlinear estimation, nonlinear regression,* and identification, although the latter sometimes is given a quite different meaning. Estimation is a statistical term and identification is an electrical engineering term.

1.1.1 Parameters, Properties, and States

A mathematical model of a dynamic process usually involves ordinary or partial differential equations. Sometimes the solution of these equations is a relatively simple set of algebraic equations. In any case, there are dependent and independent variables and also certain constants. The dependent variables are sometimes called *state* variables (or signals). The constants may be *parameters*.

In experiments the states are frequently measured directly, but the parameters are not. Approximate values of the parameters are inferred from measurements of the states. Since only approximate parameter values are found, the parameters are said to be *estimated*. This book is primarily concerned with estimating parameters. Other books concentrate on providing "best" predictions of states based on knowledge of independent variables. The two problems are quite similar. When parameters are estimated, state estimates are usually found simultaneously.

A parameter having a physical significance for a solid or fluid might also be termed a *property*. Examples of parameters that might also be properties are characteristics of materials such as density, specific heat, viscosity, thermal conductivity, electrical conductivity, Young's modulus, emittance, and electrical capacitance. A property frequently involves the concept of per unit length, area, or volume. Examples of quantities that are parameters but not properties are weight or mass of a given specimen, electrical

*Parameter estimation is not necessarily nonlinear as implied by some of these terms.

resistance of a section of wire, and drag experienced by a truck moving at a constant velocity.

The concepts of parameters, properties, and states are illustrated by the following examples.

Example 1.1.1

Newton's second law states for a system that $F_x(t) = ma_x(t)$, where $F_x(t)$ is the force in the x direction, $a_x(t)$ is the acceleration in the x direction, and m is mass. Force and acceleration, which can be functions of time t, can be considered to be states, whereas mass is a parameter. Force and acceleration are both often easily measured; usually mass is easily measured separately also, but there can be cases when the mass must be inferred, such as when determining the mass of comets and planets. If a body is homogeneous, the mass is the product of its density and volume. The density is a property. If the volume were known and measurements of F and a_x were available, the parameter to be estimated would be the density, which also happens to be a property of the material.

Example 1.1.2

Ohm's Law states that $E(t) = RI(t)$, where $E(t)$ is voltage, $I(t)$ is current, and R is electric resistance. Voltage and current are states, whereas resistance is a parameter. If the resistance of a wire of known length and diameter is being determined, one could instead estimate the electric resistivity of that type of wire, a parameter which is also a property.

Example 1.1.3

An object is thrown vertically above the earth with an initial velocity of V_0. From the solution of the appropriate differential equation, the distance s of the object above the earth is described by $s = V_0 t - gt^2/2$, where g is the acceleration of gravity. Here s would be the state and g the parameter. The independent variable is time t. V_0 could be a parameter or a state. This illustrates that the parameter and state estimation problems sometimes overlap.

Above, the term parameter was applied to what might be termed "physical" parameters in constrast with statistical parameters. Examples of statistical parameters are variances and correlation coefficients of the measurement errors. Both types of parameters may have to be estimated in some problems; in others only the physical parameters need be found.

1.1.2 Purpose of This Chapter

The purpose of this chapter is to survey some of the basic problems and concepts of parameter estimation covered in this book. By understanding some of the ideas in a simple form, the reader can better comprehend the detailed treatment in later chapters.

Much of parameter estimation can be related to five optimization problems. The first problem is the choice of the best function to extremize. The most common function chosen for minimizing is the sum of squares of deviations. This yields the method of least squares discussed in Section 1.3. The second optimization problem is the minimization of the chosen function, which is also discussed in Section 1.3. These first two optimization problems are usually what is meant by the parameter estimation problem, which is discussed further in Section 1.2.3. The third optimization problem involves optimal design of experiments to obtain the "best" parameter estimates. This is discussed in Section 1.2.4. If several competing mathematical models are known, but the true model is unknown, the (fourth) problem is the optimal design of experiments to discriminate between the models; see Section 1.2.5. (By a "known" model we mean that the mathematical structure of the equation is known even though the values of certain parameters may be unknown.) The fifth and most difficult problem is the determination of mathematical models when there is so little information that a complete, finite list of competing models is not known (Section 1.2.6).

In addition to the basic optimization problems in parameter estimation there are a number of basic concepts. One of these relates to what we call *sensitivity coefficients*. A sensitivity coefficient is formed by taking the first derivative of a dependent variable (i.e., a state variable) with respect to a parameter. This is discussed further in Section 1.4. Sensitivity coefficients are important because they can give information regarding *linearity* and *identifiability*. Linearity is concerned with the dependent variable(s) being linear or nonlinear in the parameters (Section 1.4). There are cases where unique solutions for some parameters do not exist; this relates to identifiability, which is discussed in Sections 1.3 and 1.4.

Another concept, *parsimony*, long a principle in which choices among alternative explanations of physical phenomena have been made, has recently been emphasized by Box [4] in its application to parameter estimation. Box asks that a model have a minimum number of parameters consistent with the physical basis, if there is one.

1.1.3 Related Research

A number of individuals and groups have contributed to the research in the past decade. Some of the statisticians and engineers who have made significant contributions to parameter estimation are G. E. P. Box, N. Draper, J. S. Hunter, M. J. Box, W. G. Hunter, Y. Bard, J. H. Seinfeld, and L. Lapidus. Some books on parameter estimation with a statistical emphasis are given in references 3 through 6.

Another group that has made a large contribution to estimation is the control and systems group of electrical engineers. Most of their work relates to state estimation rather than parameter estimation, which they call identification. Some books on state estimation are by Sage and Melsa [7], Sage [8], Deutsch [9], and Bryson and Ho [10]. Books by Sage and Melsa [11], Graupe [12], and Mendel [13] discuss identification. This group usually is concerned with estimating states and parameters in sets of ordinary differential equations.

Another group is composed of econometricians, that is, economists with a strong interest in statistics. One reference is Kmenta [14]. Econometricians have generally concerned themselves with models which can be approximated adequately by systems of linear algebraic equations.

Other valuable works, not identified with any of the above groups, are references 15–17.

1.1.4 Relation to Analytical Design Theory

Parameter estimation is related to analytical design theory. The concepts of choosing the best cost function and minimizing it are common to both. Because of the similarity, there may be some technology transfer from parameter estimation theory to design theory.

In addition to the design aspects related to cost functions mentioned above, parameter estimation is also concerned with the design of "best" experiments. This involves various ideas including criteria, constraints, and sensitivity. Furthermore, modern design is utilizing statistics to a greater extent than formerly to describe tolerances, life of structures, and so on.

1.2 FUNDAMENTAL PROBLEMS

A number of distinctions between various estimation problems should be understood. Some of these distinctions can be confusing because similar, but not identical problems, are encountered in control engineering.

1.2.1 Deterministic or Classical Problem

In the classical problem one mathematically models a system in a certain domain and seeks to calculate the dependent variable(s) in the domain for a known model and initial and boundary conditions. There is no uncertainity in any of these. Not only is the *structure* of the model known (e.g.,

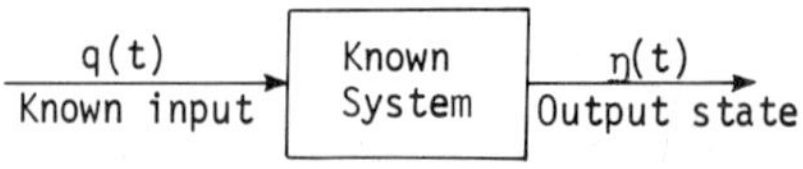

Figure 1.1 Classical problem of known input and system. The problem is to calculate the output state.

an ordinary differential equation of known order) but all the relevant parameters or properties are known. This is illustrated by Fig. 1.1, which might be visualized as the heating of a billet with $q(t)$ being the heat input and $\eta(t)$ the temperature of the billet. The model might be

$$\beta_1 \frac{d\eta}{dt} + \beta_0 \eta = q(t) \tag{1.2.1}$$

$$\eta(0) = B \tag{1.2.2}$$

where (1.2.1) is the differential equation containing the input $q(t)$ and (1.2.2) is the initial condition. All the parameters, β_0, β_1, B, and $q(t)$, are known. The objective is to calculate the *state* $\eta(t)$ for $t > 0$, in other words, to solve the differential equation.

Of all the problems listed here the classical problem is the one engineers are most often trained to solve. It is not the subject of this book.

1.2.2 State Estimation Problem

In the state estimation problem, the state is estimated using *measurements* of the input and the state. See Fig. 1.2. This problem is similar to the classical one in that $\eta(t)$ is needed and the model is known. There are extra complications, however, in that the observed input contains the noise $w(t)$ and that measurements are only available for the output state η corrupted with the noise $\varepsilon(t)$. ("Noise" means nonsystematic measurement errors.) Using the preceding example one could still write (1.2.1) but $q(t)$ would not be precisely known and neither would B in (1.2.2). In the solution of this problem the statistics of $w(t)$ and $\varepsilon(t)$ are usually assumed to be known. One seeks a "best" or optimal estimate $\eta(t)$ of the *true* system state $\eta(t)$. It is in connection with this problem that the term "filter" is used [7–12].

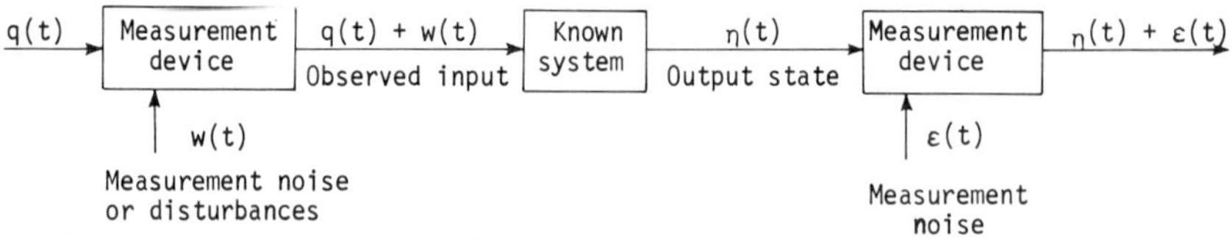

Figure 1.2 State estimation problem. The problem is to estimate $\eta(t)$.

1.2.3 Parameter Estimation Problem

In the parameter estimation problem the structure of the differential equation is known; measurements of the input $q(t)$ as well as the initial condition(s) [B in (1.2.2)] or boundary conditions are available. Some or all of the parameters may be unknown. The problem is to obtain the "best" or optimal estimate of these parameters using the measured values of input and output.

Because measurements invariably contain errors, solution of parameter estimation problems utilize concepts of probability and statistics. The requisite probability background is reviewed in Chapter 2 and the bases of the statistical methods are given in Chapter 3. The reader who has an adequate background in probability and statistics can omit these two chapters.

This estimation problem is illustrated by Fig. 1.3. The "unknown" system is modeled by a differential equation containing unknown parameters. This problem also involves state estimation because $\eta(t)$ is unknown and is usually estimated at the same time as the parameters.

The investigation of the parameter estimation problem is a primary objective of this text. The emphasis is on parameter estimation techniques that are appropriate for analysis of dynamic experiments. Methods useful for estimation involving linear and nonlinear partial differential equations are given particular attention.

1.2.4 Optimum Experiment Problem

The optimum experiment problem can be illustrated using Fig. 1.3. An objective is to adjust any inputs such as $q(t)$ or boundary and initial conditions so as to minimize the effect of errors on estimated values of the parameters. In other words the output $\eta(t)$ would be made as "sensitive" as possible to the parameters. Adjusting $q(t)$ means the selection of the time variation to accomplish the objective. Another objective would be to find the best location for sensors and the best duration for taking measure-

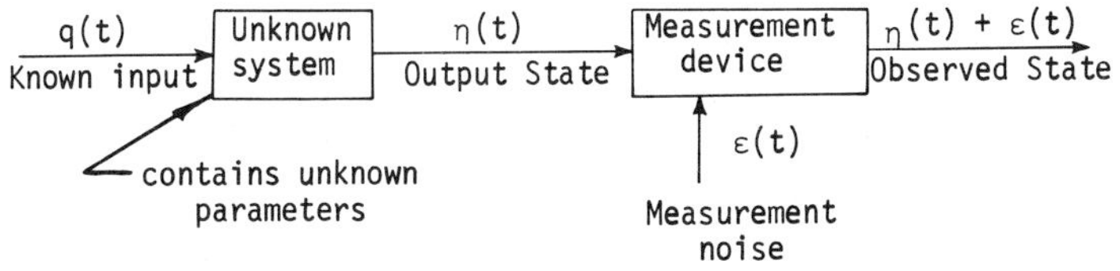

Figure 1.3 Parameter estimation problem. The problem is to estimate certain unknown parameters in the model of the system.

ments. There are certain realistic constraints that must be included when seeking these optimums, such as the maximum allowable experiment duration and maximum temperature rise. Optimum experiments are discussed mainly in Chapter 8.

1.2.5 Discrimination Problem

In the discrimination problem there are two or more possible candidates for the model, one of which is the true model. The objective is to design experiments that will enable one to decide upon the correct model. There are some similarities with the optimum experiment problem. This is also discussed in Chapter 8.

1.2.6 Identification Problem

The identification problem is similar to the parameter estimation problem in that there may be unknown parameters in the model. The problem is much more complex because the structure of the model (e.g., differential equation) is unknown. Developing models is sometimes called model building which is discussed in various connections in Chapters 5 to 8.

1.3 SIMPLE EXAMPLES

Typical problems are outlined in this section for estimation problems involving algebraic, ordinary differential, and partial differential equations. These examples are given to introduce the student to a number of parameter estimation concepts that are amplified in subsequent chapters.

1.3.1 Linear Algebraic Model

Suppose that a number of distinct experiments have been performed for a given material at different temperatures, T, and that at each T a value of the thermal conductivity k has been determined. Hence there are a number of data sets (Y_1, T_1), $(Y_2, T_2), \ldots$, where Y_i is a measured value of k and T_i is the temperature in the ith experiment. The data are shown in Fig. 1.4.

A model must now be proposed for k versus T. If there is any applicable physical law relating k and T it should be used. For this example none is known but the data of Fig. 1.4 suggest a linear relation in T,

$$k = \beta_0 + \beta_1 T \tag{1.3.1}$$

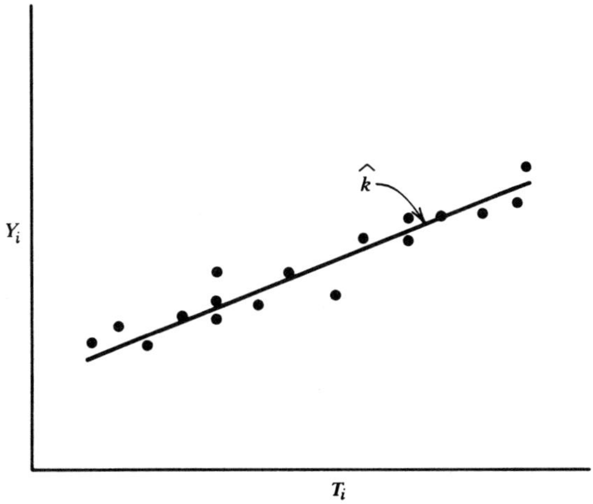

Figure 1.4 Simulated data for thermal conductivity vs. temperature.

where β_0 and β_1 are unknown parameters. The measurements Y_i are related to $k(T_i)$, abbreviated k_i, by

$$Y_i = k_i + \varepsilon_i = \beta_0 + \beta_1 T_i + \varepsilon_i, \qquad i = 1, 2, \ldots, n \tag{1.3.2a}$$

where ε_i is an unknown error. For n measurements there are n equations with two parameters and n unknown errors,

$$\begin{aligned} Y_1 &= \beta_0 + \beta_1 T_1 + \varepsilon_1 \\ Y_2 &= \beta_0 + \beta_1 T_2 + \varepsilon_2 \\ &\vdots \\ Y_n &= \beta_0 + \beta_1 T_n + \varepsilon_n \end{aligned} \tag{1.3.2b}$$

If $n = 1$, both β_0 and β_1 cannot be estimated. If $n = 2$, estimates of β_0 and β_1 can be obtained from (1.3.2b) by neglecting ε_1 and ε_2 and solving the two equations to obtain

$$\hat{\beta}_0 = \frac{Y_1 T_2 - Y_2 T_1}{T_2 - T_1}, \qquad \hat{\beta}_1 = \frac{Y_2 - Y_1}{T_2 - T_1} \tag{1.3.3}$$

where the "hat" on $\hat{\beta}_0$ and $\hat{\beta}_1$ indicates estimate. The k curve passes through both experimental points. Note that T_1 and T_2 cannot be the same temperature.

For $n>2$, a straight line (i.e., a linear curve) cannot simultaneously pass through all the points shown in Fig. 1.4. One can, however, imagine a number of strategies to place the line. For example, one could draw a line by eye through the data. After measuring the intercept and slope, β_0 and β_1 would be estimated. This has a number of advantages including simplicity and a visual check of the "fit." Moreover, all of us have had experience with this method. There are some severe shortcomings, however, including the lack of reproductivity. Different observers draw the line rather differently. Equally important is the disadvantage that the method does not lend itself to direct extension to more complex cases.

Other relatively straightforward methods, such as the method of sequential differences, are discussed by Rabinowicz [18].

The well-known method of least squares can be utilized to meet the objections noted above. The sum of squares of the errors,

$$S=\sum_{i=1}^{n}\varepsilon_i^2=\sum_{i=1}^{n}(Y_i-k_i)^2=\sum_{i=1}^{n}(Y_i-\beta_0-\beta_1T_i)^2 \tag{1.3.4}$$

is minimized with respect to the parameters β_0 and β_1. The sum of squares function S must be equal to or greater than zero simply because it is the sum of n terms, each of which is a square. S can be zero if and only if *every* measurement Y_i is on the line.

We can expand the S expression given by (1.3.4) to get (omitting, as we sometimes do, the explicit designation of limits)

$$S=\sum Y_i^2-2\beta_0\sum Y_i-2\beta_1\sum Y_iT_i+2\beta_0\beta_1\sum T_i+n\beta_0^2+\beta_1^2\sum T_i^2 \tag{1.3.5}$$

showing that in the three-dimensional coordinate system S is an elliptical paraboloid with one minimum. See Fig. 1.5. Note that (1.3.5) is of second degree in β_0 and β_1. Differentiate S with respect to β_0 and β_1 to obtain

$$\begin{aligned}\frac{\partial S}{\partial\beta_0}&=-2\sum(Y_i-\beta_0-\beta_1T_i)\\ \frac{\partial S}{\partial\beta_1}&=-2\sum(Y_i-\beta_0-\beta_1T_i)T_i\end{aligned} \tag{1.3.6}$$

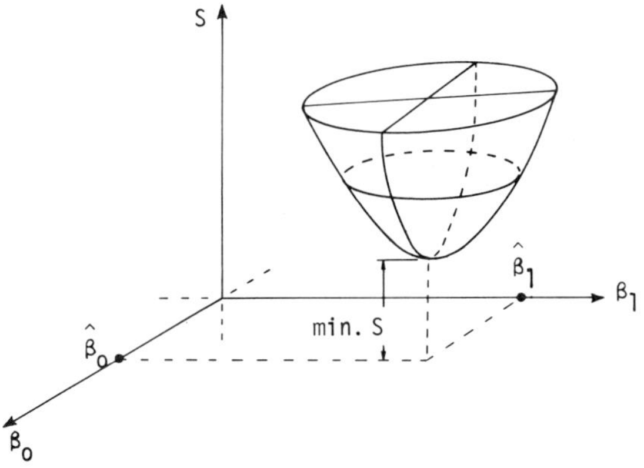

Figure 1.5 Elliptic paraboloid illustrating S given by (1.3.5).

Both derivatives of (1.3.6) are linear in β_0 and β_1. A necessary condition for a minimum is that both the derivatives in (1.3.6) be zero.

Setting these derivatives equal to zero and solving simultaneously yields

$$\hat{\beta}_0 = \frac{\left(\sum Y_i\right)\left(\sum T_i^2\right) - \left(\sum T_i\right)\left(\sum Y_i T_i\right)}{n\sum T_i^2 - \left(\sum T_i\right)^2},$$

$$\hat{\beta}_1 = \frac{n\sum Y_i T_i - \left(\sum T_i\right)\left(\sum Y_i\right)}{n\sum T_i^2 - \left(\sum T_i\right)^2} \tag{1.3.7}$$

Introduction of these values into (1.3.1) yields an estimate of k which is designated $\hat{k}$. A *residual* is defined by

$$\text{Residual} = e_i = Y_i - \hat{k}_i \tag{1.3.8}$$

which is *not* identical to the error, ε_i.

In this example it is implied that ε_i is completely unknown. When this is true, the least squares procedure just given is recommended. If, however, one knows that ε_i has a variance σ_i^2 (this term is discussed in Chapter 2), some other estimation procedure might be better.

Example 1.3.1

The thermal conductivity of air in units of W/m-K versus temperature in kelvin has been measured to be the following values:

T(K)	300	350	400	450
k(W/m-K)	0.0255	0.0309	0.0350	0.0377

Find estimates of the parameters β_0 and β_1 in (1.3.1) using least squares.

Solution

$$\sum T_i = 300 + 350 + 400 + 450 = 1500$$

$$\sum T_i^2 = 300^2 + 350^2 + 400^2 + 450^2 = 575000$$

$$\sum Y_i = 0.0255 + 0.0309 + 0.0350 + 0.0377 = 0.1291$$

$$\sum Y_i T_i = 0.0255(300) + \cdots + 0.0377(450) = 49.43$$

Then $\hat{\beta}_0$ and $\hat{\beta}_1$ from (1.3.7) are

$$\hat{\beta}_0 = \frac{0.1291(575000) - 1500(49.43)}{4(575000) - (1500)^2} = 0.00175 \text{ W/m-K}$$

$$\hat{\beta}_1 = \frac{4(49.43) - 1500(0.1291)}{4(575000) - (1500)^2} = 0.0000814 \text{ W/m-K}^2$$

Note that the parameter estimates $\hat{\beta}_0$ and $\hat{\beta}_1$ have units, each being different. The residuals are -0.00067, 0.00066, 0.00069, and -0.00068, which have a zero sum. The minimum sum of squares, which the sum of the square of each of these terms, is 1.823×10^{-6}.

In calculations such as these, at least an electronic calculator usually is needed because there frequently are small differences of large numbers.

1.3.2 Linear First Order Differential Equation Model

A case for which a fundamental law can be invoked is that of dropping a thin plate initially at temperature T_0 into a fluid at T_∞. From the first law of thermodynamics one can derive [19]

$$\frac{dT}{dt} = -\beta(T - T_\infty) \qquad (1.3.9a)$$

where β is considered the unknown parameter. Unlike the previous example β has clear physical significance; it is given by the group

$$\beta = \frac{h}{\rho c_p L}$$

where h is the heat transfer coefficient, ρ density, c_p specific heat, and L the half-thickness of the plate. It is not possible to estimate h, ρ, c_p, and L independently when given only measurements of T. One of the concepts in estimation, called *identifiability*, relates to the question of which parameter or groups of parameters can be uniquely estimated. See Section 1.5 and Appendix A.

In addition to (1.3.9a) an initial condition is needed to obtain a solution; an appropriate one is

$$T(0) = T_0 \tag{1.3.9b}$$

The solution of (1.3.9) for constant β and T_∞ is

$$T(t) = T_\infty + (T_0 - T_\infty)e^{-\beta t} \tag{1.3.10}$$

In the classical problem one stops at this point. In the estimation problem, measurements of T are used to estimate β.

Temperature data for the cooling of a plate using a single thermocouple (a temperature sensor) and uniform time spacing are shown in Fig. 1.6. Note that even though the differential equation is linear, $T(t, \beta)$ in (1.3.10) is a *nonlinear* function of β; that is, the derivative of (1.3.10) with respect to β is a function of β unlike k given by (1.3.1). This is discussed further in Section 1.4.

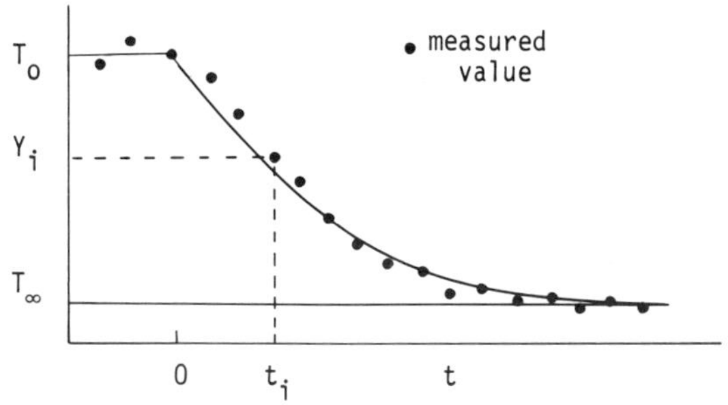

Figure 1.6 Simulated temperature measurements of a cooled thin plate.

The simplest method of estimation involves the use of *three* temperature measurements including T_0 and T_∞. Observe that at least three measurements are needed although only *one* parameter is being estimated. This is in contrast with the preceding case for which measurements at two temperatures were sufficient to estimate two parameters.

For measurements of T_0 and T_∞ and T_i at t_i, designated Y_0, Y_∞, and Y_i, an estimate of β is

$$\hat{\beta} = -\frac{1}{t_i} \ln \frac{Y_i - Y_\infty}{Y_0 - Y_\infty} \tag{1.3.11}$$

As either $t_i \to 0$, corresponding to $T_i \to T_0$, or $t_i \to \infty$, corresponding to $T_i \to T_\infty$, the error in $\hat{\beta}$ due to some small error in T_i becomes very large. Thus $\hat{\beta}$ is more sensitive to errors at some measurement times than others, which suggests the subjects of *sensitivity* and *optimum* experimental design (see Chapter 8).

If T_0 and T_∞ are not precisely known they can also be considered parameters like β. They are dissimilar from β in that (*a*) they are particular values of the dependent variable (termed the state variable in the systems literature) and (*b*) repeated measurements of these are available in this particular example.

One other parameter that could be estimated for this problem is the starting time. The starting time can be seen to be an unknown if one imagines several successive digital temperature measurements taken before the plate is dropped into the fluid. At the instant the plate contacts the fluid the plate's temperature rapidly changes; see Fig. 1.6. The time at which the plate contacts the fluid might not correspond to the instant at which any measurement was taken.

Suppose that the starting time is known to be zero and that all the measurements are for $t > 0$. For finding estimates of any combination of the parameters T_0, T_∞, and β one could start with the sum of squares for n measurements

$$S = \sum_{i=1}^{n} \left[Y_i - T_\infty - (T_0 - T_\infty) e^{-\beta t_i} \right]^2 \tag{1.3.12}$$

and minimize it with respect to the parameters. The derivatives of (1.3.12) are linear in terms of T_0 and T_∞ but nonlinear in terms of β. The nonlinearity complicates the search for a minimum.

One way to minimize S with respect to a nonlinear parameter is simply to plot S versus that parameter and graphically find the minimum. This is a slow procedure, but can give insight.

Example 1.3.2

Suppose it is known that T_∞ is equal to 100 and T_0 is equal to 300 and that two measurements of T are available, $Y_1=220$ at 5 sec and $Y_2=170$ at 10 sec. Estimate using least squares the parameter β in (1.3.10). Use a trial and error approach.

Solution

A first estimate of β can be obtained from (1.3.11) using the first observation. We obtain

$$\hat{\beta}=-\frac{1}{5}\ln\frac{220-100}{300-100}=0.1022$$

Let us then evaluate the sum of squares function S in the neighborhood of that value. From (1.3.12) we can write

$$S=(120-200e^{-5\beta})^2+(70-200e^{-10\beta})^2$$

which we evaluate at $\beta=0.1022$ to find $S=3.901$. Now another value of β must be tried. Let us try $\beta=0.1$; this gives $S=14.493$. Because this S value is bigger than the value for $\beta=0.1022$, let us try a larger value than $\beta=0.1022$. At $\beta=0.11$, $S=32.988$. Hence the minimum must be between $\beta=0.1$ and 0.11 and is probably nearer the first value. Let us try 0.103 which gives $S=2.214$. Then the minimum S must be between $\beta=0.1022$ and 0.11. A further value of $\beta=0.105$ yields $S=2.853$

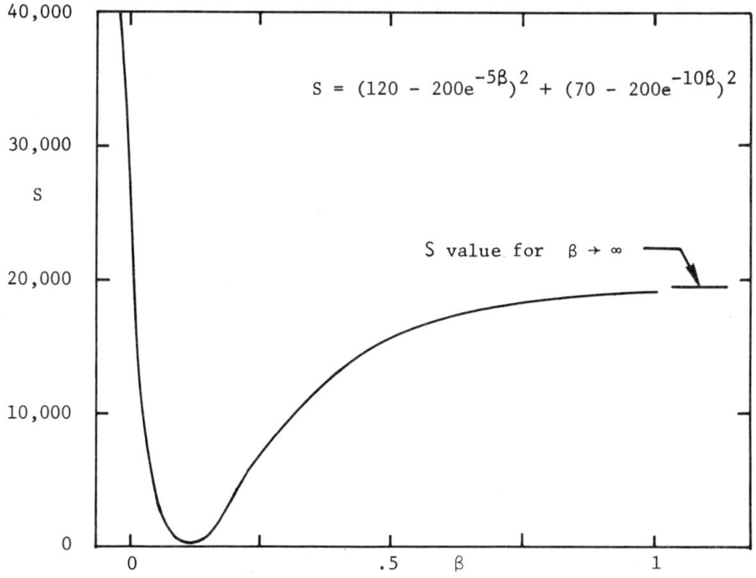

Figure 1.7 Sum of squares function for exponential example.

and thus β must be between 0.1022 and 0.105, which region could be explored further. One could continue further in this trial and error manner to estimate β more accurately. This is a possible approach but it is very tedious and time-consuming, particularly if more than one parameter is present. More direct methods of minimizing S are given in Chapter 7.

It is instructive to plot the function S for this case. See Fig. 1.7. Note that the minimum is near $\beta = 0.1$ and a local maximum is approached at large β. Thus in addition to $\partial S/\partial\beta$ being equal to zero near $\beta = 0.1$, it also approaches zero as $\beta \to \infty$. Even more ill-behaved S functions are possible. See Problem 1.5.

1.3.3 Partial Differential Equation Example

Consider again the same physical problem of a plate dropped suddenly into a fluid. Instead of negligible internal resistance ($Bi = hL/k < 0.1$) assume that there is a significant variation of temperature across the plate. The describing equations for constant properties and a plate of width $2L$ are [19]

$$k\frac{\partial^2 T}{\partial x^2} = \rho c_p \frac{\partial T}{\partial t} \qquad (0 < x < 2L) \tag{1.3.13}$$

$$-k\frac{\partial T}{\partial x}\bigg|_{x=0} = h\left[T_\infty - T(0,t)\right], \qquad -k\frac{\partial T}{\partial x}\bigg|_{x=2L} = h\left[T(L,t) - T_\infty\right] \tag{1.3.14a, b}$$

$$T(x,0) = T_0 \tag{1.3.15}$$

This is a problem which is linear in the dependent variable, T. For the estimation problem we can consider T as a function of a number of variables,

$$T = T(x, t, k, \rho, c_p, h, T_0, T_\infty, L) \tag{1.3.16}$$

In parameter estimation one must be able to solve the model repeatedly for different parameter values. For this example an exact solution is available as an infinite series but it may be easier to approximate the solution using a finite-differences representation. Such a solution can also be modified to treat nonlinearities entering in either the differential equation or the boundary conditions.

Note that in this example several different kinds of measurements are required: temperature, time, and length. Measurements of the initial conditions and boundary conditions may not be sufficient for parameter estimation; interior measurements may be needed (identifiability). The location of sensors and duration of the experiment are studied in connection with optimum experiments.

Another aspect of identifiability is the determination of what parameters or groups of parameters can be uniquely estimated. For example, (1.3.13) and (1.3.14) can be divided by k to yield the groups $\rho c_p/k$ and h/k. Since no term in these groups appears elsewhere in the problem, one would anticipate that these groups could be simultaneously estimated. That this is not always true can be proved by noting that this physical problem is identical to the one in Section 1.3.2 for which only the ratio of these two parameters could be estimated. It happens that $Bi=hL/k$ must be equal to approximately one or greater in order to estimate both. There may be other conditions that would also preclude estimation for this example. The condition for identifiability is discussed in Section 1.5.

In order to estimate the parameters one can again use the sum of squares function. Instead of a single summation over time, one could have a double summation over time and sensors located at different positions,

$$S=\sum_{i=1}^{n}\sum_{j=1}^{m}\left[Y_j(i)-T_j(i)\right]^2 \tag{1.3.17}$$

The subscript j is for position i is for time. There are m discrete locations and n different times. $Y_j(i)$ designates an observation and $T_j(i)$ a value obtained from the model.

1.4 SENSITIVITY COEFFICIENTS

In this section a brief introduction to sensitivity coefficients is given. Consider the true mathematical model to be given by $\eta(x,t,\boldsymbol{\beta})$ where x and t are independent variables and $\boldsymbol{\beta}$ is a parameter vector. The first derivative of η with respect to β_i will be called the sensitivity coefficient for β_i and designated X_i,

$$X_i=\frac{\partial\eta}{\partial\beta_i} \tag{1.4.1}$$

On some occasions the right side of (1.4.1) is multiplied by β_i and still called simply a sensitivity coefficient.

Sensitivity coefficients are very important because they indicate the magnitude of change of the response η due to perturbations in the values of the parameters. It is for this reason we have given X_i, defined by (1.4.1), the name "sensitivity coefficient." They appear in relation to many facets of parameter estimation. The reader is urged to pay particular attention to them and even to plot them versus their independent variables(s) if their shapes are not obvious. One area where the sensitivity coefficients appear is in the identifiability problem, which is briefly discussed in Section 1.5. Another area where the X_i's appear is the Gauss method of linearizing the estimation problem when the model is nonlinear in terms of parameters (see Section 7.4). In the optimal design of experiments discussed in Chapter 8, the sensitivities also play a key role.

The sensitivity coefficients also appear in a Taylor's series for $\eta(\beta_1,\ldots,\beta_p,t)$ about the neighborhood of the point $(b_1,b_2,\ldots,b_p)$ which we shall denote $\mathbf{b}$. Provided η has continuous derivatives near $\boldsymbol{\beta}=\mathbf{b}$, we can write

$$\eta(\beta_1,\ldots,\beta_p,t)=\eta(\mathbf{b},t)+\frac{\partial\eta(\mathbf{b},t)}{\partial\beta_1}(\beta_1-b_1)+\cdots+\frac{\partial\eta(\mathbf{b},t)}{\partial\beta_p}(\beta_p-b_p)$$

$$+\frac{\partial^2\eta(\mathbf{b},t)}{\partial\beta_1^2}\frac{(\beta_1-b_1)^2}{2!}+\cdots+\frac{\partial^2\eta(\mathbf{b},t)}{\partial\beta_1\partial\beta_2}(\beta_1-b_1)(\beta_2-b_2)+\cdots \quad (1.4.2)$$

If the derivatives $\partial^{r+s}\eta/\partial\beta_i^r\partial\beta_j^s$ $(i,j=1,\ldots,p)$ for $r+s>1$ are zero, then η is said to be *linear* in the parameters. For η a linear function of β_1 and β_2, we can write

$$\eta(\beta_1,\beta_2,t)=\eta(b_1,b_2,t)+X_1(\beta_1-b_1)+X_2(\beta_2-b_2) \quad (1.4.3)$$

This relation is an equality rather than an approximation if both X_1 and X_2 are not functions of the parameters. Hence η *is linear in its parameters if all the sensitivity coefficients are not functions of any parameter(s).*

Consider now some simple examples. The β_1, β_2, and β_3 sensitivity coefficients for the algebraic model

$$\eta=\beta_1+\beta_2 t+\beta_3 t^2 \quad (1.4.4)$$

are, respectively,

$$X_1=1, \qquad X_2=t, \qquad X_3=t^2 \quad (1.4.5)$$

Since each of these is independent of all the parameters, η given by (1.4.4) is linear in its parameters. Estimation involving models linear in the parameters is generally easier and more direct than estimation involving nonlinear parameters.

Another algebraic model which occurs in many fields is

$$\eta=\beta_1\exp(\beta_2 t) \quad (1.4.6)$$

The β_1 and β_2 sensitivites of this equation are

$$X_1 = \exp(\beta_2 t); \qquad X_2 = \beta_1 t \exp(\beta_2 t) \tag{1.4.7}$$

which contain the parameters and thus η given by (1.4.6) is nonlinear in terms of its parameters. If, however, the only parameter of interest is β_1, η is linear in terms of β_1.

The evaluation of sensitivity coefficients need not begin with an expression of η but could be initiated with the given differential equation. For example, if the derivative of (1.3.9a) (a *linear* differential equation) is taken with respect to β,

$$\frac{dX}{dt} = -(T - T_\infty) - \beta X, \qquad X \equiv \frac{\partial T}{\partial \beta} \tag{1.4.8a}$$

$$X(0) = 0 \tag{1.4.8b}$$

Equation 1.4.8a is termed the *sensitivity equation* for this case and, together with (1.4.8b), constitutes a statement of the sensitivity problem. In (1.4.8a) it is assumed that T (or η in the notation of this section) is a known function obtained from a previous solution of the original differential equation and initial condition. Since β appears explicitly in (1.4.8a), the sensitivity coefficient X is a function of β. Consequently the dependent variable T is nonlinear in β as can be verified from differentiating (1.3.10).

1.5 IDENTIFIABILITY

There are some models for which it is not possible to uniquely estimate all the parameters from measurements. Rather it is possible to estimate only certain functions of them. This is part of the identifiability problem. See Appendix A for a derivation of an identifiability criterion.

In this section several simple cases for which one cannot uniquely estimate all the parameters are discussed. Later an identifiability criterion utilizing sensitivity coefficients is introduced and related to some of the cases previously investigated.

A model that will not permit estimation of both β_1 and β_2 is

$$\eta_i = (\beta_1 + A\beta_2) f(t_i) \tag{1.5.1}$$

where A is a constant and $f(t)$ is any known function of t. In this case one can only estimate $\beta = \beta_1 + A\beta_2$ given measurement of η_i versus t_i. In this S, β_1, β_2 space, S does not have a unique minimum, but instead has a

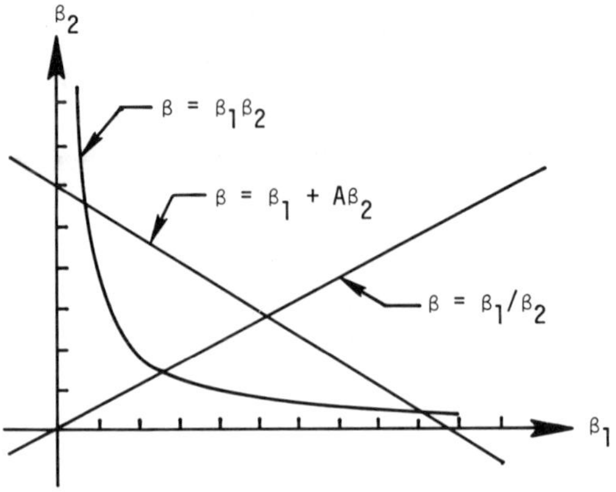

Figure 1.8 Contours of minimum S for various cases where not all the parameters can be uniquely estimated.

minimum along a line which projects into $\beta=\beta_1+A\beta_2$ in the β_1,β_2 plane; see Fig. 1.8.

Consider next the model

$$\eta_i=\beta_1\beta_2 t_i \tag{1.5.2}$$

From inspection we see that β_1 and β_2 can be replaced by the product $\beta=\beta_1\beta_2$ and that any combination of β_1 and β_2 equal to β would yield the same value of η_i for a given t_i. In terms of the three-dimensional space of S plotted versus β_1 and β_2, there is a minimum S along a curved line which projects into $\beta_2=\beta/\beta_1$ in the β_1,β_2 plane, as shown in Fig. 1.8.

A very similar case to (1.5.2) is

$$\eta_i=\frac{\beta_1}{\beta_2}t_i \tag{1.5.3}$$

where again only the ratio β is unique and various combinations of β_1 and β_2 could be given to provide $\beta=\beta_1/\beta_2=$constant. In the S, β_1, and β_2 coordinates, the minimum of S occurs along a straight line $\beta_1=\beta\beta_2$ projected into the β_1,β_2 plane. In Fig. 1.8 this line passes through the origin.

A less obvious case is for the model

$$\eta_i=\frac{\beta_1}{\beta_2+\beta_3 t_i} \tag{1.5.4}$$

Dividing by β_1 yields

$$\eta_i = (\alpha_1 + \alpha_2 t_i)^{-1}; \qquad \alpha_1 \equiv \beta_2 \beta_1^{-1}, \quad \alpha_2 = \beta_3 \beta_1^{-1} \tag{1.5.5}$$

where it is seen that η_i is a function of α_1, α_2, and t_i, and thus only α_1 and α_2 can be simultaneously estimated.

Another simple case where all three parameters cannot be uniquely estimated is for

$$\eta_i = \beta_1 e^{-(\beta_2 + \beta_3 t_i)} \tag{1.5.6a}$$

$$= \beta_1 e^{-\beta_2} e^{-\beta_3 t_i} = \alpha_1 e^{-\beta_3 t_i}; \qquad \alpha_1 = \beta_1 e^{-\beta_2} \tag{1.5.6b}$$

where only α_1 and β_3 can be found.

There are other cases where the parameters can not (easily) be uniquely estimated if measurements are made only over a certain range of the independent variable or at certain values. One example is

$$\eta_i = \beta_1 + \beta_2 (10 + t_i^4) \tag{1.5.7}$$

for $\max|t_i|$ small compared to unity. For such a model it is possible to estimate accurately only $\beta_1 + 10\beta_2$ if $|t_i|$ is small. This model is thus similar to (1.5.1) for small $|t_i|$. For sufficiently "large" t_i both β_1 and β_2 can be estimated. Another example is for the model

$$\eta_i = \beta_1 t_i + \beta_2 \sin \beta_3 t_i \tag{1.5.8a}$$

for small $\beta_3 t_i$ since then η_i can be approximated by

$$\eta_i \approx \beta_1 t_i + \beta_2 \beta_3 t_i = (\beta_1 + \beta_2 \beta_3) t_i \tag{1.5.8b}$$

Hence for small $\max|\beta_3 t_i|$, instead of being able to estimate uniquely all three parameters we can estimate only $\beta_1 + \beta_2 \beta_3$.

Many other cases could be cited that demonstrate that only certain functions of parameters can be estimated from measurements of η_i versus its independent variable(s). Some of these cases may not be at all obvious. This is particularly true where there are a number of parameters and the model is a differential equation. Rather than depending upon being able to manipulate the model so that groups of parameters appear, we would be helped by having some criterion that could be applied to the above algebraic models and also to models involving differential equations. In the latter case we imagine that the solutions of the equations and the sensitivity coefficients are available in graphical or tabular form. It turns out in the algebraic cases above, as well as for other cases involving differential equations, that the sensitivity coefficients can provide insight into the cases for which parameters can and cannot be estimated.

Parameters can be estimated if the sensitivity coefficients over the range of the observations are not linearly dependent. This is the criterion that we shall use to determine if the parameters can be simultaneously estimated without ambiguity. See Appendix A for a derivation of this criterion. Linear dependence occurs when for p parameters the relation

$$C_1\frac{\partial\eta_i}{\partial\beta_1}+C_2\frac{\partial\eta_i}{\partial\beta_2}+\cdots+C_p\frac{\partial\eta_i}{\partial\beta_p}=0 \tag{1.5.9}$$

is true for all i observations and for not all the C_j values equal to zero.

Let us illustrate the above criterion for a few examples. For (1.5.1) note that

$$\frac{\partial\eta_i}{\partial\beta_1}=f(t_i),\qquad \frac{\partial\eta_i}{\partial\beta_2}=Af(t_i)$$

and thus, if $C_1=A$ and $C_2=-1$, (1.5.9) is satisfied. Consequently, both β_1 and β_2 cannot be estimated simultaneously.

Another example involves (1.5.4) for which

$$\frac{\partial\eta_i}{\partial\beta_1}=\frac{1}{\beta_2+\beta_3t_i},\qquad \frac{\partial\eta_i}{\partial\beta_2}=\frac{-\beta_1}{(\beta_2+\beta_3t_i)^2},\qquad \frac{\partial\eta_i}{\partial\beta_3}=\frac{-\beta_1t_i}{(\beta_2+\beta_3t_i)^2}$$

It is not immediately obvious from an inspection of these sensitivity relations that there is linear dependence. It can be verified, however, that if $C_1=\beta_1$, $C_2=\beta_2$, and $C_3=\beta_3$ linear dependence exists; in equation form, we then have

$$\beta_1\frac{\partial\eta_i}{\partial\beta_1}+\beta_2\frac{\partial\eta_i}{\partial\beta_2}+\beta_3\frac{\partial\eta_i}{\partial\beta_3}=0$$

which form can occur in various cases with linear dependence. The dependent variable η and the sensitivity coefficients for the model (1.5.4) are depicted in Fig. 1.9 for $\beta_2=1$. It is strongly recommended that the sensitivity coefficients be plotted and carefully examined to see if linear dependence exists or even is approached. The relation given above between the coefficients can be approximately verified by graphically adding the three together to obtain zero at each instant of time. Furthermore, note that the β_1 and β_3 sensitivities seem to have approximately proportional magnitudes for β_3t greater than 3. This means that not only is it impossible

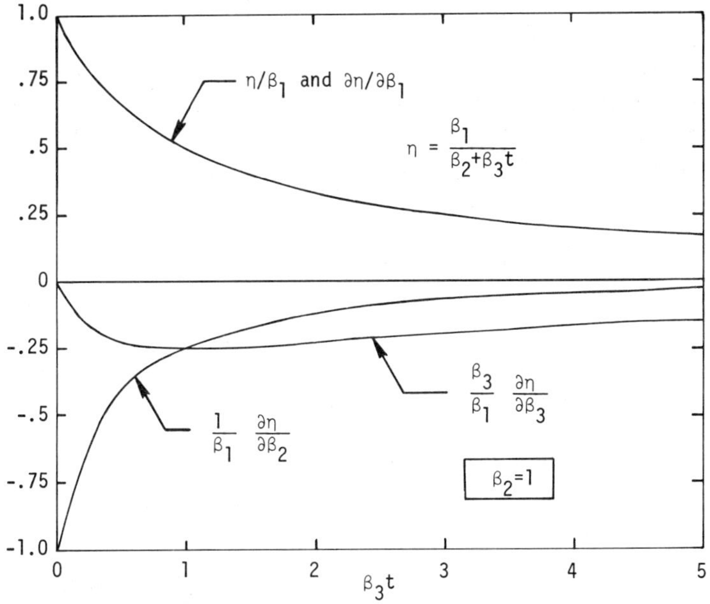

Figure 1.9 Dependent variable η and sensitivity coefficients for $\eta=\beta_1/(\beta_2+\beta_3 t)$ with $\beta_2=1$.

to estimate β_1, β_2, and β_3 simultaneously from measurements of η versus t, but it is difficult to estimate only β_1 and β_3 using data for $\beta_3 t>3$ if $\beta_2 \leqslant 1$.

1.6 SUMMARY AND CONCLUSIONS

1. Parameter estimation is a discipline that provides tools for the efficient use of data for aiding in mathematically modeling of phenomena and the estimation of constants appearing in these models. The problem of estimating parameters is that of finding constants appearing in an equation describing a system as suggested by Fig. 1.3.
2. One way to estimate the parameters for a large variety of models is to use *least squares* which involves minimizing the sum of squares of differences between measurements and model values. The minimization problem can be either linear or nonlinear.
3. One cannot always independently estimate all the parameters that appear in the model. It is clear that not all the parameters may be estimated if parameters appear in groups, but in some cases not even all these groups may be found. This is related to the subject of *identifiability*.

REFERENCES

1. Legendre, A. M., *Nouvelles Méthodes Pour la Détermination des Orbites des Comètes*, Paris, 1806.
2. Gauss, K. F., *Theory of the Motion of the Heavenly Bodies Moving about the Sun in Conic Sections*, 1809, reprinted by Dover Publications, Inc., New York, 1963.
3. Draper, N. R. and Smith, H., *Applied Regression Analysis*, John Wiley & Sons, Inc., New York, 1966.
4. Box, G. E. P. and Jenkins, G. M., *Time Series Analysis forecasting and control*, Holden-Day, Inc., San Francisco, 1970.
5. Myers, R. H., *Response Surface Methodology*, Allyn and Bacon, Inc., Boston, 1971.
6. Bard, Y., *Nonlinear Parameter Estimation*, Academic Press, New York, 1974.
7. Sage, A. P. and Melsa, J. L., *Estimation Theory with Applications to Communications and Control*, McGraw-Hill Book Co., New York, 1971.
8. Sage, A. P., *Optimum Systems Control*, Prentice-Hall, Inc., Englewood Cliffs, N. J., 1968.
9. Deutsch, R., *Estimation Theory*, Prentice-Hall, Inc., Englewood Cliffs, N. J., 1965.
10. Bryson, A. E., Jr. and Ho, Yu-Chi, *Applied Optimal Control Optimization, Estimation and Control*, Blaisdell Publishing Co., Waltham, Mass., 1969.
11. Sage, A. P. and Melsa, J. L., *System Identification*, Academic Press, New York, 1971.
12. Graupe, D., *Identification of Systems*, Van Nostrand-Reinhold Co., New York, 1972.
13. Mendel, J. M., *Discrete Techniques of Parameter Estimation The Equation Error Formulation*, Marcel Dekker, Inc., New York, 1973.
14. Kmenta, J., *Elements of Econometrics*, The Macmillan Co., New York, 1971.
15. Bevington, P. R., *Data Reduction and Error Analysis for the Physical Sciences*, McGraw-Hill Book Company, New York, 1969.
16. Wolberg, J. R., *Prediction Analysis*, D. Van Nostrand Co., Inc., Princeton, N. J., 1967.
17. Lewis, T. O. and Odell, P. L., *Estimation in Linear Models*, Prentice-Hall, Inc., Englewood Cliffs, N. J., 1970.
18. Rabinowicz, E., *An Introduction to Experimentation*, Addison-Wesley Publishing Co., Reading, Mass., 1970.
19. Kreith, F., *Principles of Heat Transfer*, 3rd ed., Intext Educational Publishers, New York, 1973.

PROBLEMS

1.1 The thermal conductivity k has been found from four independent experiments at different temperature to be given by

T_i (°C)	k_i (W/m-C)
100	90
200	98
300	111
400	121

(*a*) Estimate β_0 and β_1 in (1.3.1), using least squares.

Answer. 78.5, 0.106

(*b*) Calculate the residuals.

Answer. 0.9, −1.7, 0.7, 0.1

(*c*) For $\beta_0=80$, plot S versus β_1 in the neighborhood of the minimum.

1.2 (*a*) Derive using least squares an estimate of β for the simple model

$$\eta_i=\beta$$

for n measurements. Assume $Y_i=\eta_i+\varepsilon_i$, ε_i being the measurement error.

(*b*) Also derive estimates for β_0 and β_1 for the model

$$\eta_i=\beta_0+\beta_1\sin t_i$$

1.3 Some actual measurements for the specific heat c_p of Armco iron at room temperature are, in units of kJ/kg-C,

i	1	2	3	4	5
c_p	0.4287	0.4363	0.4451	0.4409	0.4442

i	6	7	8	9	10
c_p	0.4400	0.4400	0.4405	0.4375	0.4333

Using the model of Problem 1.2*a*, estimate c_p. Plot the residuals as a function of i. What is the sum of the residuals?

1.4 (*a*) For the model

$$\eta=e^{\beta t}$$

and the data given below, estimate β by plotting S versus β. Cover the range 0 to −20.0.

(*b*) Compare the curve with Fig. 1.7.

(*c*) Compare the residuals with the true errors ($\varepsilon_i=Y_i-\eta_i$) also given below:

t	Data, Y_i	Errors, ε_i
0.25	0.419	−0.053
0.5	0.204	−0.019
0.75	0.159	0.054
1.0	−0.106	−0.156
1.25	0.042	0.0187

1.5 Plot S versus β for the model $\eta=100\sin\beta\theta$ with $\beta\theta$ in radians and for the data $\theta_1=2.79$, $Y_1=34.2$; $\theta_2=6.98$, $Y_2=64.2$; and $\theta_3=8.38$, $Y_3=86$. Investigate the range $0<\beta<1.1$ for $\Delta\beta$ increments at least as small as 0.1. (A programmable calculator would be helpful to get the solution.) What conclusions can you draw?

1.6 How can (1.3.12) be changed to permit estimation of the starting time, t_0, β and T_0? Assume that measurements are available for t both less than and greater than t_0. Also assume that the plate has been at T_0 for a "long" time before t_0.

1.7 (*a*) For the model and data

$$\eta_i = \frac{\beta_1}{1+\beta_2 t_i}$$

t_i	Y_i
0	200
1	55
2	30
3	20

calculate the sum of squares S in the rectangular region $100<\beta_1<300$ and $2.0<\beta_2<4.0$. In particular, evaluate S at $\beta_1=100$, 200, and 300 with $\beta_2=2.0$, 3.0, and 4.

(*b*) Is η_i linear in β_1 and β_2?

(*c*) Based on the information in (*a*), estimate β_1 and β_2.

(*d*) Using the search procedure in (*a*), is it more or less than twice as much work to find two parameters as it is to estimate one?

1.8 The current i in the circuit of Fig. 1.10 after the switch S is closed satisfies the differential equation

$$L\frac{di}{dt} + Ri - E = 0$$

where L is inductance, R is resistance, and E is voltage. An initial condition is

$$i = i_0 \qquad \text{at} \quad t=0$$

Note that a solution of i is in terms of t, L, R, E, and i_0.

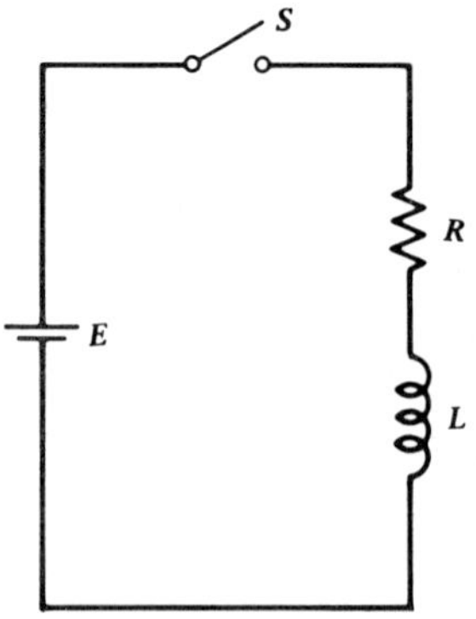

Figure 1.10 Circuit for Problem 1.8.

(*a*) What is (are) the dependent variable(s)?
(*b*) What is (are) the independent variable(s)?
(*c*) What is (are) the state(s)?
(*d*) What could be termed parameters?
(*e*) What could be termed properties?
(*f*) The solution of the problem is

$$i = \frac{E}{R}[1 - e^{(-R/L)t}] + i_0 e^{(-R/L)t}$$

Is i linear in E? R? L? i_0?
(*g*) What parameters or groups of parameters can be estimated given measurements of i?

1.9 For the following expressions of the model η, indicate for the various β_i values if η_i is linear or nonlinear in terms of them.

(*a*) $\eta_i = \beta_1 + \beta_2 \sin \pi t_i$
(*b*) $\eta_i = \beta_1 t_i^2 e^{-10t_i}$
(*c*) $\eta_i = \dfrac{\beta_1 e^{-\beta_2 t_i^2}}{1 + \beta_1 t_i + \beta_2 t_i^2}$
(*d*) $\eta_i = \sum_{j=1}^{\infty} \dfrac{1}{j}(1 - \beta_1 j) e^{-j\pi t_i}$

1.10 For the following expressions for the model η, derive expressions for the sensitivity coefficients. Also plot the sensitivity coefficients and η versus $\beta_2 t$. For parts (*b*) and (*c*) graph η/β_1, $\partial\eta/\partial\beta_1$, and $(\beta_2/\beta_1)\partial\eta/\partial\beta_2$ versus $\beta_2 t$. (If values of β_1 and β_2 are needed, let $\beta_1 = 2$ and $\beta_2 = 1$.)

(*a*) $\eta = \beta_1 + \beta_2 t$
(*b*) $\eta = \beta_1 \cos \beta_2 t \qquad (0 \leqslant \beta_2 t \leqslant 4\pi)$
(*c*) $\eta = \beta_1(1 - e^{-\beta_2 t}) \qquad (0 \leqslant \beta_2 t \leqslant 3)$

1.11 For the model

$$\eta = \beta_1 (6t - t^3) + \beta_2 \sin t$$

where t is in radians, plot the sensitivity coefficients for $-2 \leqslant t \leqslant 2$. Over what range (if any) do the parameters β_1 and β_2 appear to be linearly dependent?

1.12 Find a linear relation between the sensitivity coefficients for β_1, β_2, and β_3 for the model

$$\eta = \beta_1 \left(\frac{t}{\beta_2} \right)^{\beta_3}$$

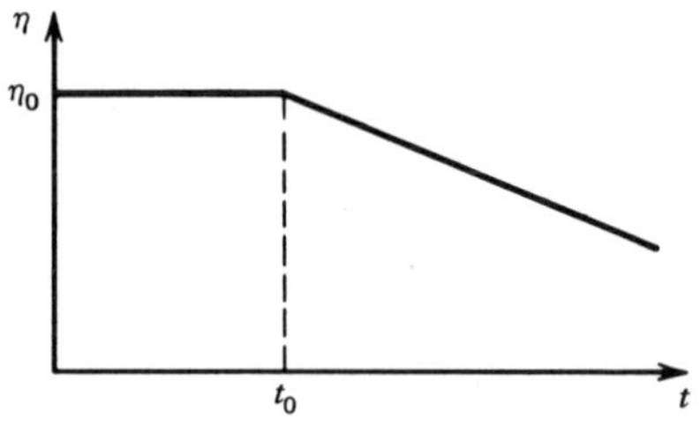

Figure 1.11 η for Problem 1.13.

1.13 Consider the model (see Fig. 1.11)

$$\eta = \eta_0 - \beta_1 (t - t_0) \qquad t \geqslant t_0$$

$$\eta = \eta_0 \qquad t < t_0$$

$$(\beta_1 \text{ is positive})$$

(*a*) Find and graph $\partial\eta/\partial\eta_0$.
(*b*) Find and graph $\partial\eta/\partial t_0$.
(*c*) Can t_0 and η_0 be simultaneously estimated using only two measurements of η if β_1 is known?

CHAPTER 2

PROBABILITY

2.1 RANDOM HAPPENINGS

If a room thermostat is set at 21°C, we do not expect the temperature throughout the room, or even right at the thermostat, to remain constant. Rather, we expect the temperature at any point to change continually and continuously while remaining very near 21°C.

If we run a test of braking distance by repeatedly bringing a car to 55 mph, then applying the brakes, we expect the distance covered after application of the brakes to differ from trial to trial no matter how we try to make sure that the road and wind conditions and pressure on the brake pedal are the same from trial to trial. We do hope to settle on some typical distance and perhaps on some measure of variability. In both cases, thermostat and braking distance, there are elements of stability and elements of randomness.

Example 2.1.1

As a simple example of randomness with an element of stability, let us observe successive determinations of percent defective in a sampling inspection of items from a production line. (The data to be exhibited were actually generated by computer simulation.) Successive items were inspected and declared to be either Good or Defective. The first two items were found to be Good, the third Defective, and so on. The results of the first 500 determinations are given in Table 2.1A. To make a long series of such determinations easier to contemplate, Table 2.1B gives the number of defectives found among the first n items inspected for various n up to 15,000. As we take observations, the fraction defective fluctuates wildly at first, but as the number of observations increases the fluctuations dampen. Note that, although the position in the sequence at which each defective occurs seems quite unpredictable, there is an element of stability in the number of defectives in large numbers of successive trials.

Table 2.1A Ordinal Numbers of Defective Items Among 500 Items Inspected

3rd	5th	55th	140th	186th
187th	380th	411th	450th	485th

Table 2.1B Cumulative Numbers of Defective Items Among 15,000 Items Inspected

Cumulative Number of Items Inspected	Cumulative Number of Defectives	Fraction Defective	Cumulative Number of Items Inspected	Cumulative Number of Defectives	Fraction Defective
1	0	0.000	1,500	40	0.027
2	0	0.000	2,000	52	0.026
3	1	0.333	3,000	80	0.027
4	1	0.250	4,000	115	0.029
5	2	0.400	5,000	144	0.0288
10	2	0.200	10,000	305	0.0305
20	2	0.100	10,500	324	0.0309
30	2	0.067	11,000	337	0.0306
40	2	0.050	11,500	350	0.0304
50	2	0.040	12,000	365	0.0304
100	3	0.030	12,500	380	0.0304
150	4	0.027	13,000	396	0.0305
200	6	0.030	13,500	404	0.0299
300	6	0.020	14,000	420	0.0300
400	7	0.018	14,500	442	0.0305
500	10	0.020	15,000	451	0.0301
1000	30	0.030			

The concept of probability was developed to describe a property of an experimental situation in which it is impossible to tell what outcome to expect for any one trial but yet, in a long enough series of trials, the fraction yielding a particular outcome seems fairly stable.

Example 2.1.2

In the example of braking distance, the observations differ in an important way from those obtained in the case of fraction defective. Since the number of defectives in a sample of n observations must be an integer, there are only a finite number, $n+1$, of possible values for fraction defective in the sample. These are discrete values and the variate, fraction defective, is called a *discrete* variate. In the braking distance example, the intrinsic variate is *continuous*, that is, for any two possible values of the variate there is another possible value between them. The recorded data by contrast are rounded, probably recorded to a fixed number of decimal places in whatever units of measurement we have chosen to use. Although this digitalization reduces the problem to one involving only a discrete variate, it is usually convenient to develop methods of analysis as though we were dealing with the intrinsic continuous variate.

Example 2.1.3

Another example of stability in randomness is that of a number of temperature sensors measuring the temperature at various (fixed) points in a room which is equipped with thermostatic control. A continuous plot of the output of three such thermocouples might look like the curves in Fig. 2.1. This figure, in one sense, constitutes a single observation on the manner in which temperature is controlled by the thermostat. A similar record beginning at another point in time would constitute another observation. We look for characteristics which are common to all records, which describe the manner in which temperature responds to the control of the thermostat.

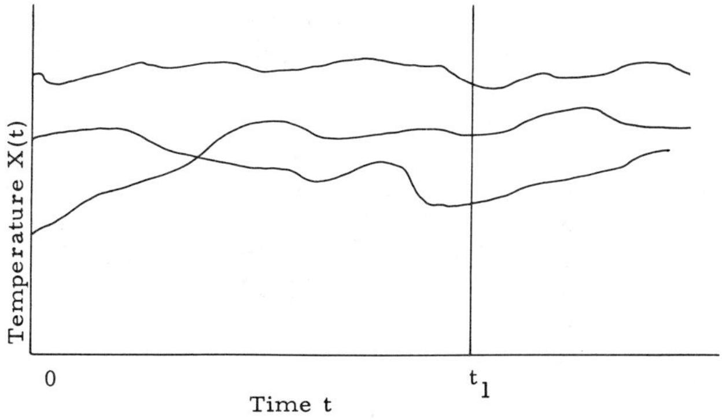

Figure 2.1 Continuous temperature measurements of air temperature.

Much of this book is concerned with analyzing data from dynamic processes with the aid of some mathematical model of the phenomena. These models contain parameters that are to be estimated with the aid of measurements. Digital data acquisition equipment used in dynamic measurements, besides recording rounded data, records only at discrete points in time. The methods of estimation discussed in this book deal with problems in which data are recorded at discrete (not necessarily equally spaced) points in time.

2.2 EVENTS

2.2.1 Events. Random Variables and Probabilities

Three probabilistic aspects of an experiment are (*a*) the *outcome*, the observed result, (*b*) the *event*, the category, of interest to the experimenter, into which the observation falls, and (*c*) the *probability* associated with the event.

Example 2.2.1

For simplicity, let us use for illustration the experiment of tossing a penny three times. For convenience, let us use the symbols HHH, HHT, HTH, HTT, THH, THT, TTH, TTT to represent the eight possible *outcomes*, aspect (*a*). Each three letter symbol represents in an obvious way the results, in order, of the three tosses. Each experiment of tossing the penny three times results in one and only one of the set of outcomes. The set of all possible outcomes constitutes the *sample space*.

As for aspect (*b*), we may be interested in whether the same face shows up on all three tosses. In this case we are interested in whether the outcome is one of the outcomes HHH and TTT. In the language of probability we say that the event "same face on all three tosses" is the subset {HHH, TTT} of the sample space. Likewise, if we are interested in whether exactly two tosses yield tails, we are interested in the event {HTT, THT, TTH}. An *event* is a subset of the sample space.

If we associate a number (or a vector) with each outcome, we form a *random variable*. Most of the events in which we shall be interested can be described in terms of random variables. In the three tosses of a penny, the number of heads is a random variable with value 0 associated with outcome TTT, 1 with THT, and so on. The set of outcomes to which we attach a given number form an event. Thus "the number of heads is one" is the event {HTT, THT, TTH}.

A summary of terms introduced in this section is given in Tables 2.2 and 2.3.

The third probabilistic aspect of an experiment is the *probability*. If we consider the eight simple events in the sample space above as equally likely, we attach a probability of $\frac{1}{8}$ to each. Sometimes our sense of symmetry leads us to such an assignment of probability but often it does not. If we ask the probability that the next item from the production line

Table 2.2 Terms Related to Events

Experiment: something which generates an observation. It may or may not require action on the part of the experimenter.

Outcome (also *simple event*, *sample point*): one of the set of possible observations which result from an experiment. One and only one outcome results from one realization of the experiment.

Sample space: the set of all possible outcomes which may result from a given experiment.

Event: a subset of the sample space. An event is said to occur if any outcome in the event occurs.

Random variable: a number (or vector) determined by an outcome, that is, a function defined on the points of the sample space.

Probability: see Table 2.3.

Table 2.3 Axioms of Probability

A function $P(E)$ defined on a set of events is called a *probability* if

(a) For any event A,

$$0 \leqslant P(A) \leqslant 1 \tag{2.2.1}$$

(b) The probability that the outcome of an experiment is some one of the outcomes which make up the sample space is 1. The probability that the outcome of an experiment is not one of the outcomes which make up the sample space is 0.

(c) If the events $A_1, A_2, \ldots$ are *disjoint*, that is, no two can occur simultaneously,

$$P(A_1 \text{ or } A_2 \text{ or } \ldots) = P(A_1) + P(A_2) + \cdots \tag{2.2.2}$$

In particular, if A_1 and A_2 are *disjoint*,

$$P(A_1 \text{ or } A_2) = P(A_1) + P(A_2) \tag{2.2.3}$$

Immediate consequences of these axioms are

(d) $P(\text{not } A) = 1 - P(A)$. (2.2.4)

(e) $P(A \text{ or } B) = P(A) + P(B) - P(A \text{ and } B)$. (2.2.5) which we inspect will be defective, we shall almost certainly *not* wish to associate a probability of $\frac{1}{2}$ with this event. Probabilists leave the word probability undefined but require that the assignment of probabilities to events satisfy certain restrictions (see Table 2.3). In essence, it is required that probabilities act like relative frequencies. The probability of a sure thing (i.e., certainty) is one. The probability of the impossible event is zero. The probability of one or the other of two disjoint events (events that cannot happen simultaneously) is the sum of the probabilities of the two events.

If events A and B are not disjoint, that is, "A and B" is a possible event, it is clear that the probability of "A and B" is not the sum of the probability of A and the probability of B. Let us see what can be said about the probability of the union of any two events, disjoint or not. Now, the event A is the union of the disjoint events "A and B" and "A and not B," and the event B is the union of the disjoint events "A and B" and "B and not A," whereas the event "A or B" is the union of the disjoint events "A and B," "A and not B," and "B and not A." Thus

$$P(A \text{ or } B) = P(A \text{ and } B) + P(A \text{ and not } B) + P(B \text{ and not } A)$$

$$P(A) = P(A \text{ and } B) + P(A \text{ and not } B)$$

$$P(B) = P(A \text{ and } B) + P(B \text{ and not } A)$$

Hence

$$P(A \text{ or } B) = P(A) + P(B) - P(A \text{ and } B) \tag{2.2.5}$$

2.2.2 Discrete and Continuous Sample Spaces and Associated Probabilities

The experiment of tossing a coin three times led us to consider a *finite* sample space consisting of eight outcomes. In contrast, the experiment of tossing a coin until a head appears has a *denumerably infinite* number of outcomes, H, TH, TTH, TTTH, ... The random variable "number of tosses" defined on these *discrete* outcomes is a discrete random variable as was "number of heads" in the earlier experiment. The probability of an event in a discrete sample space is the sum of the probabilities of the simple events which constitute the event in question. If we attach probabilities $\frac{1}{2}$, $\frac{1}{4}$, $\frac{1}{8}$, etc., respectively, to the sample points H, TH, TTH,..., the probability of the event "the number of tosses until an H appears is odd" is

$$\begin{aligned} P(\text{odd number of tosses}) &= P(\mathrm{H}) + P(\mathrm{TTH}) + P(\mathrm{TTTTH}) + \cdots \\ &= \tfrac{1}{2} + \tfrac{1}{8} + \tfrac{1}{32} + \cdots \\ &= \sum_{n=0}^{\infty} \frac{1}{2^{2n+1}} = \frac{1/2}{1-1/4} = \frac{2}{3} \end{aligned}$$

For some problems, appropriate sample spaces may be *continuous*. If the outcome is the weight of some object, the number of pounds may be any positive number (in some interval). For an example in which the physical experiment suggests probabilities, consider picking a number between 0 and 1 (including 0 but not 1) by spinning a pointer about the center of a circle of circumference 1. The distance along the perimeter in the direction of spin to the point designated by the pointer when it stops spinning is a random number. In dealing with problems involving continuous mass, the mass at a given point is zero; similarly, the probability of choosing any particular number between 0 and 1 is zero. If the probability is *uniformly* distributed over the interval $[0, 1)$, the probability of drawing a number between a and b, $0 \leqslant a \leqslant b \leqslant 1$, is $b-a$.

For another example, suppose a double-headed pointer is spun about the center of a circle of radius 1. Suppose the number we record is the distance along a tangent to the circle from the point of tangency to the indicated point, the distance being taken as positive if in one direction along the tangent and negative if in the other. The probability that the directed distance found will be between a and b, where $a \leqslant b$, might be considered to be $(\tan^{-1} b - \tan^{-1} a)/\pi$; see Fig. 2.2.

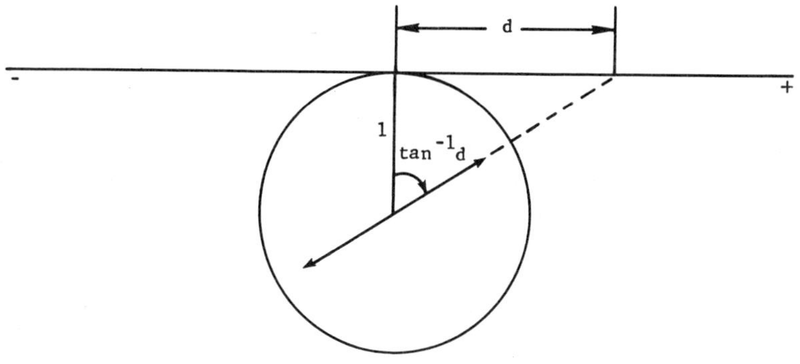

Figure 2.2 Double-headed pointer indicating a point a distance d in the positive direction from the origin of measurement.

2.2.3 Assigned Probabilities and Experience with Chance Events

Actual experiments in coin tossing have shown that, in the long run, when three coins are tossed repeatedly, HHH, HHT, HTH, HTT, THH, THT, TTH, and TTT each occurs about equally often, although not exactly equally often. The symmetry of the situation tends to make us believe that $\frac{1}{8}$ is a characteristic of the experiment. It makes some sense to say that the probability of getting three heads the next time three coins are tossed is $\frac{1}{8}$. In the experiment with stopping distances, the probability of stopping within 120 m is some number that can be approximated by observing, in a long series of similar experiments, the relative frequency of stops in less than 120 m.

2.3 PROBABILITY DISTRIBUTIONS

2.3.1 Univariate Probability Distributions. Distribution Functions

For any random variable defined on any sample space, the total probability, 1, is distributed over the possible values of the random variable. There are several ways in which a *probability distribution* may be described. For a *discrete* random variable, the probability of each possible value of the random variable may be given by formula or by table. For example, the probability of y defectives in a sample of 25 from a given process might be taken to be*

$$p(y)=P\{Y=y\}=\binom{25}{y}(.01)^{y}(.99)^{25-y}, \qquad y=0,1,\ldots,25 \quad (2.3.1)$$

or we may give the probability for each specific y as in Table 2.4 (where they are rounded to four decimal places).

For the distribution of a *continuous* random variable, we may give the probability that the value of the random variable falls in a given interval. (The probability that it takes on a particular value is zero.) Thus we may suggest that the probability that a particular piece of equipment will have a lifetime of between a and b hours might in a particular case be represented by

$$P(a \leqslant Y \leqslant b)=\int_a^b \frac{1}{12}e^{-u/12}\,du=e^{-a/12}-e^{-b/12}, \qquad 0 \leqslant a \leqslant b \quad (2.3.2)$$

$^*\binom{n}{r}\equiv\dfrac{n(n-1)\cdots(n-r+1)}{r!}=\dfrac{n!}{r!(n-r)!}, \qquad n!=1\cdot 2\cdots n,\ 0!=1.$

Table 2.4 Probabilities of Various Numbers of Defectives, y, in a Sample of 25 Taken from a Production Process Producing Defectives Randomly at a Rate of 1% [see (2.3.1)]

y	$P(Y=y)$
0	0.7778
1	0.1964
2	0.0238
3	0.0018
4	0.0001
5	0.0000
⋮	⋮
25	0.0000

The probability distribution of a continuous random variable is seldom described in this way since either of two alternative manners of description is simpler. Since $P(a \leqslant Y \leqslant b)$ for a continuous random variable can always be written as an integral with limits a and b, we can completely describe the distribution by giving the integrand. Thus for the equipment life distribution above, the integrand is

$$f_Y(y) = \begin{cases} \frac{1}{12} e^{-y/12}, & y \geqslant 0 \\ 0, & y < 0 \end{cases} \tag{2.3.3}$$

This *probability density function*, as such an integrand is called, is pictured in Fig. 2.3.

An alternative description of the distribution of a continuous random variable is that of giving $P(Y \leqslant y)$, called the *distribution function** of the random variable and symbolized by $F_Y(y)$. The subscript is often omitted when the random variable is clear. Thus

$$F_Y(y) = P(Y \leqslant y) = \int_{-\infty}^{y} f_Y(u)\,du \tag{2.3.4}$$

*The word "distribution" refers to the manner in which probabilities are associated with various events in the sample space. The phrase "distribution function" is used only in connection with random variables and is used for a particular function which describes the distribution of the random variable.

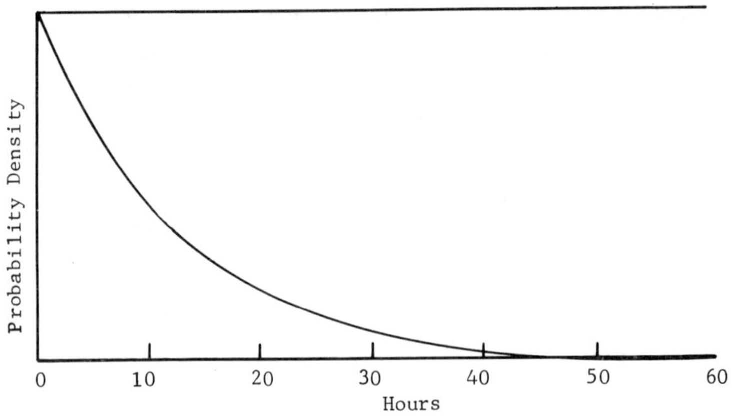

Figure 2.3 Probability density function (2.3.3).

and for the case of equipment life,

$$F_Y(y) = \begin{cases} 0, & y < 0 \\ 1 - e^{-y/12}, & y \geqslant 0 \end{cases} \tag{2.3.5}$$

This distribution function is pictured in Fig. 2.4. (When we talk of a continuous distribution we mean that the distribution function is absolutely continuous.) Discrete distributions can also be represented by distribution functions. Thus in the case of the distribution of number of

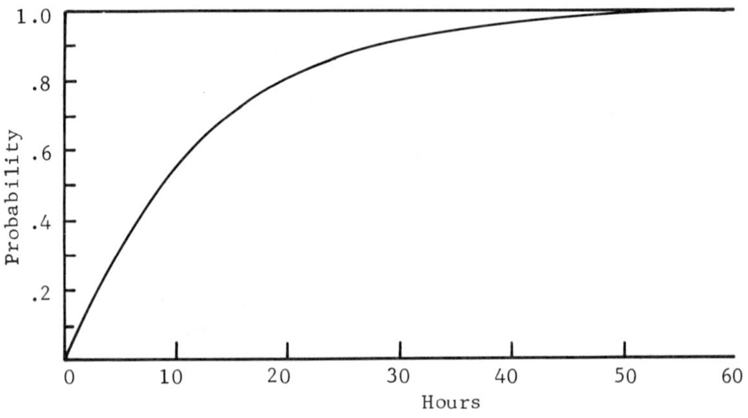

Figure 2.4 Distribution function (2.3.5).

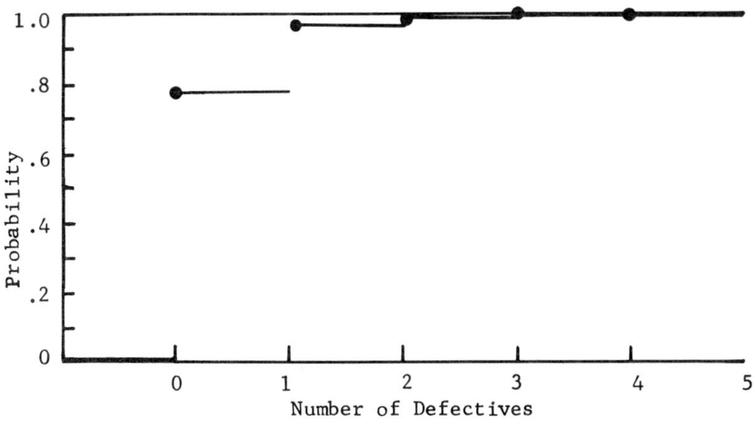

Figure 2.5 Distribution function (2.3.6) of number of defectives (Table 2.4).

defectives in a sample of 25,

$$F_Y(y) = \begin{cases} 0, & y < 0 \\ \sum_{i=0}^{[y]} \binom{25}{y} (.01)^y (.99)^{25-y}, & 0 \leqslant y < 25 \\ 1, & y \geqslant 25 \end{cases} \tag{2.3.6}$$

$[y]$ means the integral part of y. This distribution function is pictured in Fig. 2.5.

2.3.2 Multivariate Distributions

The joint distribution of two or more random variables defined on the same sample space is a multivariate distribution. An example is the distribution connected with simultaneous observations of temperature, pressure, wind direction, and wind velocity.

The *distribution function of a bivariate distribution* is $P(X \leqslant x, Y \leqslant y)$ and is usually symbolized by $F_{X,Y}(x,y)$.

$$F_{X,Y}(x,y) = P(X \leqslant x, Y \leqslant y) \tag{2.3.7}$$

with subscripts often omitted if the random variables involved are clear from context. If the two variables are absolutely continuous, there is a joint density function

$$f_{X,Y}(x,y) = \frac{\partial^2}{\partial x \, \partial y} F_{X,Y}(x,y) \tag{2.3.8}$$

If we have a bivariate distribution function $F_{X,Y}(x,y)$ we can readily get the distribution function of X or Y. Since $F_{X,Y}(x,\infty)$ is the probability that "$X \leqslant x$, $Y =$ any real number whatever,"

$$F_X(x) = F_{X,Y}(x,\infty) \tag{2.3.9}$$

and

$$F_Y(y) = F_{X,Y}(\infty,y) \tag{2.3.10}$$

If we start with a bivariate density function $f_{X,Y}(x,y)$, we find

$$f_X(x) = \int_{-\infty}^{\infty} f_{X,Y}(x,y)\,dy \tag{2.3.11}$$

and

$$f_Y(y) = \int_{-\infty}^{\infty} f_{X,Y}(x,y)\,dx \tag{2.3.12}$$

Similarly we have

$$F_{X_1,X_3}(x_1,x_3) = F_{X_1,X_2,X_3}(x_1,\infty,x_3) \tag{2.3.13}$$

and

$$f_{X_1,X_3}(x_1,x_3) = \int_{-\infty}^{\infty} f_{X_1,X_2,X_3}(x_1,x_2,x_3)\,dx_2 \tag{2.3.14}$$

and so on.

When we have a distribution involving several variables, the distribution of a proper subset of them is called a *marginal distribution*.

Table 2.5 gives a bivariate distribution with both variables discrete. Marginal distributions of X and Y are at the bottom and at the right side of the table, respectively. (Calculations were carried to more decimal places and rounded to four. Sums rounded to four places are not always identical to sums of numbers rounded to four places.) Table 2.6 gives the corresponding distribution function. Each entry is the sum of the entries in Table 2.5 which occupy the same position in the table plus all those above and to the left. Thus the entry for $X=2$, $Y=1$ is .3884 which, except for rounding, is equal to .0156+.0076+.1763+.0761+.0776+.0353 (=.3885). The marginal distribution functions are at the bottom and at the right side of the table.

Table 2.7 gives a bivariate density function with both variables continuous, the corresponding distribution function, the marginal distribution functions, and the marginal density functions. The details of deriving the other entries in the table from the joint denisty function are left as an exercise. (The fact that the joint density function changes form along the boundary of the region $0 \leqslant x \leqslant y$ must be taken into account in the integrations.)

Table 2.5 Probabilities that in a Sequence of 20 Trials, x will Result in Failure and y in Partial Success (Probability of Failure, .02; Probability of Partial Success, .1)

$$p(x,y)=P(X=x,Y=y)=\frac{20!}{x!y!(20-x-y)!}(.02)^x(.1)^y(.88)^{20-x-y}$$

	x					
y	0	1	2	3	4	Total
0	.0776	.0353	.0076	.0010	.0001	.1216
1	.1763	.0761	.0156	.0020	.0002	.2702
2	.1903	.0778	.0150	.0018	.0002	.2852
3	.1298	.0501	.0091	.0010	.0001	.1901
4	.0627	.0228	.0039	.0004		.0898
5	.0228	.0078	.0012	.0001		.0319
6	.0065	.0021	.0003			.0089
7	.0015	.0004	.0001			.0020
8	.0003	.0001				.0004
9						.0001
Total	.6676	.2725	.0528	.0065	.0006	1.0000

Table 2.6 Distribution Function of X and Y where X is the Number of Failures and Y is the Number of Partial Successes in a Series of 20 Trials (Probability of Failure, .02; Probability of Partial Success, .1)

$$F_{X,Y}(x,y)=\sum_{i=0}^{[x]}\sum_{j=0}^{[y]}\frac{20!}{i!j!(20-i-j)!}(.02)^i(.1)^j(.88)^{20-i-j}$$

	x							
y	0	1	2	3	4	5		20
0	.0776	.1128	.1204	.1215	.1216	.1216	...	.1216
1	.2538	.3652	.3884	.3914	.3917	.3917	...	.3917
2	.4441	.6334	.6716	.6765	.6769	.6769	...	.6769
3	.5739	.8133	.8606	.8665	.8670	.8670	...	.8670
4	.6366	.8987	.9499	.9562	.9568	.9568	...	.9568
5	.6593	.9293	.9817	.9881	.9887	.9887	...	.9887
6	.6658	.9378	.9906	.9970	.9976	.9976	...	.9976
7	.6673	.9397	.9925	.9990	.9995	.9996	...	.9996
8	.6676	.9400	.9929	.9993	.9999	.9999	...	.9999
9	.6676	.9401	.9929	.9994	1.0000	1.0000	...	1.0000
...	...	...	...	...	...	...	...	...
20	.6676	.9401	.9929	.9994	1.0000	1.0000	...	1.0000

Table 2.7 Example of Bivariate Density Function, Corresponding Bivariate Distribution Function, Marginal Distribution Functions, and Marginal Density Functions

$$f_{X,Y}(x,y)=\begin{cases}\frac{1}{9}e^{-y/3}, & 0\leqslant x\leqslant y\\ 0, & \text{otherwise}\end{cases}$$

$$F_{X,Y}(x,y)=\begin{cases}1-e^{-x/3}-\frac{1}{3}xe^{-y/3}, & 0\leqslant x\leqslant y\\ 1-e^{-y/3}-\frac{1}{3}ye^{-y/3}, & 0\leqslant y\leqslant x\\ 0, & \text{otherwise}\end{cases}$$

$$F_X(x)=\begin{cases}1-e^{-x/3}, & x\geqslant 0\\ 0, & \text{otherwise}\end{cases}$$

$$f_X(x)=\begin{cases}\frac{1}{3}e^{-x/3}, & x\geqslant 0\\ 0, & \text{otherwise}\end{cases}$$

$$F_Y(y)=\begin{cases}1-e^{-y/3}-\frac{1}{3}ye^{-y/3}, & y\geqslant 0\\ 0, & \text{otherwise}\end{cases}$$

$$f_Y(y)=\begin{cases}\frac{1}{9}ye^{-y/3}, & y\geqslant 0\\ 0, & \text{otherwise}\end{cases}$$

2.3.3 Sample Paths

We sometimes have continuous records of some variables such as temperature and velocity. Suppose, for example, that we are interested in several aspects of deceleration of automobiles. A driver adjusts his car to a steady speed of 55 mph, then applies his brakes. Measuring time from the moment the brake pedal is applied, we obtain a continuous record of the vehicle velocity and of the temperature of some point on the brake drum. In the sample space of this experiment, a sample point is the pair of continuous records of velocity and temperature. Since the sample point can be pictured as a curve in three-dimensional space (velocity and temperature against time), the sample point in such a case is referred to as a *sample path*.

A sample path is an infinite dimensional random variable. We sometimes conceive of our experiment as producing a sample path but actually record only periodically, say, every millisecond. We are particularly interested in random variables which are records of some continuous phenomenon taken at discrete points in time. The time points may or may not be equally spaced.

In dealing with simultaneous records over time our notation must take into account both the variable recorded and the time which is of interest; thus if we read velocity, X, at times t_1, t_3, and t_5 and temperature, Y, at t_2, t_3, and t_4, we have a six-dimensional random variable $X(t_1)$, $X(t_3)$, $X(t_5)$, $Y(t_2)$, $Y(t_3)$, $Y(t_4)$.

2.4 CONDITIONAL PROBABILITIES

2.4.1 Conditional Distributions. Discrete Case

In considering records of temperature in a room, we may be interested in finding answers to conditional questions such as "*If* the temperature at one particular point in the room is 21°C, what is the probability that, at the same time, the temperature at another particular point in the room is less than 20°C?" To handle this question we would need to build a model of the interrelations of temperature at various points in the room and at various times. Instead we illustrate concepts of conditional probability with a much simpler example, that of Tables 2.5 and 2.6. Let us ask, if there are no failures in the sample, what is the probability of two or more partial successes? Recall that our theory of probability is designed to make probabilities idealizations of relative frequencies. In this spirit let us interpret the probabilities of Table 2.6 as relative frequencies. We call attention to particular entries in Table 2.6 which we use in our illustration. From the table we read directly $P\{X=0, Y\leqslant 1\}=.2538$, $P\{X=0\}=.6676$, $P\{Y\leqslant 1\}=.3917$. From these we find $P\{X=0, Y\geqslant 2\}=P\{X=0\}-P\{X=0, Y\leqslant 1\}=.6676-.2538=.4138$, $P\{Y\geqslant 2\}=.6083$. Now if these probabilities were relative frequencies and if the total number of cases were 10,000, we would have 6676 cases in which $X=0$, of which 4138 were cases in which $X=0$ and $Y\geqslant 2$. The relative frequency of $Y\geqslant 2$ among cases of $X=0$ would be $4138/6676=.6198$. This relative frequency does not depend on the number of repetitions of the experiment and may be calculated directly from the probabilities of the event "$X=0$ and $Y\geqslant 2$" and the event "$X=0$," namely, $.4138/.6676=.6198$. It seems reasonable to call $P(X=0, Y\geqslant 2)/P(X=0)=.6198$ the conditional probability of $Y\geqslant 2$ given $X=0$. We symbolize the conditional probability of the event A given the event B by $P(A|B)$ and provisionally define

$$P(A|B)=\frac{P(A \text{ and } B)}{P(B)} \quad \text{if} \quad P(B)\neq 0 \tag{2.4.1}$$

In our example we found $P\{Y\geqslant 2|X=0\}=.6198$ which may be compared with $P\{Y\geqslant 2\}=.6083$. $P\{Y\geqslant 2|X\neq 0\}$ turns out to be .5851. Thus we see that the probability that $Y\geqslant 2$ depends on what we know about X

being equal to zero. In general,

$$\text{if} \quad P\{A|B\} > P\{A\}, \text{ then } P\{A|\text{not } B\} < P\{A\} \tag{2.4.2}$$

If we were to go through the calculations we would find that $P\{X=0|Y \geqslant 2\} \neq P\{X=0\}$ and, in general,

$$\text{if} \quad P\{A|B\} > P\{A\}, \quad \text{then} \quad P\{B|A\} > P\{B\} \tag{2.4.3}$$

Thus if the probability of A depends on whether or not B happens, the probability of B depends on whether or not A happens; we can simplify our description of the situation by saying that A and B are *dependent*.

If A and B are not dependent they are independent. If none of the probabilities involved are zero (we have not yet considered what conditional probability might mean if the denominator of the right member of (2.4.1) were zero), the inequalities of the preceding paragraphs imply that if $P\{A|B\} = P\{A\}$ then $P\{A|\text{not } B\} = P\{A\}$ and $P\{B|A\} = P\{B\} = P\{B|\text{not } A\}$. Several other equalities follow immediately. The development of these equalities is left as an exercise. A symmetric form and one that does not depend on a definition of conditional probability and does not require the exclusion of zero probabilities is used as a definition of *independence*. A and B are *independent* if

$$P\{A \text{ and } B\} = P\{A\}P\{B\} \tag{2.4.4}$$

2.4.2 Conditional Distributions. Continuous Case

If $P(a < Y \leqslant b) \neq 0$,

$$P(X \leqslant x | a < Y \leqslant b) = \frac{P(X \leqslant x, a < Y \leqslant b)}{P(a < Y \leqslant b)} \tag{2.4.5}$$

If the distribution function of Y is continuous in the interval $(a,y]$, if $a < y$, and if $f_Y(y) \neq 0$,

$$\lim_{a \to y} \frac{P(X \leqslant x, a < Y \leqslant y)}{P(a < Y \leqslant y)}$$

exists. We call the limit the conditional distribution function for X given $Y = y$ and symbolize it by $F_{X|Y}(x|y)$. By l'Hospital's rule

$$F_{X|Y}(x|y) = \frac{(\partial/\partial y)F_{X,Y}(x,y)}{f_Y(y)} \tag{2.4.6}$$

If X is a continuous random variable and y is a particular value of Y, $(\partial/\partial x)F_{X|Y}(x|y)$ exists whether Y is discrete or continuous. We call it the conditional density function and symbolize it by $f_{X|Y}(x|y)$. It is

$$f_{X|Y}(x|y)=\frac{\partial}{\partial x}F_{X|Y}(x|y)=\frac{(\partial/\partial x)(\partial/\partial y)F_{X,Y}(x,y)}{f_Y(y)}=\frac{f_{X,Y}(x,y)}{f_Y(y)} \tag{2.4.7}$$

Table 2.8 gives some conditional distributions connected with the bivariate random variable of Table 2.7.

Table 2.8 Some Conditional Distributions Related to the Bivariate Distribution of Table 2.7

$$f_{X,Y}(x,y)=\begin{cases}\frac{1}{9}e^{-y/3}, & 0\leqslant x\leqslant y\\ 0, & \text{otherwise}\end{cases}$$

$$F_{X,Y}(x,y)=\begin{cases}1-e^{-x/3}-\frac{1}{3}xe^{-y/3}, & 0\leqslant x\leqslant y\\ 1-e^{-y/3}-\frac{1}{3}ye^{-y/3}, & 0\leqslant y\leqslant x\\ 0, & \text{otherwise}\end{cases}$$

$$\begin{aligned}P(X\leqslant x,u<Y\leqslant v)&=F_{X,Y}(x,v)-F_{X,Y}(x,u)\\ &=\tfrac{1}{3}x(e^{-u/3}-e^{-v/3}), \qquad 0\leqslant x\leqslant u<v\end{aligned}$$

$$\begin{aligned}P(u<Y\leqslant v)&=F_{X,Y}(\infty,v)-F_{X,Y}(\infty,u)\\ &=\left(1+\tfrac{1}{3}u\right)e^{-u/3}-\left(1+\tfrac{1}{3}v\right)e^{-v/3}, \qquad 0\leqslant u\leqslant v\end{aligned}$$

$$P(X\leqslant x|u<Y\leqslant v)=\frac{x(e^{-u/3}-e^{-v/3})}{(3+u)e^{-u/3}-(3+v)e^{-v/3}}, \qquad 0\leqslant x\leqslant u<v$$

$$F_{X|Y}(x|y)=\lim_{u\to y}P(X\leqslant x|u<Y\leqslant y)=\frac{x}{y}, \qquad 0\leqslant x\leqslant y,\ \ y>0$$

$$f_{X|Y}(x|y)=\frac{\partial}{\partial x}\left(\frac{x}{y}\right)=\frac{1}{y} \quad \text{or} \quad \frac{f_{X,Y}(x,y)}{f_X(y)}=\frac{1}{y}, \qquad 0\leqslant x\leqslant y,\ \ y>0$$

Note that $f_{X|Y}(x|y)=0$ if $x<0<y$ or $0<y<x$ and is undefined for $y<0$.

2.4.3 Bayes's Theorem

In some problems of parameter estimation we have some information about the possible values of the parameters, and we may have some a priori idea as to the probability distribution of the possible values of the parameter. We should be able to combine this information with the information provided by our observations in estimating the actual parameter values. We assume that our model tells us the probabilities of obtaining particular observations when the parameter is known. Bayes's theorem tells us how to combine the a priori probabilities of parameter values with the probabilities of observations conditional on given parameter values to give us a posterior set of probabilities of parameter values. The theorem itself is more general, relating the probability of each of one set of disjoint events conditional on each event in a second set to the probability of each event of the second set conditional on each event in the first set. Before starting the theorem we give a simple example to illustrate the theorem in the case of discrete variates.

Example 2.4.1

We assume four factories (A_1, A_2, A_3, A_4), each with the fixed fraction of the total production of a particular product given in Table 2.9A. The product comes in three colors, (E_1, E_2, E_3). The fraction of the production of each factory devoted to each color is given in Table 2.9B. Find the fraction of the production of each color which is produced by each factory.

Table 2.9A Fraction of Total Output Produced by Each Factory

Factory	A_1	A_2	A_3	A_4
Fraction of total	0.40	0.32	0.20	0.08

Table 2.9B Fraction of Each Factory's Output in Each Color

	Factory			
Color	A_1	A_2	A_3	A_4
E_1	0.30	0.50	0.60	0
E_2	0.30	0.25	0.40	0
E_3	0.40	0.25	0	1.00
Total	1.00	1.00	1.00	1.00

Solution

The fraction of total production which falls into each of the 12 categories formed by the four factories and the three colors is given in Table 2.9C. The fraction of

Table 2.9C Fraction of Total Output in Each Factory-Color Combination

Color	Factory A_1	A_2	A_3	A_4	Totals
E_1	0.12	0.16	0.12	0	0.40
E_2	0.12	0.08	0.08	0	0.28
E_3	0.16	0.08	0	0.08	0.32
Totals	0.40	0.32	0.20	0.08	1.00

Table 2.9D Fraction of Output in Each Color Produced by Each Factory

Color	Factory A_1	A_2	A_3	A_4	Totals
E_1	0.30	0.40	0.30	0	1.00
E_2	0.43 –	0.29 –	0.29 –	0	1.00
E_3	0.50	0.25	0	0.25	1.00

total output in each color is found in the right margin of Table 2.9C. From the entries in Table 2.9C we can calculate the fraction of the production of each color which is produced by each factory. These fractions are given in Table 2.9D. A formula for obtaining the entries of Table 2.9D from those of Tables 2.9A and 2.9B is the content of Bayes's theorem.

We proceed in two steps to the statement of Bayes's theorem. First note that (2.4.1) may be rewritten as

$$P(A_i \text{ and } E_k) = P(A_i)P(E_k|A_i) \tag{2.4.8}$$

and also

$$P(A_i \text{ and } E_k) = P(E_k)P(A_i|E_k) \tag{2.4.9}$$

Thus if $P(E_k) \neq 0$,

$$P(A_i|E_k) = \frac{P(A_i)P(E_k|A_i)}{P(E_k)} \tag{2.4.10}$$

This is in essence the content of Bayes's theorem. To calculate $P(E_k)$ note that if we have a collection of disjoint events $A_1, A_2, \ldots, A_n$, the events "E_k and A_1," "E_k and A_2," etc. are disjoint and

$$\begin{aligned} P(E_k) &= P(E_k \text{ and } A_1) + P(E_k \text{ and } A_2) + \cdots + P(E_k \text{ and } A_n) \\ &= P(A_1)P(E_k|A_1) + P(A_2)P(E_k|A_2) + \cdots + P(A_n)P(E_k|A_n) \end{aligned} \tag{2.4.11}$$

and hence

$$P(A_j|E_k)=\frac{P(A_j)P(E_k|A_j)}{\sum_{i=1}^{n} P(A_i)P(E_k|A_i)} \tag{2.4.12a}$$

This is Bayes's theorem in detail. The computations required by this formula are those displayed in Tables 2.9A–D. Values of $P(A_i)$ are given in Table 2.9A; of $P(E_k|A_i)$, in Table 2.9B; of $P(A_i)P(E_k|A_i)$ and $P(E_k)=\Sigma_i P(A_i)P(E_k|A_i)$ in Table 2.9C; and of $P(A_i|E)$ in Table 2.9D.

For a continuous random variable we give the analogue of (2.4.12a) in a form we shall use later.

$$f_{\beta|X_1,\ldots,X_n}(\beta|x_1,\ldots,x_n)=\frac{g(\beta)f_{X_1,\ldots,X_n|\beta}(x_1,\ldots,x_n|\beta)}{\int_{-\infty}^{\infty} g(u)f_{X_1,\ldots,X_n|\beta}(x_1,\ldots,x_n|u)\,du} \tag{2.4.12b}$$

2.5 FUNCTIONS OF RANDOM VARIABLES

A function of a random variable is another random variable on the same sample space. It is sometimes convenient, however, to make use of relations between the two distributions. In some cases we find it convenient to find explicit descriptions of the new random variable, and in some cases we can find those properties of the new random variable which interest us in terms of the distribution of the old random variable. First let us look at some methods of describing the distribution of a random variable which is a function of a random variable whose distribution is known.

If X is a discrete random variable and if $Y=g(X)$, the probability $p_Y(y)$ is the sum of the probabilities $p_X(x)$ over those x's for which $y=g(x)$.

Example 2.5.1

Given the distribution of the random variable X defined by the first two columns of Table 2.10A, and given that

$$Y=\tfrac{1}{3}X^2(X^2-4)$$

find the distribution of Y.

Solution

For each x the value of $y=\frac{1}{3}x^2(x^2-4)$ is found, and the probabilities of the values of x which lead to each value of y are added to get the probabilities of each value of y. These probabilities are given in Table 2.10B.

Table 2.10A Probability Function for X and Values of Y

x	$p_X(x)$	y
−3	.2	15
−2	.3	0
−1	.1	−1
0	.2	0
1	.1	−1
2	.1	0

Table 2.10B Probability Function for Y

y	$p_Y(y)$
−1	.2
0	.6
15	.2

To get the distribution function of Y we need to find the probability over all the values of X such that $g(x) \leqslant y$. In some cases the distribution function of Y is simply found in terms of the distribution function of X.

If $Y = g(X)$ and $g(\cdot)$ is a continuous, strictly increasing function, that is, $x_2 > x_1$ implies $y_2 > y_1$, we have an inverse function $X = g^{-1}(Y)$ which is also monotonic strictly increasing and

$$F_Y(y) = P(Y \leqslant y) = P[g(X) \leqslant y] = P[X \leqslant g^{-1}(y)]$$

$$= F_X[g^{-1}(y)] = F_X(x) \qquad \text{where } x = g^{-1}(y) \qquad (2.5.1)$$

If X is a continuous random variable, Y is a continuous random variable and

$$f_Y(y) = \frac{d}{dy} F_Y(y) = \frac{d}{dy} F_X[g^{-1}(y)] = f_X[g^{-1}(y)] \frac{dg^{-1}(y)}{dy}$$

$$= f_X(x) \frac{dx}{dy} \qquad (2.5.2)$$

again with $x = g^{-1}(y)$.

Example 2.5.2

Let

$$f_X(x) = \begin{cases} \frac{2}{3}, & 0 < x < \frac{3}{2} \\ 0, & \text{otherwise} \end{cases}$$

or equivalently

$$F_X(x) = \begin{cases} 0, & x \leqslant 0 \\ \frac{2}{3}x, & 0 < x \leqslant \frac{3}{2} \\ 1, & \frac{3}{2} < x \end{cases}$$

Find the density function of $Y=2X$ directly from the density function of X and its distribution function from the distribution function of X.

Solution

$$f_Y(y)=f_X(x)\frac{dx}{dy}=\begin{cases}\frac{2}{3}\cdot\frac{1}{2}=\frac{1}{3}, & 0<y<3\\ 0, & \text{otherwise}\end{cases}$$

$$F_Y(y)=F_X\left(\tfrac{1}{2}y\right)=\begin{cases}0, & y\leqslant 0\\ \frac{2}{3}\left(\frac{1}{2}y\right)=\frac{1}{3}y, & 0<y\leqslant 3\\ 1, & 3<y\end{cases}$$

If $Y=g(X)$ and g is a monotonic strictly decreasing function, that is, $x_2>x_1$ implies $y_2<y_1$, we again have an inverse function and

$$F_Y(y)=P(Y\leqslant y)=P[g(X)\leqslant y]=P[X\geqslant g^{-1}(y)]$$
$$=1-F_X[g^{-1}(y)-0]$$

If X is differentiable we have

$$f_Y(y)=-f_X[g^{-1}(y)]\frac{dx}{dy}$$

If $Y=g(X)$ is monotonic but not strictly monotonic, Y may have some concentration of probability on certain values.

We give two examples in which the function $g(x)$ is not monotonic.

Example 2.5.3

Given the distribution function of the continuous random variable X, find the distribution function and the density function of

$$Y=X^2 \tag{2.5.3}$$

Solution

$$F_Y(y)=P(Y\leqslant y)=P(X^2\leqslant y)=P(-\sqrt{y}\leqslant X\leqslant\sqrt{y})$$
$$=F_X(\sqrt{y})-F_X(-\sqrt{y}-0),\qquad y>0 \tag{2.5.4}$$

$$f_Y(y)=\frac{1}{2\sqrt{y}}\left[f_X(\sqrt{y})+f_X(-\sqrt{y})\right] \tag{2.5.5}$$

Example 2.5.4

Given the density function of X

$$f_X(x)=\frac{1}{\sqrt{2\pi}}\exp\left[-\frac{(x-\mu)^2}{2}\right]$$

find the density function of $Y=X^2$.

Solution

Using (2.5.5) we find

$$f_Y(y)=\frac{1}{\sqrt{2\pi}\sqrt{y}}\exp\left[-\tfrac{1}{2}(y+\mu^2)\right]\cosh(\mu\sqrt{y})$$

2.6 EXPECTATIONS

2.6.1 Expected Value

Consider a sequence of observations on a sequence of identically distributed random variables. The average of the observed values in the sequence is the sum of those values divided by the number of observations. If we have a sufficiently long series of observations and a small enough number of different values of the random variable, it is easier to note the number of times each value appears and calculate the arithmetic average of the observed values. This can be done by calculating the sum of the products of the different values and their relative frequencies. The arithmetic average of the observed values is thus the weighted average of the possible values with weights equal to the observed relative frequencies of the possible values. If we substitute probabilities for relative frequencies, we get an idealization of arithmetic average, a weighted average of possible values with weights equal to their probabilities. This weighted average, which is not an arithmetic average, is called the *expected value* of the random variable. There are two common notations used for expected value of X, $E(X)$ and μ_X. (The μ, the Greek form of M, is an abbreviation for mean.) If there are n possible values for X: $x_1, x_2, \ldots, x_n$, the expected value for this *discrete* random variable is

$$E(X)=\sum_{i=1}^{n} x_i P(X=x_i) \tag{2.6.1}$$

Example 2.6.1

Find the expected number of defectives in a sample of 25 for a production process producing defectives randomly at a rate of 1%.

Solution

Let us use Table 2.4 and (2.6.1) to find

$$E(X)=\sum_{i=1}^{25} xP(X=x)$$

$$=0(.7778)+1(.1964)+2(.0238)+3(.0018)+4(.0001)+\cdots$$

$$=0.2498$$

(Unrounded probabilities would result in an answer $E(X)=0.25$.) Over a long period of inspecting samples of size 25 from a process producing 1% defectives, the average number of defectives per sample is expected to be near 0.25. We cannot obtain this value in a single experiment since the number of defectives must be an integer. Even the average over a long experiment will almost certainly differ from the expected value although the difference may be very small.

If X is a *continuous* random variable, we can consider the number we would get by dividing the range of values of the random variable into a number of disjoint intervals, calculating the probability of the random variable falling into each interval, picking one value of the random variable in each interval, and summing the products of each of these values with the probability of its interval. If we consider a sequence of such sums of products for sets of intervals for which the maximum length of any interval decreases to zero, the limit of the sums is the Riemann integral of the density function,

$$E(X)=\int_{-\infty}^{\infty} xf(x)\,dx \tag{2.6.2}$$

Example 2.6.2

Find the expected value of the random variable X with the density function

$$f(x)=\begin{cases} e^{-x}, & x>0 \\ 0, & \text{otherwise} \end{cases}$$

Solution

Using (2.6.2)

$$E(X)=\int_{0}^{\infty} xe^{-x}\,dx=[-(x+1)e^{-x}]_0^{\infty}=1$$

Not all probability distributions have expected values. In fact, if the angle $\tan^{-1}d$ of Fig. 2.2 is uniformly distributed between $-\pi/2$ and $\pi/2$, the random variable d does not possess an expected value. However, practical people adopt scales of measurement and models of physical situations such that it is highly unlikely that nonexistence of an expected value will cause trouble in a practical problem.

We can find the expected value of a *function of a random variable*, itself a random variable, by deriving the distribution of the values of the function and using the definition of expected values; or by associating the value of the function with each value of the original random variable, we may make use of the distribution of the original random variable. If X is a discrete random variable and $Y=g(X)$,

$$E(Y)=\sum_{\text{all } y} yP(Y=y)=\sum_{\text{all } y} yP[g(X)=y]$$

$$=\sum_{\text{all } y} y \sum_{\substack{\text{all } x \\ \text{for which} \\ g(x)=y}} P(X=x)=\sum_{\text{all } x} g(x)P(X=x) \tag{2.6.3a}$$

or

$$E[g(X)]=\sum_{\text{all } x} g(x)P(X=x) \tag{2.6.3b}$$

Example 2.6.3

Using the distributions of X and of Y of Table 2.10, find $E(Y)$ by both (2.6.1) and (2.6.3).

Solution

By (2.6.1), using Table 2.10B,

$$E(Y)=-1(.2)+0(.6)+15(.2)=-.2+3=2.8$$

By (2.6.3), using Table 2.10A,

$$E(Y)=15(.2)+0(.3)-1(.1)+0(.2)-1(.1)+0(.1)$$

$$=3-.1-.1=2.8$$

Similar to (2.6.3) we have, for continuous random variables, if $Y=g(X)$,

$$E(Y)=\int_{-\infty}^{\infty} yf_Y(y)\,dy=\int_{-\infty}^{\infty} g(x)f_X(x)\,dx \tag{2.6.4}$$

Example 2.6.4

Find $E(X^2)$ for the random variable with density function

$$f_X(x)=\begin{cases} e^{-x}, & x>0, \\ 0, & \text{otherwise} \end{cases}$$

Solution

$$E(X^2)=\int_0^{\infty} x^2e^{-x}\,dx=2$$

In the next few paragraphs we present some properties of expected value using continuous random variables as illustrations. The corresponding forms for discrete random variables may be supplied by the reader.

The expected value of a function of X and Y can be found without first finding explicitly the distribution of the function,

$$E[g(X,Y)]=\int_{-\infty}^{\infty}\int_{-\infty}^{\infty} g(x,y)f_{X,Y}(x,y)\,dx\,dy \qquad (2.6.5)$$

If, for example, $g(X,Y)=X$,

$$E(X)=\int_{-\infty}^{\infty}\int_{-\infty}^{\infty} xf_{X,Y}(x,y)\,dx\,dy \qquad (2.6.6)$$

The functional $E(\cdot)$ has the extremely important property of linearity, that is,

$$E(aX+bY)=aE(X)+bE(Y) \qquad (2.6.7)$$

To see this consider

$$\begin{aligned} E(aX+bY) &= \int_{-\infty}^{\infty}\int_{-\infty}^{\infty}(ax+by)f_{X,Y}(x,y)\,dx\,dy \\ &= a\int_{-\infty}^{\infty}\int_{-\infty}^{\infty} xf_{X,Y}(x,y)\,dy\,dx + b\int_{-\infty}^{\infty}\int_{-\infty}^{\infty} yf_{X,Y}(x,y)\,dx\,dy \\ &= aE(X)+bE(Y) \end{aligned}$$

This formula, (2.6.7), extends to linear combinations of any finite number of random variables. In words, the expected value of a linear combination of random variables defined on the same sample space is that same linear combination of expected values, that is,

$$E\left[\sum_{i=1}^{n} a_i X_i\right]=\sum_{i=1}^{n} a_i E(X_i) \qquad (2.6.8)$$

A particularly important random variable is

$$\bar{X}=\frac{1}{n}\sum_{i=1}^{n} X_i=\sum_{i=1}^{n}\frac{1}{n}X_i \qquad (2.6.9)$$

By (2.6.8)

$$E(\bar{X})=\sum_{i=1}^{n}\frac{1}{n}E(X_i) \qquad (2.6.10)$$

If $E(X_i)=\mu$ for $i=1,\ldots,n$

$$E(\bar{X})=\sum_{i=1}^{n}\frac{1}{n}E(X_i)=\mu \tag{2.6.11}$$

If X and Y are independent, possible values of X being $x_1,\ldots,x_m$, possible values of $Y_1, y_1,\ldots,y_n$,

$$\begin{aligned}
E(XY)&=\sum_{i=1}^{m}\sum_{j=1}^{n}x_i y_j P(X=x_i, Y=y_j)\\
&=\sum_{i=1}^{m}\sum_{j=1}^{n}x_i y_j P(X=x_i)P(X=y_j)\\
&=\left(\sum_{i=1}^{m}x_i P(X=x_i)\right)\left(\sum_{j=1}^{n}y_j P(Y=y_j)\right)=E(X)E(Y)
\end{aligned} \tag{2.6.12}$$

(Be warned that it is possible for $E(XY)$ to be equal to $E(X)E(Y)$ if X and Y are not independent.)

The properties of expected values which we have demonstrated are summarized in Table 2.11.

Table 2.11 Properties of Expected Value

$$E(X)=\begin{cases}\sum_{\text{all } x} xP(X=x) & \text{for discrete random variable } X\\ \int_{-\infty}^{\infty} xf_X(x)\,dx & \text{for continuous random variable } X\end{cases}$$

$$E(g(X))=\begin{cases}\sum_{\text{all } x} g(x)P(X=x) & \text{for discrete r.v.}\\ \int_{-\infty}^{\infty} g(x)f_X(x)\,dx & \text{for continuous r.v.}\end{cases}$$

$$E\left(\sum_{i=1}^{n}a_i X_i\right)=\sum_{i=1}^{n}a_i E(X_i)$$

In particular,

$$E(aX+bY)=aE(X)+bE(Y)$$

If $E(X_i)=\mu$ for all i, $E(\bar{X})=\mu$.

If X and Y are independent, $E(XY)=E(X)E(Y)$

2.6.2 Variance, Covariance, and Correlation

The *variance* of a random variable X is defined by

$$\boxed{V(X) \equiv E\left\{\left[X - E(X)\right]^2\right\} = \sigma_X^2} \tag{2.6.13}$$

The nonnegative square root of $V(X)$ is called the *standard deviation* and is symbolized σ_X. If the random variable involved is obvious from context, the subscript is often supressed.

Expansion of the square in the definition of $V(X)$ gives

$$V(X) = E(X-\mu)^2 = E(X^2 - 2X\mu + \mu^2)$$

$$= E(X^2) - 2\mu E(X) + \mu^2 = E(X^2) - \mu^2$$

or in summary

$$\boxed{V(X) = E(X^2) - \left[E(X)\right]^2} \tag{2.6.14}$$

Note that $V(X) \geqslant 0$ and is zero if and only if X is a constant.

Example 2.6.5

Using the distribution of Table 2.4, find $E(X^2)$ and $V(X)$.

Solution

$$E(X^2) = 0(.7778) + 1(.1964) + 2^2(.0238) + 3^2(.0018) + 4^2(.0001) = 0.3094$$

$$E(X) = 0(.7778) + 1(.1964) + 2(.0238) + 3(.0018) + 4(.0001) = 0.2498$$

By (2.6.14), $V(X) = .3094 - (.2498)^2 = .2470$. (If the probabilities in the table had been given exactly rather than rounded to four places, we would have found $E(X^2) = 0.31$, $E(X) = 0.25$, and $V(X) = 0.2475$.)

A dimensionless quantity known as the *coefficient of variation* is defined to be

$$\text{coefficient of variation} = \frac{\sigma_X}{\mu_X}$$

It is a measure of variation measured in terms of the size of the expected value.

Relating two random variables defined on the same sample space we have the *covariance*

$$\operatorname{cov}(X, Y) \equiv E\big[(X-\mu_X)(Y-\mu_Y)\big] \tag{2.6.15}$$

or equivalently,

$$\operatorname{cov}(X, Y) = E(XY) - \mu_X\mu_Y \tag{2.6.16}$$

The *correlation coefficient* is defined by

$$\rho_{X,Y} = \frac{\operatorname{cov}(X, Y)}{\sigma_X\sigma_Y} \tag{2.6.17}$$

Note that $\operatorname{cov}(X,X) = V(X)$.

Variance is not a linear functional. In fact,

$$V(aX) = E\big[aX - E(aX)\big]^2 = E\big[aX - aE(X)\big]^2 = E\big\{a\big[X - E(X)\big]\big\}^2$$

$$= E\Big\{a^2\big[X - E(X)\big]^2\Big\} = a^2E\big[X - E(X)\big]^2 = a^2V(X) \tag{2.6.18}$$

Table 2.12 Properties of Variance, Covariance, Correlation Coefficient

$V(X) = E\{[X-E(X)]^2\} = \sigma_X^2$	(2.6.13)
$V(X) = E(X^2) - [E(X)]^2$	(2.6.14)
$\operatorname{cov}(X, Y) = E\{[X-E(X)][Y-E(Y)]\}$	(2.6.15)
$\operatorname{cov}(X, Y) = E(XY) - [E(X)][E(Y)]$	(2.6.16)
$V(\sum_{i=1}^{n} a_iX_i) = \sum_{i=1}^{n} a_i^2V(X_i) + 2\sum_{i=1}^{n-1}\sum_{j=i+1}^{n} a_ia_j\operatorname{cov}(X_i, X_j)$	(2.6.20)
$= \sum_{i=1}^{n}\sum_{j=1}^{n} a_ia_j\operatorname{cov}(X_iX_j)$	
Specifically,	
$V(aX) = a^2V(X)$	(2.6.18)
$V(aX+k) = a^2V(X)$	(2.6.21)
$V(aX+bY) = a^2V(X) + b^2V(Y) + 2ab\operatorname{cov}(X, Y)$	(2.6.22)
If $\{X_i\}$ are independent with $V(X_i) = \sigma^2$ for $i = 1, \ldots, n$	
$V(\bar{X}) = \dfrac{\sigma^2}{n}$	(2.6.19)
$\rho_{X,Y} = \dfrac{\operatorname{cov}(X, Y)}{\sigma_X\sigma_Y}$	(2.6.17)

Other relations are found in Table 2.12. We call attention particularly to one to which repeated reference will be made. If $\{X_i\}$ are independent with identical variances σ^2, then

$$\boxed{V(\bar{X})=\frac{\sigma^2}{n}} \tag{2.6.19}$$

Example 2.6.6

For the density function of Tables 2.7 and 2.8,

$$f_{X,Y}(x,y)=\begin{cases}\frac{1}{9}e^{-y/3}, & 0\leqslant x\leqslant y\\ 0, & \text{otherwise}\end{cases}$$

find cov(X,Y) and the correlation coefficient for X and Y.

Solution

$$E(XY)=\frac{1}{9}\int_0^{\infty}\int_0^{y}xye^{-y/3}\,dx\,dy=27$$

$$E(X)=\frac{1}{9}\int_0^{\infty}\int_0^{y}xe^{-y/3}\,dx\,dy=3$$

$$E(Y)=\frac{1}{9}\int_0^{\infty}\int_0^{y}ye^{-y/3}\,dx\,dy=6$$

$$\text{cov}(X,Y)=27-(3)(6)=9$$

$$E(X^2)=18,\qquad E(Y^2)=54$$

$$\sigma_X^2=18-3^2=9,\qquad \sigma_Y^2=54-6^2=18$$

$$\rho_{X,Y}=\frac{9}{\sqrt{9\cdot 18}}=\frac{\sqrt{2}}{2}=.707$$

The variance of a function (other than linear) of two or more independent random variables is often difficult to determine in terms of the variances of the arguments of the function. We give here one example of one method of estimating such a variance, a method which has been in use for decades and which has been well described by Kline and McClintock [2]. If a random variable $Z=XY$, the product of two independent random variables, we might write X as $\mu_x+\varepsilon_x$, Y as $\mu_y+\varepsilon_y$, and Z as $\mu_z+\varepsilon_z$. We see that

$$\begin{aligned}\mu_z+\varepsilon_z&=(\mu_x+\varepsilon_x)(\mu_y+\varepsilon_y)\\&=\mu_x\mu_y+\mu_y\varepsilon_x+\mu_x\varepsilon_y+\varepsilon_x\varepsilon_y\end{aligned}$$

Since X and Y are independent, $\mu_z = \mu_x \mu_y$. Subtracting μ_z from both sides gives

$$\varepsilon_z = \mu_y \varepsilon_x + \mu_x \varepsilon_y + \varepsilon_x \varepsilon_y$$

If ε_x and ε_y are guaranteed to be very small in comparison with μ_x and μ_y, we have approximately

$$\varepsilon_z = \mu_y \varepsilon_x + \mu_x \varepsilon_z$$

$$\varepsilon_z^2 = \mu_y^2 \varepsilon_x^2 + \mu_x^2 \varepsilon_y^2 + 2\mu_x \mu_y \varepsilon_x \varepsilon_y$$

$$\sigma_z^2 = E(\varepsilon_z^2) = \mu_y^2 \sigma_x^2 + \mu_x^2 \sigma_y^2$$

or

$$\left(\frac{\sigma_z}{\mu_z}\right)^2 = \left(\frac{\sigma_x}{\mu_x}\right)^2 + \left(\frac{\sigma_y}{\mu_y}\right)^2$$

All this suggests that since, if $z = f(X, Y)$,

$$dz = \frac{\partial f}{\partial x} dx + \frac{\partial f}{\partial y} dy,$$

we might consider as an approximation to σ_z^2

$$\sigma_z^2 \cong \left(\frac{\partial f(x,y)}{\partial x}\right)^2_{\substack{x=\mu_x \\ y=\mu_y}} \sigma_x^2 + \left(\frac{\partial f(x,y)}{\partial y}\right)^2_{\substack{x=\mu_x \\ y=\mu_y}} \sigma_y^2$$

Kline and McClintock give several examples of the use of this method for two or more variates.

2.6.3 Stochastic Processes. Autocovariance, Cross-covariance

In dealing with the continuous record of temperature $X(t)$ as a function of time we may be interested in the relation between $X(t)$ at time t_1 and at time t_2 for individual records. The covariance between the joint random variables $X(t_1)$ and $X(t_2)$ representing points on the same sample path at different times, is called the *autocovariance*.

When we deal with two continuous stochastic processes, the covariance between one of them for one point in time, $X(t_1)$, and the other at a point in time possibly different, $Y(t_2)$, is called a *cross-covariance*.

Example 2.6.7

Consider a simple example which illustrates the concept of autocovariance in discrete time. Suppose the value of a random process at time t is 0.9 of its value at time $t-1$ plus the value of a random variable ϵ_t

$$X_t = 0.9X_{t-1} + \epsilon_t \tag{2.6.23}$$

where all finite sets of ϵ_i's are independent. Suppose that $X_0 = 0$, $P(\epsilon_t = -1) = .4$, $P(\epsilon_t = 0) = .2$, $P(\epsilon_t = 1) = .4$ for all t. Note that

$$E(\epsilon_t) = -1(.4) + 0(.2) + 1(.4) = 0 \qquad \text{for } t = 1, 2, \ldots$$

$$V(\epsilon_t) = 1(.4) + 0(.2) + 1(.4) = 0.8 \qquad \text{for } t = 1, 2, \ldots$$

Three realizations of this process to $t = 25$ generated by computer are shown in Fig. 2.6. Find $E(X_t)$, $V(X_t)$, $\text{cov}(X_s, X_t)$, ρ_{X_s, X_t}.

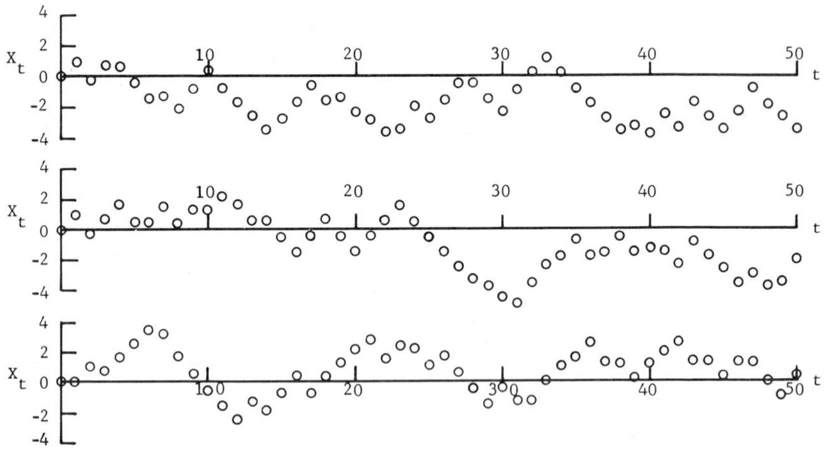

Figure 2.6 Three realizations of the process of Example 2.6.7.

Solution

$$X_1 = \epsilon_1$$

$$X_2 = 0.9\epsilon_1 + \epsilon_2$$

$$X_3 = 0.9^2\epsilon_1 + 0.9\epsilon_2 + \epsilon_3$$

$$\vdots$$

$$X_t = 0.9^{t-1}\epsilon_1 + 0.9^{t-2}\epsilon_2 + \cdots + \epsilon_t$$

$$E(X_t) = 0.9^{t-1}E(\epsilon_1) + 0.9^{t-2}E(\epsilon_2) + \cdots + E(\epsilon_t) = 0 \tag{2.6.24}$$

$$V(X_t) = E\big(0.9^{2t-2}\epsilon_1^2 + 0.9^{2t-4}\epsilon_2^2 + \cdots + \epsilon_t^2 + \text{terms each involving a product of two different } \epsilon_i\text{'s}\big)$$

$$= E\big(0.9^{2t-2}\epsilon_1^2 + 0.9^{2t-4}\epsilon_2^2 + \cdots + \epsilon_t^2\big)$$

$$= (1 + 0.9^2 + \cdots + 0.9^{2t-2})V(\epsilon_t) = \frac{1-0.9^{2t}}{1-0.9^2}(0.8)$$

$$= \frac{0.8}{0.19}(1-0.81^t)$$

$$\text{cov}(X_t, X_s | t<s) = E\big(0.9^{t+s-2}\epsilon_1^2 + 0.9^{t+s-4}\epsilon_2^2 + \cdots + 0.9^{s-t}\epsilon_t^2 + \text{terms each involving a product of two different } \epsilon_i\text{'s}\big) \tag{2.6.25}$$

$$= \frac{0.8}{0.19}\, 0.9^{s-t}(1-0.81^t)$$

$$\text{cov}(X_s, X_t) = \frac{0.8}{0.19}(1-0.81^{\min(s,t)}) \tag{2.6.26}$$

$$\rho_{X_s,X_t} = \frac{0.9^{|s-t|}(1-0.81^{\min(s,t)})}{\sqrt{(1-0.81^{\min(s,t)})(1-0.81^{\max(s,t)})}}$$

$$= 0.9^{|s-t|}\sqrt{\frac{1-0.81^{\min(s,t)}}{1-0.81^{\max(s,t)}}} \tag{2.6.27}$$

2.6.4 Stationarity

Many interesting stochastic processes are such that the distribution of $X(t)$ does not depend on t and the joint distribution of $X(t_1)$ and $X(t_2)$ for $t_2 > t_1$ depends only on the difference $t_2 - t_1$. We call such processes *stationary*. For instance, in a room with thermostatic control, the probability that the temperature will be between 21 and 22°C at any particular time in the future will be the same as for any other future instant. The probability that the temperature will be between 21 and 22°C at some particular time and between 22 and 24°C half an hour later will be the same as for any other two times half an hour apart. We call a process on

$0 \leqslant t < \infty$ a *stationary process* if the distribution of $X(t_1+h)$, $X(t_2+h), \ldots, X(t_n+h)$ is the same as the distribution of $X(t_1)$, $X(t_2), \ldots, X(t_n)$ for all n and all $t_i > 0$, $i = 1, 2, \ldots, n$ and $h > 0$.

Processes that are not stationary may possess some of the aspects of stationarity. If $E[X(t)] = E[X(t+h)]$ for all $h > 0$, $t > 0$ and $\sigma_{XX}(t_1, t_1+h)$ is a function of h only, the process is said to be *wide sense stationary*. Since a process which is stationary in the wide sense is not necessarily stationary, the stationarity of a process is sometimes emphasized by calling a stationary process strictly stationary.

For stationary processes, notation can be simplified unambiguously

$$\mu_{X(t)} = \mu_X \tag{2.6.28}$$

$$\operatorname{cov}\left[X(t_1), X(t_1+h)\right] = C_X(h) \tag{2.6.29}$$

$$\operatorname{cov}\left[X_1(t_1), X_2(t_1+h)\right] = C_{X_1, X_2}(h) \tag{2.6.30}$$

2.7 LAW OF LARGE NUMBERS. CENTRAL LIMIT THEOREM

2.7.1 Chebyshev's Inequality

We may ask why we think estimates based on observations should be any more likely to be near the quantity being estimated than pure guesses. Chebyshev's inequality tells us why we might hope. In this section we talk about the Chebyshev inequality itself. In the next section we show how it applies to averages.

In essence, Chebyshev's inequality says that, if we use the standard deviation of a random variable as a unit of measurement, the probability of being far from the expected value is small. To be precise, for any random variable X which possesses a standard deviation σ, the probability of X being at least $k\sigma$ from the expected value μ cannot be more than $1/k^2$. The Chebyshev inequality is expressed in symbols as

$$P(|X-\mu| \geqslant k\sigma) \leqslant \frac{1}{k^2} \tag{2.7.1}$$

Before giving a proof, let us give the Chebyshev inequality in three other forms for ease in reference.

$$P(|X-\mu| < k\sigma) \geqslant 1 - \frac{1}{k^2} \tag{2.7.2}$$

$$P(|X-\mu| \geqslant \epsilon) \leqslant \frac{\sigma^2}{\epsilon^2} \tag{2.7.3}$$

$$P(|X-\mu| < \epsilon) \geqslant 1 - \frac{\sigma^2}{\epsilon^2} \tag{2.7.4}$$

We give a proof for the case of a discrete random variable. For a continuous random variable replace sums with integrals.

By definition

$$\sigma^2 = \sum_{\text{all } x} (x-\mu)^2 p(x)$$

We may divide the possible x's into two sets, set A in which $|x-\mu| \geqslant k\sigma$ and set B in which $|x-\mu| < k\sigma$. We have

$$\sigma^2 = \Sigma_A (x-\mu)^2 p(x) + \Sigma_B (x-\mu)^2 p(x)$$

Now

$$\Sigma_B (x-\mu)^2 p(x) \geqslant 0$$

and

$$\Sigma_A (x-\mu)^2 p(x) \geqslant k^2\sigma^2 \Sigma_A p(x) = k^2\sigma^2 P(|X-\mu| \geqslant k\sigma)$$

Thus

$$\sigma^2 \geqslant k^2\sigma^2 P(|X-\mu| \geqslant k\sigma)$$

which, when both sides are divided by $k^2\sigma^2$, is the Chebyshev inequality. Notice that no particular probability distribution is assumed.

2.7.2 Weak Law of Large Numbers

We have based our axioms of probability on the way relative frequencies work, expecting near agreement between probabilities and relative frequencies if the number of trials is large enough. We now look at a theorem in probability which sounds as though it ensures such agreement. Actually, we are only showing some sort of consistency in the theory.

Consider an experiment, an event A with probability θ, and a random variable X which is 1 if A happens and 0 otherwise. Independent repetitions of the experiment constitute a super experiment which yields a sequence of independent random variables. Let Y be the average of the values of X over n independent repetitions of the experiment; that is, Y is the fraction of the experiments in which A happens. Since Y is the average of the X's, $E(Y) = E(X) = \theta = P(A)$. Since X has finite variance, σ^2, say, we have by (2.6.19), $V(Y) = \sigma^2/n$. Using this variance in (2.7.4) we get

$$P(|Y-\theta| < \epsilon) \geqslant 1 - \frac{\sigma^2}{n\varepsilon^2} \tag{2.7.5}$$

However small ϵ, we can find an n large enough that $P(|Y-\theta|<\epsilon)$ is as close as we please to 1. Thus

$$\lim_{n\to\infty} P(|Y-P(A)|<\epsilon)=1 \tag{2.7.6}$$

More generally, using the same type of argument, we have the *weak law of large numbers*; that is, if $\{X_i\}$ is a set of independent identically distributed random variables with expected value μ and variance σ^2,

$$P\left(|\overline{X}-\mu|<\epsilon\right) \geqslant 1-\frac{\sigma^2}{n\epsilon^2} \tag{2.7.7}$$

and

$$\lim_{n\to\infty} P\left(|\overline{X}-\mu|<\epsilon\right)=1 \tag{2.7.8}$$

2.7.3 Central Limit Theorem

We state the *central limit theorem* without proof, since a proof seems to be of no aid in understanding the theorem.

If X_i, $i=1,2,\ldots$ are independent identically distributed random variables, each with expected value μ and variance σ^2, then

$$Y=\frac{X_1+X_2+\cdots+X_n-n\mu}{\sigma\sqrt{n}} \tag{2.7.9}$$

has approximately the standard normal distribution, with density function

$$f(y)=\frac{1}{\sqrt{2\pi}}\exp\left(\frac{-y^2}{2}\right) \tag{2.7.10}$$

The approximation is as accurate as may be desired if n is sufficiently large. In this sense

$$\overline{X}=\frac{X_1+X_2+\cdots+X_n}{n} \tag{2.7.11}$$

may be said to be approximately distributed in accordance with the density function

$$f_{\overline{X}}(\overline{x})=\frac{\sqrt{n}}{\sigma\sqrt{2\pi}}\exp\left\{-\frac{n(\overline{x}-\mu)^2}{2\sigma^2}\right\} \tag{2.7.12}$$

Of course, the distribution of Y will be exactly normal if the X's are normally distributed.

2.8 EXAMPLES OF DISTRIBUTIONS

2.8.1 Bernoulli Distributions

The distribution which consists of 0 with probability $1-\theta$ and 1 with probability θ is known as the *Bernoulli distribution.*

A Bernoulli random variable is one which has (1) two possible values, 0 and 1, and for which (2) the probability that it takes on the value 1 is θ.

x	$P(X=x)$
0	$1-\theta$
1	θ

or

$$P(X=x)=\theta^{x}(1-\theta)^{1-x}, \qquad x=0,1 \tag{2.8.1}$$

The expected value and variance of a Bernoulli random variable are

$$\mu=\theta, \qquad \sigma^2=\theta(1-\theta) \tag{2.8.2}$$

The Bernoulli distribution was used implicitly earlier in connection with the weak law of large numbers.

2.8.2 Binomial Distributions

The sum of a fixed number of independent Bernoulli random variables is said to have a *binomial distribution.** Since a binomial random variable is the sum of independent identically distributed Bernoulli random variables, the expected value of a binomial random variable is $n\theta$ and its variance is $n\theta(1-\theta)$. The individual binomial probabilities are somewhat more difficult to compute. They turn out to be

$$P(X=x)=\frac{n!}{x!(n-x)!}\theta^{x}(1-\theta)^{n-x} \tag{2.8.3}$$

*The probabilities of various values for the random variable are the terms of the binomial expansion of $[(1-\theta)+\theta]^n$.

A binomial random variable is

1. The sum of
2. a fixed number n of
3. independent
4. Bernoulli random variables which have
5. a common value of θ.

The expected value and variance of a binomial random variable are

$$\mu = n\theta \tag{2.8.4}$$

$$\sigma^2 = n\theta(1-\theta) \tag{2.8.5}$$

2.8.3 Poisson Distributions

If arrivals are random in time, that is, if $P\{X(t)=x\}$ is the probability of x arrivals between time zero and time t, and if

$$\lim_{\Delta t \to 0} \frac{P\{X(t+\Delta t)-X(t)=1\}}{\Delta t} = \lambda, \qquad t>0, \quad \Delta t>0, \qquad x=0,1,\ldots \tag{2.8.6}$$

and

$$\lim_{\Delta t \to 0} \frac{P\{X(t+\Delta t)-X(t)\geqslant 2\}}{\Delta t} = 0, \qquad t>0, \quad \Delta t>0, \qquad x=0,1,2,\ldots \tag{2.8.7}$$

then X has a Poisson distribution

$$\{X(t)=x\} = \frac{(\lambda t)^x e^{-\lambda t}}{x!}, \qquad t>0, \quad x=0,1,\ldots \tag{2.8.8}$$

The distribution of time between arrivals is an exponential distribution (see Section 2.8.7). Note that if we are concerned with only one time, we can take this time as a unit and the symbol t may be omitted from the formula above.

The distribution function is

$$F_X(x) = \sum_{y=0}^{x} \frac{(\lambda t)^y e^{-\lambda t}}{y!} = \frac{1}{x!}\int_{\lambda t}^{\infty} u^x e^{-u}\,du, \qquad x=0,1,2,\ldots \tag{2.8.9}$$

$$\mu_X = \lambda t \tag{2.8.10}$$

$$\sigma_X^2 = \lambda t \tag{2.8.11}$$

If $X_1,\dots,X_n$ are independently distributed with Poisson distributions with expected values $\lambda t_1, \lambda t_2, \dots, \lambda t_n$, respectively, $X_1+\cdots+X_n$ has a Poisson distribution with expected value $\lambda(t_1+\cdots+t_n)$.

2.8.4 Uniform Distributions

If the probability that X is in an interval (X_L, X_U) is proportional to the length of the interval for X_L and X_U both inside an interval (a,b), then X is said to have a *uniform distribution* over the interval (a,b). Its probability density function is

$$f_X(x) = \begin{cases} 0, & x < a \\ \dfrac{1}{b-a}, & a < x \leqslant b \\ 0, & x > b \end{cases} \tag{2.8.12}$$

and its distribution function

$$F_X(x) = \begin{cases} 0, & x < a \\ \dfrac{x-a}{b-a}, & a < x \leqslant b \\ 1, & x > b \end{cases} \tag{2.8.13}$$

$$\mu = \frac{a+b}{2}, \qquad \sigma^2 = \frac{(b-a)^2}{12} \tag{2.8.14}$$

Figures 2.7 and 2.8 illustrate the approach to the normal distribution of the distribution of averages of independent observations from a uniform population. In Fig. 2.7 the horizontal scale is the average of n observations. In Fig. 2.8 the scale is $\sqrt{n}$ times the average, a scale which makes the standard deviation of the distributions equal.

2.8.5 Normal Distributions

The distribution that appears most frequently in this book both as a distribution in its own right and as an approximation to other distributions is the *normal* distribution. We have seen the normal distribution in the central limit theorem (Section 2.7.3). A random variable with a normal distribution with mean μ and variance σ^2, [abbreviated $N(\mu,\sigma^2)$], has the density function

$$f(x) = \frac{1}{\sqrt{2\pi}\,\sigma} \exp\left[-\frac{1}{2}\left(\frac{x-\mu}{\sigma}\right)^2 \right] \tag{2.8.15}$$

$$F(x) = \frac{1}{\sqrt{2\pi}\,\sigma} \int_{-\infty}^{x} \exp\left[-\frac{1}{2}\left(\frac{v-\mu}{\sigma}\right)^2 \right] dv \tag{2.8.16}$$

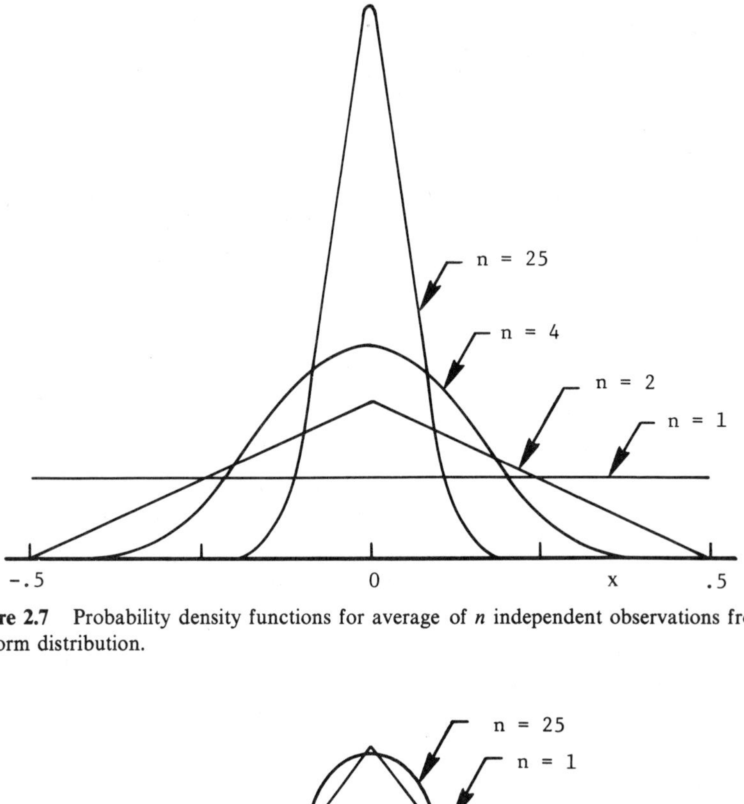

Figure 2.7 Probability density functions for average of n independent observations from a uniform distribution.

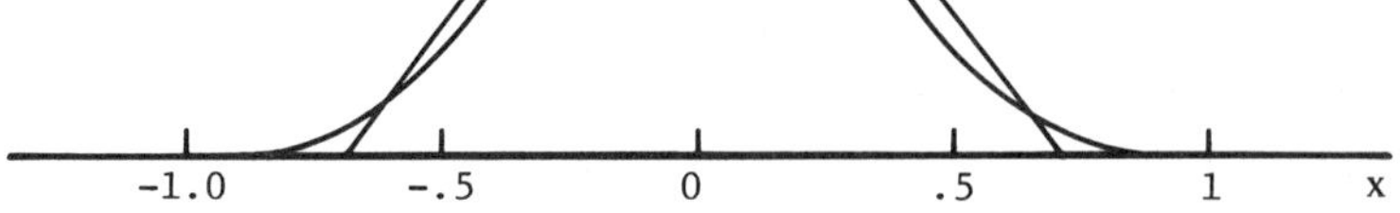

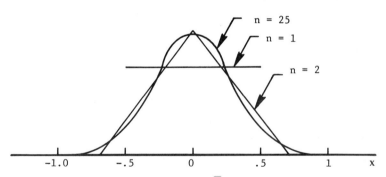

Figure 2.8 Probability density functions for $\sqrt{n}$ times the average of n independent observations from a uniform distribution.

The standard normal density $\varphi(z)$ and the standard normal distribution function $\Phi(z)$ are defined to be

$$\varphi(z)=\frac{1}{\sqrt{2\pi}}\exp\left(\frac{-z^2}{2}\right), \qquad \Phi(z)=\int_{-\infty}^{z}\varphi(v)\,dv \tag{2.8.17}$$

respectively. z is referred to as a standard normal deviate. We can express $f(x)$ and $F(x)$ in terms of $\varphi(x)$ and $\Phi(x)$. Thus

$$f(x)=\frac{1}{\sigma}\varphi\left(\frac{x-\mu}{\sigma}\right), \qquad F(x)=\Phi\left(\frac{x-\mu}{\sigma}\right) \tag{2.8.18}$$

Normal density functions with $\mu=0$ and $\sigma=0.4$, 1.0, and 2.5 are shown in Fig. 2.9. $\varphi(x)$ is pictured in Fig. 2.10 and $\Phi(z)$ in Fig. 2.11.

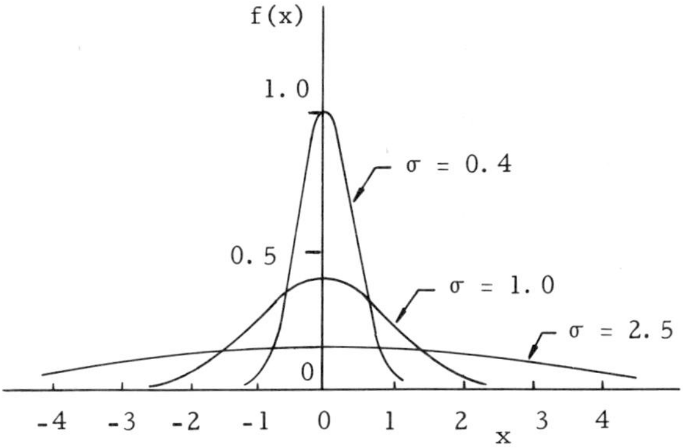

Figure 2.9 Normal probability density functions.

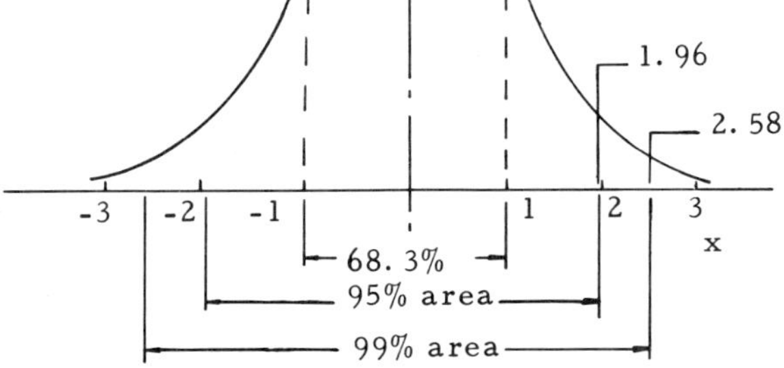

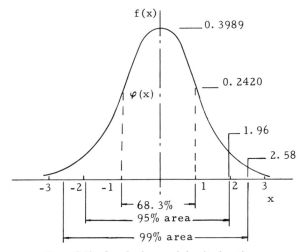

Figure 2.10 Standard normal density function.

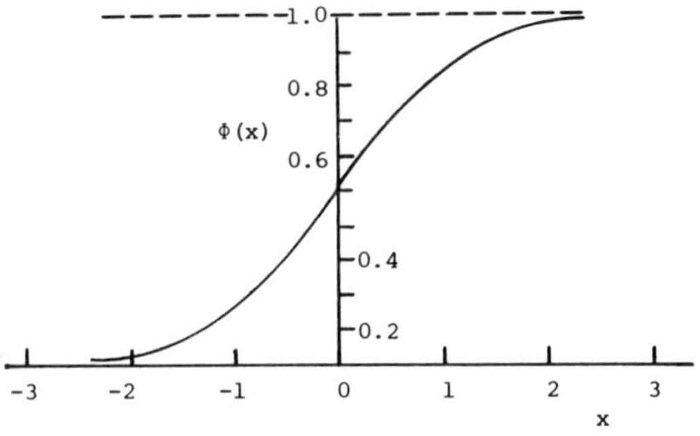

Figure 2.11 Standard normal distribution function.

Table 2.13 gives selected values of $\Phi(z)$. More extensive tables are readily available.

Table 2.13 Standard Normal Distribution Function

z	$\Phi(z)$	z	$\Phi(z)$
0	.5	1.282	.90
0.1	.5398	1.5	.9332
0.2	.5793	1.645	.95
0.3	.6179	1.960	.975
0.4	.6554	2.0	.9772
0.5	.6915	2.326	.99
0.6	.7257	2.5	.9938
0.8	.7881	2.576	.995
1.0	.8413	3.0	.9987

If X has a normal distribution with mean μ and variance σ^2 the probability of X being between a and b is

$$P\{a \leqslant X \leqslant b\} = \Phi\left(\frac{b-\mu}{\sigma}\right) - \Phi\left(\frac{a-\mu}{\sigma}\right) \tag{2.8.19}$$

If $X_1, X_2, \ldots, X_n$ are *independent* random variables, each normally distributed with mean μ and variance σ^2,

$$\bar{X} = \frac{X_1 + X_2 + \cdots + X_n}{n} \tag{2.8.20}$$

is normally distributed with mean μ and variance σ^2/n. In other words,

$$z = \frac{\bar{X}-\mu}{\sigma}\sqrt{n} \tag{2.8.21}$$

has distribution function $\Phi(z)$.

2.8.6 Multivariate Normal Distributions

Of the many possible generalizations of the normal distribution, the one which is called the bivariate normal is that with density function

$$f_{X,Y}(x,y) = \frac{1}{2\pi\sigma_x\sigma_y\sqrt{1-\rho^2}}$$
$$\times \exp\left\{-\frac{1}{2(1-\rho^2)}\left[\left(\frac{x-\mu_x}{\sigma_x}\right)^2 - 2\rho\left(\frac{x-\mu_x}{\sigma_x}\right)\left(\frac{y-\mu_y}{\sigma_y}\right)+\left(\frac{y-\mu_y}{\sigma_y}\right)^2\right]\right\} \tag{2.8.22}$$

Not only are the marginal distributions of the variates normal but all conditional distributions of one variate for fixed value of the other are normal. The choice of symbols for the parameters indicates exactly the role these parameters play in the joint distribution. The correlation coefficient between X and Y is ρ.

The marginal distribution of X has the density function

$$f_X(x) = \frac{1}{\sigma_x\sqrt{2\pi}}\exp\left[-\frac{1}{2}\left(\frac{x-\mu_x}{\sigma_x}\right)^2\right] \tag{2.8.23}$$

The conditional distribution of $Y|X$ is

$$f_{Y|X}(y|x) = \frac{1}{\sqrt{2\pi}\,\sigma_y\sqrt{1-\rho^2}}\exp\left[-\frac{1}{2\sigma_y^2(1-\rho^2)}\left[y-\mu_y-\rho\frac{\sigma_y}{\sigma_x}(x-\mu_x)\right]^2\right] \tag{2.8.24}$$

showing incidentally that the expected value of Y conditional on $X=x$ is

$$\mu_y + \rho\frac{\sigma_y}{\sigma_x}(x-\mu_x) \tag{2.8.25}$$

and its conditional variance is

$$\sigma_y^2(1-\rho^2) \tag{2.8.26}$$

A little algebra will disclose that, if X and Y have a bivariate normal distribution, the random variable $aX+bY$ is normal with mean

$$a\mu_x + b\mu_y \tag{2.8.27}$$

and variance

$$a^2\sigma_x^2 + b^2\sigma_y^2 + 2ab\rho\sigma_x\sigma_y \tag{2.8.28}$$

In general, an n-variate normal has density function of the form

$$k\exp\left[-\frac{1}{2}\sum_{i=1}^{n}\sum_{j=1}^{n}a_{ij}(x_i-\mu_{x_i})(x_j-\mu_{x_j})\right] \tag{2.8.29}$$

The relation between the a_{ij}'s and the variances and covariances of the X_i's is difficult to present without using matrix concepts. In Chapter 6, (6.1.66) gives the multivariate normal density function in terms of the variance–covariance matrix of the X_i's.

2.8.7 Gamma Distributions

A random variable Y with density function

$$f_Y(y)=\begin{cases}\dfrac{y^{\alpha-1}}{\lambda^\alpha\Gamma(\alpha)}e^{-y/\lambda}, & y>0;\quad \alpha>0,\quad \lambda>0\\ 0, & \text{otherwise}\end{cases} \tag{2.8.30}$$

is said to have a gamma distribution with parameters α and λ. Its distribution function is

$$F_Y(y)=\frac{1}{\lambda^\alpha\Gamma(\alpha)}\int_0^y u^{\alpha-1}e^{-u/\lambda}\,du=\frac{1}{\Gamma(\alpha)}\int_0^{y/\lambda}u^{\alpha-1}e^{-u}\,du \tag{2.8.31}$$

$$\mu_Y=\alpha\lambda \tag{2.8.32}$$

$$\sigma_Y^2=\alpha\lambda^2 \tag{2.8.33}$$

The parameter α need not be an integer. One application in which it is restricted to integral values is that in which Y is the time to the αth arrival under the conditions described in Section 2.8.3.

If $Y_1,\ldots,Y_k$ are independently distributed with gamma distributions with parameters (α_1,λ), $(\alpha_2,\lambda),\ldots,(\alpha_k,\lambda)$, then $Y_1+\cdots+Y_k$ has a gamma distribution with parameters $(\alpha_1+\cdots+\alpha_k,\lambda)$.

A gamma distribution with parameters $(1,\lambda)$ is said to have an exponential distribution with expected value λ.

2.8.8 Chi-squared Distributions

If Y has a gamma distribution with parameters $(\nu/2,2)$ it is said to have a χ^2 (chi, *ch* as in Christmas) distribution with ν degrees of freedom. Its density and distribution functions are

$$f_{\chi^2}(y)=\begin{cases}\dfrac{1}{2^{\nu/2}\Gamma(\nu/2)}y^{(\nu/2)-1}e^{-y/2}, & y>0\\ 0, & \text{otherwise}\end{cases} \tag{2.8.34}$$

and

$$F_{\chi^2}(y)=\frac{1}{2^{\nu/2}\Gamma(\nu/2)}\int_0^y u^{(\nu/2)-1}e^{-(u/2)}\,du \tag{2.8.35}$$

respectively.

If $X_1,\ldots,X_n$ are independently normally distributed with expected values $\mu_1,\ldots,\mu_n$, resepectively, and variances $\sigma_1^2,\ldots,\sigma_n^2$, respectively, the sum

$$\sum_{i=1}^{n}\left(\frac{X_i-\mu_i}{\sigma_i}\right)^2=\left(\frac{X_1-\mu_1}{\sigma_1}\right)^2+\cdots+\left(\frac{X_n-\mu_n}{\sigma_n}\right)^2 \tag{2.8.36}$$

has a χ^2 distribution with $\nu=n$ degrees of freedom.

If $X_1,\ldots,X_n$ are independently normal with common expected value μ and common variance σ^2, and if

$$\bar{X}=\frac{X_1+\cdots+X_n}{n}$$

then

$$\frac{1}{\sigma^2}\sum_{i=1}^{n}\left(X_i-\bar{X}\right)^2=\frac{1}{\sigma^2}\left\{\left(X_1-\bar{X}\right)^2+\cdots+\left(X_n-\bar{X}\right)^2\right\} \tag{2.8.37}$$

has a χ^2 distribution with $\nu=n-1$ degrees of freedom. Sums such as (2.8.37) frequently occur in parameter estimation.

If $Y_1,\ldots,Y_n$ are independently distributed as χ^2 with $\nu_1,\ldots,\nu_n$ degrees of freedom, respectively, $Y_1+\cdots+Y_n$ is distributed as χ^2 with $\nu_1+\cdots+\nu_n$ degrees of freedom.

The distribution of χ^2/ν and the distribution of $\chi^2\sigma^2$ are frequently of interest. They are so closely related to χ^2 that it is usual to use tables of $F_{\chi^2}^{-1}(\cdot)$ in calculations related to them.

The density functions of χ^2 for 1, 2, 3, 4, 5, and 10 degrees of freedom are given in Fig. 2.12.

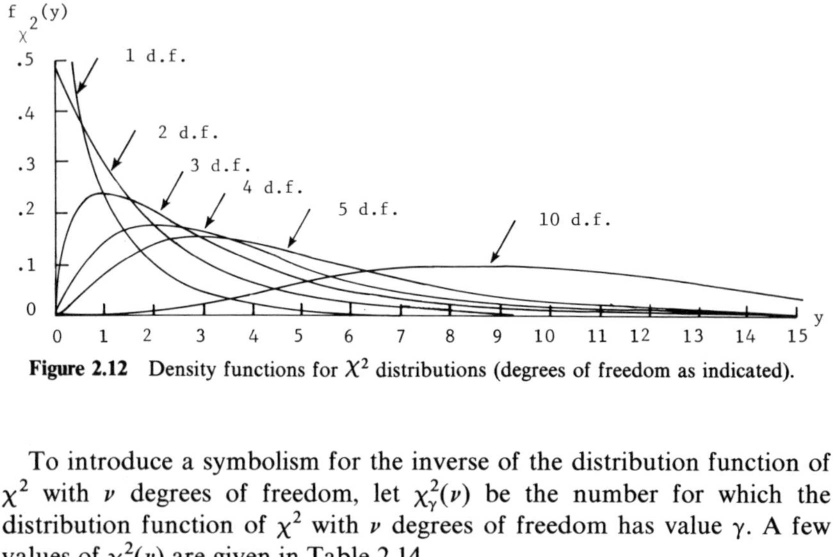

Figure 2.12 Density functions for X^2 distributions (degrees of freedom as indicated).

To introduce a symbolism for the inverse of the distribution function of χ^2 with ν degrees of freedom, let $\chi^2_\gamma(\nu)$ be the number for which the distribution function of χ^2 with ν degrees of freedom has value γ. A few values of $\chi^2_\gamma(\nu)$ are given in Table 2.14.

Table 2.14 Some Values of $\chi^2_\gamma(\nu)$

Degrees of Freedom, ν	γ 0.01	0.025	0.05	0.1	0.5	0.9	0.95	0.975	0.99
1	0.00	0.00	0.00	0.02	0.45	2.71	3.84	5.02	6.63
10	2.56	3.25	3.94	4.87	9.34	15.99	18.31	20.48	23.21
15	5.23	6.26	7.26	8.55	14.34	22.31	25.00	27.49	30.58
30	14.95	16.79	18.49	20.60	29.34	40.26	43.77	46.98	50.89
60	37.5	40.5	43.2	46.5	59.3	74.4	79.1	83.3	88.4

2.8.9 *t* Distributions

If X has a normal distribution with mean 0 and variance 1, and Y is independently distributed as χ^2 with ν degrees of freedom, the statistic T,

$$T = \frac{X}{\sqrt{Y/\nu}} \tag{2.8.38}$$

has a distribution known as a *t distribution* with ν degrees of freedom. The density function of T is

$$f_T(t) = \frac{\Gamma[(\nu+1)/2]}{\sqrt{\pi\nu}\,\Gamma(\nu/2)} \cdot \frac{1}{(1+t^2/\nu)^{(\nu+1)/2}}, \qquad -\infty < t < \infty \tag{2.8.39}$$

As ν increases without limit, the distribution of t approaches the normal distribution as a limit.

Density functions for t with 1, 5, and ∞ degrees of freedom are displayed in Fig. 2.13.

To introduce a symbolism for the inverse of the distribution function of t with ν degrees of freedom, let $t_\gamma(\nu)$ be the number for which the distribution function of t with ν degrees of freedom has value γ. A few values of $t_\gamma(\nu)$ are given in Table 2.15.

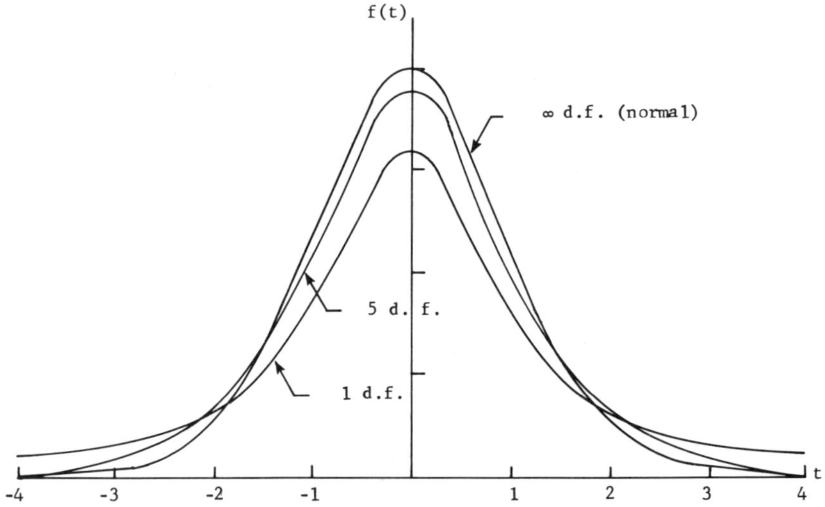

Figure 2.13 Density functions for t distributions (degrees of freedom as indicated).

Table 2.15 Some Values of $t_\gamma(\nu)$

Degrees of Freedom, ν	γ 0.5	0.8	0.9	0.95	.975	.99	.995
1	0.0	1.38	3.08	6.31	12.71	31.82	63.66
10	0.0	0.88	1.37	1.81	2.23	2.76	3.17
15	0.0	0.87	1.34	1.75	2.13	2.60	2.95
30	0.0	0.85	1.31	1.70	2.04	2.46	2.75
60	0.0	0.85	1.30	1.67	2.00	2.39	2.66
120	0.0	0.84	1.29	1.66	1.98	2.36	2.62
∞	0.0	0.84	1.28	1.64	1.96	2.33	2.58

The statistic

$$\frac{\sqrt{n}\,(\bar{Y}-\mu)}{S} \tag{2.8.40}$$

where $Y_1,\ldots,Y_n$ are independently normally distributed with common variance and where

$$\mu = E(Y_i) \qquad \text{and} \qquad S^2 = \frac{\sum_{i=1}^{n}(Y_i-\bar{Y})^2}{n-1} \tag{2.8.41}$$

has a t distribution with $\nu = n-1$ degrees of freedom. Although somewhat unusual, the capital S is used in this section to clearly indicate that S is a random variable.

Many statistics of importance in parameter estimation have t distributions.

Under the conditions stated in connection with (2.8.40), the statistic

$$\frac{\bar{Y}-a}{S}\sqrt{n} \tag{2.8.42}$$

is said to have a noncentral t distribution with noncentrality parameter $(\mu-a)\sqrt{n}/\sigma$.

2.8.10 *F* Distributions

If U has a χ^2 distribution with ν_1 degrees of freedom (d.f.) and V has an independent χ^2 distribution with ν_2 d.f., $(U/\nu_1)/(V/\nu_2)$ has an F distribution with ν_1 and ν_2 d.f.

$$f(F) = \frac{\Gamma[(\nu_1+\nu_2)/2]}{\Gamma(\nu_1/2)\Gamma(\nu_2/2)}\,\frac{\nu_1^{\nu_1/2}\nu_2^{\nu_2/2}F^{\nu_1/2-1}}{(\nu_1 F+\nu_2)^{(\nu_1+\nu_2)/2}} \tag{2.8.43}$$

The F distribution is used frequently in parameter estimation concerned with model building.

To introduce a symbolism for the inverse of the distribution function of F with ν_1 and ν_2 degrees of freedom, let $F_\gamma(\nu_1, \nu_2)$ be the number for which the distribution function of F with ν_1 degrees of freedom in the numerator sum of squares and ν_2 degrees of freedom in the denominator sum of squares has value γ. A few values of $F_{.95}(\nu_1, \nu_2)$ are given in Table 2.16.

Table 2.16 $F_{.95}(\nu_1, \nu_2)$

	ν_1					
ν_2	1	10	30	60	120	∞
1	161.45	241.88	250.10	252.20	253.25	254.31
10	4.96	2.98	2.70	2.62	2.58	2.54
30	4.17	2.16	1.84	1.74	1.68	1.62
60	4.00	1.99	1.65	1.53	1.47	1.39
120	3.92	1.91	1.55	1.43	1.35	1.25
∞	3.84	1.83	1.46	1.32	1.22	1.00

If X has an F distribution with ν_1 and ν_2 degrees of freedom, X^{-1} has an F distribution with ν_2 and ν_1 degrees of freedom.

2.8.11 Tables and Computer Programs for Commonly Used Statistics

Programs for calculating values of the distribution functions of normal, t, χ^2, and F distributions are available for many computers including some microcomputers. Expansions which are tabled in Abramowitz and Stegun [1] are the basis for most such programs.

Many collections of tables of percentiles of normal, t, χ^2, and F distributions are available. All of the collections of references [3] through [9] contain these and more. The two handbooks also contain extensive collections of formulas used in statistical computation.

REFERENCES

1. Abramowitz, M. and Stegun, I. A., editors, *Handbook of Mathematical Functions with Formulas, Graphs, and Mathematical Tables*, National Bureau of Standards Applied Mathematics Series No. 55, Washington, D.C., 1964.
2. Kline, S. J. and McClintock, F. A., "Describing Uncertainties in Single-sample Experiments," *Mech. Eng.*, **75** (1953) 2–8.
3. Owen, D. B., *Handbook of Statistical Tables*, Addison-Wesley Publishing Co., Reading, Mass., 1962.

4. Pearson, E. S. and Hartley, H. O., editors, *Biometrika Tables for Statisticians*, 3rd ed., University Press, Cambridge 1966.
5. Fisher, R. A. and Yates, F., *Statistical Tables for Biological, Agricultural and Medical Research*, 6th rev. ed., Oliver and Boyd London, 1963; reprinted by Hafner Publishing Co., Darien, Conn., 1974.
6. Beyer, W. H., editor, *CRC Handbook of Tables for Probability and Statistics*, 2nd ed., Chemical Rubber Co., Cleveland, 1968.
7. Burington, R. S. and May, D. C., Jr., *Handbook of Probability and Statistics with Tables*, 2nd ed., McGraw-Hill Book Company, New York, 1970.
8. Hald, A., *Statistical Tables and Formulas*, John Wiley & Sons, Inc., New York, 1952.
9. Arkin, H. and Colton, R. R., *Tables for Statisticians*, 2nd ed., Barnes & Noble Inc., New York, 1963.

PROBLEMS

2.1 Three parts are subject to large stresses in starting a machine at the beginning of the working day. Let A be the event that Part A fails, B the event that Part B fails, C the event that Part C fails. $P(A)=.04$, $P(B)=.02$, $P(C)=.03$, $P(A \text{ and } B)=.008$, $P(A \text{ and } C)=.007$, $P(B \text{ and } C)=.01$, $P(A \text{ and } B \text{ and } C)=.0001$. Find $P(A \text{ or } B)$, $P(A \text{ or not } B)$, $P(A \text{ or } B \text{ or } C)$, $P(\text{neither } A \text{ nor } B)$.

2.2 If

$$f_Y(y)=\begin{cases} 1, & 0 \leqslant y < 1 \\ 0, & \text{otherwise} \end{cases}$$

find $F_Y(y)$.

2.3 Consider the joint probability function given below

	x_1		
x_2	0	1	2
0	h	$2h$	$3h$
1	$2h$	$4h$	$6h$
2	$3h$	$6h$	$9h$
3	$4h$	$8h$	$12h$

(*a*) What is h?
(*b*) Give the joint distribution function.
(*c*) Find the marginal probability functions.
(*d*) Are X_1 and X_2 independent?

Answer. (partial). For $x_1=0,1,2; p(X_1=x_1)=\frac{1}{6},\frac{1}{3},\frac{1}{2}$ respectively

2.4 If

$$f_{X,Y}(x,y)=\begin{cases} 8xy, & 0\leqslant x\leqslant y\leqslant 1 \\ 0, & \text{otherwise} \end{cases}$$

find $F_{X,Y}(x,y)$, $F_X(x)$, $F_Y(y)$.

2.5 For the joint probability function of Problem 2.4, give the conditional probability density function $f_{X|Y}(x|y)$.

Answer. $2x/y^2$ if $0 \leq x \leq y \leq 1$; 0 if $0 \leq y \leq 1$ and $x < 0$ or $x > y$; undefined otherwise

2.6 If

$$f_{X,Y}(x,y)=\begin{cases} 8xy, & 0\leqslant x\leqslant y\leqslant 1 \\ 0, & \text{otherwise} \end{cases}$$

find $f_{Y|X}(y|x)$.

2.7 A test for the presence of a pollutant which causes the quality of output to deteriorate is not infallible. It falsely indicates the presence of the pollutant with probability .05, and it fails to detect the presence of the pollutant with probability .01. From our experience, we would judge the probability of the pollutant being present is .01. If we get a positive indication from the test, what is the probability that the pollutant is present?

2.8 If the actual diameters of shot vary uniformly over the interval (.200, .205), what is the distribution of volume?

2.9 Define the new random variable Z which is related to the random variable X by

$$Z=2+4X+2X^2$$

where X has the probability function $P(X=1)=\frac{1}{2}$, $P(X=2)=\frac{1}{4}$, $P(X=4)=\frac{1}{4}$ and $P(X=x)=0$ otherwise. Find the mean of Z.

Answer. 21

2.10 Consider the joint probability density for X, Y, and Z,

$$f(x,y,z)=Ae^{-a_1x-a_2y-a_3z}, \qquad \begin{aligned} x>0;\ \ a_1>0 \\ y>0;\ \ a_2>0 \\ z>0;\ \ a_3>0 \end{aligned}$$

(*a*) Find A in terms of a_1, a_2, and a_3.
(*b*) Find $E(X)=\mu_X$.
(*c*) Are X, Y, and Z independent?
(*d*) Give the joint distribution function for X, Y, and Z.

2.11 Compute the mean and variance for the discrete probability functions given below:

(*a*)

$$P(X=0)=\tfrac{1}{3}$$

$$P(X=1)=\tfrac{2}{3}$$

$$P(X=x)=0 \quad \text{for} \quad x \neq 0,1$$

Answer. $\frac{2}{3}$, $\frac{2}{9}$

$$(b)\ P(X=x)=\frac{6!}{x!(6-x)!}\left(\frac{2}{3}\right)^x\left(\frac{1}{3}\right)^{6-x} \qquad \text{for } x=0,1,2,\ldots,6$$
$$=0 \qquad \text{otherwise}$$

Answer. 4, $\frac{4}{3}$

$$(c)\ P(X=x)=e^{-2}\frac{2^x}{x!} \qquad \text{for } x=0,1,2,\ldots$$
$$=0 \qquad \text{otherwise}$$

Answer. 2, 2

2.12 Compute the mean and variance of the distributions whose density function are

(*a*)

$$f(x)=\begin{cases} 2x, & 0<x<1 \\ 0, & \text{otherwise}\end{cases}$$

Answer. $\frac{2}{3}$, $\frac{1}{18}$

(*b*)

$$f(x)=\begin{cases} |x|, & |x|\leqslant 1 \\ 0, & \text{otherwise}\end{cases}$$

Answer. 0, $\frac{1}{2}$

(*c*)

$$f(x)=\begin{cases} (2/\pi)^{1/2}e^{-x^2/2}, & x>0 \\ 0, & \text{otherwise}\end{cases}$$

Answer. $\sqrt{2/\pi}$, $1-\dfrac{2}{\pi}$

2.13 Find the variance of Z of Problem 2.9. (How does it compare with the variance of $Z-2$?)

2.14 For the joint probability function given below, find the covariance between X and Y.

	x			
y	1	2	3	4
0	0	0.1	0	0.1
1	0.1	0.2	0	0.1
2	0.2	0.1	0.1	0

Answer. $-.34$

2.15 If $F=\Sigma_{i=1}^{n} a_i Y_i$, and

$$E\{[Y_i-E(Y_i)][Y_j-E(Y_j)]\}=\begin{cases}\sigma^2 & \text{for } i=j\\ 0 & \text{for } i\neq j\end{cases}$$

prove that $V(F)=\sigma^2\Sigma_{i=1}^{n} a_i^2$.

2.16 For the distribution of Problem 2.4 find the covariance of X and Y. Find the coefficient of correlation between X and Y.

2.17 If $Y_i = Y_{i-1}+\epsilon_i$, $i=1,2,\ldots, Y_0=0$ where ϵ_i are normal, independently distributed with zero expected value and variance 5, find the correlation between Y_i and Y_{i-1}.

Answer. $\sqrt{(i-1)/i}$

2.18 An autoregressive model is defined by

$$\epsilon_i=\rho\epsilon_{i-1}+u_i, \qquad i=1,2,\ldots,n$$

where $\epsilon_0\equiv 0$ and ρ is a given constant, $-1<\rho<1$

$$u_i\sim N(0,\sigma^2) \quad i=1,2,\ldots,n, \quad E(u_iu_j)=0 \qquad \text{for} \quad i\neq j.$$

(*a*) Show that $E(\epsilon_i)=0$ for all i.
(*b*) Show that

$$E(\epsilon_1\epsilon_i)=\rho^{i-1}\sigma^2, \qquad i=1,2,\ldots,n$$

(*c*) Show that

$$E(\epsilon_2\epsilon_i)=\rho^{i-2}(1+\rho^2)\sigma^2, \qquad i=2,\ldots,n$$

(*d*) Show that

$$E(\epsilon_i^2)=[1+\rho^2+\rho^4+\cdots+\rho^{2(i-1)}]\sigma^2, \qquad i=1,2,\ldots,n$$

(*e*) What is the limiting value of $E(\epsilon_i^2)$ as $i\to\infty$?

(*f*) Show that

$$E\left[\sum_{i=1}^{n} \epsilon_i^2\right] = \frac{\sigma^2}{1-\rho^2}\left[n - \frac{\rho^2(1-\rho^{2n})}{1-\rho^2}\right]$$

and thus

$$\lim_{n\to\infty} \frac{1}{n} E\left[\sum_{i=1}^{n} \epsilon_i^2\right] = \frac{\sigma^2}{1-\rho^2}$$

2.19 Toss two pennies 12 times and record your results for the number of heads. Repeat this three times. Compare the actual numbers of heads with the expected number.

2.20 What are the expected value and variance of the number of successes in 100 independent trials in each of which the probability of success is $\frac{1}{2}$? What are the expected value and the variance of the fraction of trials which are successes?

2.21 If Y is normally distributed with expected value 100 and standard deviation 5, find (*a*) $P(Y>100)$, (*b*) $P(Y<95)$, (*c*) $P(|Y-100|<10)$, (*d*) $P(Y\geqslant 95)$, (*e*) $P(|Y-100|\geqslant 10)$, (*f*) the y such that $P(Y\geqslant y)=.95$, (*g*) the c such that $P(|Y-100|<c)=.90$.

2.22 If Y is normally distributed with expected value 100 and standard deviation 5, and if the average of a sample of 25 is $\bar{Y}$, find (a) $P(\bar{Y}>101)$, (b) $P(\bar{Y}>101)$, (c) $P(|\bar{Y}-100|>2)$, (d) the c such that $P(\bar{Y}>c)=.95$, (e) the c such that $P(|\bar{Y}-100|>c)=.05$.

2.23 If Y has a Poisson distribution with $\lambda t=4$, find $P(3\leqslant Y\leqslant 5)$ and compare with the normal approximation, $P(2.5<Y\leqslant 5.5)$ for Y normal with $E(Y)=4$ and $V(Y)=4$.

2.24 Prove that μ and σ^2 are indeed mean and variance respectively of the random variable whose density function is given by (2.8.15).

2.25 Derive (2.8.24).

2.26 Derive (2.8.26).

2.27 One of the convenient properties of the normal density functions is the following identity. Verify algebraically that for any real numbers x, μ_1, μ_2, σ_1, and σ_2 $(\sigma_1>0,\sigma_2>0)$

$$\exp\left[-\frac{1}{2}\left(\frac{x-\mu_1}{\sigma_1}\right)^2\right]\exp\left[-\frac{1}{2}\left(\frac{x-\mu_2}{\sigma_2}\right)^2\right]$$

$$=\exp\left[-\frac{1}{2}\left(\frac{x-\mu}{\sigma}\right)^2\right]\exp\left[-\frac{1}{2}\frac{(\mu_1-\mu_2)^2}{\sigma_1^2+\sigma_2^2}\right]$$

where

$$\mu = \frac{\mu_1\sigma_2^2 + \mu_2\sigma_1^2}{\sigma_1^2 + \sigma_2^2}, \qquad \sigma^2 = \frac{\sigma_1^2\sigma_2^2}{\sigma_1^2 + \sigma_2^2}$$

2.28 Using (2.8.22), derive (2.8.23). Note that

$$\int_{-\infty}^{\infty} e^{-ax^2}\,dx = \sqrt{\pi/a}$$

$$\int_{-\infty}^{\infty} x\exp\left[-a(x-\mu)^2 - b(x-\mu)\right]dx = e^{b^2/4a}\left(\mu - \frac{b}{2a}\right)\sqrt{\pi/a}$$

2.29 Eleven independent observations were taken from a normal population,

$$\{32.2,\ 32.7,\ 31.9,\ 32.9,\ 32.3,\ 31.7,\ 32.6,\ 32.5,\ 32.5,\ 32.2,\ 32.2\}.$$

(*a*) Find the sample variance.

(*b*) What is the probability that an s^2 for a random sample of size 11 from a normal population with $\sigma^2 = 6$ would be larger than the value found in (*a*)?

2.30 In addition to the sample of Problem 2.29, two observations from another normal population with the same variance but not necessarily identical mean were obtained. The sample variance of this sample of two is to be calculated. What value will be exceeded with probability 5% by the ratio of this sample variance to a random sample of 11 such as that of Problem 2.29?

CHAPTER 3

INTRODUCTION TO STATISTICS

In Chapter 2 we developed the idea of a *probabilistic model*. We studied how a knowledge of the model may be used to derive probabilities of the various possible outcomes in the sample space and of the various values of a random variable defined on the sample space. We now turn to problems in which the values of one or more parameters in the model are unknown and we are to use observations to estimate them.

For example, assuming that the heat equation for an infinite slab adequately represents the physical aspects of the experiment, we may gather data on temperature changes on the face of an actual slab and estimate thermal conductivity. We may assume exponential decay of acoustical energy in a recording studio, measure acoustical energy as a function of time at various points in the studio, and estimate the rate of decay.

In studying methods of deriving estimates from data, it is worthwhile to use different words to distinguish between a method of estimating and the result of applying the method. An *estimator* is a formula or procedure for deriving an estimate from a sample. An estimator is a random variable. An *estimate*, by contrast, is a number. The word *statistic* is used both for a formula for deriving a number or numbers, not necessarily an estimate, from a sample and for the number obtained by applying the formula to a particular sample.

In this chapter we consider what can be expected of estimators and how to choose the best possible estimator for our immediate purposes. After we look at one or two more or less specific problems, we begin a somewhat systematic consideration of characteristics that we would like estimators to have and investigate ways of devising estimators that possess these characteristics.

3.1 SOME EXAMPLES OF ESTIMATORS

Before beginning a formal study of methods of developing and evaluating estimators, we look at a few particular problems. These serve to introduce some basic ideas and raise some questions that are studied in the remainder of the chapter.

3.1.1 Two Estimators of the Center of a Symmetric Distribution

Suppose that we have a sample of n independent observations $X_i, i = 1, 2, \ldots, n$ from the same distribution and that the one thing we know about this distribution is that it is symmetric. Suppose that we wish to estimate the center of symmetry. A random variable X is symmetrically distributed if $P\{X \leqslant \gamma - t\} = P\{X \geqslant \gamma + t\}$ for some γ and for all t; γ is the center of symmetry. We might look for symmetry in the sample but samples, even from symmetric distributions, are unlikely to be symmetric. It is fairly clear, however, that, if γ is a center of symmetry, it is also the expected value of X if X possesses an expected value and, in addition, it is the median of X. By definition, λ is a *median* of X if

$$P\{X \leqslant \lambda\} \geqslant \tfrac{1}{2} \qquad \text{and} \qquad P\{X \geqslant \lambda\} \geqslant \tfrac{1}{2} \tag{3.1.1}$$

The *median of a sample*, M, is similarly defined as a number such that at least half of the sample values are at least as large and at least half of the sample values are at least as small. If there is an interval of medians, it is quite common to name the average of the largest and the smallest median as *the* median.

Two of the many possible estimators of the center of a symmetric probability distribution are the mean and the median of a sample of independent observations.

If a continuous random variable Y has distribution function $F_Y(y)$ and probability density function $f_Y(y)$, the probability density function of the median of a sample of an odd number n of independent observations—and for simplicity, we deal only with the case of an odd number of observations—can be shown to be

$$f_M(x) = \left[F_Y(x)\right]^{(n-1)/2}\left[1 - F_Y(x)\right]^{(n-1)/2} f_Y(x) \tag{3.1.2}$$

If

$$f_Y(y) = \frac{1}{\sqrt{2\pi}} \exp\left[-\tfrac{1}{2}(y-\mu)^2\right] \tag{3.1.3}$$

then

$$F_Y(y) = \Phi(y-\mu)$$

where the notation of (2.8.18) is used. The probability density function for the median of a sample of n (n odd) is given by (3.1.2). The density function for the mean of a sample of n is [by (2.8.21)],

$$f_{\bar{Y}}(\bar{y}) = \sqrt{\frac{n}{2\pi}} \exp\left(-\frac{n}{2}(\bar{y}-\mu)^2\right) \tag{3.1.4}$$

Figure 3.1 pictures the probability density function of mean and of median for samples of size 101 from the distribution of (3.1.3). It is clear that the mean is the better estimator from any reasonable point of view. The probability that the mean of a sample of 101 will be within ε of the center of symmetry of the distribution is, for all $\varepsilon > 0$, greater than the probability that the median of 101 will be within ε of the center. For $\varepsilon = .12$, $P(\mu - .12 < \bar{Y} < \mu + .12) = .77$; $P(\mu - .12 < M < \mu + .12) = .66$.

The comparison of mean and median of samples of 101 independent observations on a double exponential random variable with probability density function

$$f_Y(y) = \tfrac{1}{2}\exp(-|y-\mu|) \tag{3.1.5}$$

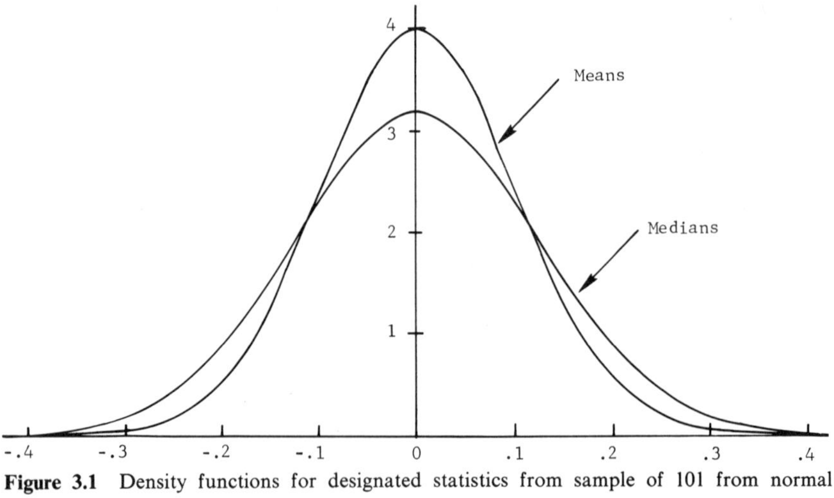

Figure 3.1 Density functions for designated statistics from sample of 101 from normal population.

is quite different. The probability density functions for the two estimators are pictured in Fig. 3.2. The probability that the mean of a sample of 101 will be within ε of the center of the distribution is, for all $\varepsilon > 0$, less than the probability that the median will be within ε of the center. $P(\mu - .12 < \bar{Y} < \mu + .12) = .61$; $P(\mu - .12 < M < \mu + .12) = .75$.

The choice of an estimator is not always as clear-cut even when the form of the distribution is as fully known. It is possible, for example, that for two different ε's, ε_1 and ε_2, and for two different estimators of θ, T_1 and T_2,

$$P(|T_1 - \theta| < \varepsilon_1) < P(|T_2 - \theta| < \varepsilon_1)$$

and

$$P(|T_1 - \theta| < \varepsilon_2) > P(|T_2 - \theta| < \varepsilon_2)$$

3.1.2 Estimating a Variance

If we wish to estimate the variance $V(Y) = \sigma^2$ of a random variable Y—see Section 2.6.2—using a sample of n independent observations, we might consider the statistic

$$\frac{\sum_{i=1}^{n} (Y_i - \mu)^2}{n} \tag{3.1.6}$$

To see that the expected value of this statistic is σ^2 (if σ^2 is finite), we note

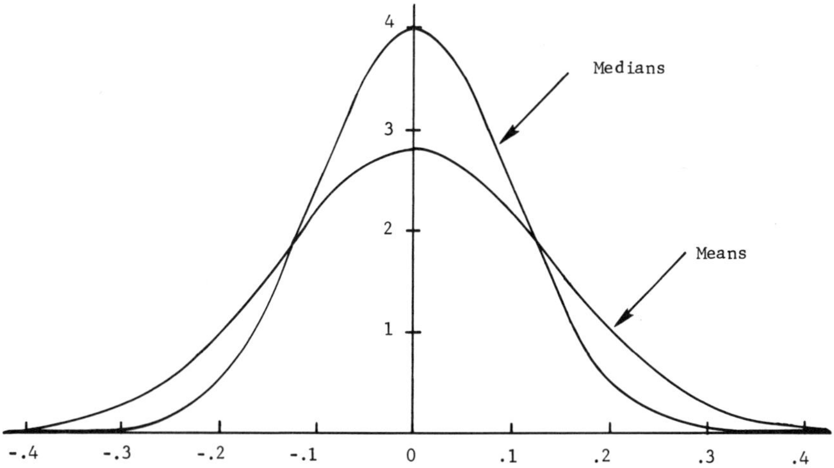

Figure 3.2 Density functions for designated statistics from sample of 101 from double exponential distribution.

that

$$E(Y_i-\mu)^2=\sigma^2 \tag{3.1.7}$$

by definition and

$$E\left[\frac{\sum_{i=1}^{n}(Y_i-\mu)^2}{n}\right]=\frac{1}{n}\sum_{i=1}^{n}E(Y_i-\mu)^2=\frac{1}{n}\sum_{i=1}^{n}\sigma^2=\sigma^2 \tag{3.1.8}$$

Unfortunately, to calculate (3.1.6) we must know μ.

It seems reasonable to ask whether

$$A=\frac{\sum_{i=1}^{n}\left(Y_i-\bar{Y}\right)^2}{n} \tag{3.1.9}$$

might be used as an estimator of σ^2. It can, of course, but it will tend, on the average, to underestimate σ^2. This can be seen by noting that

$$\sum_{i=1}^{n}(Y_i-\mu_Y)^2=\sum_{i=1}^{n}\left[\left(Y_i-\bar{Y}\right)+\left(\bar{Y}-\mu_Y\right)\right]^2$$

$$=\sum_{i=1}^{n}\left(Y_i-\bar{Y}\right)^2+\sum_{i=1}^{n}\left(\bar{Y}-\mu_Y\right)^2+2\left(\bar{Y}-\mu_Y\right)\sum_{i=1}^{n}\left(Y_i-\bar{Y}\right)$$

$$=\sum_{i=1}^{n}\left(Y_i-\bar{Y}\right)^2+n\left(\bar{Y}-\mu_Y\right)^2 \tag{3.1.10}$$

Dividing the left member by n and taking expected value, we get σ^2 [by (3.1.8)]. Dividing the right side by n and using (3.1.9) and (2.6.19), we get

$$E(A)+E\left(\bar{Y}-\mu_Y\right)^2=E(A)+\frac{\sigma^2}{n} \tag{3.1.11}$$

Solving for $E(A)$,

$$E(A)=\frac{n-1}{n}\sigma^2 \tag{3.1.12}$$

Rather than using A as an estimator of σ^2, we use

$$S^2 = \frac{1}{n-1} \sum_{i=1}^{n} \left(Y_i - \bar{Y} \right)^2 \tag{3.1.13}$$

which is an unbiased estimator of σ^2; that is, its expected value is σ^2. We use a capital S, as we did in Section 2.8.9, to emphasize the fact that we are here talking about a random variable. We see that an estimator may sometimes be modified to give another estimator the properties of which please us more. Similar modifications of estimators of variance are used throughout the following chapters.

It is remarkable that we can derive the expected value of a sample variance for independent observations with no information about the distribution of the random variable except that it possesses a variance. In fact, any unbiased estimator of the standard deviation depends on much more detailed information about the distribution of the random variable.

If $\{Y_i;\ i=1,2,\ldots,n\}$ are independent normally distributed with common variance σ^2, the statistic

$$\frac{(n-1)S^2}{\sigma^2} \tag{3.1.14}$$

has a χ^2 distribution (Section 2.8.8) with $n-1$ degrees of freedom. The distribution function of S^2 is the distribution function of $[\sigma^2/(n-1)]\chi^2$, that is, a χ^2 distribution with a scale change in the argument of the function.

3.2 PROPERTIES OF ESTIMATORS

3.2.1 Unbiasedness

A statistic T is said to be an *unbiased* estimator of a parameter θ if

$$E(T) = \theta \tag{3.2.1}$$

Unbiasedness sounds like a good property for an estimator to have. It is a simple property to describe. Roughly, an estimator is unbiased if, on the average, it yields the correct value of the parameter.

Other properties are almost always more important to us, however. Fortunately, we can sometimes find an unbiased estimator which also has the other properties we want for a particular application.

The standard deviation of a sample from a normal population with standard deviation σ is biased, whether standard deviation of a sample is defined with a divisor of n or, as we have defined it, with an $n-1$. In fact, reference to the χ^2 distribution (Section 2.8.8) yields

$$E(S)=\frac{\Gamma\left(\frac{n}{2}\right)}{\sqrt{\frac{n-1}{2}}\;\Gamma\left(\frac{n-1}{2}\right)}\sigma \tag{3.2.2}$$

For $n=2, E(S)=.80\sigma$; for $n=5, .94\sigma$; for $n=10, .97\sigma$.

Example 3.2.1

What, if any, multiple of the average of n independent observations on a gamma random variable with known fixed α—see Section 2.8.7—will be an unbiased estimator of λ?

Solution

Using the properties of gamma random variables described in Section 2.8.7, we see that, if each observation has parameters (α,λ), the sum of n has the parameters $(n\alpha,\lambda)$. The expected value of the sum is $n\alpha\lambda$ and the expected value of the average is therefore $\alpha\lambda$. The sample average multiplied by $1/\alpha$ will be an unbiased estimator of λ.

3.2.2 Consistency

Unless a sample is actually the whole population being sampled, an estimate of a population parameter cannot be expected to be equal to that parameter. If we have a sequence of estimators, one for each sample size, we would hope that larger samples would tend to give better estimates.

A *consistent sequence of estimators*, $T(n), n=1,2,\ldots,$ of a parameter, θ, is one for which

$$\lim_{n\to\infty} P(|T(n)-\theta|<\delta)=1 \qquad \text{for every } \delta>0 \tag{3.2.3}$$

Thus a consistent sequence of estimators is one for which a sufficiently large sample is almost certain to produce an estimate close to the parameter value.

Example 3.2.2

Show that sample means form a consistent sequence of estimators of the population mean if the distribution sampled possesses a variance.

Solution

If the distribution possesses a variance, the weak law of large numbers (see Section 2.7.2) says

$$\lim_{n \to \infty} P\left(|\bar{X} - \mu| < \varepsilon\right) = 1 \tag{3.2.4}$$

Hence, the sequence is consistent.

3.2.3 Efficiency. Minimum Variance Unbiased Estimators

Recall Figure 3.1 picturing the distribution of the mean $\bar{Y}$ and of the median M for samples of 101 from a normal population. The distribution of $\bar{Y}$ is, of course, normal (see Section 2.8.5). It appears that the distribution of M is approximately normal. A mathematical analysis confirms the approximate normality of the distribution of M for large sample size and yields the additional information that the variance of M is approximately $\pi\sigma^2/2n$ compared with σ^2/n for the variance of $\bar{Y}$. In both cases σ^2 is the variance of the normal population being sampled. If we wish a given probability that our estimate will not differ from μ by more than a specified amount, we could attain our objective using either $\bar{Y}$ or M but we would need only $2/\pi$ as large a sample if we choose to use $\bar{Y}$ rather than M. We say that M has an efficiency of $2/\pi$ relative to $\bar{Y}$.

If $T_1(n)$ and $T_2(n)$ are unbiased estimators of θ, the relative efficiency of $T_1(n)$ relative to $T_2(n)$ is

$$\text{relative efficiency} = \frac{V[T_2(n)]}{V[T_1(n)]} \tag{3.2.5}$$

In certain cases there is a smallest possible variance for unbiased estimators based on n independent observations. Under certain regularity conditions, the variance of any unbiased estimator $T(n)$ from a distribution with probability function or probability density function $f(y|\theta)$ is

$$V[T(n)] \geqslant \frac{1}{nE\left[\left(\dfrac{\partial \ln f(y|\theta)}{\partial \theta}\right)^2\right]} \tag{3.2.6}$$

This inequality is the Cramér–Rao or Cramér–Frechet–Rao inequality. (The regularity conditions are not satisfied if the range of possible values of Y depends on θ.)

If a lower bound on the variance of an unbiased estimator of θ is calculable by (3.2.6), the *efficiency* of an *unbiased* estimator $T(n)$ of θ is defined to be

$$\frac{1}{nE\left[\left(\frac{\partial \ln f(Y|\theta)}{\partial\theta}\right)^2\right]V[T(n)]} \tag{3.2.7}$$

Example 3.2.3

Find a lower bound to the variance of an unbiased estimator, using a sample of n, of the mean of a normal random variable with known variance. Making use of the fact that the variance of the median is approximately $\pi\sigma^2/2n$ for large n, find the efficiency of the median as an estimator of the mean of a normal population.

Solution

Develop forms (3.2.6) and (3.2.7) for this specific case.

$$f(Y|\mu) = \frac{1}{\sqrt{2\pi}\,\sigma}\exp\left(-\frac{(Y-\mu)^2}{2\sigma^2}\right) \tag{3.2.8}$$

$$\ln f(Y|\mu) = -\frac{1}{2}\ln(2\pi) - \ln\sigma - \frac{(Y-\mu)^2}{2\sigma^2} \tag{3.2.9}$$

$$\frac{\partial \ln f(Y|\mu)}{\partial\mu} = \frac{Y-\mu}{\sigma^2} \tag{3.2.10}$$

$$E\left[\left(\frac{\partial \ln f(Y|\mu)}{\partial\mu}\right)^2\right] = \frac{E(Y-\mu)^2}{\sigma^4} = \frac{\sigma^2}{\sigma^4} = \frac{1}{\sigma^2} \tag{3.2.11}$$

$$\frac{1}{nE\left[\left(\frac{\partial \ln f(Y|\mu)}{\partial\mu}\right)^2\right]} = \frac{1}{n(1/\sigma^2)} = \frac{\sigma^2}{n} \tag{3.2.12}$$

A lower bound to the variance of an unbiased estimator of the mean of a normal distribution with known variance using a sample of size n is σ^2/n.

Since the variance of the median is $\pi\sigma^2/2n$, the efficiency of the median is

$$\frac{\sigma^2/n}{\pi\sigma^2/2n} = \frac{2}{\pi}$$

Whether or not the conditions under which (3.2.5) may be used are satisfied, an estimator which has the minimum possible variance among unbiased estimators is called a *minimum variance unbiased estimator.* Many commonly used estimators are minimum variance unbiased estimators.

Considerations of efficiency become much more complex if we do not insist on unbiased estimators. Minimum variance is a meaningless criterion since we can easily attain zero variance with a worthless estimator. For example, we can always attain zero variance by always estimating the parameter to be 100.

One useful generalization of minimum variance to situations in which biased estimators are allowed is minimum expected square deviations. A *minimum expected squared deviation estimator* is an estimator T of the parameter θ for which $E(T-\theta)^2$ is the smallest possible for any estimator of θ. If $E(T)$ is θ, a minimum expected squared deviation estimator is a minimum variance unbiased estimator.

3.2.4 Sufficiency

Roughly, a *sufficient statistic* is one which contains all the information from a sample which is relevant to the estimation of any property of the random variable being sampled. A sufficient statistic need not be an estimator; it need only contain all the information necessary for an estimator with the properties of any estimator we might care to devise.

Before giving a definition of sufficient statistic, let us consider the relative information contained in two estimators of the mean of a normal distribution with known variance. If we are using a sample of size n to estimate μ, we might ask whether knowing the sample median as well as the sample mean would make it possible to better estimate the population mean. It turns out that the conditional distribution of the sample median given the sample mean does not depend on μ. Furthermore, it turns out that the conditional distribution of the observations given the sample mean (a distribution of an $n-1$ dimensional random variable) does not depend on μ and hence, if the sample mean is known, no further information about the sample is relevant to μ. We call $\bar{X}$ a sufficient statistic for the family of normal distributions with known variance.

Finding conditional distributions of observations given statistics can be tedious. Fortunately, we can determine whether the conditional distribution depends on the parameter without finding the conditional distribution explicitly.

Definition

T is a *sufficient statistic* for a family of distributions whose members are identified by values of a parameter θ if the joint probability function or the joint probability density function can be factored into two factors, one of which is the probability function or the probability density function of T and the other does not depend on θ.

Example 3.2.4

For the family of distributions represented by the density function

$$f(x|\theta)=\frac{1}{\sqrt{2\pi}}e^{-\frac{1}{2}(x-\theta)^2}, \qquad -\infty<x<\infty \tag{3.2.13}$$

we have the joint density function for n independent observations

$$f(x_1,\ldots,x_n|\theta)=\frac{1}{(2\pi)^{n/2}}\exp\left(-\frac{1}{2}\sum_{i=1}^{n}(x_i-\theta)\right)^2, \qquad -\infty<x_i<\infty,$$

$$i=1,2,\ldots,n \tag{3.2.14}$$

The density function of $\bar{X}$ is

$$g(\bar{x}|\theta)=\sqrt{\frac{n}{2\pi}}\;e^{-\frac{n}{2}(\bar{x}-\theta)^2}, \qquad -\infty<\bar{x}<\infty \tag{3.2.15}$$

$$f(x_1,\ldots,x_n|\theta)=g(\bar{x}|\theta)\left[\frac{1}{\sqrt{n}\,(2\pi)^{(n-1)/2}}\exp\left(-\frac{1}{2}\sum_{i=1}^{n}(x_i-\bar{x})^2\right)\right] \tag{3.2.16}$$

since

$$\sum_{i=1}^{n}(x_i-\theta)^2=\sum_{i=1}^{n}\left[(x_i-\bar{x})+(\bar{x}-\theta)\right]^2$$

$$=\sum_{i=1}^{n}(x_i-\bar{x})^2+n(\bar{x}-\theta)^2 \tag{3.2.17}$$

Thus we see that $\bar{X}$ is a sufficient statistic in the family of normal distributions with unit standard deviation.

For the normal family, $\bar{x}$ and s are jointly sufficient for μ and σ.

3.2.5 Maximum Likelihood Estimators

In the estimators so far considered, we have generally looked for some meaning for the parameter and have sought an estimator which has a similar meaning with respect to the sample. There are some more general methods for deriving estimators. One of these is the method of maximum likelihood. In essence this method consists of choosing from among the possible values for the parameter, the value which maximizes the probability of obtaining the sample which was obtained.

In dealing with a family of probability distributions, we have found it convenient to use the symbol $f(x,\ldots,x_n|\theta)$ to represent the joint probability function or the joint probability density functions for discrete or continuous random variables, respectively. For each possible value of θ, $f(x_1,\ldots,x_n|\theta)$ defines the distribution of $X_1,\ldots,X_N$. We may wish to consider more than one member of the family at a time; in this case θ is a parameter. When θ is fixed, we know which function of $x_1,\ldots,x_n$ we are dealing with. If we fix θ and $x_1,\ldots,x_n$, $f(x_1,\ldots,x_n|\theta)$ represents a number. In the preceding paragraph we posed the problem of finding that member of a family of probability distributions for which the probability of getting the observations we got (or for continuous random variables, the probability density for the observation we got) is greatest. Here we look at $f(x_u,\ldots,x_n|\theta)$ as a function of θ for fixed values of $x_1,\ldots,x_n$. This function is *not* a probability density function. We need a new name. If $f(x_1,\ldots,x_n|\theta)$ is a probability function or probability density function for the random variable $x_1,\ldots,x_n$ for fixed θ, it is the likelihood function of θ for fixed $x_1,\ldots,x_n$. For example, the family of binomial probability functions for a fixed n can be described by

$$f(x|\theta)=\binom{n}{x}\theta^x(1-\theta)^{n-x}, \qquad x=0,1,\ldots,n, \quad 0<\theta<1$$

Although the random variable X is discrete, the possible values of θ are continuous. That $f(x|\theta)$ is not a probability density function for θ for fixed x is easily seen when we note that

$$\int_0^1\binom{n}{x}\theta^x(1-\theta)^{n-x}\,d\theta=\frac{1}{n+1}$$

and not one.

We use the letter L for likelihood function. (Some authors use L for the logarithm of likelihood.)

Example 3.2.5

Consider an example involving the binomial distributions. If all values of θ between 0 and 1 are possible and if we observe x successes in the n trials, we choose as our estimate that value of θ which maximizes $L=\binom{n}{x}\theta^x(1-\theta)^{n-x}$. The θ which maximizes L also maximizes the natural logarithm of L or

$$\ln L=\ln\binom{n}{x}+x\ln\theta+(n-x)\ln(1-\theta) \tag{3.2.18}$$

which is easier to work with than L in this case. Now take the first and second derivatives of $\ln L$ to find

$$\frac{\partial \ln L}{\partial\theta}=\frac{x}{\theta}-\frac{n-x}{1-\theta} \qquad \frac{\partial^2\ln L}{\partial\theta^2}=-\frac{x}{\theta^2}-\frac{n-x}{(1-\theta)^2} \tag{3.2.19}$$

Since the second derivative is always negative, the value of θ which makes the first derivative zero maximizes $\ln L$. Designating the estimator of θ by $\hat{\theta}$, we have

$$x(1-\hat{\theta})-\hat{\theta}(n-x)=0 \quad \text{or} \quad \hat{\theta}=\frac{x}{n} \tag{3.2.20}$$

Example 3.2.6

For the rectangular distribution with range $(0,\theta)$

$$L=\frac{1}{\theta^n},\ 0\leqslant x_i\leqslant\theta, \qquad i=1,2,\ldots,n \tag{3.2.21}$$

to maximize L subject to $0\leqslant x_i\leqslant\theta$ we make θ as small as possible under these restrictions. This smallest possible value is $\max(x_1,x_2,\ldots,x_n)$. Hence

$$\hat{\theta}=\max(x_1,x_2,\ldots,x_n) \tag{3.2.22}$$

Example 3.2.7

Find maximum likelihood estimators of μ and σ for a normal distribution using a sample $(X_1,\ldots,X_n)$.

Solution

We find $\ln L$, differentiate with respect to each of the parameters, and set derivatives equal to zero, finding second derivatives to make sure we have a maximum.

$$L=\frac{1}{(2\pi)^{n/2}\sigma^n}\exp\left(-\frac{1}{2\sigma^2}\sum_{i=1}^{n}(x_i-\mu)^2\right) \tag{3.2.23}$$

$$\ln L=-\frac{n}{2}\ln(2\pi)-n\ln\sigma-\frac{\sum_{i=1}^{n}(x_i-\mu)^2}{2\sigma^2} \tag{3.2.24}$$

$$\frac{\partial\ln L}{\partial\mu}=\frac{\sum_{i=1}^{n}(x_i-\mu)}{\sigma^2}, \qquad \frac{\partial^2\ln L}{\partial\mu^2}=-\frac{n}{\sigma^2} \tag{3.2.25}$$

$$\frac{\partial\ln L}{\partial\sigma}=-\frac{n}{\sigma}+\frac{\sum_{i=1}^{n}(x_i-\mu)^2}{\sigma^3}, \qquad \frac{\partial^2\ln L}{\partial\sigma^2}=\frac{n}{\sigma^2}-\frac{3\sum_{i=1}^{n}(x_i-\mu)^2}{\sigma^4} \tag{3.2.26}$$

$$\sum_{i=1}^{n}(x_i-\hat{\mu})=0, \qquad \hat{\sigma}^2=\frac{\sum_{i=1}^{n}(x_i-\hat{\mu})^2}{n} \tag{3.2.27}$$

where (3.2.27a) comes from setting the expression at (3.2.25a) equal to zero and (3.2.27b) from setting the expression at (3.2.26a) equal to zero. The maximum likelihood estimators of μ and σ are then

$$\hat{\mu} = \frac{\sum_{i=1}^{n} x_i}{n} = \bar{x}, \qquad \hat{\sigma} = \left(\frac{\sum_{i=1}^{n} (x_i - \bar{x})^2}{n} \right)^{1/2} \tag{3.2.28}$$

Now the second derivative with respect to μ is always negative. The second derivative with respect to σ at the point at which the two first derivatives are zero is

$$\frac{n}{\hat{\sigma}^2} - \frac{3n\hat{\sigma}^2}{\hat{\sigma}^4} = -\frac{2n}{\hat{\sigma}^2} \tag{3.2.29}$$

which is negative. Hence we have maximum likelihood estimators.

3.2.6 Estimators a posteriori

If we know something about which values our parameter is likely to have, we should be able to obtain better estimates by using this information. If our parameter is a random variable, a value of which has been chosen in accordance with its distribution, this value being an unknown constant throughout our experiment, Bayes's theorem gives us a means of combining this prior information with the results of our experiment. In cases in which our prior knowledge amounts to much less than a prior distribution of parameter values, we may still find it useful to form one or two hypothetical prior distributions and see what estimators are suggested by these prior distributions.

For purposes of discussion here, we assume that a prior distribution of our parameter is available. If this prior distribution has density function $g(\theta)$ and the conditional probability function or probability density function of our observations given θ is $f(x_1, x_2, \ldots, x_n|\theta)$, Bayes's theorem tells us that the posterior density function of our parameter is

$$g(\theta|x_1, \ldots, x_n) = \frac{g(\theta)f(x_1, \ldots, x_n|\theta)}{\int_{-\infty}^{\infty} g(u)f(x_1, \ldots, x_n|u)\,du} \tag{3.2.30}$$

If we have a prior probability function for the parameter, a sum appears in place of the integral in this form. For some applications of (3.2.30) it is convenient to note that θ appears in the right member only in the numerator.

Example 3.2.8

If we think that the parameter of the Bernoulli distribution which we are investigating is likely to be near 2/3 with probability density falling off to zero at $\theta = 0$ and $\theta = 1$ and with expected value about .6, we might be interested in using the prior probability density function $12\theta^2(1-\theta)$. We run our experiment and obtain x successes in n trials. Let us find the posterior density. Now

$$g(\theta)=12\theta^2(1-\theta), \qquad 0\leqslant\theta\leqslant 1 \tag{3.2.31}$$

$$f(x_1,\ldots,x_n|\theta)=\theta^x(1-\theta)^{n-x}, \qquad x=0,1,\ldots,n \tag{3.2.32}$$

$$12\int_0^1 \theta^{x+2}(1-\theta)^{n-x+1}\,d\theta=\frac{12\Gamma(x+3)\Gamma(n-x+2)}{\Gamma(n+5)}$$

$$g(\theta|x_1,\ldots,x_n)=\frac{12\theta^{x+2}(1-\theta)^{n-x+1}}{\dfrac{12\Gamma(x+3)\Gamma(n-x+2)}{\Gamma(n+5)}}=\frac{\Gamma(n+5)}{\Gamma(x+3)\Gamma(n-x+2)}\theta^{x+2}(1-\theta)^{n-x+1} \tag{3.2.33}$$

If we use a body of data in conjunction with an initial prior distribution to get a posterior distribution, then use the posterior as a prior in conjunction with a second body of data to get a second posterior distribution, we end up with the same posterior as though we combined the two bodies of data and used the combined data in conjunction with the initial prior.

Fortunately, as data accumulate, the initial prior distribution matters less and less. We may therefore look among functions which are easier to work with for one which fits our feeling as to the correct prior. (We do want to make sure that we do not initially rule out any possible value, since once ruled out, it cannot later show up among the possible values no matter what the data are.)

We may find a suitable prior among functions $g(\theta)$ which are proportional to likelihood functions. In this case, $g(\theta|x_1,\ldots,x_n)$ is also proportional to a possible likelihood function.

For an investigator who knows what family of distributions his data come from but has no idea whatever from which member of the family they come, a concept of "*noninformative prior*" distribution has been developed. Noninformative priors have impressive properties and may in many cases be the most suitable choice. There are however no universally acceptable criteria for choosing a prior distribution.

3.2.7 Bayes Squared Error Loss Estimators. MAP Estimators

Some people feel uncomfortable with a distribution of parameter values rather than a single value. For such people there are, of course, a variety of ways of using a parameter of the posterior distribution as a point estimate.

One possible estimator is the *mode* of the posterior distribution. This estimator is called the *maximum a posteriori estimator* or MAP estimator.

The value of the parameter which maximizes the posterior density function also maximizes the joint density function for parameter and observations, the numerator of the right member of (3.2.30).

Example 3.2.9

Find the MAP estimator for the estimation problem of Example 3.2.8.

Solution

Finding the first and second derivatives of the logarithm of $g(\theta|x_1,\ldots,x_n)$ [or of $g(\theta)f(x_1,\ldots,x_n|\theta)$] with respect to θ, we have

$$\frac{\partial \ln g(\theta|x_1,\ldots,x_n)}{\partial\theta}=\frac{x+2}{\theta}-\frac{n-x+1}{1-\theta} \tag{3.2.34}$$

$$\frac{\partial^2 \ln g(\theta|x_1,\ldots,x_n)}{\partial\theta^2}=-\frac{x+2}{\theta^2}-\frac{n-x+1}{(1-\theta)^2} \tag{3.2.35}$$

we see that the second derivative is always negative and hence the estimator may be found by setting the first derivative equal to zero.

$$(x+2)(1-\hat{\theta})-\hat{\theta}(n-x+1)=0 \tag{3.2.36}$$

$$\hat{\theta}=\frac{x+2}{n+3} \tag{3.2.37}$$

If $n=0$ and therefore $x=0$, this posterior estimate reduces, as it should, to the value which maximizes the prior distribution. As n increases, the prior distribution of θ matters less and less and the estimate of θ approaches the maximum likelihood estimate, which is x/n.

Another possible estimator is the expected value of the posterior distribution. This estimator is called the *Bayes estimator for squared error loss.*

Example 3.2.10

Find the Bayes squared error loss estimator for the estimation problem of Example 3.2.8.

Solution

$$E(\theta)=\frac{\Gamma(n+5)}{\Gamma(x+3)\Gamma(n-x+2)}\int_0^1\theta^{x+3}(1-\theta)^{n-x+1}\,d\theta$$

$$=\frac{\Gamma(n+5)\Gamma(x+4)\Gamma(n-x+2)}{\Gamma(x+3)\Gamma(n-x+2)\Gamma(n+6)}=\frac{x+3}{n+5} \tag{3.2.38}$$

If $n=0$, the expected value reduces, as it should, to the expected value of the prior distribution. As n increases, the prior distribution matters less and less and the estimator θ approaches the minimum variance estimator.

If, for the Bernoulli family we take a prior distribution of θ,

$$g(\theta)=\frac{\Gamma(\alpha+\beta)}{\Gamma(\alpha)\Gamma(\beta)}\theta^{\alpha-1}(1-\theta)^{\beta-1}, \qquad \alpha>0,\beta>0,0\leqslant\theta\leqslant 1$$

and observe x successes in n trials, the posterior distribution of θ will be

$$g(\theta|x_1,\ldots,x_n)=\frac{\Gamma(n+\alpha+\beta)}{\Gamma(x+\alpha)\Gamma(n-x+\beta)}\theta^{x+\alpha-1}(1-\theta)^{n-x+\beta-1}, \qquad 0\leqslant\theta\leqslant 1$$

The MAP estimator of θ is $(\alpha+x-1)/(\alpha+\beta+n-2)$ unless $\alpha<1$, $x=0$ or $\beta<1$, $x=n$, in which case $\hat{\theta}$ is 0 or 1 respectively. The Bayes estimator for squared error loss is $(\alpha+x)/(n+\alpha+\beta)$. Note that the MAP estimator is the maximum likelihood estimator if $\alpha=\beta=1$. The minimum variance unbiased estimator of θ cannot be the Bayes estimator for squared error loss since neither α nor β can be zero. If $\alpha=\beta=$ ½, the prior is the noninformative prior and the MAP and Bayes squared error loss estimator are on either side of the minimum variance unbiased estimator.

3.2.8 Bayes Intervals

If we have a probability distribution for our parameter, θ, whether an a priori distribution or an a posteriori distribution (see Section 2.4.3), we can find the probability that the parameter is in any interval of possible values. If the distribution is continuous, we can find intervals for any desired probability. For many purposes an interval of shortest length, that is, one concentrated in the region of greatest probability density, is probably most desirable. If the probability density function is unimodal and symmetric, such an interval is relatively easy to find. If the density function is not symmetric, we may be satisfied to use an interval which leaves out extreme values at each extreme with equal probability. Thus for a distribution with distribution function $F(\theta)$ we might choose the interval (θ_1,θ_2) where $F(\theta_1)=1-F(\theta_2)$. For a symmetric, unimodal distribution the two types of intervals are identical.

We present herewith a *Bayes interval* for the case of normal prior of normal expected value when the prior distribution of μ has expected value μ_0 and variance σ_0^2 $(=\sigma^2/k$, say) and when the conditional distribution of

X given μ has variance σ^2. For n independent observations, the posterior distribution of μ is normal with expected value $(k\mu_0 + n\overline{X})/(k+n)$ and variance $\sigma^2/(k+n)$. A *Bayes interval* with probability γ and μ is

$$\left[\frac{k\mu_0 + n\overline{X}}{k+n} - \frac{z_{(1+\gamma)/2}\sigma}{(k+n)^{1/2}}, \quad \frac{k\mu_0 + n\overline{X}}{k+n} + \frac{(z_{(1+\gamma)/2})\sigma}{(k+n)^{1/2}}\right]$$

where $\Phi[z_{(1+\gamma)/2}] = (1+\gamma)/2$. See Section 2.8.5.

3.2.9 Minimizing Expected Cost

Presumably, in a practical situation a precise knowledge of the parameter or parameters of the model of our system is useful. If we act on the basis of an estimate of the parameter we fail to attain this utility. In comparing estimators it is convenient to consider as a loss the decrease in utility owing to using an estimate t instead of the true value θ of the parameter. The loss as a function of parameter and estimate is called a *loss function* and denoted $\mathcal{L}(\theta, t)$. Although in almost every case we cannot know the value taken on by $\mathcal{L}(\theta, t)$ since we do not know θ, the properties of the random variable $\mathcal{L}(\theta, T)$, where T is an estimator, may be amenable to investigation. The expected value of $\mathcal{L}(\theta, T)$ for given θ is particularly interesting. It is called the *risk* and is denoted $r(\theta, T)$.

$$r(\theta, T) = E_\theta(\mathcal{L}(\theta, T))$$

$$= \int_{-\infty}^{\infty} \cdots \int_{-\infty}^{\infty} \mathcal{L}[\theta, T(x_1, \ldots, x_n)] f(x_1, \ldots, x_n | \theta) \, dx_1 \ldots dx_n \quad (3.2.39)$$

If we have to choose among estimators, we may be able to calculate $r(\theta, T)$ for each estimator for values which are important to us and choose that estimator for which the pattern of $r(\theta, T)$ looks best.

If we have a prior distribution of θ, $g(\theta)$ say, we can go further in selecting an estimator. For each estimator T, the random variable $r(g(\theta), T)$ has a distribution. We can look at the distributions of those estimators we are considering and choose the estimator whose distribution we like best. Many investigators seem content to choose that estimator for which the expected risk, $E_g[r(\theta, T)] = E_g E_\theta[\mathcal{L}(\theta, T)]$ is least.

For the popular loss function $\mathcal{L}(\theta, t) = k(t-\theta)^2$, minimum expected risk is attained by taking T as the posterior expected value of θ, that is, the expected value of the conditional distribution of θ given the observations.

3.3 CONFIDENCE INTERVALS

If we have no prior distribution of our parameter, we may still give, along with an estimate of the parameter, some indication of how far the estimate may be expected to be from the true parameter value. In the following discussion, it is important to remember the distinction between estimates and estimators. An *estimate* is a number which is likely to differ from the numerical value of the parameter being estimated. The estimator, the formula by which an estimate is calculated, is a random variable and thus has a probability distribution. The probability that it will take on a value, the estimate, within 0.02 of the parameter being estimated may be calculable without knowledge of the value of the parameter.

We may know that the estimator has probability .95 of yielding an estimate within 0.03 of the value of the parameter. If our *point estimate* is 52.26, for example, we may report it as 52.26 ± 0.03. We do not thereby mean to imply that the parameter value is between 52.23 and 52.29 but merely that the estimator we have used is such that the probability is .95 that an interval constructed in this way will include the actual parameter value. The interval itself may be called a *confidence interval*. It is clear that if we had chosen a smaller probability than .95, the corresponding interval would in all probability have been smaller. Larger probabilities go with larger intervals. We must balance the advantage of being more *definite* against the advantage of being more *sure*.

The idea of confidence interval is developed in the following sections for particular cases but in a manner which, it is hoped, suggests how confidence intervals may be constructed in other cases. We start with a case in which the construction of the interval is simple and we can concentrate on the properties of confidence intervals. We then move to more complex constructions and more practical assumptions.

3.3.1 Confidence Intervals for the Mean of a Normal Population when the Population Standard Deviation is Known

Let us apply the principles of the preceding section to find a confidence interval for the mean μ of a normal population whose standard deviation is known. Let $\{X_i\}$ be normal and independent, identically distributed (i.i.d.) with $V(X_i) = \sigma^2$, a known constant, and let us use the distribution of $\bar{X}$ as a basis for construction of a confidence interval. We begin by noting that

$$\frac{\bar{X} - \mu}{\sigma / \sqrt{n}} \tag{3.3.1}$$

has a normal distribution with mean 0 and variance 1 [see (2.8.21)].

Thus if

$$P\left(-z \leqslant \frac{\bar{X}-\mu}{\sigma}\sqrt{n} \leqslant z\right) = \Phi(z) - \Phi(-z) = \gamma \tag{3.3.2}$$

we have

$$P\left(\mu - z\frac{\sigma}{\sqrt{n}} \leqslant \bar{X} \leqslant \mu + z\frac{\sigma}{\sqrt{n}}\right) = \gamma \tag{3.3.3}$$

and

$$P\left(\bar{X} - z\frac{\sigma}{\sqrt{n}} \leqslant \mu \leqslant \bar{X} + z\frac{\sigma}{\sqrt{n}}\right) = \gamma \tag{3.3.4}$$

and thus a γ confidence interval for μ of a normal population with known variance σ is the interval based on a sample of n

$$\left(\bar{X} - z\frac{\sigma}{\sqrt{n}}, \bar{X} + z\frac{\sigma}{\sqrt{n}}\right) \tag{3.3.5}$$

where

$$\Phi(z) - \Phi(-z) = 2\Phi(z) - 1 = \gamma \tag{3.3.6}$$

Example 3.3.1

Sixteen independent observations on normally distributed random variables with $\sigma = 5$ show $\bar{x} = 52.26$. Find a 95% confidence interval for μ.

Solution

For $\gamma = .95$, $z = 1.96$ (see Table 2.13); hence the end points of the confidence interval are

$$\bar{x} + z\frac{\sigma}{\sqrt{n}} = 52.26 \pm 1.96\frac{5}{\sqrt{16}} = 52.26 \pm 2.45 = (49.81, 54.71)$$

The steps from (3.3.3) to (3.3.4) may be more meaningful if a graphical method of construction is considered. In Fig. 3.3 we picture the $(\mu, \bar{x})$ plane. On the vertical line representing a particular value of μ we plot the points representing the end points of the interval whose probability appears in (3.3.3), $\mu - z(\sigma/\sqrt{n})$ and $\mu + z(\sigma/\sqrt{n})$. If we were to plot corresponding points for each possible μ, the loci of these points would be the lowest and the uppermost sloping lines of the figure. The observed $\bar{x}$ is

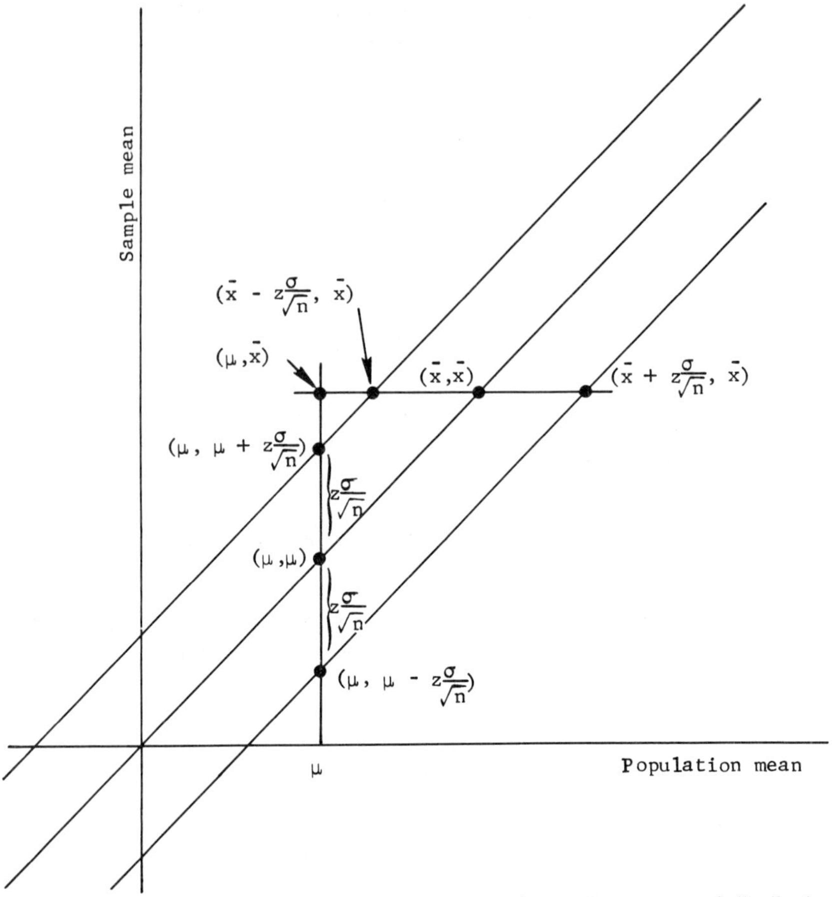

Figure 3.3 Construction of confidence intervals for μ given x from a normal distribution with known σ.

represented as a point $(\mu, \bar{x})$. We do not know μ but we can draw a horizontal line at height $\bar{x}$. Although we do not know μ, we do know that $(\mu, \bar{x})$ will fall between the two 45° lines just constructed with probability γ. Viewing $\bar{x}$ as a random variable, the probability that the horizontal line segment at height $\bar{x}$ between the two lines crosses the vertical line at μ is γ. The range of μ represented by such a line segment for a specific $\bar{x}$ is called a confidence interval for μ. To avoid crowding in the figure, we picture a case in which the sample did not lead to a confidence interval which contained the parameter value.

A confidence interval need not be symmetric. Indeed, since for a random sample from a normal population with known σ^2,

$$P\left(\overline{X} > \mu - z\frac{\sigma}{\sqrt{n}}\right) = \Phi(z) \tag{3.3.7}$$

$$\left(-\infty, \bar{x} + z\frac{\sigma}{\sqrt{n}}\right) \tag{3.3.8}$$

is a $\Phi(z)$ confidence interval for the mean. So also is

$$\left(x - z\frac{\sigma}{\sqrt{n}}, \infty\right) \tag{3.3.9}$$

One-sided confidence intervals should be used more often than they are. Still, two-sided intervals are usually wanted and we seldom, if ever, refer to one sided confidence intervals.

3.3.2 Confidence Intervals for the Standard Deviation of a Normal Population

Suppose $\{X_i\}$ are normal and independent with mean μ and variance σ^2. To find a confidence interval for σ we begin by noting that

$$\frac{\Sigma_{i=1}^n \left(X_i - \overline{X}\right)^2}{\sigma^2} \tag{3.3.10}$$

has a χ^2 distribution with $\nu = n-1$ degrees of freedom. For γ between 0 and 1, we can find a pair (in fact, many pairs) for which

$$P\left[\chi_L^2 \leqslant \frac{\Sigma_{i=1}^n \left(X_i - \overline{X}\right)^2}{\sigma^2} \leqslant \chi_U^2\right] = F_{\chi^2}(\chi_U^2) - F_{\chi^2}(\chi_L^2) = \gamma \tag{3.3.11}$$

We solve the inequalities in the argument of the probability and obtain

$$P\left\{\left[\frac{\Sigma_{i=1}^n \left(X_i - \overline{X}\right)^2}{\chi_U^2}\right]^{1/2} \leqslant \sigma \leqslant \left[\frac{\Sigma_{i=1}^n \left(X_i - \overline{X}\right)^2}{\chi_L^2}\right]^{1/2}\right\} = \gamma \tag{3.3.12}$$

which defines a γ confidence interval for the standard deviation of a normal population.

If we wish the probability that the entire interval is above the population standard deviation to be equal to the probability that the entire interval is below the population standard deviation, we find χ_L^2 and χ_U^2 by solving

$$F_{\chi^2}(\chi_L^2) = 1 - F_{\chi^2}(\chi_U^2) = \frac{1-\gamma}{2} \tag{3.3.13}$$

This interval is not the shortest possible interval for the specified confidence coefficient.

Example 3.3.2

For a sample of size 11 from a normal population, we have found

$$\sum_{i=1}^{n} (x_i - \bar{x})^2 = 2304$$

Find a 95% confidence interval for σ.

Solution

Since $\chi^2_{.025}(10) = 3.25$ and $\chi^2_{.975}(10) = 20.48$, using (3.3.12) we see that

$$\left(\sqrt{\frac{2304}{20.48}}, \sqrt{\frac{2304}{3.25}}\right) = (10.6, 26.6)$$

is a 95% confidence interval for σ.

Figure 3.4 shows the graphical construction of confidence intervals for the standard deviation of a normal distribution using a sample of 10 (using nine degrees of freedom). For fixed σ we lay off ordinates of $\sigma(\chi^2_{9,.025}/9)^{1/2}$ and $\sigma(\chi^2_{9,.975}/9)^{1/2}$. To get the confidence interval corresponding to s for nine degrees of freedom, draw a horizontal line at height s. The values of σ at the points of intersection with the two slanting lines give the end points of a 95% confidence interval.

The distribution of $(n-1)s^2/\sigma^2$ is χ^2 if the sample is from independent normal distributions with the same mean and variance. Unfortunately, the distribution of $(n-1)s^2/\sigma^2$ is sensitive to departures from normality in the population being sampled. These confidence intervals must be used with discretion.

3.3.3 Confidence Intervals for the Mean of a Normal Population when the Population Standard Deviation is Unknown

In sampling from a normal population with known variance, we developed confidence intervals for μ based on

$$Z = \frac{\bar{X} - \mu}{\sigma} \sqrt{n} \tag{3.3.14}$$

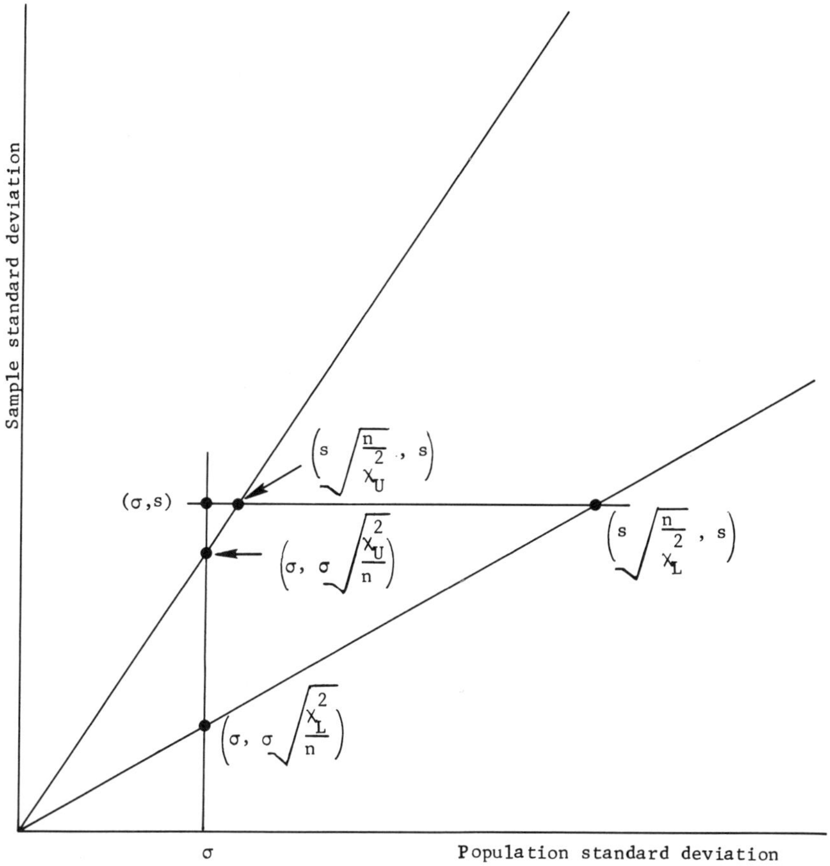

Figure 3.4 Construction of confidence intervals for σ given s from a normal distribution.

If σ is not known, we might consider using

$$S^2 = \frac{\sum_{i=1}^{n}\left(X_i - \bar{X}\right)^2}{n-1} \tag{3.3.15}$$

in its place. The resulting

$$T = \frac{\bar{X} - \mu}{S}\sqrt{n} \tag{3.3.16}$$

has a t distribution (see Section 2.8.9). The added variability introduced by S causes T to have a greater variance than Z, a variance which decreases to that of Z as n increases.

Analogous to (3.3.2) we have

$$P\left(-t < \frac{\bar{X} - \mu}{S}\sqrt{n} < t\right) = \gamma \tag{3.3.17}$$

if

$$2F_T(t) - 1 = \gamma \tag{3.3.18}$$

Rearranging the inequalities, we get

$$P\left(\bar{X} - t\frac{S}{\sqrt{n}} < \mu < \bar{X} + t\frac{S}{\sqrt{n}}\right) = \gamma \tag{3.3.19}$$

and thus we see that

$$\left(\bar{X} - t\frac{S}{\sqrt{n}}, \bar{X} + t\frac{S}{\sqrt{n}}\right) \tag{3.3.20}$$

is a confidence interval with confidence coefficient γ.

Again, we easily get one-sided confidence intervals

$$\left(-\infty, \bar{X} + t\frac{S}{\sqrt{n}}\right) \tag{3.3.21}$$

or

$$\left(\bar{X} - t\frac{S}{\sqrt{n}}, \infty\right) \tag{3.3.22}$$

in each of which the confidence coefficient is $F_T(t)$. It should be noted that although the symmetric interval is relatively insensitive to departures from normality in the observations, the one-sided intervals are sensitive to skewness in that population.

3.4 HYPOTHESIS TESTING

If we have competing hypotheses about the correct model for an experiment, we need some means of deciding among the hypotheses. We could reduce the problem to one of estimation by attaching an index number to each hypotheses. Our problem would then be one of estimating the index number of the true hypothesis. However, there are advantages to taking a fresh point of view, especially when the choice is between two hypotheses.

In Section 3.4.1, we consider a choice between two simple hypotheses, that is, hypotheses which specify completely all parameters. In this case it is possible to develop a method of attack leading to procedures which simultaneously satisfy a wide variety of criteria for good decision procedures. In Section 3.4.2 we see that some decisions between compound hypotheses can be treated as decisions between simple hypotheses.

In Section 3.4.3 we broaden our scope to include other compound hypotheses. We shall find it convenient to consider these problems as requiring decisions as to whether our sample came from a distribution which is a member of a particular subclass of all the distributions we admit as possibilities. Examples of such problems are (1) deciding whether our sample came from a normal distribution with known mean and unknown variance or from some other normal distribution, and (2) deciding whether a regression curve is first or second degree.

The structure of a decision making procedure should depend on the costs of gathering data and of making wrong decisions. To a first approximation, the expected costs of wrong decisions may be taken to depend only on the probabilities of wrong decisions and the cost of gathering data to depend only on sample size. We consider only sampling plans for which the sample size is fixed before sampling begins.

3.4.1 Two Simple Hypotheses

A *simple hypothesis* is one in which the model is completely specified (with no unknown parameters). Consider a case in which we are to choose between two simple hypotheses. We can find the distribution of possible samples for each hypothesis. If the cost of choosing the wrong hypothesis depends only on which hypothesis is true, the development of a test falls neatly into two parts. First, we can order the possible samples with respect to the degree to which each favors one of the hypotheses, H_2, say, over the other, H_1, say. Secondly, we determine a critical point, a point in the ordering such that, if our sample is on one side of this point we choose to consider H_1 as true and if on the other we choose to consider H_2 as true. (For those samples which have the same position in the ordering as the critical point, we may choose to include them with one side or the other, or divide them between the two sides.)

The ordering of possible samples can be accomplished by computing, for each possible sample, the ratio of (1) the probability of that sample if H_2 were true to (2) the probability of that sample if H_1 were true:

$$\frac{P(x_1,\ldots,x_n|H_2)}{P(x_1,\ldots,x_n|H_1)} \tag{3.4.1}$$

or if we are dealing with continuous random variables and therefore with density function, we compute the ratio

$$\frac{f_{X_1,\ldots,X_n|H_2}(x_1,\ldots,x_n)}{f_{X_1,\ldots,X_n|H_1}(x_1,\ldots,x_n)} \tag{3.4.2}$$

If the observations are independent, both the numerator and denominator can be written as products of factors involving the individual observations. The samples are ordered from those most favorable to H_1 (with ratio closest to 0), to those most favorable to H_2 (the largest possible ratio). To decide where in the ordering to make the division between those which cause us to accept one hypothesis and those which cause us to accept the other, we compute for each hypothesis and for each possible point of division the probability of making a wrong decision. Changing the point of division cannot simultaneously decrease both the probability of deciding H_1 in case H_2 is true and the probability of deciding H_2 in case H_1 is true. Among the pairs of probabilities, we choose the pair most acceptable to us.

Example 3.4.1

Let both hypotheses specify independent normal observations with $\sigma = 16$. Let $n = 25$. Let H_1 specify $\mu = 10$ and H_2 specify $\mu = 15$. Find a decision procedure if the probability of deciding H_2 when H_1 is true is to be twice the probability of deciding H_1 is true when H_2 is true.

Solution

Under H_1

$$f_{X|H_1}(x) = \frac{1}{16\sqrt{2\pi}} \exp\left[-\frac{1}{2}\left(\frac{x-10}{16}\right)^2\right] \tag{3.4.3}$$

and under H_2

$$f_{X|H_2}(x) = \frac{1}{16\sqrt{2\pi}} \exp\left[-\frac{1}{2}\left(\frac{x-15}{16}\right)^2\right] \tag{3.4.4}$$

Let us find the ratio given by (3.4.2).

$$\frac{f_{X_1,\ldots,X_n|H_2}(x_1,\ldots,x_n)}{f_{X_1,\ldots,X_n|H_1}(x_1,\ldots,x_n)} = \frac{\dfrac{1}{16^n(2\pi)^{n/2}} \exp\left[-\dfrac{1}{2\cdot 256}\Sigma(x_i-15)^2\right]}{\dfrac{1}{16^n(2\pi)^{n/2}} \exp\left[-\dfrac{1}{2\cdot 256}\Sigma(x_i-10)^2\right]}$$

$$= \exp\left[-\frac{1}{512}(\Sigma x_i^2 - 30\Sigma x_i + 25\cdot 225 - \Sigma x_i^2 + 20\Sigma x_i - 25\cdot 100)\right]$$

$$= \exp\left(\frac{10}{512}\Sigma x_i - \frac{25\cdot 125}{512}\right) \tag{3.4.5}$$

As Σx_i or $\bar{x}$ increases, the ratio increases. Hence the ordering is in accord with the values of $\bar{x}$. Our decision rule will be: decide H_2 if $\bar{x} \geqslant k$ where k is still to be determined. To find k we use

$$P(\bar{x} \geqslant k | H_1) = 1 - \Phi\left(\frac{k-10}{16}\sqrt{25}\right) \tag{3.4.6}$$

$$P(\bar{x} \leqslant k | H_2) = \Phi\left(\frac{k-15}{16}\sqrt{25}\right) \tag{3.4.7}$$

By trial and error, we find the k which makes

$$1 - \Phi\left(\frac{k-10}{3.2}\right) = 2\Phi\left(\frac{k-15}{3.2}\right) \tag{3.4.8}$$

to be $k = 11.68$.

3.4.2 Problems Reducible to Problems of Two Simple Hypotheses

Example 3.4.2

As a purchaser of electric light bulbs, we wish to make reasonably sure, by sampling inspection, that the average life of the bulbs in each consignment is not too low. We agree with the manufacturer on an accelerated life test. We believe that the distribution of life under the accelerated test will, for production under stable conditions, be sufficiently close to normal that we can safely assume normality in our analysis. We believe that the standard deviation of results of our accelerated life test will be very close to 200 hr. We wish a good chance, .95, of rejecting a consignment if the expected life is as low as 1000 hr and a reasonably good chance, .90, of accepting an expected value as good as 1100 hr. How large a sample is needed and what values of $\bar{x}$ will cause us to reject a consignment?

Solution

We shall reject a consignment if $\bar{x}$, for a sample of size n is less then k, where n and k are to be found. The constraints we have introduced are

$$P(\bar{x} < k | \mu \leqslant 1000) \geqslant .95 \qquad \text{and} \qquad P(\bar{x} \geqslant k | \mu \geqslant 1100) \geqslant .90$$

which is equivalent to

$$P(\bar{x} < k | \mu = 1000) = .95 \qquad \text{and} \qquad P(\bar{x} \geqslant k | \mu = 1100) = .90$$

Using (2.8.21) we see that these are equivalent to

$$P\left(\frac{\bar{x}-1000}{200}\sqrt{n} < \frac{k-1000}{200}\sqrt{n}\right) = .95$$

and

$$P\left(\frac{\bar{x}-1100}{200}\sqrt{n} \geqslant \frac{k-1100}{200}\sqrt{n}\right) = .90$$

or

$$\frac{k-1000}{200}\sqrt{n} = 1.645 \qquad \text{and} \qquad \frac{k-1100}{200}\sqrt{n} = -1.282$$

Solving for n and k we find that we should use a sample of size 35 and reject the consignment if $\bar{x} \leqslant 1056$.

We have just seen how a problem in test construction may sometimes be solved by methods designed to construct decision procedures for deciding between two simple hypotheses.

3.4.3 Generalized Likelihood Ratio Tests. Power

For tests of hypothesis which do not reduce to the case of choice between two simple hypotheses, we describe only one type of test, the *generalized likelihood ratio test*. Suppose that observations come from one of a (broad) class of distributions and we want to test the hypothesis, called the *null hypothesis* and symbolized H_0, that the observations come from a distribution belonging to a particular subclass. We form a likelihood ratio, λ. For the numerator, we use the maximum of the likelihood over all distributions belonging to the subclass. For the denominator, we use the maximum of the likelihood over all distributions of the (broad) class.

$$\lambda = \frac{\max\limits_{H_0} L}{\max L} \tag{3.4.9}$$

The ratio must be between zero and one; the smaller the ratio, the less we are inclined to accept the null hypothesis. The decision rule will be: reject H_0 if $\lambda < \lambda_0$ where λ_0 is determined so that

$$P(\lambda < \lambda_0 | H_0) = \alpha \tag{3.4.10}$$

where α is the *level of significance* of the test.

Example 3.4.3

Assuming that we have n observations from a normal distribution, test the hypothesis that $\mu = \mu_0$ using a 5% level of significance.

Solution

The broad class we are considering is that of all normal distributions. The subclass is that of all normal distributions with $\mu = \mu_0$. In either case the likelihood function has the form

$$L = \frac{1}{(2\pi)^{n/2}\sigma^n} \exp\left(-\frac{1}{2\sigma^2} \sum_{i=1}^{n} (x_i - \mu)^2 \right) \tag{3.4.11}$$

L is maximized in the (broad) class by replacing μ by $\bar{x}$ and σ^2 by $[\Sigma_{i=1}^n (x_i - \bar{x})^2]/n$, giving us

$$\max L = \frac{1}{(2\pi)^{n/2}\left[\frac{\Sigma_{i=1}^n (x_i - \bar{x})2}{n}\right]^{n/2}} e^{-n/2} \tag{3.4.12}$$

L is maximized under the null hypothesis, (i.e., in the subclass), by replacing μ by μ_0 and σ^2 by $[\Sigma_{i=1}^n (x_i - \mu_0)^2]/n$ giving us

$$\max_{H_0} L = \frac{1}{(2\pi)^{n/2}\left[\frac{\Sigma_{i=1}^n (x_i - \mu_0)^2}{n}\right]^{n/2}} e^{-n/2} \tag{3.4.13}$$

The likelihood ratio, λ, is given by

$$\lambda = \left[\frac{\Sigma_{i=1}^n (x_i - \bar{x})^2}{\Sigma_{i=1}^n (x_i - \mu_0)^2} \right]^{n/2} \tag{3.4.14}$$

Since $\Sigma_{i=1}^n (x_i - \mu_0)^2$ of the last form can be written

$$\sum_{i=1}^{n} (x_i - \mu_0)^2 = \sum_{i=1}^{n} (x_i - \bar{x})^2 + n(\bar{x} - \mu_0)^2 \tag{3.4.15}$$

we can write

$$\begin{aligned} \lambda^{2/n} &= \frac{\Sigma_{i=1}^n (x_i - \bar{x})^2}{\Sigma_{i=1}^n (x_i - \bar{x})^2 + n(\bar{x} - \mu_0)^2} \\ &= \frac{1}{1 + \frac{1}{n-1} \cdot \frac{n(\bar{x} - \mu_0)^2}{\left[\Sigma_{i=1}^n (x_i - \bar{x})^2\right]/(n-1)}} \\ &= \frac{1}{1 + t^2/(n-1)} \end{aligned} \tag{3.4.16}$$

where t is the familiar t of Section 2.8.9. Since small values of λ correspond to large values of t^2, the decision rule will read: reject H_0 if $|t| > t_0$ where t_0 is the solution of $F_{T,n-1\text{ d.f.}}(t_0)=1-(\alpha/2)$. If $n=11$, we have 10 degrees of freedom. Entering Table 2.15 with $n=10, F_T(t)=.975$, we find $t_0=2.23$.

The *power* of the test to detect deviations from the null hypothesis is defined to be the probability of rejecting the null hypothesis as a function of the specific distributions in the broad class of alternatives.

REFERENCE

1. Box, G. E. P. and Tiao, G. C. *Bayesian Inference in Statistical Analysis*, Addison-Wesley Publishing Co., Reading, Mass., 1973.

PROBLEMS

3.1 For the following family of probability distributions, find all possible samples of 3 (with replacement), find the mean, the median, and the probability of each, and find $P(|\bar{X}-\theta|<|M-\theta|)$ and $P(|\bar{X}-\theta|>|M-\theta|)$ when $\bar{X}$ is the sample mean and M is the sample median.

x	$\theta-2$	$\theta-1$	θ	$\theta+1$	$\theta+2$
$P(X=x)$	.1	.2	.4	.2	.1

Answer. Working through this problem may be somewhat tedious but certain short cuts are possible. There are 35 possible samples if order of observations is not considered. The different possible orders of observations for each possible sample must be taken into account in calculating probabilities. You will find that

$$P(|\bar{x}-\theta|<|M-\theta|)=.294 \qquad \text{and} \qquad P(|M-\theta|<|\bar{x}-\theta|)=.408.$$

3.2. A sample of 10 independent observations on a random variable is given below. Using the usual unbiased estimators, find estimates of the expected value and of the variance of the population sampled.

137.2, 138.4, 136.8, 137.5, 137.4, 137.2, 137.9, 136.9, 137.4, 137.6.

3.3. Is the average of n independent observations x_i an unbiased estimator of θ if the density function for each X is $\begin{cases} \theta^{-1}e^{-x/\theta} & 0<x<\infty \\ 0 & \text{otherwise} \end{cases}$

3.4 If X has a binomial distribution with parameters n and p, show that

$$E\left[\frac{\frac{X}{n}\left(1-\frac{X}{n}\right)}{n-1}\right]=\frac{p(1-p)}{n} \qquad \left[\text{Remember } E(X^2)=\sigma_X^2+[E(X)]^2.\right]$$

3.5 For the random variable $Y, P(Y=y)=e^{-\lambda}\lambda^y/y!$ find an unbiased estimator of λ^2 based on one observation.

Answer. $Y(Y-1)$.

3.6 An estimator of the least upper bound of possible values of a given random variable is the maximum of a sample of independent observations. If the random variable has a rectangular distribution with density θ^{-1} over the interval $(0,\theta)$ and if we use sample size n, (1) find an unbiased estimator of θ of the form k times the maximum observation, (2) find an unbiased estimator of θ of the form k times the average of the observations, (3) find the relative efficiencies of the two estimators.

3.7 For the random variable Y distributed uniformly over $(0,\theta)$, is $\max(y_1,\dots,y_n)$ an unbiased estimator of θ?

3.8 Find the minimum possible variance of an unbiased estimator of p when $P(Y=y)=\binom{n}{y}p^y(1-p)^{n-y}$.

Answer. $p(1-p)/n$.

3.9 For a binomial random variable generated by n independent trials, each with probability p, show that $x=$number of successes/number of trials is a sufficient statistic.

3.10 If

$$f(x)=\begin{cases}\theta e^{-\theta x} & x>0\\ 0 & \text{otherwise}\end{cases}$$

find the maximum likelihood estimator of θ given a sample of n independent observations.

3.11 Find the maximum likelihood estimator of λ from a sample of n independent observations from the Poisson probability function $P(Y=y)=e^{-\lambda}\lambda^y/y!$.

3.12 Find the maximum likelihood estimator of θ from a sample of n independent observations from the uniform distribution over the interval $(-\theta,\theta)$.

Answer. $\max\{|x_1|,|x_2|,\dots,|x_n|\}$.

3.13 If p is a random variable with density function $6p(1-p)$ for $0<p<1$ and 0 otherwise, and if X is a binomial random variable with parameters n and p, (1) find the posterior distribution of p for a given value of x. (2) Find the expected value of the posterior distribution, and (3) find the maximum posterior likelihood estimator. [Remember $\int_0^1 x^{m-1}(1-x)^{n-1}dx = \Gamma(m)\Gamma(n)/\Gamma(m+n)$.]

Answer. (partial). (2) $(x+2)/(n+4)$; (3) $(x+1)/(n+2)$.

3.14 Show that the likelihood ratio test of the hypothesis that $\lambda=\lambda_0$ where λ is the parameter of the exponential distribution with density function

$$\begin{array}{ll} \frac{1}{\lambda}e^{-x/\lambda} & x>0 \\ 0 & \text{otherwise} \end{array}$$

will have a rejection region which is the solution of an inequality of the form $(\bar{x}/\lambda_0)\exp(-x/\lambda_0) \leqslant$ constant.

3.15 If $f_{X|p}(x|p)=p^x(1-p)^{1-x}$ $x=0,1$, $\mathcal{L}(p,\hat{p})=(\hat{p}-p)^2$,

(*a*) find $r[p,\hat{p}_1(x)]$ for $\hat{p}_1(x)=(x+1)/3$;
(*b*) find $r[p,\hat{p}_0(x)]$ for $\hat{p}_0(x)=\frac{1}{2}$.
(*c*) If $g(p)=2(1-p), 0 \leqslant p \leqslant 1$, find $E\{r[g(p),\hat{p}_1(x)]\}$.

3.16 If X has a binomial distribution with $n=4$, if $\hat{p}=(x+1)/6$, and if $\mathcal{L}(p,\hat{p})=(\hat{p}-p)^2$, find $r(p,\hat{p})$.

Answer. $\frac{1}{36}$.

3.17 From a normal distribution with standard deviation 5 a sample of 16 independent observations was obtained. $\bar{x}$ was calculated and found to be 31.5. Find a 90% confidence interval for μ.

Answer. (29.44, 33.56).

3.18 From a normal distribution a sample of 16 was obtained. $\bar{x}$ and s were calculated and found to be 31.5 and 5, respectively. Find a 90% confidence interval for μ.

Answer. (29.32, 33.68).

3.19 From a normal distribution with standard deviation 5 a sample of 16 independent observations was obtained. $\bar{x}$ was calculated and found to be 31.5. Test at the 5% level of significance the hypothesis that $\mu=30$ against alternatives that $\mu \neq 30$.

3.20 If $f(x|\theta)=1+\theta^2(x-\frac{1}{2}), 0 \leqslant x \leqslant 1, 0 \leqslant \theta \leqslant \sqrt{2}$, using one observation, a rejection region $x>.9$ was decided upon. Find the power function of this test.

Answer. $.1+.045\theta^2$.

CHAPTER 4

PARAMETER ESTIMATION METHODS

4.1 INTRODUCTION

In this chapter we introduce some of the concepts which we shall develop in the remainder of the book. Some canonical forms for the models of problems of parameter estimation are presented, least squares estimators are described, and modifications suggested by various criteria for good estimators are mentioned. We close the chapter with a short discussion of simulation techniques for comparing methods of estimation.

4.2 RELATIONS BETWEEN OBSERVED RANDOM VARIABLES AND PARAMETERS

We assume a functional relationship among several measurable variables, $(Y, X_1, \ldots, X_k)$, one or more parameters $(\beta_1, \beta_2, \ldots, \beta_p)$, and particular values of one or more random variables $(\varepsilon, \varepsilon^{(1)}, \ldots, \varepsilon^{(m)})$. (In this book we deal almost exclusively with cases in which each observation involves only one random variable; thus we have no need for the index indicated by the superscript.) The measurement of the measurable variables provides the observations. The parameters are unknown and we wish to estimate at least some of them. The particular values of the random variables ε's are unknown. We may estimate the ε's themselves in order to get a picture of how well the estimates fit with our preconception of the distribution of the ε's.

It is convenient to pick one of the measurable variables and express this variable in terms of the others. It is the picked variable with which we associate the pronoun Y. Thus we write

$$Y=f(X_1,\ldots,X_k;\ \beta_1,\ldots,\beta_k;\ \varepsilon) \tag{4.2.1}$$

The variable to be called Y is traditionally chosen because we are interested in how its value is affected by the values assumed by $X_1,\ldots,X_k$. It is traditionally called the dependent variable. The variables $X_1,\ldots,X_k$ are called the independent variables. (Be clear in noting that independent in this context does not mean independence in the statistical sense.) The X's may be thought of as the causes of the Y, as when Y represents the yield in a chemical process into which amounts $X_1,\ldots,X_k$ of material from sources $1,\ldots,k$ are combined. The X's may merely describe the physical environment, as when Y represents temperature at point (X_1,X_2,X_3) in space at time X_4.

We are fortunate if the ε's can be combined into one ε and especially fortunate if the errors are *additive*, that is, if we can write

$$Y=(X_1,\ldots,X_k;\ \beta_1,\ldots,\beta_p)+\varepsilon \tag{4.2.2}$$

in which the distribution of ε does not depend on the unknown β's although it may depend on parameters which do not appear in the other term of (4.2.2). It will sometimes be convenient to index the β's by integers beginning with 0; thus $\beta_0,\beta_1,\ldots,\beta_{p-1}$. It will also be convenient to use vector notation to abbreviate; thus

$$\mathbf{X}=(X_1,\ldots,X_k)$$

$$\boldsymbol{\beta}=(\beta_1,\ldots,\beta_p) \quad \text{or} \quad (\beta_0,\beta_1,\ldots,\beta_{p-1}) \text{ as appropriate}$$

and

$$Y=\eta(\mathbf{X},\boldsymbol{\beta})+\varepsilon \tag{4.2.3}$$

Thus the ith observation will be signified by adding a subscript i to $(Y,X_1,\ldots,X_k)$; that is, $Y_i,X_{i1},\ldots,X_{ik}$. ε_i is found by (4.2.2) or (4.2.3) to be such that

$$Y_i=\eta(X_{i1},\ldots,X_{ik};\ \beta_1,\ldots,\beta_p)+\varepsilon_i=\eta(\mathbf{X}_i,\boldsymbol{\beta})+\varepsilon_i=\eta_i+\varepsilon_i \tag{4.2.4}$$

to introduce in the last relation one further abbreviation. And yet one more abbreviation is to use $\mathbf{Y}$ for $(Y_1,\ldots,Y_n)$, $\boldsymbol{\varepsilon}$ for $(\varepsilon_1,\ldots,\varepsilon_n)$, and $\mathbf{X}$ for the matrix

$$(\mathbf{X}_1,\mathbf{X}_2,\ldots,\mathbf{X}_k)=\begin{bmatrix} X_{11} & X_{12} & \cdots & X_{1k} \\ X_{21} & X_{22} & \cdots & X_{2k} \\ \vdots & \vdots & & \vdots \\ X_{n1} & X_{n2} & \cdots & X_{nk} \end{bmatrix} \tag{4.2.5}$$

Note that the $\mathbf{X}_i$'s are column vectors.

The distribution of the ε's is generally unknown. If the ε_i's are correlated, estimation of parameters may be much more difficult, the estimators less reliable, and the reliability difficult to assess. We therefore deal first with cases of independent ε_i's, later investigating estimation under various assumptions regarding correlation. For those methods of estimation which require some assumption about the form of the distribution of the ε_i's or when the evaluation of the method requires assumptions of the form, we shall invariably investigate first under the assumption of normality. Usually we go no further. Fortunately many aspects of the relationships between sample moments and moments of the random variable being investigated do not depend on details of the form of the distribution.

One trouble arises from too great a preoccupation with cases in which the ε's may be assumed to be normally distributed. Different criteria for evaluating estimators may be expected to lead to different choices of estimators to be used. We seem frequently to deal with criteria which in general do suggest somewhat different estimators but which, when applied to a case in which normal ε's are assumed, lead to identical estimators. The casual student is sometimes misled to assume that if the estimators are the same the criteria must be essentially equivalent. Beware. We have used and shall use assumptions other than normality not only to illuminate the difference but also to give insight into the effect of wrongly assuming normality.

To estimate parameters we must first gather data and then analyze them. It is essential that experimental procedures be such that the data can be analyzed. The form of the function to be used in (4.2.1) and the experimental procedure must be developed together. "Design of experiments" deals, for the most part, with choosing values of the X's to facilitate analysis and to improve accuracy of estimation.

4.3 EXPECTED VALUES, VARIANCES, COVARIANCES

With a sequence of random variables, $\varepsilon_1,\ldots,\varepsilon_n$, we associate expected values $E(\varepsilon_1),\ldots,E(\varepsilon_n)$, variances $V(\varepsilon_1),\ldots,V(\varepsilon_n)$, and covariances $\mathrm{cov}(\varepsilon_1,\varepsilon_2)$, $\mathrm{cov}(\varepsilon_1,\varepsilon_3),\ldots,\mathrm{cov}(\varepsilon_{n-1},\varepsilon_n)$. In Chapter 6 we use vector and matrix forms to save space and to increase clarity:

$$E(\boldsymbol{\varepsilon})=\left[E(\varepsilon_1),E(\varepsilon_2),\ldots,E(\varepsilon_n)\right] \tag{4.3.1}$$

$$\mathrm{cov}(\boldsymbol{\varepsilon})=\begin{bmatrix} V(\varepsilon_1) & \mathrm{cov}(\varepsilon_1,\varepsilon_2) & \cdots & \mathrm{cov}(\varepsilon_1,\varepsilon_n) \\ \mathrm{cov}(\varepsilon_1,\varepsilon_2) & V(\varepsilon_2) & \cdots & \mathrm{cov}(\varepsilon_2,\varepsilon_n) \\ \vdots & \vdots & & \vdots \\ \mathrm{cov}(\varepsilon_1,\varepsilon_n) & \mathrm{cov}(\varepsilon_2,\varepsilon_n) & \cdots & V(\varepsilon_n) \end{bmatrix} \tag{4.3.2}$$

In $\mathrm{cov}(\boldsymbol{\varepsilon})$ it is convenient to think of each covariance as occupying two positions symmetrically situated with respect to the main diagonal.

4.4 LINEAR PROBLEMS

If we can write $\eta(\mathbf{X},\boldsymbol{\beta})$ in the form

$$\eta(\mathbf{X},\boldsymbol{\beta})=\beta_1X_1+\beta_2X_2+\cdots+\beta_pX_p \tag{4.4.1}$$

$$Y=\beta_1X_1+\beta_2X_2+\cdots+\beta_pX_p+\varepsilon \tag{4.4.2}$$

we find the problem of estimating $\boldsymbol{\beta}$ simpler than otherwise. At the same time, if η is not linear in its parameters, we may find a form such as (4.4.1) a useful approximation to $\eta(\mathbf{X},\boldsymbol{\beta})$ in the neighborhood of some particular value of $\mathbf{X}$ and $\boldsymbol{\beta}$. Chapters 5 and 6 deal with linear estimation.

4.5 LEAST SQUARES

In forms (4.2.2) or (4.2.3) or (4.4.2) we shall be interested in $E(Y|\mathbf{X}_i)$. It is

$$E(Y|\mathbf{X}_i)=\eta(\mathbf{X}_i,\boldsymbol{\beta})+E(\varepsilon_i) \tag{4.5.1}$$

If $E(\varepsilon_i)=\mu$, we can easily rewrite

$$E(Y|\mathbf{X}_i)=\left[\eta(\mathbf{X}_i,\boldsymbol{\beta})+\mu\right]+E(\varepsilon_i-\mu) \tag{4.5.2}$$

We see that (4.5.2) has the same form as (4.5.1) with $\eta(\mathbf{X}_i,\boldsymbol{\beta})$ of (4.5.1) replaced by $\eta(\mathbf{X}_i,\boldsymbol{\beta})+\mu$ of (4.5.2) and ε_i replaced by a new random variable $\varepsilon_i-\mu$. If μ is known, it is just a number in (4.5.2). If μ is unknown it plays the role of one of the β's. We shall lose nothing and gain simplicity if we assume ε_i has expected value 0. We deal further with this question in Section 5.10.

If $\eta(\mathbf{X},\boldsymbol{\beta})$ is a constant, say, $\eta(\mathbf{X},\boldsymbol{\beta})=\mu$ for all $\mathbf{X}$, the problem of estimation is one we handled in Chapter 3. Our estimate of μ is $\bar{Y}$ and if the ε's are normally distributed $\bar{Y}$ is the best estimator of μ from many points of view. In looking for some property of $\bar{Y}$ which is at the same time simple and generalizable, mathematicians a couple of centuries ago turned to the fact that, if we have a set of numbers $Y_1, Y_2,\ldots, Y_n$, $\bar{Y}$ is the value of the variable $\hat{\mu}$ which minimizes $\Sigma_{i=1}^n(Y_i-\hat{\mu})^2$. As estimator of $\boldsymbol{\beta}$ of (4.2.2) they chose that $\hat{\boldsymbol{\beta}}$ that minimizes

$$S=\|Y-\eta(\mathbf{X},\hat{\boldsymbol{\beta}})\|^2=\sum_{i=1}^{n}\left[Y_i-\eta(\mathbf{X}_i,\hat{\boldsymbol{\beta}})\right]^2 \tag{4.5.3}$$

with respect to changes in $\hat{\boldsymbol{\beta}}$. This method of estimation is known as the ordinary least squares method. For (4.4.1) or (4.4.2) the computation of the least squares estimates of $\boldsymbol{\beta}$ can be described in a straightforward manner without using successive approximations or iterations. If $\hat{\boldsymbol{\beta}}$ is unique, it is called the least squares estimator of $\boldsymbol{\beta}$.

4.6 GAUSS–MARKOV ESTIMATION

If in (4.2.2) or (4.4.2) the ε's are independent but do not all have the same variance and if we know the proportion $\sigma_1^2:\sigma_2^2:\cdots:\sigma_n^2$ we would almost certainly wish to consider in place of $\Sigma_{i=1}^n(Y_i-\eta_i)^2$,

$$S(\boldsymbol{\beta})=\sum_{i=1}^{n}\frac{(Y_i-\eta_i)^2}{\sigma_i^2} \tag{4.6.1}$$

in order that more accurate measurements be counted more heavily. The minimization of S with respect to $\boldsymbol{\beta}$ does not depend on the size of any σ_i^2 but only on their proportion. In Chapter 5 we expand on this idea.

If the ε's are not independent, the weighting that suggests itself involves the covariances among the ε's as well as the variances. We shall deal with some such cases in later chapters. The sum of squares to be minimized is

$$S(\boldsymbol{\beta})=\sum_{i=1}^{n}\sum_{j=1}^{n}(Y_i-\eta_i)W_{ij}(Y_j-\eta_j) \tag{4.6.2}$$

where the W_{ij}'s are elements of the inverse of the variance–covariance matrix of the ε's. See Section 6.1.7, Gauss–Markov theorem.

4.7 SOME OTHER ESTIMATORS

If we know the form of the joint distribution of the ε_i's and therefore of the Y's, we can seek joint maximum likelihood estimators of the β_i's. We may sometimes need to estimate parameters of the distribution of the ε_i's in the process. Maximum likelihood estimators are discussed in Chapters 5 and 6 under various sets of assumptions about the distribution of the ε's. We shall find that, if the ε_i's are normally independently distributed with zero mean and common variance, the maximum likelihood estimators turn out to be the ordinary least squares estimators. If the ε_i's have a multivariate normal distribution, not necessarily with equal variances or zero covariances, but possessing known proportions among the variances and covariances, the maximum likelihood estimators are generalizations of least squares estimates.

If the conditional distribution of $Y_1,\ldots,Y_n$ given $\beta_1,\ldots,\beta_p$ can be described by a probability density function (or by a discrete probability function) which has continuous second partial derivatives with respect to each β_i and jointly with respect to each pair, then the first partial derivatives will be zero at the maximum likelihood estimate $\mathbf{b}_{\mathrm{ML}}$ of $\boldsymbol{\beta}$, that is,

$$\left.\frac{\partial f(\mathbf{Y}|\boldsymbol{\beta})}{\partial \beta_i}\right|_{\mathbf{b}_{\mathrm{ML}}} = 0 = \left.\frac{\partial \ln f(\mathbf{Y}|\boldsymbol{\beta})}{\partial \beta_i}\right|_{\mathbf{b}_{\mathrm{ML}}} \quad \text{for } i=1,2,\ldots,p \tag{4.7.1}$$

If the form of the distribution of the random variable $\boldsymbol{\varepsilon}$ is known and it is known that the parameter(s), (σ, for example) of the distribution of $\boldsymbol{\varepsilon}$ are chosen in accordance with a known probability distribution and if we know we can adequately approximate the prior distribution of the parameters, we can use Bayes's theorem to obtain the posterior distribution, the MAP estimators, and the squared error loss estimators.

The MAP estimates are found by maximizing $f(\boldsymbol{\beta}|\mathbf{Y})$. If there is a unique $\boldsymbol{\beta}$ which maximizes $f(\boldsymbol{\beta}|\mathbf{Y})$, that is, if there is a unique mode, the MAP estimate of $\boldsymbol{\beta}$ is the mode of the posterior distribution of $\boldsymbol{\beta}$. If the distribution of $\boldsymbol{\beta}$ is described by a probability density function which has continuous second partial derivatives with respect to each β_i and jointly with respect to each pair of β_i's, the first partial derivative will be zero at the mode, that is,

$$\left.\frac{\partial f(\boldsymbol{\beta}|Y)}{\partial \beta_i}\right|_{\mathbf{b}_{\text{MAP}}} = 0 \qquad \text{for } i=1,2,\ldots,p \tag{4.7.2}$$

The estimator which minimizes the expected value of the square of the deviation of the estimator from the parameter being estimated is called the squared error loss Bayes estimator and we symbolize it by $\mathbf{b}_{\text{SEL}}$. For scalar β, if the expected value of the posterior distribution of β given $\mathbf{Y}$ exists, it is b_{SEL}; that is,

$$b_{\text{SEL}} = \int_{-\infty}^{\infty} \beta f(\beta|\mathbf{Y})\, d\beta \tag{4.7.3}$$

Another possible estimator is the median of the posterior distribution. This estimator is associated with minimizing the expected absolute deviation of estimator from estimated.

If the posterior distribution of β is symmetric the median and mean coincide. Some symmetric densities for scalar $\boldsymbol{\beta}$ are shown in Fig. 4.1. For such cases the b_{SEL} vector defined by (4.7.3) is given by the mean or

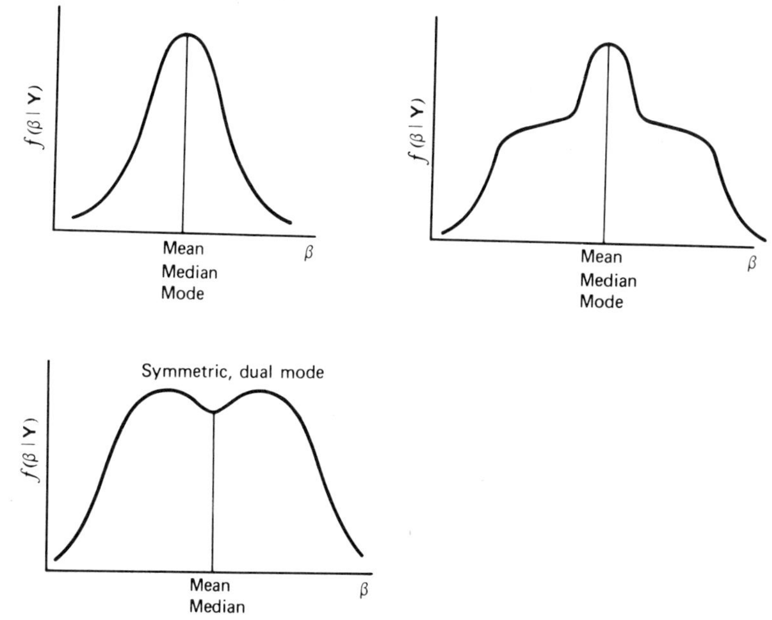

Figure 4.1 Some symmetric conditional probability densities.

median value of the conditional distribution of β given $\mathbf{Y}$, $f(\beta|\mathbf{Y})$. If the density $f(\beta|\mathbf{Y})$ in addition to being symmetric is also unimodal, the mean, median, and mode will all be at the *same* location. Hence when $f(\beta|\mathbf{Y})$ is symmetric about the parameter vector β and is also unimodal, b_{SEL} is b_{MAP}. When the distribution is *not* symmetric or *not* unimodal b_{SEL} and b_{MAP} are rarely the same. Some nonsymmetric unimodal probability densities are depicted in Fig. 4.2. Note that the modes do not coincide with the means. This causes the parameters b_{SEL} given by (4.7.3) and associated with the *mean* to be *not* equivalent to those given by the *mode* which are indicated by (4.7.2).

The conditional probability density $f(\boldsymbol{\beta}|\mathbf{Y})$ used in (4.7.2) can be written in terms of other densities using the form of Bayes's theorem written as

$$f(\boldsymbol{\beta}|\mathbf{Y})=\frac{f(\mathbf{Y}|\boldsymbol{\beta})f(\boldsymbol{\beta})}{f(\mathbf{Y})} \tag{4.7.4}$$

The probability density $f(\boldsymbol{\beta})$ contains the prior information known regarding the parameter vector $\boldsymbol{\beta}$. Notice that the parameters appear only in the numerator of the right side of (4.7.4); this numerator can also be written as

$$f(\mathbf{Y},\boldsymbol{\beta})=f(\mathbf{Y}|\boldsymbol{\beta})f(\boldsymbol{\beta}) \tag{4.7.5}$$

Then the necessary conditions given by (4.7.2) can be written equivalently as

$$\left.\frac{\partial \ln[f(\mathbf{Y},\boldsymbol{\beta})]}{\partial \beta_i}\right|_{\mathbf{b}_{\mathrm{MAP}}}=\left.\frac{\partial \ln[f(\mathbf{Y}|\boldsymbol{\beta})]}{\partial \beta_i}\right|_{\mathbf{b}_{\mathrm{MAP}}}+\left.\frac{\partial \ln[f(\boldsymbol{\beta})]}{\partial \beta_i}\right|_{\mathbf{b}_{\mathrm{MAP}}}=0 \tag{4.7.6}$$

since the maximum of $f(\mathbf{Y},\boldsymbol{\beta})$ exists at the same location as the maximum of its natural logarithm.

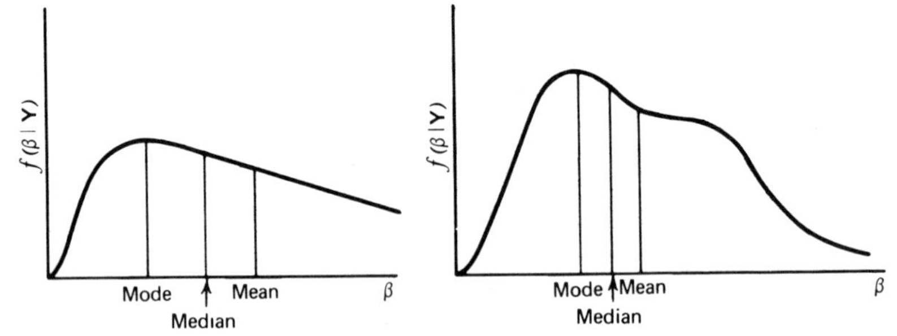

Figure 4.2 Some nonsymmetric conditional probability densities.

The estimators $\mathbf{b}_{MAP}, \mathbf{b}_{SEL}$ are described without reference to the linearity or nonlinearity of the expected value of Y in the β's nor to the independence of the Y_i's. Under some assumptions about the structure of η_i and under some assumptions about the prior distribution of the β's, the MAP and SEL procedures are equivalent in arithmetic to certain least squares or Gauss–Markov procedures.

4.8 COST

Methods of collecting data and analyzing them must be coordinated. If observations are expensive, sophisticated methods of analysis to extract all pertinent information are justified. Sometimes more expensive methods of collecting data yield net returns by drastically reducing the cost of analysis. Increased costs due to collecting more data or using more sophisticated methods of analysis may or may not reduce the cost occasioned by the degree to which the estimate is incorrect. Some remarks in Chapter 3 were directed to these matters.

4.9 MONTE CARLO METHODS

One method for investigating the effects of nonlinearity or various other effects that are difficult to analyze otherwise is called the Monte Carlo method. Actually, what we describe is sometimes referred to as the "crude" Monte Carlo method. More sophisticated Monte Carlo methods often provide the same amount of information as the crude method but at a lower cost [1].

The Monte Carlo method can be used to investigate analytically the properties of a proposed estimation method. To simulate a series of experiments on the computer we proceed as follows:

1. Define the system by prescribing (*a*) the model equation, also called regression function, (*b*) the way in which "errors" are incorporated in the model of the observations, (*c*) the probability distribution of all the errors and, where applicable, (*d*) a prior distribution. Assign "true" values to all the parameters ($\boldsymbol{\beta}$) in the regression function and to those in the distribution of error.
2. Select a set of values of the independent variables. Then calculate the associated set of "true" values of $\boldsymbol{\eta}$ from the regression equations.
3. Use the computer to produce a set of errors $\boldsymbol{\varepsilon}$ drawn from the prescribed probability distribution. For most computers programs are

available which can generate a stream of numbers that have all the important characteristics of successive independent observations on a population uniform over the interval (0, 1). Since they are generated by a deterministic scheme, they are not actually random. Such numbers are called *pseudorandom numbers*. Suitable transformations are used to obtain samples for any other distribution.

To obtain a sequence of pseudorandom observations on a normal population with expected value 0 and variance 1, we can make use of the Box–Muller transformation [2]. If u_{2i-1} and u_{2i} are independent (0, 1) random numbers,

$$x_{2i-1} = (-2\ln u_{2i-1})^{1/2}\cos(2\pi u_{2i}) \tag{4.9.1a}$$

and

$$x_{2i} = (-2\ln u_{2i-1})^{1/2}\sin(2\pi u_{2i}) \tag{4.9.1b}$$

are independent random observations on a normal distribution with expected value 0 and variance 1. The normal random numbers are then adjusted to have the desired variances and covariances.

The simulated measurements are obtained by combining the errors with the regression values. For additive errors, the ith error is simply added to the ith η value. This then provides simulated measurements.

4. Acting as though the parameters are unknown, we estimate the parameters, denoting the estimates $\boldsymbol{\beta}^*$.
5. Replicate the series of simulated experiments N times by repeating steps 3 and 4, each time with a new set of errors.
6. We use appropriate methods to estimate properties of the distribution of parameter estimates. (We consider the estimates actually obtained by our pseudorandom number scheme to be a random sample from the distribution of all possible estimates.) The expected value of our parameter estimator is estimated by the mean of our parameter estimates,

$$\bar{\beta}_j^* = \frac{1}{N}\sum_{i=1}^{N}\beta_{ji}^* \tag{4.9.2}$$

where β_{ji}^* is the jth component of the $\boldsymbol{\beta}^*$ found on the ith replication. If $\boldsymbol{\beta}^*$ may be a biased estimator, $\boldsymbol{\beta}^* - \boldsymbol{\beta}$ is an estimate of the bias. If it is not clear whether or not $\boldsymbol{\beta}^*$ is biased the size of $\boldsymbol{\beta}^* - \boldsymbol{\beta}$ needs to be compared with an estimate of its variance–covariance matrix.

The variances and covariances of the distribution of $\boldsymbol{\beta}^*$ may be estimated by

$$\text{est. cov}(\beta_j^*, \beta_k^*) = \frac{1}{N-1} \sum_{i=1}^{N} \left(\beta_{ji}^* - \bar{\beta}_j^*\right)\left(\beta_{ki}^* - \bar{\beta}_k^*\right) \qquad (4.9.3a)$$

If $\boldsymbol{\beta}^*$ is known to be unbiased, we can make use of our knowledge of $\boldsymbol{\beta}$ and use a slightly more efficient estimator

$$\text{est. cov}(\beta_j^*, \beta_k^*) = \frac{1}{N} \sum_{i=1}^{N} (\beta_{ji}^* - \beta_j)(\beta_{ki}^* - \beta_k) \qquad (4.9.3b)$$

If $\boldsymbol{\beta}^*$ is biased, the right side of (4.9.3b) which are estimates of mean square error and corresponding product moments, may be more interesting than variances and covariances. If we use actual experiments rather than simulated ones (4.9.3b) will not be available although (4.9.2) and (4.9.3a) are.

The flexibility of the above simulation procedure is great. We can estimate the sample properties for any model, linear or nonlinear, and for any parameter values. We can estimate the effect of different probability distributions upon ordinary least squares estimation or other estimation methods. Many other possibilities also exist. An example of a Monte Carlo simulation is given below and another one is given in Section 6.9. These simulations can be accomplished on a modern high-speed computer at a small fraction of the cost, in time and money, of a comparable set of physical experiments.

The great power of the Monte Carlo procedure is that we can investigate the properties of estimators in cases for which the character of the estimators cannot be derived. To demonstrate the validity of a Monte Carlo procedure an example is considered which is simple enough to be analyzed without recourse to simulation. We investigate estimating β in the model $\eta_i = \beta X_i$ for the case of additive, zero mean, constant variance, uncorrelated errors; that is

$$y_i = \eta_i + \varepsilon_i, \qquad E(\varepsilon_i) = 0, \qquad V(\varepsilon_i) = \sigma^2, \qquad E(\varepsilon_i \varepsilon_j) = 0 \qquad \text{for } i \neq j$$

The distribution of ε_i is uniform in the interval $(-.5, .5)$; each ε_i is found using a pseudorandom number generator. There are no errors in X_i and there is no prior information.

The X_i values are $X_i = i$ for $i = 1, 2, \ldots, 10$ and $\beta = 1$. For the kth set of simulated measurements, β_k^* is found using the ordinary least squares estimator,

$$\beta_k^* = \left[\sum_{i=1}^{10} X_i Y_{i,k}\right]\left[\sum_{i=1}^{10} X_i^2\right]^{-1}$$

The estimated expected value of β_k^*, (4.9.2), the estimated variance of β_k^*, (4.9.3a), and the estimated mean square error of β_k^*, (4.9.3b), are obtained by using

$$\bar{\beta}^* = \frac{1}{10}\sum_{k=1}^{10} \beta_k^*, \text{ est. } V(\beta^*) = \frac{1}{9}\sum_{k=1}^{10}\left(\beta_k^* - \bar{\beta}^*\right)^2$$

$$\text{est. mean square error } (\beta^*) = \frac{1}{10}\sum_{k=1}^{10}(\beta_k^* - 1)^2$$

For independent sets of errors, estimates were calculated for $N=5$, 25, 50, 100, 200, and 500. The results are shown in Table 4.1 where the estimated standard deviation and estimated root mean square error are given rather than their squares. In Table 4.2 comparable results for a simulation involving normal errors are given. The variance of ε_i in this case was taken as $1/12$, the same as the variance for the uniform case.

In both Tables 4.1 and 4.2 the sample mean $\bar{\beta}^*$ tends to approach the true value of 1 as N becomes large. Hence β^* is an unbiased estimator of β. Also the estimated standard error of β^* and estimated root mean square error tend to their common exact value

$$\left\{\sigma^2\left[\sum X_i^2\right]^{-1}\right\}^{1/2} = \left\{\frac{1/12}{385}\right\}^{1/2} = 0.014712$$

Table 4.1 Monte Carlo Simulation for $\eta_i = \beta X_i$, with $\beta = 1$ and $X_i = i$, $i = 1, 2, \ldots, 10$. Uniform Distribution of Errors

Sample Size	$\bar{\beta}^*$	Est. Std Dev (β^*)	Est. Root Mean Square Error (β^*)
5	1.0044	0.00950	0.00958
25	1.0014	0.01616	0.01589
50	0.9992	0.01350	0.01339
100	0.9996	0.01425	0.01418
200	1.0018	0.01440	0.01448
500	0.9987	0.01415	0.01419

Table 4.2 Monte Carlo Simulation for $\eta_i = \beta X_i$, with $\beta = 1$ and $X_i = i$, $i = 1, 2, \ldots, 10$. Normal Distribution of Errors

Sample Size	$\bar{\beta}^*$	Est. Std Dev (β^*)	Est. Root Mean Square Error (β^*)
5	1.0021	0.01156	0.01055
25	0.9969	0.01608	0.01606
50	0.9972	0.01496	0.01507
100	0.9973	0.01486	0.01502
200	0.9995	0.01410	0.01407
500	0.9997	0.01480	0.01478

This example shows that the number of simulations N must be quite large in order to provide accurate estimates of the variance of the parameter estimate. Such simulations are still inexpensive compared to actual experiments to determine the variance. Moreover, methods are available for making the simulation procedure more efficient [1].

REFERENCES

1. Hammersley, J. M. and Handscomb, D. C., *Monte Carlo Methods*, Methuen & Co. Ltd., London, 1964.
2. Box, G. E. P. and Muller, M. E., "A Note on the Generation of Random Normal Deviates," *Ann. Math. Stat.*, **29** (1958), 610–611.

CHAPTER 5

Introduction to Linear Estimation

5.1 MOTIVATION, MODELS, AND ASSUMPTIONS

5.1.1 Motivation

One of the basic principles in engineering is to start analysis with simple cases. For that reason estimation of parameters in several simple linear algebraic models is studied in this chapter. Many of the estimation ideas can be introduced in connection with these models without the added complexities introduced by nonlinear algebraic models or by models described by differential equations.

In addition to the pedagogic value of simple algebraic cases, there are numerous physical situations for which the regression function is linear in the parameters. Moreover, when the regression function is unknown and cannot be derived from first principles, simple models are usually proposed.

Simple linear models have been widely studied by statisticians, economists, and others. Various terms designating certain parts of the study of estimation of parameters in statistical models have also been used to refer to much larger segments of that study. When the models are linear in the parameters, *regression analysis* and *analysis of variance* are sometimes used interchangeably. However, regression analysis also specifically refers to the analysis of the dependence of the expected value of a random variable on the conditions* under which the experiment is conducted; the method of least squares is frequently used to estimate the parameters. Analysis of variance refers to the breakdown of the variability of the observed values of the dependent variable into a part which is the sum of squares about the fitted regression function and other parts due to the exclusion of parameters or groups of parameters from the regression function. Those using analysis of variance methods when the independent variables are limited in possible values to 0 and 1 (presence or absence) tend to be unaware that a model is implied [1, p. 243]. *Analysis of covariance* uses a combination of techniques which are specially adapted to 0 or 1 independent variables and techniques needed in more general cases.

5.1.2 Models

Certain aspects of models are discussed in this section. First considered is the model functional form, which is termed the *regression function*. Some restrictions on designs for these functions are also given. Second, two error models are discussed. In one there are measurement errors and in the other the random component is in the equation describing the system. Third, in the next subsection various standard assumptions relating to the statistics of the errors are given.

The regression functions for the cases used are considered to have the correct functional forms, that is, not empirical approximations or best guesses. The functions considered in this chapter are linear in the parameters and contain at most two parameters. For convenience in later references, the regression functions used in this chapter are listed and labeled as follows:

$$\text{Model 1,} \quad \eta_i = \beta_0 \tag{5.1.1a}$$

$$\text{Model 2,} \quad \eta_i = \beta_1 X_i \tag{5.1.1b}$$

$$\text{Model 3,} \quad \eta_i = \beta_0 + \beta_1 X_i \tag{5.1.1c}$$

$$\text{Model 4,} \quad \eta_i = \beta_0' + \beta_1 \left(X_i - \bar{X}\right); \quad \bar{X} = \sum_{i=1}^{n} \frac{X_i}{n} \tag{5.1.1d}$$

$$\text{Model 5,} \quad \eta_i = \beta_1 X_{i1} + \beta_2 X_{i2} \tag{5.1.1e}$$

The variable η is sometimes called the dependent variable†; X_i, X_{i1}, and X_{i2} are independent variables that might represent time, position, tempera-

*"Conditions" refer, for example, to the X_i values in (5.1.1c).

†In the statistical literature Y is called the dependent variable.

ture, velocity, cost, and so on. Clearly some of these models are related. For example, Model 2 reduces to Model 1 if $X_i=1$. Also, Model 5 includes both Models 3 and 4.

In each case there is a restriction related to the measurements. Assume that there are n observations. For Model 1 the restriction is simply that there is at least one observation or $n \geqslant 1$. For Models 2, 3, 4, and 5 the respective restrictions are as follows:

$$\sum_{i=1}^{n} |X_i| \neq 0 \qquad \text{(at least one } X_i \neq 0 \text{ needed)} \tag{5.1.2a}$$

$$\sum_{i=1}^{n} \left(X_i - \overline{X}\right)^2 \neq 0 \qquad \text{(at least 2 different } X_i \text{ values needed)} \tag{5.1.2b}$$

$$\sum_{i=1}^{n} \left(X_i - \overline{X}\right)^2 \neq 0 \qquad \text{(at least 2 different } X_i \text{ values needed)} \tag{5.1.2c}$$

$$\left(\sum_{i=1}^{n} X_{i1}^2\right)\left(\sum_{j=1}^{n} X_{j2}^2\right) - \left(\sum_{i=1}^{n} X_{i1}X_{i2}\right)^2 \neq 0$$

$$\text{(at least 2 different sets of } X_{i1}, X_{j2}) \tag{5.1.2d}$$

where $\overline{X} \equiv \sum_{i=1}^{n} X_i / n$.

In each of the models, except the first, the independent variables X_i or X_{ij} could represent a number of equally or unequally spaced values. Alternately, X_i might represent values of various functions of time, t, such as

$$t_i,\ t_i^2,\ 3t_i^3 - t_i,\ \sin at_i,\ \cos at_i,\ e^{-at_i},\ \ln t_i$$

or some combination of them. The quantity a is here assumed to be known.

In most of this chapter the errors are considered to be additive. Then for Model 3

$$Y_i = \beta_0 + \beta_1 X_i + \varepsilon_i \tag{5.1.3}$$

where ε_i is the unknown error and Y_i is the measurement at X_i. The model given by (5.1.3) can, however, represent the following two cases:

Error Model A. Errors in Measurements

$$\eta_i = \beta_0 + \beta_1 X_i$$

$$Y_i = \eta_i + \varepsilon_i \tag{5.1.4}$$

Error Model B. Errors (Noise) in Process

$$\eta_i = \beta_0 + \beta_1 X_i + \varepsilon_i$$

$$Y_i = \eta_i \tag{5.1.5}$$

where η_i represents the quantity being measured and Y_i is its measurement. Implicit in these models is the assumption that there is no error in X_i; that is, X_i is not a random variable as are Y_i and ε_i. In Error Model B, η_i is also a random variable.

In Error Model A there are errors in the measurements but there is none in η. In order to quantify ε_i one can study the error characteristics of the measuring devices be they thermocouples, hot-wire anemometers, micrometers, etc. These errors can be reduced by more precise devices. As technology improves, one would expect ε_i in Error Model A to decrease. The system model itself is assumed to be errorless or noiseless. This implies that the physics is well-understood and that there is no stochastic noise entering in η. This would be the case for many physical measurements. Consider, for example, the steady state temperature distribution in a flat plate which is linear with position. The randomness in observed temperatures for repeated measurements would be the result of measurement noise rather than some physical phenomenon causing the fluctuation.

In Error Model B the measurements are assumed errorless; but the model (η) contains "noise"; that is, the variable being measured deviates by some stochastic component from its expected value. An example is turbulent flow between two parallel plates. Part of the universal velocity profile for turbulent flow is described by the expression

$$u^+ = \beta_0 + \beta_1 \ln y^+$$

where the dependent variable u^+ is a dimensionless velocity and y^+, the independent variable, is a dimensionless distance. In this case instantaneous velocity measurements fluctuate about the mean value u^+ owing more to the turbulence phenomenon than to measurement inaccuracies. Hence this is an Error Model B case. For Error Model B ε_i would not be expected to decrease with time (that is, with improved measurement capability). Also a study of the sensor would not yield any information regarding ε_i.

Regardless of whether Error Model A or B is correct, the estimation problem is formally the same for the physical models considered in this chapter. The meaning of η and ε is different, however, as are the statistics for ε. We shall visualize Error Model A as the model considered in this chapter.

5.1.3 Statistical Assumptions Regarding the Measurement Errors

Assumptions regarding the measurement errors should be carefully stated in each estimation problem. If the assumptions do not accurately describe the data, then one can at least pinpoint the assumption(s) which are not satisfied. The mere identification of the incorrect assumptions may lead to more realistic assumptions and thus better estimators.

Different assumptions lead to different estimation methods. In this chapter we consider three commonly used methods: ordinary least squares (OLS), maximum likelihood (ML), and maximum a posteriori (MAP). The following conditions given in terms of Error Model A and Model 3 are termed the *standard statistical assumptions* for $i=1, 2, \ldots, n$:

1. $Y_i = E(Y_i|\beta_0, \beta_1) + \varepsilon_i = \eta_i + \varepsilon_i$ (additive errors) (5.1.6)
2. $E(\varepsilon_i) = 0$ (zero mean errors) (5.1.7)
3. $V(Y_i|\beta_0, \beta_1) = \sigma^2$ (constant variance errors, homoskedasticity) (5.1.8)
   [Note $E(\varepsilon_i^2) = \sigma^2$ if $E(\varepsilon_i) = 0$.]
4. $E\{[\varepsilon_i - E(\varepsilon_i)][\varepsilon_j - E(\varepsilon_j)]\} = 0$ for $i \neq j$ (uncorrelated errors) (5.1.9)
   [or $E(\varepsilon_i \varepsilon_j) = 0$ if $E(\varepsilon_i) = 0$ and $i \neq j$.]
5. ε_i has a normal probability distribution (5.1.10)
6. Known statistical parameters (5.1.11)
7. $V(X_i) = 0$ (nonstochastic independent variable) (5.1.12)
8. No prior information regarding β_0 and β_1 and parameters nonrandom (5.1.13)

In order to describe the assumptions concisely and explicitly, we assign a 1 or 0 to the above assumptions where 1 means yes and 0 no. For a case when all the assumption are satisfied we designate them as 11111111 where the first 1 on the left refers to the additive error assumption, the second 1 refers to the zero mean assumption, etc. In some cases additional numbers are used to indicate more information than a simple no. For example, for the uncorrelated error condition, 2 designates first-order autoregressive errors. See Section 6.1.5 for a more complete list of possibilities other than 1 or 0. If an assumption is not used then a dash will be used in lieu of a 1 or 0.

Assumptions 2, 3, 4, and 7 are sometimes referred to as the Gauss–Markov assumptions.

5.2 ORDINARY LEAST SQUARES ESTIMATORS (OLS)

In ordinary least squares estimation the sum of squares function to be minimized with respect to the parameters is simply

$$S=\sum_{i=1}^{n}\left[Y_i-\eta_i\right]^2 \tag{5.2.1}$$

where η_i is a function of the parameters such as β_0 and β_1.

It is important to observe that *no* statistical assumptions are used in obtaining OLS parameter estimates, that is, the assumptions are --------. In order to make statistical statements regarding the estimators it is necessary to possess information regarding the measurement errors, however.

In derivations to be given we may need the variance of $\sum d_i Y_i$ where d_i is not a random variable. Assume that the errors in Y_i are additive, have zero mean, and are uncorrelated (assumptions 1, 2, and 4, respectively). Then

$$V\left(\sum_{i=1}^{n} d_i Y_i\right)=V\left[\sum_{i=1}^{n} d_i(\eta_i+\varepsilon_i)\right]$$

$$=E\left\{\left[\sum_{i=1}^{n} d_i(\eta_i+\varepsilon_i)-\sum_{i=1}^{n} d_i\eta_i\right]^2\right\}$$

$$=V\left(\sum_{i=1}^{n} d_i\varepsilon_i\right)=\sum_{i=1}^{n} d_i^2 V(\varepsilon_i)=\sum_{i=1}^{n} d_i^2\sigma_i^2 \tag{5.2.2}$$

where the first assumption is used on the first line of (5.2.2), second assumption on the second line, and fourth on the third line. (5.2.2) is a special case of (2.6.20).

5.2.1 Models 1 and 2 ($\eta_i=\beta_0$ and $\eta_i=\beta_1 X_i$)

Both Models 1 and 2 are covered in this section. Since Model 2 is the more general, we start with it and then apply the results to Model 1. For Model 2 ($\eta_i=\beta_1 X_i$), (5.2.1) can be written

$$S=\sum_{i=1}^{n}\left[Y_i-\beta_1 X_i\right]^2 \tag{5.2.3}$$

Differentiating S with respect to β_1, replacing β_1 by the estimator b_1, and setting equal to zero give the *normal equation*,

$$\Sigma Y_i X_i - b_1 \Sigma X_i^2 = 0 \tag{5.2.4}$$

whose solution for Model 2 is

$$\boxed{b_1 = \left(\sum_{i=1}^{n} Y_i X_i \right) \left(\sum_{j=1}^{n} X_j^2 \right)^{-1}} \tag{5.2.5}$$

By setting $X_i = 1$ in (5.2.5) the Model 1 estimator is

$$\boxed{b_0 = \frac{\Sigma_{i=1}^{n} Y_i}{n} \equiv \bar{Y}} \tag{5.2.6}$$

which is the average Y_i. For these two estimators, no statistical assumptions are used but at least one observation must be made, and in the case of Model 2, at least one X_i must not be zero.

The *predicted*, *regression*, or *smoothed* value is denoted $\hat{Y}_i$ and is called "Y_i hat." For Models 1 and 2, respectively, $\hat{Y}_i$ is

$$\hat{Y}_i = b_0 = \bar{Y}, \qquad \hat{Y}_i = b_1 X_i \tag{5.2.7a,b}$$

The *residual* e_i is the measured value of Y_i minus the predicted value or

$$e_i \equiv Y_i - \hat{Y}_i \tag{5.2.8}$$

The residual e_i is not equal to the error ε_i but it can be used to estimate ε_i.

5.2.1.1 Mean and Variances of Estimates

Using the standard statistical assumptions of additive, zero mean errors and nonstochastic X_i, β_0, and β_1 (11----11), we get for the expected value of the Model 2 parameter

$$E(b_1) = \Sigma E(Y_i) X_i \left[\Sigma X_j^2 \right]^{-1} = \Sigma(\beta_1 X_i) X_i \left[\Sigma X_j^2 \right]^{-1} = \beta_1$$

One can also show for Model 1 that $E(b_0) = \beta_0$. Hence the least squares estimators b_0 and b_1 are unbiased for the stated assumptions (see Section 3.2.1).

Suppose that all the standard assumptions are valid except that ε_i need not possess a normal density and σ^2 may or may not be known (assumptions 1111--11); then the variance of b_0 using (5.2.6) and (5.2.2) is

$$V(b_0) = \sum_{i=1}^{n} n^{-2}\sigma^2 = \frac{\sigma^2}{n} \tag{5.2.9}$$

From (5.2.5) and (5.2.2) the variance of b_1 is

$$V(b_1) = \left[\sum_{i=1}^{n} X_i^2\sigma^2\right]\left[\sum_{j=1}^{n} X_j^2\right]^{-2} = \sigma^2\left[\sum_{i=1}^{n} X_i^2\right]^{-1} \tag{5.2.10}$$

Notice that (5.2.9) and (5.2.10) both indicate that estimates as accurate as desired can be obtained by simply taking a sufficiently large number of observations. This naturally requires that the underlying assumptions be valid. If the measurements were correlated, for example, this conclusion might not be true.

Also note that for Model 2 ($\eta_i = \beta_1 X_i$) there is optimum placement of observations. Suppose that n observations are to be obtained and it is desired to obtain a minimum variance estimate by selecting the X_i so that $|X_i| \leqslant |X_m|$. Then the variance of b_1 is minimized if all the measurements are concentrated at X_m giving $V(b_1) = \sigma^2/nX_m^2$. This would be the best choice of the X_i values provided there is no uncertainty in the model (i.e., functional form of η_i).

Suppose that all the standard assumptions are valid except there may or may not be normality and σ^2 is unknown (1111-011). Then the variances of b_0 and b_1 are estimated by replacing σ^2 by an estimate which is designated s^2. The square roots of $V(b_0)$ and $V(b_1)$ with this replacement are called the *estimated standard errors* (or standard deviations),

$$\text{est. s.e.}(b_0) = sn^{-1/2} \tag{5.2.11}$$

$$\text{est. s.e.}(b_1) = s\left[\sum_{i=1}^{n} X_i^2\right]^{-1/2} \tag{5.2.12}$$

5.2.1.2 *Expected Value of* $S_{\min}$

An estimator for σ^2 is not directly obtained using OLS as it is using ML estimation. One can, however, for the assumptions 1111-011 relate the expected value of the minimum sum of squares, designated $S_{\min}$, to σ^2.

Since $E(Y_i - \hat{Y}_i) = 0$,

$$E\left[\left(Y_i - \hat{Y}_i\right)^2\right] = V\left(Y_i - \hat{Y}_i\right) = V(e_i) \tag{5.2.13a}$$

and thus the expected value of $S_{\min}$ is

$$E(S_{\min}) = E\left[\Sigma\left(Y_i - \hat{Y}_i\right)^2\right] = \Sigma E\left[\left(Y_i - \hat{Y}_i\right)^2\right] = \Sigma V(e_i) \tag{5.2.13b}$$

(5.2.13b) is valid for any number of parameters. It still remains to find $V(e_i)$ in terms of σ^2. It is always true that

$$V(e_i) = V\left(Y_i - \hat{Y}_i\right) = V(Y_i) + V\left(\hat{Y}_i\right) - 2\,\text{cov}\left(Y_i, \hat{Y}_i\right) \tag{5.2.14}$$

The $V(Y_i)$ term is simply σ^2. The other two terms are considered below.

For the one-parameter models we can write $\hat{Y}_i = b_1 X_i = X_i \, \Sigma \, d_j Y_j$ so that

$$V\left(\hat{Y}_i\right) = X_i^2 \Sigma d_j^2 \sigma_j^2 \tag{5.2.15}$$

using (5.2.2). For constant error variance σ^2 the variance of $\hat{Y}_i$ for Model 2 is

$$V\left(\hat{Y}_i\right) = X_i^2 \sigma^2 \left[\Sigma X_j^2\right]^{-1} \tag{5.2.16}$$

and then letting $X_i = 1$ we have for Model 1 ($\eta_i = \beta_0$),

$$V\left(\hat{Y}_i\right) = \frac{\sigma^2}{n.} \tag{5.2.17}$$

Observe that the variance of the predicted value of Y_i is a constant for Model 1 but increases with X_i^2 for Model 2.

The third term on the right side of (5.2.14) for assumptions 1111--11 and Model 2 is

$$-2\text{cov}\left(Y_i, \hat{Y}_i\right) = -2X_i d_i \sigma^2 = -2X_i^2 \left[\Sigma X_k^2\right]^{-1} \sigma^2 \tag{5.2.18}$$

Combining the above results yields for Models 1 and 2, respectively,

$$V(e_i) = \sigma^2\left[1 - \frac{1}{n}\right]; \qquad V(e_i) = \sigma^2\left[1 - \frac{X_i^2}{\Sigma X_j^2}\right] \tag{5.2.19a,b}$$

which are both less than $V(\varepsilon_i) = \sigma^2$. In both cases the expected value of

$S_{\min}$ is found using (5.2.13) and (5.2.19) to be

$$E(S_{\min}) = (n - 1)\sigma^2 \tag{5.2.20}$$

and thus an unbiased estimator for σ^2, designated s^2 or $\hat{\sigma}^2$, is

$$s^2 = \hat{\sigma}^2 = \frac{S_{\min}}{(n-1)} = \frac{\Sigma(Y_i - \hat{Y}_i)^2}{(n-1)}, \qquad n > 1 \tag{5.2.21}$$

This expression is valid for *one* parameter with assumptions 1111-011 and can be used in (5.2.11) or (5.2.12). For one parameter, s^2 can be estimated by only using two or more observations.

Example 5.2.1

An automobile is traveling at a constant speed and the distances traveled at the end of 1, 2, and 3 min are measured to be 1.01, 2.03, and 3.00 km. Assume that distance is the dependent variable and time the independent variable. The regression function for this case is that the distance traveled, h, is equal to the velocity, v, times the duration traveled, t; in symbols, $h = vt$. Use OLS to estimate v.

Solution

This is a Model 2 case with v being the parameter. Using (5.2.5) with Y_i being the h_i measurement, we find

$$\hat{v} = \left[\sum_{i=1}^{3} Y_i t_i\right]\left[\sum_{j=1}^{3} t_j^2\right]^{-1}$$

$$= [1.01(1) + 2.03(2) + 3(3)][1 + 4 + 9]^{-1} = 1.005 \text{ km/min}$$

where Y_i is the observation of h_i.

Example 5.2.2

An object is dropped in a vacuum and the position h is observed at various times t_i. The observations of h_i, designated Y_i, are given as

t_i(sec)	0.1	0.2	0.3	0.4
Y_i(m)	0.05	0.2	0.4	0.8

The measurements are to be used to estimate the local gravitational constant g.

The position h is described by the differential equation $\ddot{h} = g$ and the initial conditions $h = \dot{h} = 0$ at $t = 0$; the solution for h is $h = gt^2/2$.

(a) Using ordinary least squares, find an estimate of g.

(b) Using the standard assumptions except that σ^2 is unknown and ε_i need not be normal, give an estimate of the standard error of $\hat{g}$.

Solution

(a) The given model is the same as Model 2 with g being β and X_i being $t_i^2/2$. The estimator for OLS is (5.2.5) which can be written as

$$\hat{g} = \left(\frac{\sum_{i=1}^{4} Y_i t_i^2}{2} \right) \left(\frac{\sum_{j=1}^{4} t_j^4}{4} \right)^{-1}$$

Then the numerator and denominator are, respectively,

$$\frac{1}{2} \sum_{i=1}^{4} Y_i t_i^2 = \frac{1}{2} \{ .05(.1)^2 + .2(.2)^2 + \cdots + .8(.4)^2 \} = 0.08625$$

$$\frac{1}{4} \sum_{j=1}^{4} t_j^4 = \frac{1}{4} \left[(.1)^4 + (.2)^4 + (.3)^4 + (.4)^4 \right] = 0.00885$$

and thus the estimate is $\hat{g} = 0.08625/0.00885 = 9.7458$ m/sec^2.

(b) The residuals, $e_i = Y_i - \hat{Y}_i$, are, respectively, 0.00127, 0.00508, -0.03855, and 0.02034 and the sum of squares of these terms is $S_{\min} = 0.001928$. From (5.2.21) the estimated standard deviation is

$$s = \left[\frac{S_{\min}}{(n-1)} \right]^{1/2} = \left[\frac{.001928}{(4-1)} \right]^{1/2} = 0.02535$$

and then from (5.2.12) the estimated standard error of $\hat{g}$ is

$$\text{est. s.e.}(\hat{g}) = s \left[\frac{\sum_{i=1}^{4} t_i^4}{4} \right]^{-1/2} = .02535[.00885]^{-1/2}$$

$$= 0.2695 \text{ m/sec}^2$$

which can be compared with the estimate of 9.7458 m/sec^2.

5.2.2 Two-Parameter Models

5.2.2.1 Model 5, $\eta_i = \beta_1 X_{i1} + \beta_2 X_{i2}$

In order to simplify the presentation of the two-parameter cases, the general two-parameter case, Model 5, is considered first. Using the sum of squares function, (5.2.1), with Model 5, (5.1.1e), we have

$$S=\sum_{i=1}^{n}\left[Y_i-\beta_1X_{i1}-\beta_2X_{i2}\right]^2 \tag{5.2.22}$$

We differentiate S with respect to β_1, setting the derivative equal to zero, and replace β_1 by its estimator b_1 and β_2 by b_2. Repeating the same procedure for β_2 then yields the two normal equations

$$b_1c_{11}+b_2c_{12}=d_1 \tag{5.2.23a}$$

$$b_1c_{12}+b_2c_{22}=d_2 \tag{5.2.23b}$$

where

$$c_{k\ell}\equiv\sum_{i=1}^{n}X_{ik}X_{i\ell};\qquad d_k\equiv\sum_{i=1}^{n}Y_iX_{ik} \tag{5.2.23c}$$

Notice that the coefficient c_{12} appears in a symmetric manner in (5.2.23a, b). Solving (5.2.23a, b) for b_1 and b_2 yields (for Model 5)

$$b_1=\frac{d_1c_{22}-d_2c_{12}}{\Delta};\qquad b_2=\frac{d_2c_{11}-d_1c_{12}}{\Delta} \tag{5.2.24a}$$

$$\Delta\equiv c_{11}c_{22}-c_{12}^2 \tag{5.2.24b}$$

No statistical assumptions were necessary to derive the estimators given in (5.2.24a). Using the three standard assumptions of additive, zero mean errors and nonstochastic X_i it can be shown that b_1 and b_2 are unbiased estimates of β_1 and β_2.

The variance of b_1 can be readily found by writing b_1 as

$$b_1=\Sigma(f_i-g_i)Y_i;\qquad f_i=\frac{c_{22}X_{i1}}{\Delta},\qquad g_i=\frac{c_{12}X_{i2}}{\Delta} \tag{5.2.25}$$

Then using the standard statistical assumptions 1111--11 and (5.2.2) the variance of b_1 is

$$V(b_1)=\Sigma(f_i-g_i)^2\sigma^2=\Sigma(f_i^2-2f_ig_i+g_i^2)\sigma^2$$
$$=\left[c_{22}^2c_{11}-2c_{22}c_{12}^2+c_{12}^2c_{22}\right]\sigma^2/\Delta^2$$

or simplifying gives

$$V(b_1)=\frac{c_{22}\sigma^2}{\Delta} \tag{5.2.26a}$$

In a similar manner it can be shown that $V(b_2)$ and $\mathrm{cov}(b_1, b_2)$ are given by

$$V(b_2) = \frac{c_{11}\sigma^2}{\Delta}, \qquad \mathrm{cov}(b_1, b_2) = \frac{-c_{12}\sigma^2}{\Delta} \tag{5.2.26b,c}$$

The predicted value of Y_i is $\hat{Y}_i$,

$$\hat{Y}_i = b_1 X_{i1} + b_2 X_{i2} \tag{5.2.27}$$

The variance of $\hat{Y}_i$ is then

$$\begin{aligned} V(\hat{Y}_i) &= X_{i1}^2 V(b_1) + X_{i2}^2 V(b_2) + 2X_{i1}X_{i2}\,\mathrm{cov}(b_1, b_2) \\ &= \left[X_{i1}^2 c_{22} + X_{i2}^2 c_{11} - 2X_{i1}X_{i2}c_{12}\right]\sigma^2/\Delta \end{aligned} \tag{5.2.28}$$

where (5.2.26) is used. It can also be shown that $\mathrm{cov}(Y_i, \hat{Y}_i)$ is equal to the same value or

$$\mathrm{cov}(Y_i, \hat{Y}_i) = V(\hat{Y}_i) \tag{5.2.29}$$

From (5.2.14), (5.2.28), and (5.2.29) the variance of the residual e_i $(= Y_i - \hat{Y}_i)$ is equal to

$$V(e_i) = \left[\Delta - X_{i1}^2 c_{22} - X_{i2}^2 c_{11} + 2X_{i1}X_{i2}c_{12}\right]\sigma^2/\Delta \tag{5.2.30}$$

Then using the result that $E(S_{\min})$ is equal to $\Sigma V(e_i)$ given by (5.2.13b), we find that

$$\begin{aligned} E(S_{\min}) &= \left[n\Delta - c_{11}c_{22} - c_{22}c_{11} + 2c_{12}^2\right]\sigma^2/\Delta \\ &= (n-2)\sigma^2 \end{aligned} \tag{5.2.31}$$

since $\Delta \equiv c_{11}c_{22} - c_{12}^2$. Consequently, for the two-parameter case with Model 5 and assumptions 1111-011, an unbiased estimator for σ^2 is

$$s^2 = S_{\min}/(n-2) \qquad (n > 2) \tag{5.2.32}$$

which differs from (5.2.21) in that there is a factor of $n-2$ rather than $n-1$. Observe that (5.2.32) is properly meaningless for $n=2$. For two parameters and two observations the two residuals must be zero also giving $S_{\min} = 0$. Consequently, for two parameters, σ^2 can be estimated only if $n > 2$.

5.2.2.2 *Model 3,* $\eta_i = \beta_0 + \beta_1 X_i$

Model 3 results can be found from those of Model 5 by replacing in Model 5 β_1 by β_0, b_1 by b_0, β_2 by β_1, b_2 by b_1, X_{i1} by 1, and X_{i2} by X_i. This gives

$$c_{11} = n, \qquad c_{22} = \Sigma X_i^2, \qquad c_{12} = \Sigma X_i \tag{5.2.33}$$

$$\Delta = n(\Sigma X_i^2) - (\Sigma X_i)^2, \qquad d_1 = \Sigma Y_i, \qquad d_2 = \Sigma Y_i X_i \tag{5.2.34}$$

One must be careful where the squares are placed in Δ; note that ΣX_i^2 means the sum of X_i^2 whereas $(\Sigma X_i)^2$ means the square of the sum of the X_i values. It also can be shown that Δ is also equal to

$$\Delta = n \sum_{i=1}^{n} \left(X_i - \bar{X}\right)^2, \qquad \bar{X} \equiv \sum_{j=1}^{n} \frac{X_j}{n} \tag{5.2.35a,b}$$

From the above relations b_1, the estimator of β_1 in Model 3, which is $\eta_i = \beta_0 + \beta_1 X_i$, can be found from b_2 in (5.2.24a) to be

$$b_1 = \frac{n(\Sigma Y_i X_i) - (\Sigma Y_i)(\Sigma X_i)}{\Delta} \tag{5.2.36}$$

Using (5.2.35a) this expression can also be written (Model 3)

$$\boxed{b_1 = \frac{\Sigma\left(X_i - \bar{X}\right) Y_i}{\Sigma\left(X_i - \bar{X}\right)^2} = \frac{\Sigma\left(X_i - \bar{X}\right)\left(Y_i - \bar{Y}\right)}{\Sigma\left(X_i - \bar{X}\right)^2}} \tag{5.2.37}$$

where $\bar{Y} = \Sigma Y_i / n$ and the range of each summation is from $i = 1$ to n. The estimator for b_0 can also be found from (5.2.24a) by using the expression for b_1. Instead we shall use (5.2.23a) divided by n (and $b_1 \to b_0$ and $b_2 \to b_1$) to get

$$\boxed{b_0 = \bar{Y} - b_1 \bar{X}} \tag{5.2.38}$$

Hence if $\bar{X} = \Sigma X_i / n$ is equal to zero, b_0 is simply $\bar{Y}$. For this reason and the resulting simplifications in (5.2.37), a transformation sometimes used in hand calculations redefines X_i so that $\bar{X} = 0$.

As mentioned several times above, no statistical assumptions are used to obtain the estimators for b_0 and b_1 given respectively by (5.2.38) and (5.2.37). Suppose now that the standard assumptions are valid. A number

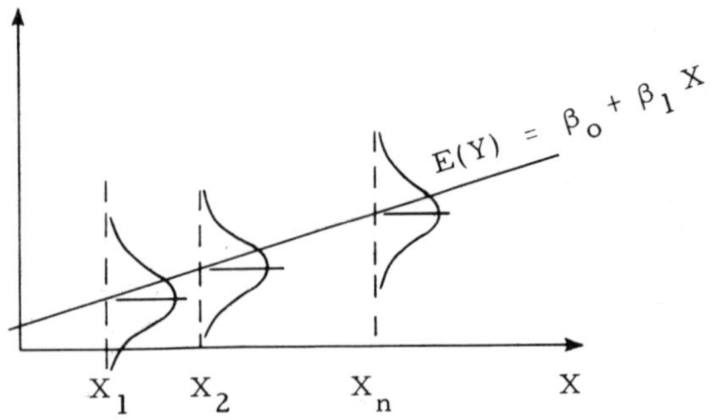

Figure 5.1 Linear model with Y being a random variable with constant σ^2 and normal probability distribution.

of these are illustrated by Fig. 5.1 for Model 3, $\eta_i = \beta_0 + \beta_1 X_i$. The normal probability density is superimposed upon the curve for several X_i values. The first two assumptions of additive, zero means errors are implied in Fig. 5.1. The third assumption of constant variance is depicted explicitly, as is the normality assumption (number 5). The nonstochastic X_i assumption (number 7) is implied by the lack of a probability density in the X_i direction.

Mean and Variances for Model 3

The OLS estimates of β_0 and β_1 are unbiased for additive, zero mean errors as was demonstrated for the more general case, Model 5.

From (5.2.26), (5.2.33), and (5.2.34) the variances and covariance of b_0 and b_1 are

$$V(b_0) = \sigma^2 \Sigma X_i^2 / \Delta, \; V(b_1) = \frac{n\sigma^2}{\Delta} \tag{5.2.39a,b}$$

$$\text{cov}(b_0, b_1) = -n\bar{X}\sigma^2/\Delta \tag{5.2.40}$$

where Δ is given by (5.2.35a). Assumptions 1111-111 are used.

From (5.2.28) and (5.2.30) the variances of the predicted value $\hat{Y}_i$ and the residual e_i can be written

$$V(\hat{Y}_i) = \left[n^{-1} + n\left(X_i - \bar{X}\right)^2 \Delta^{-1} \right] \sigma^2 \tag{5.2.41a}$$

$$V(e_i) = \left[1 - n^{-1} - n\left(X_i - \bar{X}\right)^2 \Delta^{-1} \right] \sigma^2 \tag{5.2.41b}$$

Unlike the variances of b_0 and b_1, the variances of $\hat{Y}_i$ and e_i are functions of i. Note that $V(\hat{Y}_i)$ has a minimum at $X_i = \bar{X}$ and maximum value at the smallest or largest value of X_i. The variance of the residual e_i is different in that it has a maximum at $X_i = \bar{X}$.

The estimated standard errors of b_0 and b_1 are found from (5.2.39a,b) to be for assumptions 1111-011,

$$\text{est. s.e.}(b_0) = s\left[\frac{\Sigma X_i^2}{\Delta}\right]^{1/2} \tag{5.2.42a}$$

$$\text{est. s.e.}(b_1) = s\left[n/\Delta\right]^{1/2}, \qquad \Delta \equiv n\Sigma\left(X_i - \bar{X}\right)^2 \tag{5.2.42b}$$

where from (5.2.32), $s = [S_{\min}/(n-2)]^{1/2}$

For Model 3 the sum of the residuals is equal to zero or

$$\sum_{i=1}^{n} e_i = 0 \tag{5.2.43}$$

This interesting result can be used to check the accuracy of calculations for the parameters. This result is true for *any* linear or nonlinear model provided there is a β_0 term in the model, that is, a parameter not multiplied by a function of an independent variable, and provided OLS is used.

Example 5.2.3

Experiments have been performed for the heat transfer to air flowing in a pipe. A dimensionless group related to the heat flow rate is the Nusselt number, designated Nu. This is a function of the Reynolds number, denoted Re, which is proportional to the average velocity in the tube. Below are some values for the turbulent fluid flow range.

Re	10^4	2×10^4	4×10^4	5×10^4
Nu	32	60	90	119

The suggested model is $\text{Nu} = a_0\text{Re}^{a_1}$ where the parameters are a_0 and a_1. Reduce to a linear form and estimate a_0 and a_1 using ordinary least squares with log Nu being the dependent variable.

Solution

Take the logarithm to the base 10 to get

$$\log \text{Nu} = \log a_0 + a_1 \log \text{Re}$$

For convenience write the model in the Model 3 form, $\eta_i = \beta_0 + \beta_1 X_i$, with

$$\log \mathrm{Nu} \to \eta_i, \qquad \log a_0 \to \beta_0, \qquad a_1 \to \beta_1, \qquad \log \mathrm{Re} \to X_i$$

The tabulated values of Nu are used to obtain log Nu which is now Y_i as given below

X_i (=log Re)	4.0	4.3010	4.6021	4.6990
Y_i	1.5051	1.7782	1.9542	2.0755

The estimates of b_0 and b_1 are found using (5.2.37) and (5.2.38). In these equations the following are needed,

$$\bar{X} = \frac{\Sigma X_i}{n} = \frac{[4.0 + 4.301 + 4.6021 + 4.699]}{4}$$

$$= 4.400525$$

$$\bar{Y} = \frac{\Sigma Y_i}{n} = \frac{1.5051 + 1.7782 + \cdots}{4} = 1.82825$$

$$\Sigma(X_i - \bar{X})^2 = (4 - 4.400525)^2 + \cdots + (4.699 - 4.400525)^2 = 0.3000453$$

$$\Sigma(X_i - \bar{X})Y_i = (4 - 4.400525)(1.5051) + (4.301 - 4.400525)(1.7782)$$

$$+ \cdots = 0.2335972$$

Then (5.2.37) gives

$$b_1 = \frac{\Sigma(X_i - \bar{X})Y_i}{\Sigma(X_i - \bar{X})^2} = \frac{.2335972}{.3000453} = 0.7785397$$

and from (5.2.38) b_0 is

$$b_0 = \bar{Y} - b_1\bar{X} = -1.597734$$

The estimate of a_0 is

$$\hat{a}_0 = 10^{b_0} = 0.252503$$

Thus the prediction equation for Nu is

$$\widehat{\mathrm{Nu}} = .0253\mathrm{Re}^{.779}$$

where some of the decimal places have been dropped.

Example 5.2.4

Normal random error terms [2] with a mean of zero and unit variance have been added to the model $\eta_i = \beta_0 + \beta_1 X_i$ with β_0 set equal to 1 and β_1 set equal to 0.1. The "data" are tabulated in Table 5.1.

(*a*) Estimate the parameters β_0 and β_1 using ordinary least squares.

(*b*) Find the estimated standard errors for b_0, b_1, and $\hat{Y}_i$ using the standard assumptions except that the errors need not be normal and that σ^2 is unknown (1111-011).

Table 5.1 Data for Example 5.2.4

Observation	X_i	η_i	ϵ_i	Y_i
1	0	1.0	−0.742	0.258
2	10	2.0	−0.034	1.966
3	20	3.0	1.453	4.453
4	30	4.0	0.963	4.963
5	40	5.0	0.040	5.040
6	50	6.0	0.418	6.418
7	60	7.0	1.792	8.792
8	70	8.0	−0.374	7.626
9	80	9.0	−0.222	8.778
	$360 = \Sigma X_i$		$3.294 = \Sigma \varepsilon_i$	$48.294 = \Sigma Y_i$

Solution

(*a*) The OLS estimators for b_0 and b_1 are given by (5.2.37) and (5.2.38). In these equations $\bar{X}$ and $\bar{Y}$ are needed,

$$\bar{X} = \frac{1}{n}\sum_{i=1}^{n} X_i = \frac{1}{9}(0 + 10 + 20 + \cdots + 80) = \frac{360}{9} = 40$$

$$\bar{Y} = \frac{1}{n}\sum_{i=1}^{n} Y_i = \frac{1}{9}(.258 + 1.966 + \cdots + 8.778) = \frac{48.294}{9} = 5.366$$

Additional required calculations are given in the second, third, and fourth columns of Table 5.2. Then the estimates of β_1 and β_0 are

$$b_1 = \frac{\Sigma(X_i - \bar{X})Y_i}{\Sigma(X_i - \bar{X})^2} = \frac{611.93}{6000} = 0.10198833$$

$$b_0 = \bar{Y} - b_1\bar{X} = 5.366 - (0.10198833)(40) = 1.2864667$$

which happen to be about 2% and 29% larger than the true values.

Table 5.2 Calculations for Example 5.2.4

X_i	$X_i-\bar{X}$	$(X_i-\bar{X})^2$	$Y_i(X_i-\bar{X})$	$\hat{Y}_i$	$Y_i-\hat{Y}_i=e_i$
0	−40	1600	−10.32	1.28647	−1.02847
10	−30	900	−58.98	2.30635	−0.34035
20	−20	400	−89.06	3.32623	1.12677
30	−10	100	−49.63	4.34612	0.61688
40	0	0	0	5.36600	−0.32600
50	10	100	64.18	6.38588	0.03212
60	20	400	175.84	7.40577	1.38623
70	30	900	228.78	8.42565	−0.79965
80	40	1600	351.12	9.44553	−0.66753
360 = ΣX_i		6000 = $\Sigma(X_i-\bar{X})^2$	611.93 = $\Sigma(X_i-\bar{X})Y_i$		0.00000 = $\Sigma(Y_i-\hat{Y}_i)$

All eight significant figures given in these estimates are not needed, but it is usually wise to carry a couple of extra significant digits in the calculations because there can be small differences of large numbers.

The predicted value of the dependent variable, $\hat{Y}_i$, can be found from

$$\hat{Y}_i=\bar{Y}+b_1\left(X_i-\bar{X}\right)=5.366+0.10198833(X_i-40)$$

and is also given in Table 5.2. The residuals e_i are also given. Note that the sum is zero.

(*b*) In order to find the estimated standard errors it is necessary to evalute s^2 which in turn needs $S_{\min}=\Sigma e_i^2$ which is 5.937718. Then from (5.2.32)

$$s=\left[\frac{S_{\min}}{n-2}\right]^{1/2}=\left[\frac{5.937718}{9-2}\right]^{1/2}=0.921002$$

which is an estimate of the standard deviation. Compared with the true value of unity this is only about 8% too low.

From (5.2.42a) the standard error of b_0 is

$$\text{est. s.e.}(b_0)=\left[\frac{\Sigma X_i^2}{n\Sigma\left(X_k-\bar{X}\right)^2}\right]^{1/2}s=\left[\frac{20400}{9(6000)}\right]^{1/2}(0.921002)=0.56608$$

and the standard error of b_1 is obtained from (5.2.42b)

$$\text{est. s.e.}(b_1)=s\left[\Sigma\left(X_i-\bar{X}\right)^2\right]^{-1/2}=\frac{0.921002}{(6000)^{1/2}}=0.011890$$

Notice that $b_0\pm$est. s.e.(b_0) is 1.286 ± 0.566 which includes the true value of $\beta_0=1$. $b_1\pm$est. s.e.(b_1) is 0.10199 ± 0.0119 which also includes the true value of 0.1.

Statistical statements regarding the accuracy of the estimates are discussed in Chapter 6 in connection with the confidence region.

The estimated standard error of the predicted (or smoothed) value of $\hat{Y}_i$ using (5.2.41a) is

$$\text{est. s.e.}(\hat{Y}_i)=\left[\frac{1}{n}+\frac{(X_i-\bar{X})^2}{\Sigma(X_k-\bar{X})^2}\right]^{1/2} s=\left[\frac{1}{9}+\frac{(X_i-40)^2}{6000}\right]^{1/2}(0.921002)$$

which varies from a minimum at $X_i=40$ of 0.307 to maximums of 0.566 at $X_i=0$ and 80. This latter value is the same as for est. s.e.(b_0) because b_0 in this case is also the $X_i=0$ value of $\hat{Y}_i$.

5.2.2.3 *Estimators for Model 4,* $\eta_i=\beta_0'+\beta_1(X_i-\bar{X})$

Model 4 is interesting because a number of the results have simple forms. Without any statistical assumptions the OLS estimator for β_0' is

$$b_0'=\bar{Y}=\frac{\Sigma Y_i}{n} \tag{5.2.44}$$

and the OLS estimator for β_1 is the same as that given for Model 3.

Using the assumptions 1111-011 the variance of b_0' is

$$V(b_0')=\frac{\sigma^2}{n} \tag{5.2.45}$$

and that of b_1 is given by (5.2.39b). The covariance of b_0' and b_1 is simply

$$\text{cov}(b_0',b_1)=0=\text{cov}(\bar{Y},b_1) \tag{5.2.46}$$

The variance of $\hat{Y}_i$ and e_i are equal to those given for Model 3, (5.2.41a,b).

5.2.2.4 *Optimal Experiments for Models 3 and 4*

If one has the freedom of taking the observations at any X_i values for estimating parameters in Models 3 and 4, then one should select the X_i values so that the most accurate estimates of parameter values are produced. Such designs of experiments are termed optimal and yield optimal parameter estimates. Our criterion of optimality in this section is that of minimum variance of b_1. A more general criterion and analysis is given in Chapter 8.

Models 3 and 4 provide exactly the same OLS $\hat{Y}_i$ values. For that reason we consider the variances for Model 4 for assumptions 1111-011. The variance of b_0' is independent of X_i and the covariance of b_0' and b_1 is zero. Hence only the variance of b_1 which is given by (5.2.39b) need be considered. Note that $V(b_1)$ is minimized by maximizing $\Sigma(X_i-\bar{X})^2$. Let the maximum permissible range of X_i be between X_{min} and X_{max}. Then it can be rigorously shown that $V(b_1)$ is minimized if one half the measurement are made at X_{min} and the other half at X_{max}. No intermediate measurements are taken. The optimal case is illustrated by Fig. 5.2.

The variances of b_1 with uniform spacing of the X_i values given by

$$X_i=(i-1+c)\delta, \qquad i=1,2,\ldots,n \tag{5.2.47}$$

for various models are given in the fifth column of Table 5.3 which is a summary of the results of this section. The spacing between the X_i values is δ and the first X_i value is $X_1=c\delta$ where c is a factor locating X_1. The largest X_i value is $X_n=(n-1+c)\delta$. For this uniform spacing the variance of b_1 is

$$V_u(b_1)=\frac{12\sigma^2}{n(n^2-1)\delta^2} \tag{5.2.48}$$

If one half the observations were located at $X_{\text{min}}=c\delta$ and the other half at $X_{\text{max}}=(n-1+c)\delta$, the variance of b_1 is (for this nonuniform spacing)

$$V_n(b_1)=\frac{4\sigma^2}{n(-1)^2\delta^2} \tag{5.2.49}$$

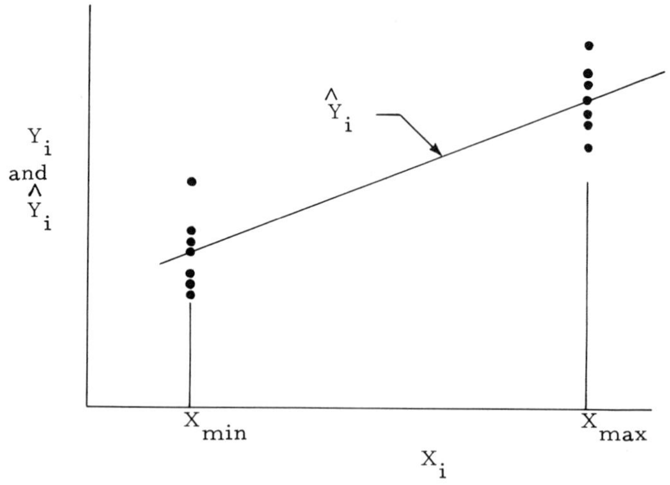

Figure 5.2 Recommended location of measurements when model is *known* to be a straight line in X.

The ratio of $V_u(b_1)/V_n(b_1)$ is

$$\frac{V_u(b_1)}{V_n(b_1)}=\frac{3(n-1)}{n+1} \tag{5.2.50}$$

which is equal to 1 for $n=2$ and monotonically increases to 3 as $n\to\infty$. Hence for large n, there is a factor 3 in the ratio of variances of b_1 for the uniform spaced case and the case of placement of the observations at the extremes.

In using the next to last column of Table 5.3 one should note that

$$X_{\min}=c\delta, \qquad X_{\max}=(n-1+c)\delta \tag{5.2.51}$$

and thus

$$\delta=\frac{X_{\max}-X_{\min}}{n-1}, \qquad c=\frac{X_{\min}}{\delta}=\frac{X_{\min}(n-1)}{X_{\max}-X_{\min}} \tag{5.2.52}$$

In this discussion of optimal design of experiments it is important to note that the standard assumptions of 1111-011 are assumed. Also there should be *no* uncertainity regarding the validity of the model. If the model is in question then one would be better advised to choose equal spacing of the X_i values or equal spacing in "time" if X_i is a function of time such as $t^2/2$.

5.2.3 Comments Regarding Definitions

In this section a number of definitions are given. Some of these can be confusing. There are, for example, several expressions related to Y_i. We have

$Y_i=\eta_i+\varepsilon_i$, measured value of Y_i

$E(Y_i)=\eta_i$, expected value of Y_i or model or dependent variable

$\hat{Y}_i=b_0+b_1X_i$, predicted value of Y_i for Model 3

$\bar{Y}=\dfrac{\Sigma Y_i}{n}$, average value of Y_i for $i=1$ to $i=n$

Also used is the symbol ε_i for measurement error or noise. This should not be confused with the residual e_i which is $Y_i-\hat{Y}_i$. The independent variable X_i is assumed to be errorless and has an average value given by $\bar{X}=\Sigma X_i/n$. All these terms are illustrated in Fig. 5.3. Modified definitions for $\bar{X}$ and $\bar{Y}$ may be used in subsequent sections when σ_i^2 is not a constant.

Table 5.3 Summary of Estimators, Variances, and Covariances for Five Simple Linear Models. Standard Assumptions of 1111-111 Apply.

Model No.	Model	Estimators	Variances and Covariance	Variances and Covariances for Uniformly Increasing X_i; $X_i=(i-1+c)\delta$; $i=1,2,\ldots$		Variances and Covariances for 1/2 Measurements at $X=c\delta$ and Rest at $X=(n-1+c)\delta$	
				$n=2,3,\ldots$	large n	$n=2,4,6,\ldots$	large n
1	$\eta_i=\beta_0$	$b_0=\overline{Y}$	$V(b_0)=\frac{\sigma^2}{n}$	$\frac{\sigma^2}{n}$	$\frac{\sigma^2}{n}$	$\frac{\sigma^2}{n}$	$\frac{\sigma^2}{n}$
2	$\eta_i=\beta_1 X_i$	$b_1=\frac{\Sigma Y_i X_i}{\Sigma X_j^2}$	$V(b_1)=\frac{\sigma^2}{\Sigma X_j^2}$	$\frac{6\sigma^2}{n(n+1)(2n+1)\delta^2}$ [b] for $c=1$	$\frac{3\sigma^2}{n^3\delta^2}$ for $c=1$	$\frac{2\sigma^2}{n\left[c^2+(n-1+c)^2\right]\delta^2}$ [a]	$\frac{2\sigma^2}{n^3\delta^2}$ [a] for $n\gg c$
3	$\eta_i=\beta_0+\beta_1 X_i$	$b_0=\overline{Y}-b_1\overline{X}$	$V(b_0)=\frac{\sigma^2\Sigma X_i^2}{n\Sigma\left(X_j-\overline{X}\right)^2}$	$\frac{2(2n+1)\sigma^2}{n(n-1)}$ for $c=1$	$\frac{4\sigma^2}{n}$ for $c=1$	$\frac{2\left[c^2+(n-1+c)^2\right]\sigma^2}{n(n-1)^2}$	$\frac{2\sigma^2}{n}$
		$b_1=\frac{\Sigma\left(X_i-\overline{X}\right)Y_i}{\Sigma\left(X_i-\overline{X}\right)^2}$	$V(b_1)=\frac{\sigma^2}{\Sigma\left(X_j-\overline{X}\right)^2}$	$\frac{12\sigma^2}{n(n^2-1)\delta^2}$	$\frac{12\sigma^2}{n^3\delta^2}$	$\frac{4\sigma^2}{n(n-1)^2\delta^2}$	$\frac{4\sigma^2}{n^3\delta^2}$
			$\text{cov}(b_0,b_1)=\frac{-\overline{X}\sigma^2}{\Sigma\left(X_i-\overline{X}\right)^2}$	$-\frac{6(n-1+2c)\sigma^2}{n(n^2-1)\delta}$	$\frac{-6\sigma^2}{n^2\delta}$ for $n\gg c$	$-\frac{2(n-1+2c)\sigma^2}{n(n-1)^2\delta}$	$\frac{-2\sigma^2}{n^2\delta}$ for $n\gg c$

4	$\eta_i = \beta_0' + \beta_1(X_i - \bar{X})$	$b_0' = \bar{Y}$	$V(b_0') = \frac{\sigma^2}{n}$	$\frac{\sigma^2}{n}$	$\frac{\sigma^2}{n}$	$\frac{\sigma^2}{n}$	$\frac{\sigma^2}{n}$
		$b_1 = \frac{\Sigma(X_i - \bar{X})Y_i}{\Sigma(X_i - \bar{X})^2}$	$V(b_1) = \frac{\sigma^2}{\Sigma(X_j - \bar{X})^2}$	$\frac{12\sigma^2}{n(n^2-1)\delta^2}$	$\frac{12\sigma^2}{n^3\delta^2}$	$\frac{4\sigma^2}{n(n-1)^2\delta^2}$	$\frac{4\sigma^2}{n^3\delta^2}$
			$\mathrm{cov}(b_0', b_1) = 0$	0	0	0	0
5	$\eta_i = \beta_1 X_{i1} + \beta_2 X_{i2}$	$b_1 = \frac{d_1 c_{22} - d_2 c_{12}}{\Delta}$	$V(b_1) = \frac{c_{22}\sigma^2}{\Delta}$				
		$b_2 = \frac{d_2 c_{11} - d_1 c_{12}}{\Delta}$	$V(b_2) = \frac{c_{11}\sigma^2}{\Delta}$				
		$c_{kl} = \Sigma_i X_{ik} X_{il}$ $d_k = \Sigma_i Y_i X_{ik}$ $\Delta = c_{11}c_{22} - c_{12}^2$	$\mathrm{cov}(b_1, b_2) = -\frac{c_{12}\sigma^2}{\Delta}$				

[a]If the measurements are only at $X = (n-1+c)\delta$ and $n \gg c$, the variances are one half of those indicated.

[b]For uniform spacing, that is, $X_i = (i-1+c)\delta$, $i = 1, 2, \ldots, n$, we have

$$\bar{X} = \frac{1}{n}\Sigma X_i = \frac{1}{2}(n-1+2c)\delta, \quad \Sigma(X_i - \bar{X})^2 = \frac{1}{12}n(n^2-1)\delta^2$$

$$\Sigma X_i^2 = \frac{\delta^2}{6}\{(n+c)(n+c-1)(2n+2c-1) - c(c-1)(2c-1)\}$$

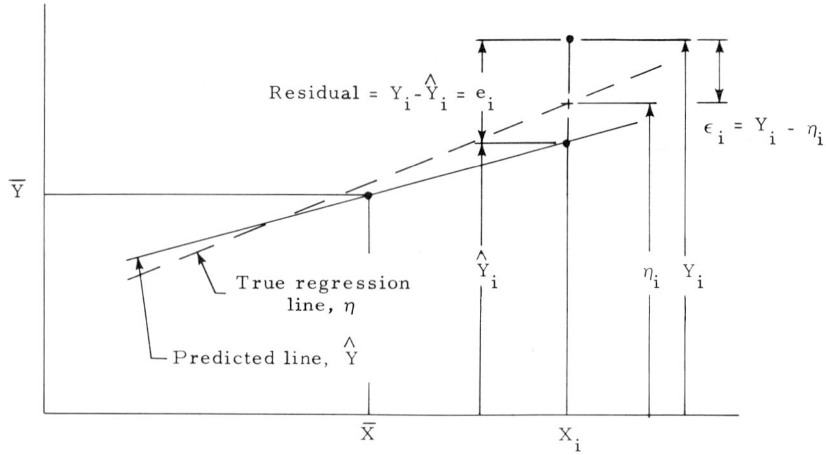

Figure 5.3 Figure showing some terms used in Section 5.2.

5.3 MAXIMUM LIKELIHOOD (ML) ESTIMATION

Maximum likelihood estimates make use of whatever information we have about the distribution of the observations. We illustrate ML estimation for the case of additive errors, $Y_i = \eta(X_i, \beta) + \varepsilon_i$, and when the errors ε_i have zero mean, are independent, are normal, and have known variances σ_i^2. The X_i's are errorless and the parameters are nonrandom. These assumptions are designated 11-11111. This information can be used to obtain estimates of parameter variances.

The natural logarithm of the normal probability density for independent measurements is given by

$$\ln f(Y_1, Y_2, \ldots, Y_n | \beta_0, \beta_1, \ldots) = -\frac{1}{2}\left[n \ln 2\pi + \sum_{i=1}^{n} \ln \sigma_i^2 + S_{ML}\right] \tag{5.3.1}$$

where the "physical" parameters are only contained in S_{ML},

$$S_{\text{ML}} = \sum_{i=1}^{n} \left[\frac{Y_i - \eta_i}{\sigma_i}\right]^2 \tag{5.3.2}$$

The one- and two-parameter cases are considered briefly in this section. It is pointed out that the ML estimators for Models 2 and 5 can be given in a similar form to those given by OLS.

5.3.1 One-Parameter Cases

Consider the linear model of $\eta_i = \beta_1 X_i$ (Model 2) and introduce this η_i expression in (5.3.2). The function $\ln f(Y_1,\ldots,Y_n|\beta_1)$ is maximized with respect to β_1 by minimizing S_{ML} since β_1 appears only in S_{ML}. Differentiating with respect to β_1, replacing β_1 by its estimator b_1, and setting the derivative equal to zero yields the normal equation

$$b_1 \Sigma\left(\frac{X_i}{\sigma_i}\right)^2 = \Sigma\left(\frac{Y_i}{\sigma_i}\right)\left(\frac{X_i}{\sigma_i}\right) \tag{5.3.3}$$

which can be solved for b_1 to obtain (for Model 2)

$$b_1 = \left[\Sigma\left(\frac{Y_i}{\sigma_i}\right)\left(\frac{X_i}{\sigma_i}\right)\right]\left[\Sigma\left(\frac{X_j}{\sigma_j}\right)^2\right]^{-1} \tag{5.3.4}$$

Note that this expression reduces to exactly the same one as given by (5.2.5) for OLS if $\sigma_i^2 = \sigma^2$, a constant. Also note that by defining

$$F_i \equiv \frac{Y_i}{\sigma_i}, \qquad Z_i \equiv \frac{X_i}{\sigma_i} \tag{5.3.5}$$

(5.3.4) can be written as

$$b_1 = (\Sigma F_i Z_i)(\Sigma Z_j^2)^{-1} \tag{5.3.6}$$

which is also similar to the OLS expression, (5.2.5); here F_i is analogous to Y_i and Z_i to X_i. In terms of F_i and Z_i, S_{ML} is a sum of squares of terms which have constant variance and has the same form as for OLS. Finally note that the variance of F_i is unity.

From the analogies given above between Y_i and F_i, X_i and Z_i, and σ^2 and unity, the variance of b_1 can be found from (5.2.10) to be

$$V(b_1) = (\Sigma Z_i^2)^{-1} = \left[\Sigma\left(\frac{X_i}{\sigma_i}\right)^2\right]^{-1} \tag{5.3.7}$$

For Model 1, $\eta_i = \beta_0$, the estimator b_0 and the variance of b_0 are found by letting $X_i = 1$ in the above two equations,

$$b_0 = \bar{Y}; \qquad \bar{Y} \equiv (\Sigma Y_i \sigma_i^{-2})(\Sigma \sigma_j^{-2})^{-1} \tag{5.3.8a,b}$$

$$V(b_0) = \left[\Sigma \sigma_i^{-2}\right]^{-1} \tag{5.3.8c}$$

5.3.2 Two-Parameter Cases

For the general model, Model 5, given by $\eta_i = \beta_1 X_{i1} + \beta_2 X_{i2}$, the estimators for β_1 and β_2 and their variances can be obtained by letting

$$X_{i1} \to Z_{i1} \equiv \frac{X_{i1}}{\sigma_i}, X_{i2} \to Z_{i2} \equiv \frac{X_{i2}}{\sigma_i}$$

$$Y_i \to F_i, \qquad \sigma^2 \to 1 \tag{5.3.9}$$

and thus (5.2.24) and (5.2.26) could be used for the estimators b_1 and b_2, their variances, and covariance.

For Model 3, $\eta_i = \beta_0 + \beta_1 X_i$, with assumptions 11-11111 (5.3.9) can be used to find

$$b_1 = \frac{\Sigma Y_i \sigma_i^{-2}\left(X_i - \bar{X}\right)}{\Sigma \sigma_j^{-2}\left(X_j - \bar{X}\right)^2}, \qquad b_0 = \bar{Y} - b_1 \bar{X} \tag{5.3.10a,b}$$

$$\bar{X} = \frac{\Sigma X_k \sigma_k^{-2}}{\Sigma \sigma_j^{-2}}, \bar{Y} = \frac{\Sigma Y_k \sigma_k^{-2}}{\Sigma \sigma_j^{-2}} \tag{5.3.11a,b}$$

$$V(b_0) = \frac{\left(\Sigma X_j^2 \sigma_j^{-2}\right)/\Sigma \sigma_i^{-2}}{\Sigma\left(X_k - \bar{X}\right)^2 \sigma_k^{-2}}, \qquad V(b_1) = \frac{1}{\Sigma\left(X_k - \bar{X}\right)^2 \sigma_k^{-2}} \tag{5.3.12a,b}$$

Note the new definition of $\bar{X}$ given by (5.3.11a). The same definition of $\bar{Y}$ is given in (5.3.8b) and (5.3.11b). For constant σ^2, these definitions for $\bar{X}$ and $\bar{Y}$ reduce to those given in Section 5.2.

Example 5.3.1

Simple harmonic motion can be described by $\eta_i = \beta_0 + \beta_1 \sin t_i$ where β_0 is a shift of the axis and β_1 is the amplitude of the motion. Measurements and their standard deviations vary as indicated in the following table.

i	t_i (°)	σ_i	Y_i
1	0	0.01	0.4926
2	30	0.05	0.9985
3	90	0.1	1.3547
4	150	0.05	0.9519
5	180	0.01	0.4996

(*a*) Estimate the parameters using ML. Let the standard assumptions apply except that we do not assume that σ_i^2 equals a constant, σ^2.

(*b*) Find the standard errors for b_0 and b_1.

Solution

(*a*) For this example, the model is Model 3 and the estimators are given by (5.3.10) and (5.3.11). Note that $X_i = \sin t_i$. Some of the required detailed calculations are given below.

i	X_i	σ_i^{-2}	$X_i\sigma_i^{-2}$	$X_i - \bar{X}$	$(X_i - \bar{X})^2\sigma_i^{-2}$	$Y_i\sigma_i^{-2}$	$Y_i\sigma_i^{-2}(X_i - \bar{X})$
1	0	10,000	0	−0.0239	5.723	4926.0	−117.847
2	0.5	400	200	0.4761	90.660	399.4	190.145
3	1	100	100	0.9761	95.273	135.47	132.229
4	0.5	400	200	0.4761	90.660	380.76	181.271
5	0	10,000	0	−0.0239	5.723	4996.0	−119.522
		20,900	500		288.039	10837.63	266.276

In addition to the sums indicated in the above table, $\bar{X}$ and $\bar{Y}$ are found from (5.3.11) to be

$$\bar{X} = \frac{500}{20900} = 0.0239234, \qquad \bar{Y} = \frac{10837.63}{20900} = 0.518547$$

Then from (5.3.10)

$$b_1 = \frac{266.276}{288.039} = 0.924449$$

$$b_0 = \bar{Y} - b_1\bar{X} = 0.518547 - 0.924449\ (0.0239234) = 0.496431$$

(*b*) The standard errors are found from the square roots of (5.3.12a,b)

$$\text{s.e.}(b_0) = \left[\frac{\Sigma X_j^2\sigma_j^{-2}/\Sigma\sigma_j^{-2}}{\Sigma\left(X_k - \bar{X}\right)^2\sigma_k^{-2}}\right]^{1/2} = \left[\frac{200/20900}{288.039}\right]^{1/2} = 0.0057639$$

$$\text{s.e.}(b_1) = \left[\Sigma\left(X_k - \bar{X}\right)^2\sigma_k^{-2}\right]^{-1/2} = (288.039)^{-1/2} = 0.05892$$

Least squares estimates of the parameters for this example are $b_0 = 0.510329$ and $b_1 = 0.872829$. The b_0 value is outside the $b_0 \pm \text{s.e.}(b_0)$ interval found using maximum likelihood.

5.3.3 Estimating σ^2 Using Maximum Likelihood

When the error variance is a constant, that is, $\sigma_i^2 = \sigma^2$, an estimator for σ^2 can be obtained by differentiating (5.3.1) with respect to σ^2 and setting the result equal to zero. The result is

$$-\frac{1}{2}\left[\frac{n}{\hat{\sigma}^2}-\frac{1}{(\hat{\sigma}^2)^2}\Sigma(Y_i-\hat{Y}_i)^2\right]=0 \tag{5.3.13}$$

or

$$\hat{\sigma}^2=\frac{1}{n}\Sigma(Y_i-\hat{Y}_i)^2 \tag{5.3.14}$$

This is unfortunately a biased estimator for σ^2. For one parameter, the denominator should be $n-1$ to provide an unbiased estimator. For that and other reasons use (5.2.21) to estimate σ^2 for one parameter and use (5.2.32) for two parameters when the assumptions 1111-011 are valid.

5.3.4 Maximum Likelihood Estimation Using Information from Prior Experiments

After one set of data has been used to estimate the parameters, a second set of data may become available. If the second set of observations is independent of the first and parameter estimates based on all the data are needed, then the first set of data can provide prior information for analysis of the second set. A method is given below whereby the number of calculations in simultaneously analyzing all the data can be reduced by taking advantage of the results of the analysis of the first set of data.

For simplicity let us derive the method for one parameter. The ML estimator for one set of data when the standard assumptions 11-11111 are valid is given by (5.3.6); assume that there are n_1 observations and write (5.3.6) as

$$b_{,1}=\left(\sum_{i=1}^{n_1}F_iZ_i\right)\left(\sum_{j=1}^{n_1}Z_j^2\right)^{-1}=V_{b_{,1}}\sum_{i=1}^{n_1}F_iZ_i \tag{5.3.15}$$

where $V_{b_{,1}}$ is the variance of $b_{,1}$

$$V_{b_{,1}}\equiv V(b_{,1})=\left(\sum_{i=1}^{n_1}Z_i^2\right)^{-1} \tag{5.3.16}$$

Consider now a combined analysis of $n=n_1+n_2$ observations. Then (5.3.6) becomes

$$b_{,2}=\left(b_{,1}V_{b_{,1}}^{-1}+\sum_{i=n_1+1}^{n}F_iZ_i\right)V_{b_{,2}} \tag{5.3.17a}$$

where $V_{b_{,2}}$ (the variance of $b_{,2}$) is given by

$$V_{b_{,2}}=\left(V_{b_{,1}}^{-1}+\sum_{j=n_1+1}^{n} Z_j^2\right)^{-1} \tag{5.3.17b}$$

We point out that (5.3.17) uses only the previously calculated $b_{,1}$ and $V_{b_{,1}}$ values; no other information regarding the first n_1 observations is needed to calculate improved values of b and V. The same procedure can be used for more than one parameter.

5.4 MAXIMUM A POSTERIORI (MAP) ESTIMATION

There are several ways to introduce prior information. One of these is given in Section 5.3.4 above for ML estimation. In this method, information from previous tests is included in such a way that exactly the same estimates are obtained as if all the data were analyzed together. This ML method also assumed that the parameters were nonrandom.

Another way to include prior information utilizes the maximum a posteriori (MAP) method. The MAP estimators are based on Bayes's theorem and are therefore called bayesian estimators. In the MAP method the parameters either are random or are conceived as being random. Hence there are two situations when MAP estimators might be used: (1) when the parameters are random and (2) when there is subjective information. What is meant by random parameters is discussed further below.

In this section the standard assumptions of additive, zero mean, uncorrelated, normal errors as well as known statistical parameters and nonstochastic independent variables are considered to be valid. Also, there is information about a prior distribution of values of the parameters (β). We assume this prior distribution to be normal with known mean and variance. We assume throughout our *experiment* that the β's are constant, that is, nonrandom. These assumptions are designated 11011110. (In Chapter 6 where a more detailed set of standard assumptions are given, two particular sets of MAP assumptions considered are designated 11--1112 and 11--1113.)

5.4.1 Random Parameter Case

In the random parameter case the parameter for a *particular* experiment or set of experiments is considered to be constant (or *non*random). This may be clarified by an example. A particular steel is occasionally produced by a plant. The thermal conductivity is known to vary from batch to batch. The long-run room-temperature average thermal conductivity (the parameter, β, of interest) is 20 W/m-°C with the standard deviation among batch averages being 0.1 W/m-°C. The distribution is normal. Then this information regarding the random nature of β from batch to batch is described by the probability density of

$$f(\beta)=\left[(2\pi)^{1/2}(0.1)\right]^{-1}\exp\left[-\frac{1}{2}\left(\frac{\beta-20}{0.1}\right)^2\right] \tag{5.4.1}$$

The standard deviation of measurements Y_i for a given batch is known to be 0.4. For a single normal measurement the probability of this measurement given the true conductivity β of the batch is

$$f(Y|\beta)=\left[(2\pi)^{1/2}(0.4)\right]^{-1}\exp\left[-\frac{1}{2}\left(\frac{Y-\beta}{0.4}\right)^2\right] \tag{5.4.2}$$

Let us use Bayes's theorem in the form

$$f(\beta|Y)=\frac{f(Y|\beta)f(\beta)}{f(Y)} \tag{5.4.3}$$

where $(\beta|Y)$ is the posterior distribution of β given Y. It includes information both from a large number of batches, $f(\beta)$, and from a given batch, $f(Y|\beta)$. If additional measurements Y_i are made, *they are also considered to be from this given batch.*

Since the parameter β appears only in the numerator of (5.4.2) and since it is convenient to take the logarithm of (5.4.2), we find that $f(\beta|Y)$ is maximized by minimizing

$$\left(\frac{\beta-20}{.1}\right)^2+\left(\frac{Y-\beta}{.4}\right)^2$$

with respect to β.

Notice in this example that the conductivity of a batch chosen at random is a random parameter. Once the batch is chosen, however, all our specimens are from this batch and thus the expected value of each is the same.

If we examine the conductivity as a function of temperature, instead of having a single parameter corresponding to room temperature conductivity we have a regression function containing a number of parameters. These parameters vary from batch to batch but our estimates are estimates of the specific values of this particular batch.

Let us now develop an estimator for the parameter β_1 in Model 2, $\eta_i = \beta_1 X_i$, β_1 being chosen at random from a given population. With the assumptions mentioned above and that β_1 is independent of ε_i, we have

$$Y_i = \beta_1 X_i + \varepsilon_i, \qquad \beta_1 \sim N(\mu_\beta, V_\beta) \tag{5.4.4a}$$

$$\varepsilon_i \sim N(O, \sigma_i^2), \qquad E(\varepsilon_i \beta_1) = 0 \tag{5.4.4b}$$

and thus the (prior) probability density of the random parameter β_1 is

$$f(\beta_1) = (2\pi V_\beta)^{-1/2} \exp\left[-\frac{1}{2} \frac{(\beta_1 - \mu_\beta)^2}{V_\beta} \right] \tag{5.4.5}$$

and that of $Y_1, \ldots, Y_n$ given β_1 is

$$f(Y_1, \ldots, Y_n | \beta_1) = \left\{ \Pi (2\pi\sigma_i^2)^{-1/2} \right\} \exp\left[-\frac{1}{2} \sum_{i=1}^{n} (Y_i - \beta_1 X_i)^2 \sigma_i^{-2} \right] \tag{5.4.6}$$

Introducing (5.4.5) and (5.4.6) into (5.4.3) and then taking the logarithm of $f(\beta_1 | Y_1, \ldots, Y_n)$ gives

$$\ln[f(\beta_1 | Y_1, \ldots, Y_n)] = -\frac{1}{2} \left[(n+1)\ln 2\pi + \ln V_\beta + \Sigma \ln \sigma_i^2 \right.$$

$$\left. + \frac{(\beta_1 - \mu_\beta)^2}{V_\beta} + \Sigma (Y_i - \beta_1 X_i)^2 \sigma_i^{-2} \right] - \ln f(Y_1, \ldots, Y_n) \tag{5.4.7}$$

Note that $f(Y_1, \ldots, Y_n)$ is not a function of the parameter β_1.

In (5.4.7) we are effectively considering the joint probability of each random choice of (both) β_1 and the subsequent collection of observations. We concentrate our attention on those possible choices which include the observations we actually obtained and hunt among them for that β_1 for which the probability is greatest. This β_1 we use as an estimate of the particular value for the batch chosen. Note that we are dealing with a *random variable*, β_1, a collection of possible values, and a *constant* β_1, the value actually chosen, that is, the parameter for the particular batch used in the experiment.

Taking the derivative of (5.4.7) with respect to β_1 yields the normal equation,

$$(b_1 - \mu_\beta)(V_\beta)^{-1} - \Sigma (Y_i - b_1 X_i) X_i \sigma_i^{-2} = 0 \tag{5.4.8}$$

which, after the addition and subtraction of $\mu_\beta X_i$ within the summation, can be written as

$$(b_1-\mu_\beta)(V_\beta)^{-1}-\Sigma\left[F_i-(b_1-\mu_\beta)Z_i-Z_i\mu_\beta\right]Z_i=0 \tag{5.4.9}$$

where

$$F_i=\frac{Y_i}{\sigma_i}, \qquad Z_i=\frac{X_i}{\sigma_i} \tag{5.4.10a,b}$$

Solving (5.4.9) for b_i then yields

$$b_1=\mu_\beta+\frac{\Sigma(F_i-\mu_\beta Z_i)Z_i}{\Sigma Z_j^2+V_\beta^{-1}}=\frac{\Sigma F_i Z_i+\mu_\beta V_\beta^{-1}}{\Sigma Z_j^2+V_\beta^{-1}} \tag{5.4.11a,b}$$

The expected value of b_1 given by (5.4.11) is μ_β. Hence the MAP estimator for b_1 is biased since it is not β_1, the value for the particular batch.

The variance of b_1 is affected not only by the errors in the measurements, Y_i, but by the variability of β_1 from batch to batch. For measurements involving a particular batch we are interested in the variability of b_1 compared to the value of the batch (β_1). Hence we are interested in the variance of the difference, $b_1-\beta_1$. Using (5.4.11b) we can show that

$$b_1-\beta_1=\left(\Sigma\varepsilon_i Z_i\sigma_i^{-1}-\beta_1 V_\beta^{-1}+\mu_\beta V_\beta^{-1}\right)\left(\Sigma Z_j^2+V_\beta^{-1}\right)^{-1} \tag{5.4.12}$$

Then the variance of the difference, $b_1-\beta_1$, is given by

$$V(b_1-\beta_1)=\left(\Sigma Z_j^2+V_\beta^{-1}\right)^{-1} \tag{5.4.13}$$

where $V(\beta_1)=V_\beta$ is used. Notice that as more observations are taken, the relative effect of the prior information regarding the random parameter diminishes. As the number of measurements becomes arbitrarily large, $\Sigma Z_j^2\to\infty$ and thus $V(b_1-\beta_1)\to 0$. This means that the variability of estimators obtained using (5.4.11) approaches zero for a particular batch if a very large number of measurements are taken for this batch.

Equations for the two-parameter cases involving Model 5 are given in Problem 5.21.

5.4.2 Subjective Prior Information

Some authors such as Box and Tiao [3] regard the prior probability distribution as a mathematical expression of degree of belief with respect to a certain proposition. In this context the concept of developing probabilities utilizing repeated observations is regarded merely as a means of calibrating a subjective attitude. In this view to say that one thinks the probability is one half that candidate A will be elected president means that we have the same *belief* in the proposition "candidate A will be elected president" as we would in the proposition "a toss of fair coin will produce a head." We need not imagine an infinite series of elections in half of which A is elected, and in half of which he is defeated.

This view can also be applied to the estimation of a physical property. The following is an example given in reference 3. Two physicists, A and B, are concerned with obtaining more accurate estimates of some physical constant β, known only approximately. Imagine physicist A is very familiar with previous measurements of β and thus can make a moderately good guess of the true β value; let his prior opinion about β be approximately represented as a normal density centered at 900 and having a standard deviation of 20,

$$f_A(\beta)=\left[(2\pi)^{1/2}20\right]^{-1}\exp\left[-\frac{1}{2}\left(\frac{\beta-900}{20}\right)^2\right] \tag{5.4.14a}$$

This implies that A believes that the chance of β being outside the interval of 860 to 940 is only about one in 20. By contrast, suppose that physicist B has little experience regarding values of β and that his rather vague prior beliefs can be represented by a normal density with mean of 800 and standard deviation of 200,

$$f_B(\beta)=\left[(2\pi)^{1/2}200\right]^{-1}\exp\left[-\frac{1}{2}\left(\frac{\beta-800}{200}\right)^2\right] \tag{5.4.14b}$$

We can see that B is much less certain of the true β value because any value between 400 and 1200 is considered plausible.

Suppose that one of the physicists performs an experiment and an observation of β is made. Further assume that this measurement contains an additive, zero mean, normal error with a standard deviation of 40. The probability density of Y is the same as given by (5.4.2) with the 0.4 replaced by 40.

To make the results more general let us use the notation $f(\beta_1|\mu)$ for the prior subjective information for β_1; for a normal distribution we have

$$f(\beta_1|\mu)=\left[(2\pi)^{1/2}\sigma_\mu\right]^{-1}\exp\left[-\frac{1}{2}\left(\frac{\beta_1-\mu}{\sigma_\mu}\right)^2\right] \tag{5.4.15}$$

The conditional probability density of $f(Y_1,\ldots,Y_n|\beta_1)$ is given by (5.4.6). For this case the use of Bayes's theorem leads to maximizing the natural logarithm of the product $f(\beta|\mu)\,f(Y_1,\ldots,Y_n|\beta)$, or

$$\ln\left[f(\beta_1|\mu)f(Y_1,\ldots,Y_n|\beta_1)\right]=-\frac{1}{2}\left[(n+1)\ln 2\pi+\ln\sigma_\mu^2+\Sigma\ln\sigma_i^2 + \frac{(\beta_1-\mu)^2}{\sigma_\mu^2}+\frac{\Sigma(Y_i-\beta_1X_i)^2}{\sigma_i^2}\right] \tag{5.4.16}$$

which is quite similar to (5.4.7). The estimate for β_1 is

$$b_1=\mu+\frac{\Sigma(F_i-\mu Z_i)Z_i}{\Sigma Z_j^2+\sigma_\mu^{-2}}=\frac{\Sigma F_iZ_i+\mu\sigma_\mu^{-2}}{\Sigma Z_j^2+\sigma_\mu^{-2}} \tag{5.4.17a,b}$$

which is identical to (5.4.11a,b), with μ being μ_β and V_β being σ_μ^2. It is also very similar to (5.3.17a,b) which give ML estimations for a combined analysis of two sets of observations.

As for the random parameter case the expected value of b_1 and the variance $b_1-\beta_1$ are

$$E(b_1)=\mu,\qquad V(b_1-\beta_1)=\left[\Sigma Z_j^2+\sigma_\mu^{-2}\right]^{-1} \tag{5.4.18a,b}$$

Note that though the estimators given by (5.4.11) and (5.4.17) are identical in form, the meanings attached to the quantities μ_β, V_β, and σ_μ^2 are different.

Let us return to the example of the two physicists. For one measurement $Y=850$ the estimator b and its variance for physicist A are (since $X_i=1$ for $\eta=\beta$)

$$b_A=\frac{\left(Y\sigma^{-2}+\mu\sigma_\mu^{-2}\right)}{\sigma^{-2}+\sigma_\mu^{-2}}=\frac{(850)(40)^{-2}+900(20)^{-2}}{40^{-2}+20^{-2}}=890$$

$$V(b_A)=\left(\sigma^{-2}+\sigma_\mu^{-2}\right)^{-1}=320$$

Repeating the same calculation for physicist B gives $b_B=848$ and $V(b_B)=1538$. Note that though the observation was the same for both physicists, the different normal prior distributions resulted in physicist A having the posterior distribution of $n(890,\,17.9^2)$ and physicist B having $n(848,\,39.2^2)$. Hence physicists A and B have different estimates and different standard deviations of 17.9 and 39.2, respectively.

We see that after the single observation the ideas of A and B about β (represented by the posterior distributions) are much closer than before using the observation. Note that A did not learn as much from the experiment as did B. The reason is that for A the uncertainity in the measurement indicated by $\sigma=40$ was larger than that indicated by the prior standard deviation, $\sigma_\mu=20$. In contrast, for B the uncertainty in the measurement was considerably smaller than that of B's prior ($\sigma_\mu=200$). For A the greater influence on the posterior distribution is the prior whereas for B the measurement has greater effect. As, however, more and more Y_i measurements are used for estimating β, (5.4.17) and (5.4.18) indicate that the prior information has less and less effect upon the estimate and its standard deviation.

5.4.3 Comparison of Viewpoints

Three different types of prior information have been discussed. First, in Section 5.3.4 prior information from actual experiments is combined with that from a new set of experiments. Only maximum likelihood need be used and the ideas are relatively straightforward. In the MAP cases, which use Bayes's theorem, the ideas are less clear and have been the subject of controversy. In the first case, the parameters are random, as in the case of the thermal conductivities of different batches of steel in the example above. In the second MAP case the parameters are not random but our prior *belief* can be incorporated into a subjective prior.

For each viewpoint the form of the parameter estimators are identical. The only differences are in symbols and meanings of the terms for the prior mean and variance. In each case, the variance of $b_1-\beta_1$ gives the same mathematical expression.

Problem 5.21 gives the estimators for the two-parameter model (Model 5).

Example 5.4.1

A scientist has measured a certain physical phenomenon and obtained the data given below. From knowledge of his measuring device, the variances of the measurements are also given. From his previous experience he feels that he can give a prior normal distribution with a mean of 1.01 and a variance of 0.001 for the parameter.

i	X_i	Y_i	σ_i
1	0.01	0.02	0.01
2	0.1	0.12	0.05
3	1	0.8	0.1
4	10	13	2

The regression function is $\eta_i = \beta_1 X_i$ and the assumptions regarding the data are

$$Y_i = \eta_i + \varepsilon_i, \qquad \varepsilon_i \sim N(0, \sigma_i^2), \qquad E(\varepsilon_i \varepsilon_j) = 0 \qquad \text{for} \, i \neq j$$

$V(X_i) = 0$, σ_i^2 values are known. Estimate β_1 using (*a*) OLS, (*b*) ML, and (*c*) MAP estimators. Also find the variance of the estimate in each case.

Solution

The assumptions given above can be designated 11011110. Various sets of assumptions are used in the different estimator methods.

(*a*) The OLS estimator does not use any statistical assumption. Using (5.2.5) the estimate is

$$b_{1,\text{OLS}} = [\Sigma Y_i X_i][\Sigma X_i^2]^{-1} =$$

$$\frac{0.01(0.02) + 0.1(0.12) + 1(0.8) + 10(13)}{0.0001 + 0.01 + 1 + 100} = 1.2950$$

The calculation of the variance of $b_{1,\text{OLS}}$ does require some assumptions; we use those designated 1101-11-. With the nonconstant σ^2, (5.2.10) is not valid for finding the variance. Instead the reader should derive

$$V(b_{1,\text{OLS}}) = [\Sigma X_i^2 \sigma_i^2][\Sigma X_i^2]^{-2}$$

$$= \frac{0.0001(0.01)^2 + 0.01(0.05)^2 + 1(0.1)^2 + (100)(4)}{(101.0101)^2} = 0.0392$$

(*b*) For ML estimation the assumptions needed are those given above. Prior information is not used. From (5.3.4) and (5.3.7) we find

$$b_{1,\text{ML}} = \frac{\Sigma Y_i X_i \sigma_i^{-2}}{\Sigma X_j^2 \sigma_j^{-2}} = \frac{0.02(0.01)(0.01)^{-2} + \cdots + 13(10)(2)^{-2}}{(0.01)^2(0.01)^{-2} + (0.1)^2(0.05)^{-2} + 1^2(0.1)^{-2} + 10^2(2)^{-2}}$$

$$= \frac{119.3}{130} = 0.91769$$

$$V(b_{1,\text{ML}}) = [\Sigma X_j^2 \sigma_j^{-2}]^{-1} = (130)^{-1} = 0.00769$$

(*c*) For MAP estimation the subjective prior information is included. Using the assumptions given above permits the use of (5.4.17b) and (5.4.18b) to get

$$b_{1,\text{MAP}} = \frac{\Sigma F_i Z_i + \mu \sigma^{-2}}{\Sigma Z_j^2 + \sigma_\mu^{-2}} = \frac{119.3 + 1.01(.001)^{-1}}{130 + (.001)^{-1}} = 0.99938$$

$$V(b_{1,\text{MAP}}) = [\Sigma Z_j^2 + \sigma_\mu^{-2}]^{-1} = (1130)^{-1} = 0.000885$$

For the OLS estimation no statistical assumptions are used; this implies that no information is used regarding the errors. Maximum likelihood estimation uses information regarding the measurement errors. MAP estimation uses the prior information regarding the parameter in addition to the information used in ML estimation. This suggests that the parameter variance for ML would be less than that of OLS and that of MAP would be the smallest. This is indeed what occurs in this example. However, if many additional measurements are given, the effect of the prior information is to reduce the disparity in values given by ML and MAP. If the errors do not have constant variance, the OLS values could be different from those given by ML and MAP even for a large number of observations.

5.5 MULTIPLE DATA POINTS

One way to gain insight into the assumption of the constant error variance (that is, $\sigma_i^2 = \sigma^2$) is to use repeated measurements. For Models 2, 3, and 4, this means to have more than one measurement of Y at each X_i. For Model 5 repeated measurements occur for more than one Y_i value at each combination of X_{i1}, X_{i2}. Repeated measurements are not always possible to obtain, but whenever possible they should be obtained for each new problem until the nature of the dependence of σ_i^2 on i is understood. Furthermore, multiple data points could be useful in investigating the validity of other assumptions such as those of zero mean, uncorrelated, and normal errors.

In some cases repeated measurements can be simply obtained by investigating another specimen at the "same" conditions. In other cases, repeated measurements can be obtained by using several sensors attached to the same specimen. An example of the latter is for temperature measurements in solids and fluids; the thermocouples (if they are used) might be all placed to measure the same temperature. The same could be true for other sensors as well.

It is important to distinquish between repeated measurements and taking repeated readings of the same measurement. A failure to do so may lead to inefficient design of experiments and to erroneous statements regarding accuracy of the parameters. The difference between repeated measurements and those that are essentially repeated readings can be illustrated by an example involving the temperature history of a solid copper block that is initially hot and then allowed to cool in open air. Several thermocouples are attached to it. Because of the high thermal conductivity of the copper the temperature of the block is quite uniform throughout it at any given time. The temperature of the block gradually decreases with time, however.

Consider first a given thermocouple. At any time the thermocouple would yield a temperature measurement which is in error owing to a number of different factors. Perhaps the largest factor is that due to calibration errors. Over the whole calibration temperature range the average error is nearly zero but at most temperatures the calibration error is not zero. Hence if several temperature measurements are made with only a short time interval between them, the "same" calibration errors would be in each measurement. Very nearly the same measurements would be obtained so that these could be considered repeated readings of the same measurement. These repeated readings may contain random components but the variance would be small compared to the calibration error.

A repeated measurement of the temperature at a specified time is more appropiately given by another thermocouple embedded in the specimen. It too would have a calibration error but the error would be independent of that of the first one (provided the calibrations are independently made for each sensor).

If a measurement is taken at some later time when the temperature has dropped considerably, the calibration error in the temperature measurement will be nearly independent of the early measurements for the same sensor.

It is also possible to obtain repeated measurements involving thermocouples (or other sensors) using the *same* sensor. This would occur in the above example if the calibration were very good and the associated variance were small compared to fluctuations in the readings due to electronic noise. For example, it might be that unbiased measurements of the temperature of a stirred water–ice mixture would produce values of 0.11, −0.06, −0.01, 0.03,...,0.05°C when the correct value is 0°C. The same type of random measurements might be produced for small or large time spacing between the measurements. In this case the errors are random with zero mean. These measurements can be considered repeated values even if the "same" specimen and sensor are used. The above examples illustrate that it is necessary to be careful to distinguish between repeated measurements and repeated readings.

5.5.1 Sum of Squares

The case of ordinary least squares is first considered. One can always number the observations so that we can write

$$S = \sum_{i=1}^{n} (Y_i - \eta_i)^2 \tag{5.5.1}$$

if there are any repeated values, the estimators given in Section 5.2 still apply. Some saving in effort, however, can be sometimes achieved by denoting the observations Y_{ij} and the regression function η_{ij}. There might be m_1 measurements of Y at X_1, m_2 measurements at $X_2,\ldots,$and m_r at X_r. Typically the Y values will be designated Y_{ij} for location X_i with $j=1, 2,\ldots,m_i$. Then (5.5.1) can be written

$$S=\sum_{i=1}^{r}\sum_{j=1}^{m_i}(Y_{ij}-\eta_i)^2 \qquad \text{where} \sum_{j=1}^{r} m_j=n \tag{5.5.2}$$

Let us now derive another expression for S that is frequently easier to use than (5.5.2). It applies equally well for both linear and nonlinear cases and shows that minimizing S need involve only means of the Y_{ij}'s for each i. Consider first the identity,

$$Y_{ij}-\eta_i=\left(Y_{ij}-\bar{Y}_i\right)+\left(\bar{Y}_i-\eta_i\right) \tag{5.5.3}$$

where $\bar{Y}_i$ and another mean (to be used later) are

$$\bar{Y}_i\equiv\frac{1}{m_i}\sum_{j=1}^{m_i}Y_{ij}, \qquad \bar{Y}\equiv\frac{1}{n}\sum_{i=1}^{K}m_i\bar{Y}_i \tag{5.5.4}$$

Squaring and summing (5.5.3) over i and j gives

$$S=\sum_{i,j}(Y_{ij}-\eta_i)^2=\sum_{i,j}\left(Y_{ij}-\bar{Y}_i\right)^2+\sum_i m_i\left(\bar{Y}_i-\eta_i\right)^2$$

$$+2\sum_{i,j}\left(Y_{ij}-\bar{Y}_i\right)\left(\bar{Y}_i-\eta_i\right) \tag{5.5.5a}$$

$$=\sum_{i,j}\left(Y_{ij}-\bar{Y}_i\right)^2+\sum_i m_i\left(\bar{Y}_i-\eta_i\right)^2 \tag{5.5.5b}$$

The cross-product sum in (5.5.5a) is zero because the summation on j is equal to zero. Note that the first summation in (5.5.5b) is not a function of the parameters. Hence for *linear* and *nonlinear* parameter estimation problems with repeated measurements the same parameters will be found if we start with the function

$$S_1=\sum_{i=1}^{r}m_i\left(\bar{Y}_i-\eta_i\right)^2 \tag{5.5.6}$$

rather than (5.5.2). Note that (5.5.6) requires less computation, however. If the measurement errors are independent, but have variances dependent only on i, maximum likelihood estimation (with the assumptions 11-11111) can be performed by minimizing

$$S_1 = \sum_{i=1}^{r} (F_i - H_i)^2 \tag{5.5.7}$$

where

$$F_i \equiv \bar{Y}_i m_i^{1/2} \sigma_i^{-1}, \qquad H_i \equiv \eta_i m_i^{1/2} \sigma_i^{-1} \tag{5.5.8}$$

When estimating parameters using repeated measurements, it is necessary that $r \geqslant p$ where p is the number of parameters. In Model 3, for example, estimates of β_0 and β_1 would require measurements at no less than two different X_i values regardless how large n is.

5.5.2 Parameter Estimates

Parameters can be estimated by minimizing (5.5.7) for various models given in this chapter. Economy in obtaining estimators can be obtained by utilizing previous results. Consider first Model 2 ($\eta_i = \beta_1 X_i$) and ML estimation with the assumptions 11-11111. Then b_1 is given by (5.3.6) with F_i as defined by (5.5.8) and Z_i by

$$Z_i \equiv X_i m_i^{1/2} \sigma_i^{-1} \tag{5.5.9}$$

The variance of b_1 is given by (5.3.7) with Z_i defined by (5.5.9).

For Model 5 given by $\eta_i = \beta_1 X_{i1} + \beta_2 X_{i2}$, the estimators b_1 and b_2, their variances, and covariance can be obtained from (5.2.24) and (5.2.26) by letting

$$X_{i1} \to Z_{i1} \equiv X_{i1} m_i^{1/2} \sigma_i^{-1}, X_{i2} \to Z_{i2} \equiv X_{i2} m_i^{1/2} \sigma_i^{-1} \tag{5.5.10a,b}$$

$$Y_i \to_{fi} \equiv \bar{Y}_i m_i^{1/2} \sigma_i^{-1}, \qquad \sigma^2 \to 1 \tag{5.5.10c,d}$$

For Model 3, $\eta_i = \beta_0 + \beta_1 X_i$, the ML results can be obtained from the above procedure more simply from (5.3.10–12) by replacing σ_i^2 by σ_i^2 / m_i and Y_i by $\bar{Y}_i$.

The number of terms related to Y and η has increased in this section. In addition to the observed value Y_{ij}, there is the value $\bar{Y}_i$, which is the average of the Y_{ij} values at a given X_i. $\hat{Y}_i$ is the predicted regression value at X_i; η_i is the actual regression value at X_i, that is, by definition $E(Y_{ij})$ and thus the expected value of $\bar{Y}_i$ and of $\hat{Y}_i$ also; and $\bar{Y}$ is the weighted average of the Y_{ij} values over all the X_i values. These symbols are illustrated by Fig. 5.4.

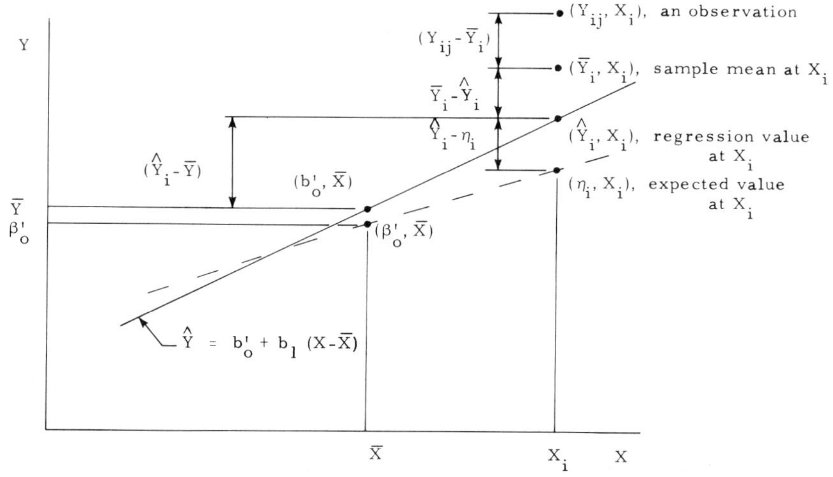

Figure 5.4 Relationships among observations, etc. for repeated measurements.

Example 5.5.1

Four measurements are made for both $X_1=0$ and $X_2=80$ with the same errors ε_i as in Example 5.2.4 except the fifth error is not used. Then the Y_{1j} measurements at X_1 are 0.258, 0.966, 2.453, and 1.963, whereas at $X_2=80$, Y_{2j} is 9.418, 10.792, 8.626, and 8.778. The assumptions of additive, zero mean, constant variance, uncorrelated, normal errors, and errorless X_i are valid. There is no prior information and σ^2 is unknown.

(*a*) Using expressions developed in this section, estimate the parameters β_0 and β_1 in Model 3.

(*b*) Find the estimated standard errors of b_0 and b_1.

Solution

(*a*)With the assumptions given, 11111011, the estimates can be obtained using OLS or ML. The simplest expressions to use are those given by (5.3.10,–12) by replacing σ_i^2 by σ^2/m_i and Y_i by $\bar{Y}_i$. Since σ_i^2 is a constant (5.3.10) and (5.3.11) can be written

$$b_1 = \frac{\Sigma_i^r \bar{Y}_i m_i (X_i - \bar{X})}{\Sigma_j^r m_j (X_j - \bar{X})^2} \qquad (a)$$

$$b_0 = \bar{Y} - b_1 \bar{X} \qquad (b)$$

$$\bar{X} = \frac{\Sigma_k^r X_k m_k}{\Sigma_j^r m_j}, \qquad \bar{Y} = \frac{\Sigma_k^r \bar{Y}_k m_k}{\Sigma_j^r m_j} \qquad (c)$$

In the above equations $r=2$, $m_1=m_2=4$, $X_1=0$, and $X_2=80$; also

$$\bar{X}=\frac{\sum_{k=1}^{2} 4X_k}{8}=\frac{1}{8}[0+4(80)]=40$$

$$\bar{Y}_1=\frac{1}{m_1}\sum_{j=1}^{m_1} Y_{1j}=\frac{1}{4}[0.258+\cdots+1.963]=1.410$$

$$\bar{Y}_2=\frac{1}{m_2}\sum_{j=1}^{m_2} Y_{2j}=\frac{1}{4}[9.418+\cdots+8.778]=9.4035$$

$$\bar{Y}=\frac{\sum_{k=1}^{2} \bar{Y}_k 2}{4}=\frac{1}{2}(1.410+9.4035)=5.40675$$

Then using the expression (a) for b_1, we obtain

$$b_1=\left[\sum_{i=1}^{2} \bar{Y}_i 4(X_i-40)\right]\Big/\left[\sum_{j=1}^{2} 4(X_j-40)^2\right]$$

$$=[1.41(4)(-40)+9.4035(4)(40)]/[4(1600)+4(1600)]$$

$$=0.09991875$$

(*b*) The expressions for the estimated standard errors can be obtained from (5.3.12) by replacing σ_i^{-2} by $m_i s^{-2}$ to get

$$\text{est. s.e.}(b_0)=s\left[\frac{(\Sigma X_j^2 m_j)/\Sigma m_i}{\Sigma(X_k-\bar{X})^2 m_k}\right]^{1/2} \qquad (d)$$

$$\text{est. s.e.}(b_1)=s\left[\Sigma(X_k-\bar{X})^2 m_k\right]^{-1/2} \qquad (e)$$

Since there are *two* X_i values and *two* parameters, the predicted line passes through $\bar{Y}_1$ and $\bar{Y}_2$. Then the minimum sum of squares resulting from (5.5.2) is the first term on the right side of (5.5.5),

$$S_{\min}=\sum_{i,j}\left(Y_{ij}-\bar{Y}_i\right)^2=2.9179+2.9239=5.8418$$

so that the estimated variance of the errors is

$$s^2=S_{\min}/(n-2)=5.8418/6=0.9736$$

and the estimated standard error is $s=0.9867$.Then using (d) and (e)

$$\text{est. s.e.}(b_0)=0.9867\left[\frac{6400(4)/8}{2(1600)(4)}\right]^{1/2}=0.4933$$

$$\text{est. s.e.}(b_1)=0.9867[2(1600)(4)]^{-1/2}=0.00872$$

Though the value of b_0 is less accurate than that given in Example 5.2.4, the variances are smaller in this example than in Example 5.2.4. These estimated variances corroborate the theoretical result that smaller estimated variances are generally obtained for Models 3 and 4 by concentrating the measurements at the minimum and maximum X_i values.

5.6 COEFFICIENT OF MULTIPLE DETERMINATION (R^2)

In this section the sum of squares are compared for two different models applied to the same data. Ordinary least squares is used as the estimation procedure. The analysis will start in sufficient generality to permit the models to be linear or nonlinear in the parameters. Later the results are specialized to Models 1 and 4. In the following discussion we consider two models, designated A and B. Frequently Model B has the same functional form and parameters as Model A except there is an additional parameter in Model B. Many authors restrict the meaning of R^2 to the case where Model A is Model 1.

Let ${}_A\hat{Y}_i$ be the predicted value of Y_i for Model A and ${}_B\hat{Y}_i$ for Model B. We start with the identity

$$Y_i-{}_A\hat{Y}_i=\left(Y_i-{}_B\hat{Y}_i\right)+\left({}_B\hat{Y}_i-{}_A\hat{Y}_i\right) \tag{5.6.1}$$

which can be also written as

$${}_Ae_i={}_Be_i+\left({}_B\hat{Y}_i-{}_A\hat{Y}_i\right) \tag{5.6.2}$$

for which the residuals for Models A and B are defined by

$${}_Ae_i\equiv Y_i-{}_A\hat{Y}_i \qquad \text{and} \qquad {}_Be_i\equiv Y_i-{}_B\hat{Y}_i \tag{5.6.3}$$

Let us square and sum (5.6.2) over i to get

$$\Sigma_A e_i^2=\Sigma_B e_i^2+\Sigma\left({}_B\hat{Y}_i-{}_A\hat{Y}_i\right)^2+2\Sigma_B e_i\left({}_B\hat{Y}_i-{}_A\hat{Y}_i\right) \tag{5.6.4a}$$

$$\text{SST} = \text{SSE} + \quad \text{SSR} \quad + \quad 2\text{SC} \tag{5.6.4b}$$

Each term in (5.6.4b) corresponds to the term in (5.6.4a) directly above.

Note that SST is the minimum sum of squares for Model A and SSE is the minimum sum of squares for Model B. Let us specify Models A and B so that

$$\text{SST} = \Sigma_A e_i^2 \geqslant \Sigma_B e_i^2 = \text{SSE} \tag{5.6.5}$$

which would be always true if Model A could be obtained from Model B by making a certain parameter in Model B equal to zero.

Divide (5.6.4) by the left side and rearrange to the form

$$R^2 = 1 - \frac{\Sigma_B e_i^2}{\Sigma_A e_j^2} = 1 - \frac{\text{SSE}}{\text{SST}} \tag{5.6.6}$$

where R^2 is called the coefficient of multiple determination and is defined by

$$R^2 = \frac{\text{SSR} + 2\text{SC}}{\text{SST}} \tag{5.6.7}$$

Because of condition (5.6.5), an examination of (5.6.6) reveals that $0 \leqslant R^2 \leqslant 1$ where $R^2 \approx 0$ corresponds to both models being nearly as effective and $R^2 \approx 1$ corresponds to Model B being much better than Model A. Then R^2 can be used to say something about the improvement in the "goodness of fit," $R^2 = 0$ being the poorest and $R^2 = 1$ being the best improvement in using Model B rather than Model A.

For nonlinear problems, the parameter estimates and sum of squares can be found separately for Models A and B and then R^2 would be evaluated using (5.6.6). For the simple linear models given next a simplified form of (5.6.7) is frequently used.

A classical case considered in connection with R^2 is for Models 1 and 4 being A and B, respectively,

$$\text{Model A} \quad {}_AY_i = \beta_0 + \varepsilon_i, \qquad {}_A\hat{Y}_i = \bar{Y} \tag{5.6.8}$$

$$\text{Model B} \quad {}_BY_i = \beta_0' + \beta_1\left(X_i - \bar{X}\right) + \varepsilon_i, \qquad {}_B\hat{Y}_i = \bar{Y} + b_1\left(X_i - \bar{X}\right) \tag{5.6.9}$$

The term SC in (5.6.4b) and (5.6.7) is then

$$\text{SC} = \Sigma_B e_i\left[b_1\left(X_i - \bar{X}\right)\right] = 0 \tag{5.6.10}$$

where the normal equation for Model 4 and parameter β_1 was used. Hence

R^2 can be calculated from (5.6.7) which becomes

$$R^2 = \frac{\text{SSR}}{\text{SST}} = \frac{\Sigma(\hat{Y}_i - \bar{Y})^2}{\Sigma(Y_i - \bar{Y})^2} = \frac{b_1^2\Sigma(X_i - \bar{X})^2}{\Sigma(Y_i - \bar{Y})^2} = \frac{b_1\Sigma(X_i - \bar{X})Y_i}{\Sigma(Y_i - \bar{Y})^2} \tag{5.6.11}$$

where $\bar{Y}$ is associated with Model A (Model 1 in this case) and $\hat{Y}_i$ with Model B (i.e., 4). If $\hat{Y}_i = Y_i$, that is, the prediction is perfect, then $R^2 = 1$. If $\hat{Y}_i = \bar{Y}$, that is, $b_1 = 0$ or the model $Y = \beta_0 + \varepsilon$ alone fits the data, $R^2 = 0$. Thus R^2 is a measure of the usefulness of the term $\beta_1(X_i - \bar{X})$ in the model, it being not needed for $R^2 \approx 0$ and needed for $R^2 \approx 1$. R^2 as given by (5.6.11) is the correlation coefficient of (2.6.17).

Example 5.6.1

Investigate the goodness of fit as indicated by R^2 for Example 5.2.4.

Solution

Using (5.6.11) and values given in Example 5.2.4 gives

$$R^2 = \frac{(0.101988)^2(6000)}{\Sigma(Y_i - 5.366)^2} = 0.9131$$

which is nearly unity, indicating that the $\beta_1(X_i - \bar{X})$ term may be needed in the model.

5.7 ANALYSIS OF VARIANCE ABOUT THE SAMPLE MEAN

The subject of analysis of variance is a broad one and contains many different facets. In this section only certain aspects of the analysis of variance (ANOVA) are considered.

The preceding section employed no statistical information and thus no probabilistic statements could be made. This section uses many of the standard assumptions. Assume that the errors are additive, uncorrelated, and normal and have zero mean and constant variance. The σ^2 value is unknown and there is no prior information regarding the constant parameters. The X_i values are nonstochastic (i.e., errorless). These assumptions are designated 11111011.

For models 1 and 4 given by (5.6.8) and (5.6.9), equation (5.6.4a) can be written

$$\Sigma(Y_i - \bar{Y})^2 = \Sigma(Y_i - \hat{Y}_i)^2 + \Sigma(\hat{Y}_i - \bar{Y})^2 \tag{5.7.1a}$$

$$\text{SST} \quad = \quad \text{SSE} \quad + \quad \text{SSR} \tag{5.7.1b}$$

$\bar{Y}$ is for Model A (or 1) and $\hat{Y}_i$ is for Model B (or 4). The sum of squares on the left side of (5.7.1a) is sometimes called the *total* sum of squares and designated SST. The first term on the right of (5.7.1a) is called the *error* sum of squares, SSE. The remaining term is (5.7.1a), called the *regression* sum of squares, SSR. It can be proved that SSE and SSR are independent.

Any sum of squares has associated with it a number called its *degrees of freedom*. Let the sum of squares be written as a sum of the squares of independent linear forms. (A linear form, for example, is $\Sigma\, a_i\, Y_i$ where the a_i's are constants and the Y_i's are variables.) Then the number of independent linear forms is the number of degrees of freedom. The sum of squares of $Y_i - \hat{Y}_i$ for the assumptions 11111011 is $n-p$ for n being the number of observations and p the number of independent parameters. Hence SST has $n-1$ degrees of freedom and SSE has $n-2$. Since SSE and SSR are independent, we know from Cochran's theorem [4] that the sum of the degrees of freedom of SSE and of SSR is equal to the degrees of freedom of SST. This information can be used to obtain that which is displayed in Table 5.4.

Table 5.4 ANOVA Table for Partition of Variance About $\bar{Y}$, (5.7.1)

Source of Variation	Sum of Squares	Degrees of Freedom	Mean Square
1. Deviation about regression line (residuals)	$\text{SSE} = \Sigma(Y_i - \hat{Y}_i)^2$	$n-2$	$s^2 = \text{SSE}/(n-2)$
2. Deviation between the regression line and mean	$\text{SSR} = \Sigma(\hat{Y}_i - \bar{Y})^2$	1	SSR
3. Total deviation between data and mean	$\text{SST} = \Sigma(Y_i - \bar{Y})^2$	$n-1$	

We now wish to employ an F test to obtain an indication if the $\beta_1(X_i - \bar{X})$ term in Model 4 is needed. For the assumptions indicated by 11111011, an F statistic can be given. Recall that an F statistic is the ratio of two independent random variables, each having a χ^2 distribution and each divided by its respective degrees of freedom. One χ^2 statistic can be formed by SSE divided by σ^2 and another independent χ^2 statistic is SSR/σ^2. Then an F statistic is

$$F = \frac{(\text{SSR}/\sigma^2)/1}{(\text{SSE}/\sigma^2)/(n-2)} \tag{5.7.2}$$

This statistic can provide a measure of how much the additional parameter β_1 (i.e., using the model $Y_i = \beta_0' + \beta_1(X_i - \overline{X}) + \varepsilon_i$ rather than $Y_i = \beta_0 + \varepsilon_i$) is needed. If F is near unity [corresponding to $R^2 \approx 0$ in (5.6.11)], then the two-parameter model (Model 4) does not significantly improve the fit compared to the one parameter model (Model 1). The other extreme is large F [which corresponds to $R^2 \approx 1$ in (5.6.11)]; in this case we can be confident that the β_1 parameter is needed.

A probability statement can be made utilizing the F statistic and a table of its distribution which could be used to obtain the value of $F_{1-\alpha}(1, n-2)$. See Section 2.8.10. The probability of F being less than $F_{1-\alpha}(1, n-2)$ is $1-\alpha$ or

$$P\left[F < F_{1-\alpha}(1, n-p)\right] = 1 - \alpha \tag{5.7.3}$$

Alternatively we can write

$$P\left[F_{1-\alpha}(1, n-p) < F\right] = \alpha \tag{5.7.4}$$

In words, if the null hypothesis $H_0: \beta_1 = 0$ is true, the probability that the calculated value F exceeds the tabulated value is α. If F is greater than $F_{1-\alpha}(1, n-p)$, we reject the null hypothesis at the given significance level α. If the calculated F value is less than $F_{1-\alpha}(1, n-p)$, we say that we *cannot reject* the null hypothesis — that is, it may be that $\beta_1 = 0$.

Example 5.7.1

Using the data of Example 5.2.4 develop an analysis of variance table and determine if the β_1 parameter is needed. Make the probability 1% of falsely deciding that β_1 is needed.

Solution

Using the data from Example 5.2.4 the following ANOVA table is constructed.

Source	Sum of Squares	Degrees of Freedom	Mean Square	Calculated F
1. Residual	5.9377	7	0.92100	
2. Deviation between line and mean	62.4097	1	62.4097	$\frac{62.4097}{0.92100} =$ 67.763
3. Total	68.3474	8		

From a table of the F distribution, we find

$$F_{1-\alpha}(1, n-2) = F_{0.99}(1, 7) = 12.25$$

Since $F > F_{0.99}(1,7)$, we reject the null hypothesis that $\beta_1 = 0$. If β_1 is not needed, our method has only a 1% chance of causing us to use the model $\eta_i = \beta_0' + \beta_1(X_i - \overline{X})$ rather than $\eta_i = \beta_0' = \beta_0$.

The use of the F test for model building is considered further in Chapter 6.

5.8 ANALYSIS OF VARIANCE ABOUT THE REGRESSION LINE FOR MULTIPLE MEASUREMENTS AT EACH X_i

Consider the case of partitioning the variation about the predicted regression line for multiple measurements at each X_i. From (5.5.7), which applies for linear and nonlinear parameter estimation, we have

$$\sum_{i=1}^{r} \sum_{j=1}^{m_i} \left(Y_{ij} - \hat{Y}_i\right)^2 = \sum_{i=1}^{r} \sum_{j=1}^{m_i} \left(Y_{ij} - \overline{Y}_i\right)^2 + \sum_{i=1}^{r} m_i \left(\overline{Y}_i - \hat{Y}_i\right)^2 \tag{5.8.1}$$

or

SS_t	=	SS_e	+	SS_r
Total sum of squares* between data and regression line; "residuals" (d.f. $= n - p$)	=	Sum of squares within data sets; "pure error sum of squares" (d.f. $= n - r$)	+	Sum of squares of local mean about regression line; "lack of fit sum of squares" (d.f. $= r - p$)

where d.f. stands for degrees of freedom. The number of degrees of freedom on the left has been discussed previously; it is the total number of points minus the number of parameters. The first term on the right has the contribution from $i = 1$ of

$$\sum_{j=1}^{m_1} \left(Y_{1j} - \overline{Y}_1\right)^2$$

which has $m_1 - 1$ degrees of freedom; the second contribution ($i = 2$) would have $m_2 - 1$ degrees of freedom. Hence for the first term on the right hand side of (5.8.1), the number of degrees of freedom is

$$\text{d.f.} = \sum_{i=1}^{r} (m_i - 1) = n - r \tag{5.8.2}$$

*SS_t is our former SSE.

The number of degrees of freedom of the last term are given by subtraction. The various terms are labeled SS_t, SS_e, and SS_r; note that the terms are *not* completely analogous to those in (5.7.1), but are similarly labeled. In fact, (5.8.1) can be used in (5.7.1) to get

$$\begin{aligned} \text{SST} &= \text{SSE} + \text{SSR} \\ &= \left[\text{SS}_e + \text{SS}_r\right] + \text{SSR} \end{aligned} \tag{5.8.3}$$

where an additional summation is used in (5.7.1), and then

$$\text{SST} = \sum_{i=1}^{r} \sum_{j=1}^{m_i} \left(Y_{ij} - \overline{Y}\right)^2 \tag{5.8.4}$$

$$\text{SSR} = \sum_{i=1}^{r} m_i \left(\hat{Y}_i - \overline{Y}\right)^2 \tag{5.8.5}$$

where $\overline{Y}$ is defined by (5.5.4).

Table 5.5 shows the analysis of variance table for (5.8.1) in lines 2 and 3; the table as a whole illustrates (5.8.3).

The mean square s_e^2, which is defined by

$$s_e^2 = \frac{\text{SS}_e}{n - r} \tag{5.8.6}$$

Table 5.5 ANOVA Table for Partition of Variance About $\hat{Y}_i$, (5.8.1), and About Y, (5.8.3)

Source of Variation	Sum of Squares	Degrees of Freedom	Mean Square
1. Pure error sum of squares	$\text{SS}_e = \Sigma\Sigma(Y_{ij} - \overline{Y}_i)^2$	$n - r$	$s_e^2 = \text{SS}_e/(n - r)$
2. Lack of fit sum of squares	$\text{SS}_r = \Sigma m_i(\overline{Y}_i - \hat{Y}_i)^2$	$r - p$	$s_r^2 = \text{SS}_r/(r - p)$
3. Residual sum of squares	$\text{SS}_t = \Sigma\Sigma(Y_{ij} - \hat{Y}_i)^2$	$n - p$	$s^2 = \text{SS}_t//(n - p)$
4. Sum of squares between line and mean	$\text{SSR} = \Sigma m_i(\hat{Y}_i - \overline{Y})^2$	$p - 1$	SSR
5. Sum of squares between data and mean	$\text{SST} = \Sigma\Sigma(Y_{ij} - \overline{Y})^2$	$n - 1$	

is an unbiased estimate of σ^2 even if the true model is not used or if the model is nonlinear. Hence this estimate of σ^2 is said to arise from "pure error." On the other hand, s^2,

$$s^2 = \frac{SS_t}{n-2} \tag{5.8.7}$$

is *not* an unbiased estimate of σ^2 if the model is incorrect.

5.8.1 Expected Values of s^2 for Incorrect Model

Let us investigate the effect upon s^2 of an incorrect mathematical model. We recall that e_{ij} is the residual for the jth measurement at X_i; it "contains all available information on the ways in which the fitted model fails to properly explain the observed variation in the dependent variable Y" [1, p. 26]. Recalling $\eta_i = E(Y_{ij})$ and writing

$$\begin{aligned} e_{ij} = Y_{ij} - \hat{Y}_i &= (Y_{ij} - \hat{Y}_i) - E(Y_{ij} - \hat{Y}_i) + E(Y_{ij} - \hat{Y}_i) \\ &= \{(Y_{ij} - \hat{Y}_i) - [\eta_i - E(\hat{Y}_i)]\} + [\eta_i - E(\hat{Y}_i)] & (5.8.8) \\ &= q_{ij} + B_i & (5.8.9) \end{aligned}$$

where

$$q_{ij} = \{(Y_{ij} - \hat{Y}_i) - [\eta_i - E(\hat{Y}_i)]\}, \qquad B_i = \eta_i - E(\hat{Y}_i) \tag{5.8.10}$$

B_i is called the *bias error* at X_i; it is zero if the model is correct ($E[\hat{Y}_i] = \eta_i$). The random variable q_{ij} has a zero mean *whether the model is correct or not* since $E(Y_{ij}) = \eta_i$ is true in any case. These statements regarding B_i and q_{ij} are true for *nonlinear* as well as linear models.

For Model 5 with the assumptions denoted 1111--11 (except that $E(\hat{Y}_i) = \eta_i - B_i$) it can be shown for OLS and ML estimation that

$$E(s^2) = \frac{1}{n-2} \Sigma\Sigma \left\{ V(Y_{ij} - \hat{Y}_i) + [\eta_i - E(\hat{Y}_i)]^2 \right\} \tag{5.8.11}$$

which reduces to

$$E(s^2) = \sigma^2 + \frac{1}{n-2} \sum_{i=1}^{n} B_i^2 \tag{5.8.12}$$

where (5.2.31) is used. If the model is correct, the last term in (5.8.12) disappears.

When the model is incorrect, the residuals contain both random (q_{ij}) and systematic or biased components (B_i) which are respectively called *variance* and *bias* error components of the residuals. An incorrect model results in an inflated residual mean square.

5.8.2 F Test with Repeated Data

For this case of repeated observations, an F statistic is (for $p=2$)

$$F_r = \frac{\left[\dfrac{\mathrm{SS}_t}{\sigma^2} - \dfrac{\mathrm{SS}_e}{\sigma^2}\right]\dfrac{1}{r-2}}{\dfrac{1}{\sigma^2}\dfrac{\mathrm{SS}_e}{n-r}} = \frac{s_r^2}{s_e^2} \tag{5.8.13}$$

where numerator and denominator contain χ^2 distributions if the model is correct; s_r^2 is called the mean square due to lack of fit. This F_r value should be compared with $F_{1-\alpha}(r-2, n-r)$. If $F_r > F_{1-\alpha}(r-2, n-r)$, we say that F_r is significant and we mean that the model is inadequate. An estimate of σ^2 using s_e^2 would be unbiased, but using s^2 or s_r^2 would be biased and tend to yield too large an estimate. If, on the other hand, $F_r < F_{1-\alpha}(r-2, n-r)$, F_r is said to be not significant; there is no reason to doubt the adequacy of the model and both the pure error and lack of fit mean squares (s_e^2 and s_r^2) can be used as estimates of σ^2. Moreover, s^2 is a pooled estimate of s^2. See Fig. 5.5 for a schematic diagram summarizing the steps for checking for lack of fit with repeated observations.

The use of the F_r statistic as given by (5.8.13) does not preclude the use of the F statistic given by (5.7.2). They give different information. F, (5.7.2), can be used whether there are repeated measurements or not; it tells whether β_1 is needed and can be generalized to investigate the validity of adding another or several parameters to the model. For cases where there are repeated measurements, the F_r test can indicate if the model is satisfactory (with no reference to adding another parameter) and can tell if σ^2 can be estimated from s^2. For repeated measurements both tests should be used.

With the two F tests we can have four combinations associated with (*a*) significant (or not significant) lack of fit and (*b*) significant (or not significant) linear regression. These combinations are illustrated in Fig. 5.6 and the results are summarized in Table 5.6. In each case the model

$$Y = \beta_0 + \beta_1 X + \epsilon = \beta_0' + \beta_1 (X - \bar{X}) + \epsilon$$

is used.

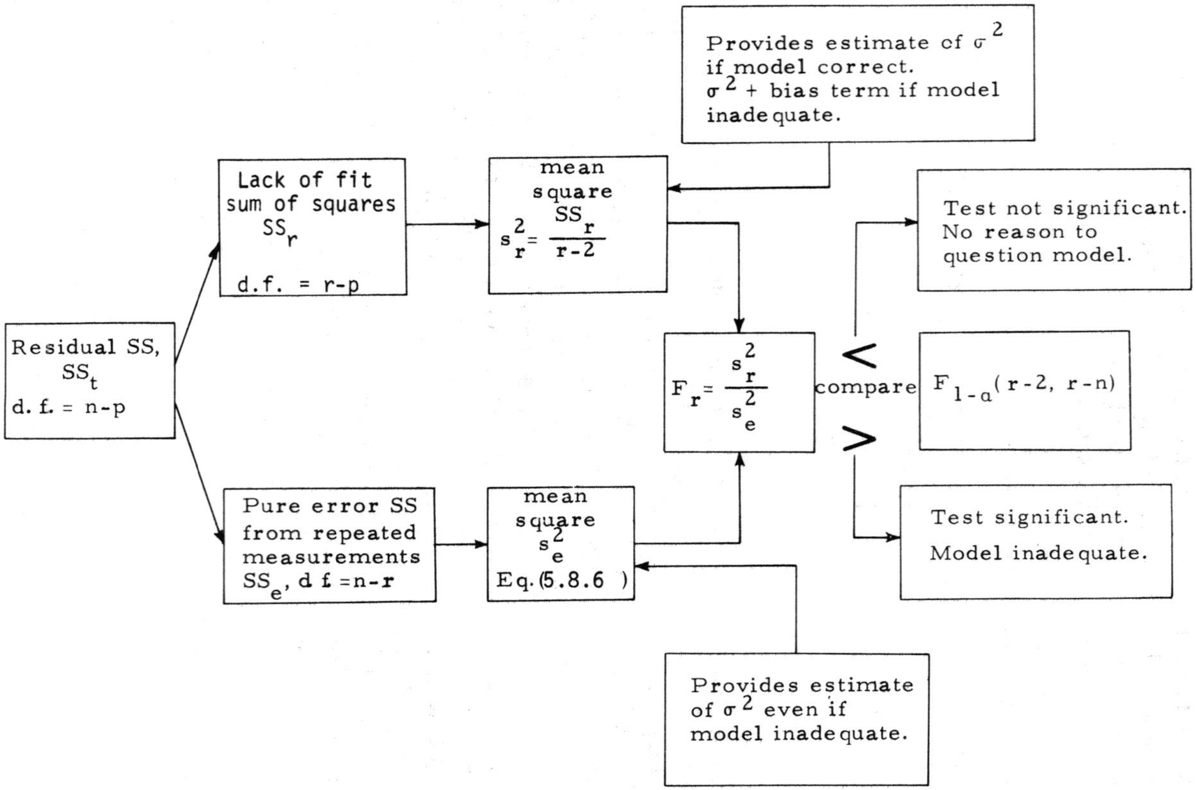

Figure 5.5. Schematic diagram for checking lack of fit with repeated observations. (Adapted from Applied Regression Analysis by Norman R. Draper and Harry Smith, John Wiley & Sons.)

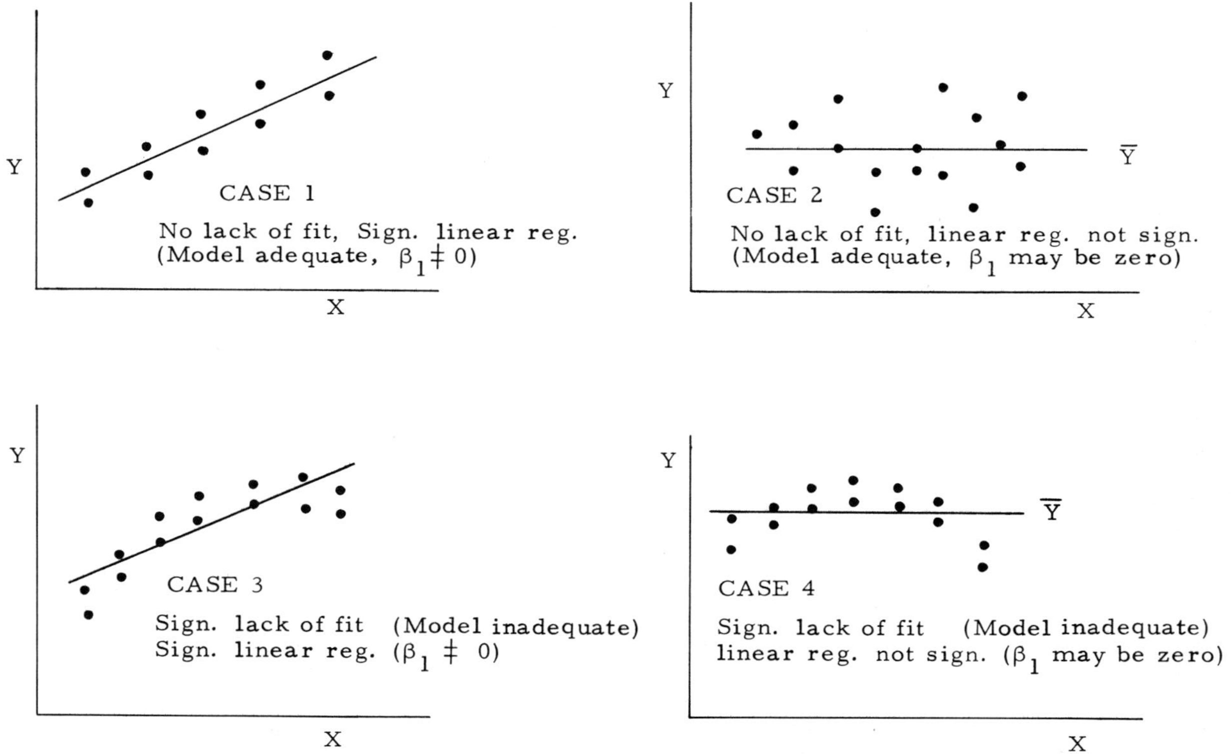

Figure 5.6. Typical straight line situations. (Adapted from Applied Regression Analysis by Norman R. Draper and Harry Smith, John Wiley & Sons.)

Table 5.6 Summary of Observations from Figure 5.6

Observation	Case 1	Case 2	Case 3	Case 4
Significant lack of fit $F_r > F_{1-\alpha}(r-2, n-r)$			X	X
Significant linear regression $F > F_{1-\alpha}(1, n-2)$	X		X	

For case 1 the linear model is adequate since there is *no* lack of fit and there is significant linear regression. For case 2 the linear regression is not significant; hence the model $\hat{Y} = \bar{Y}$ would be recommended. For case 3 there is lack of fit, but the linear regression is significant; thus one might try $Y = \beta_0 + \beta_1 X + \beta_{11} X^2 + \epsilon$. In case 4 there is a significant lack of fit and not significant linear regression. A model such as $Y = \beta_0 + \beta_1 X + \beta_{11} X^2 + \epsilon$ would be recommended even though there is not significant linear regression. (Why?)

Both tests need not be limited to testing the adequacy of the simple linear model $Y_i = \beta_0 + \beta_1 X_i + \epsilon_i$, but can be applied to linear estimation with more parameters and even to nonlinear parameter estimation; this can be done if there are repeated observations for the standard conditions of zero mean, independent, constant variance, and normal errors.

After saying the above, it should be emphasized that considerable insight can sometimes be gained in unfamiliar cases if the residuals are plotted and inspected visually.

5.9 CONFIDENCE INTERVAL ABOUT THE POINTS ON THE REGRESSION LINE

Let us consider a confidence interval about any point on the regression line

$$\hat{Y}_k = b_0' + b_1\left(X_k - \bar{X}\right) \tag{5.9.1}$$

This requires the variance of $\hat{Y}_k$, which is given by (5.2.41a). Using this expression with σ replaced by s the estimated standard error is

$$\text{est. s.e.}\left(\hat{Y}_k\right) = s\left[\frac{1}{n} + \frac{\left(X_k - \bar{X}\right)^2}{\Sigma\left(X_i - \bar{X}\right)^2}\right]^{1/2} \tag{5.9.2}$$

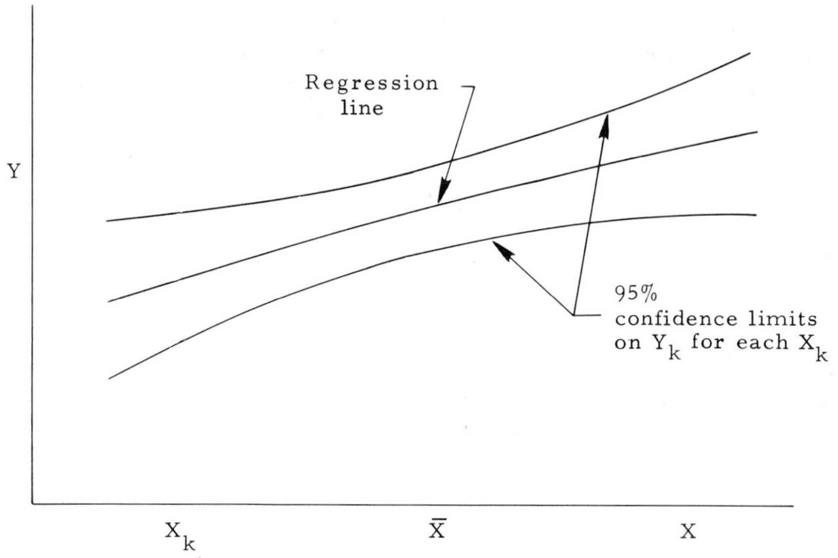

Figure 5.7 Confidence intervals about points on the regression line.

which is clearly a minimum at $X_k = \overline{X}$ and becomes larger toward the extremities; (5.9.2) implies that we do not know σ. The confidence limits for $\hat{Y}_k$ are

$$Y_k \pm t_{1-\alpha/2}(n-p)\left[\text{est. s.e.}\left(\hat{Y}_k\right)\right] \tag{5.9.3}$$

for n observations of Y_i, p parameters, and $100(1-\alpha)$ confidence. Figure 5.7 shows the 95%, say, confidence limits for the model (5.9.1); the curved, hyperbolic lines about the straight regression line give the confidence limits.

These limits can be interpreted as follows. Suppose that repeated sets of measurements of Y are taken at the same X values as were used to find the confidence limits given in Fig. 5.7. Then, of all the 95% confidence intervals constructed for $\eta_k = E(Y_k)$ at X_k, 95% of these intervals will contain $E(Y_k)$.

Confidence intervals and regions for parameters are discussed in Chapter 6.

5.10 VIOLATION OF THE STANDARD ASSUMPTION OF ZERO MEAN ERRORS

In the next few sections violations of the basic assumptions are considered. One of the easiest to treat is the case of additive errors that do not have a zero mean. The assumptions then are 10111111.

We are concerned here with nonzero mean errors that remain after any appropriate corrections have been made. Suppose, however, after all known corrections have been made, the errors still do not have a zero mean so that

$$E(\varepsilon_i) = f_i \tag{5.10.1}$$

where $f_i \neq 0$. Let ε_i be written as two terms one of which has a zero mean,

$$\varepsilon_i = f_i + v_i, \qquad E(v_i) = 0 \tag{5.10.2}$$

Consider several functions of f_i in connection with Model 2, $\eta_i = \beta_1 X_i$, with X_i not being the same for all i. The first function that we consider is $f_i = c$, constant. Then Y_i for Model 2 can be written

$$Y_i = \eta_i + \varepsilon_i = \beta_1 X_i + f_i + v_i = c + \beta X_i + v_i \tag{5.10.3}$$

where now the bias c is a parameter to be estimated in addition to β_1. In this case a one-parameter Model 2 problem becomes a two-parameter Model 3 problem.

If f_i happens to be proportional to X_i or $f_i = cX_i$ then instead of (5.10.3) we write

$$Y_i = \eta_i + \varepsilon_i = \beta_1 X_i + cX_i + v_i = (\beta_1 + c)X_i + v_i \tag{5.10.4}$$

and thus it is possible to estimate only the sum $\beta_1 + c$.

Another case is when $f_i = cZ_i$ is some known function which is not proportional to X_i. This reduces to a Model 5 estimation problem which involves two parameters.

5.11 VIOLATION OF THE STANDARD ASSUMPTION OF NORMALITY

If the standard assumptions excluding that of normality are valid (11110111), ordinary least squares estimation can still be used. The resulting least squares estimators are unbiased and have minimum variance among all linear unbiased estimators, but they are not efficient. A consequence of the central limit theorem is that the least squares estimators are consistent and asymptotically efficient almost regardless of the distribution of the errors, however. Hence when the normality assumption is not justified, least squares estimators still retain most of their desirable properties.

We note that the previously used estimators of the variances of the parameters are unchanged. Confidence intervals and tests for significance given in this chapter are based on the assumption of normal errors, however; for small numbers of observations the intervals and tests could be substantially in error. Fortunately, for larger sample sizes and provided the distribution is not radically nonnormal, the confidence limits and tests of significance can be used as reasonable approximations.

If the form of the underlying probability density of the errors is known, then the maximum likelihood and maximum a posteriori methods can be used. For example assume that all the standard assumptions apply except that the probability of ε_i is given by

$$f(\varepsilon_i) = \frac{1}{2\alpha}\exp(-|\varepsilon_i|\alpha^{-1}) \tag{5.11.1}$$

Then the ML function to minimize is

$$S_{\text{ML}} = \sum_{i=1}^{n} |Y_i - \eta_i| \tag{5.11.2}$$

Unfortunately, minimizing S_{ML} is not as simple as it would be for normal measurement errors.

Example 5.11.1

For Model 1, $\eta_i = \beta_0$, estimate β_0 for the data as given below. Assume that the assumptions 11110111 are valid and that $f(\varepsilon_i)$ is given by (5.11.1).

(*a*) $Y_1 = 0$, $Y_2 = 1$.
(*b*) $Y_1 = 0$, $Y_2 = 0.5$, $Y_3 = 1$.
(*c*) $Y_1 = 0$, $Y_2 = 0.25$, $Y_3 = 0.5$, $Y_4 = 1$.
(*d*) Generalize the results.

Solution

(*a*) For the observations $Y_1 = 0$ and $Y_2 = 1$,

$$S_{\text{ML}} = |\beta_0| + |1 - \beta_0|$$

A plot of S_{ML} versus β_0 shows that S_{ML} has a minimum between 0 and 1. In that range S_{ML} is equal to 1. Thus there is neither unique minimum nor parameter estimate.

(*b*) For the three observations of 0, 0.5, and 1, a plot of S_{ML} versus β_0 gives a minimum value of S_{ML} also equal to 1 at $b_0 = 0.5$.

(*c*) For this case a plot of S_{ML} versus β_0 shows that a minimum occurs between $b_0 = 0.25$ and 0.5.

(*d*) From the pattern of the answers obtained, it appears that there are two possibilities; one is for an even number of observations n and the other is for an odd number. Let the Y_i values be ordered so that the smallest Y_i value is Y_1, the next larger value is Y_2, etc. Then for n even, the b_0 value is located between $Y_{n/2}$ and $Y_{n/2+1}$. For n odd, b_0 is equal to $Y_{(n+1)/2}$.*

Another example with other than the normal distribution is given in Section 4.9 in connection with Monte Carlo methods.

5.12 VIOLATION OF THE STANDARD ASSUMPTION OF CONSTANT VARIANCE

When $V(\varepsilon_i)=\sigma_i^2$ varies with i, ordinary least squares estimation does not yield minimum variance estimators. Minimum variance estimators can be obtained, however, using maximum likelihood. These estimators for one- and two-parameter cases are given in Sections 5.2 and 5.3.

The effect upon the estimator(s) can be investigated for many σ_i^2 functions. Assume that the standard assumptions (11011111) apply in this section where two possible functions are considered. For illustrative purposes, the one-parameter case, Model 2, which is $\eta_i=\beta_1 X_i$, is used. The OLS and ML estimators and variances are

$$b_{1,\text{OLS}}=(\Sigma Y_i X_i)(\Sigma X_k^2)^{-1}, \qquad V(b_{1,\text{OLS}})=(\Sigma X_i^2\sigma_i^2)(\Sigma X_i^2)^{-2} \tag{5.12.1}$$

$$b_{1,\text{ML}}=\frac{\Sigma Y_i X_i \sigma_i^{-2}}{\Sigma X_j^2 \sigma_j^{-2}}, \qquad V(b_{1,\text{ML}})=(\Sigma X_i^2\sigma_i^{-2})^{-1} \tag{5.12.2}$$

In the case of the ML estimator and variance the quantity $Z_i=X_i/\sigma_i$ can be considered as a modified sensitivity coefficient; Z_i plays the same role as X_i when OLS is used with *all* the standard assumptions being valid.

Before investigating some cases of nonuniform σ_i^2, some situations are suggested where nonuniform σ_i^2 might arise. Error variances tend to increase with the amplitude of signal (or observation). When the response of Y_i varies over several orders of magnitude–say, from 0.001 to 100–the accuracy of the measuring device(s) is rarely constant. For small signals the errors usually are even smaller; for the large signals the standard deviation of the errors may be the same small fraction of the signal, but the actual error may be many times the value of the smallest signal. For

*The estimator b_0 conforms to the definition of the median given in Section 3.1.1.

example, suppose the voltage of some device, such as heat flow meter, varies from 0.00001 to 0.1 V in a series of observations. (Another device having large variations in output is a thermistor, for which the electric resistance varies greatly with temperature.) In order to measure such a range, a digital voltmeter with several full scale settings could be used. One range might go up to 0.001 V, another range might be used for 0.001 to 0.01 V, and so on. Then for readings near 0.001 and 0.01 V the percent accuracy might be the same; note that this infers a varying σ_i^2 that is approximately proportional to η_i^2.

5.12.1 Variance of ε_i Given by $\sigma_i^2=(X_i/\delta)^2\sigma^2$

One possible variation of σ_i^2 is $\sigma_i^2=(X_i/\delta)^2\sigma^2$ where δ is some quantity with the same units as X_i. The OLS estimator is unaffected, but the variance of $b_{1,\mathrm{OLS}}$ becomes

$$V(b_{1,\mathrm{OLS}})=\frac{\left[\Sigma X_i^4/(\Sigma X_i^2)^2\right]\sigma^2}{\delta^2} \tag{5.12.3}$$

The $b_{1,\mathrm{ML}}$ estimator and variance becomes

$$b_{1,\mathrm{ML}}=\frac{1}{n}\Sigma\frac{Y_i}{X_i}, \qquad V(b_{1,\mathrm{ML}})=\frac{\sigma^2}{n\delta^2} \tag{5.12.4}$$

Note that the variance of $b_{1,\mathrm{ML}}$ is a simple expression, but that for OLS is not. In order to make a comparison let $X_i=i\delta$. One can derive the following summation expressions

$$\sum_{i=1}^{n} i^2=\frac{n(n+1)(2n+1)}{6}$$

$$\sum_{i=1}^{n} i^4=\frac{n(n+1)(2n+1)(3n^2+3n-1)}{30}$$

which yield for the stipulated σ_i^2 the expression for $V(b_{1,\mathrm{OLS}})$ of

$$V(b_{1,\mathrm{OLS}})=\frac{6(3n^2+3n-1)\sigma^2}{5n(n+1)(2n+1)\delta^2}$$

For large values of n this expression reduces to $9\sigma^2/5n\delta^2$. Hence for large values of n, the OLS estimate for this Model 2 case with $\sigma_i^2=(X_i/\delta)^2\sigma^2$ has a variance of b_1 which is 80% larger than that of b_1 given by ML. This means that ML estimation is substantially superior in this case to OLS estimation.

One further benefit of the maximum likelihood (ML) method of estimation is that it can be used to provide an estimate of σ^2. This can be accomplished by replacing σ_i^2 in (5.3.1) and (5.3.2) by $(X_i/\delta)^2\sigma^2$, differentiating (5.3.1) with respect to σ^2, and then replacing σ^2 by $\hat{\sigma}^2$ and η_i by $\hat{Y}_i$ to get

$$\hat{\sigma}^2 = \frac{\delta^2}{n}\Sigma\left(\frac{Y_i - \hat{Y}_i}{X_i}\right)^2 \tag{5.12.5}$$

which is a consistent, asymptotically efficient, and biased estimate.

5.12.2 Variance of ε_i Equal to $\sigma^2\eta_i^2$

A commonly occurring case is for the standard deviation of the error to be proportional to the dependent variable η_i. In terms of the variance of ε_i, this can be expressed by

$$\sigma_i^2 = \sigma^2 E^2(Y_i) = \sigma^2\eta_i^2 \tag{5.12.6}$$

The OLS estimator is the same as usual, but the variance can only be approximated. For our purposes it is permissible to replace $E(Y_i) = \eta_i$ by $\hat{Y}_i$, the regression value for OLS; then let

$$\sigma_i^2 = \sigma^2\hat{Y}_i^2 \tag{5.12.7}$$

In ML estimation the $\sigma_i^2 = \sigma^2\eta_i^2$ relation makes the problem nonlinear because the parameters appear in both the denominator and numerator of S_{ML} given by (5.3.2) and also in the $\ln\sigma_i^2$ term contained in (5.3.1). A suggested procedure to get approximate ML values is to first solve for the parameter(s) using OLS and so obtain approximate values of $\hat{Y}_{i,\text{OLS}}$. These are then used to approximate σ_i^2 as $\sigma^2\hat{Y}_{i,\text{OLS}}$ in the ML estimators such as (5.12.2a).

5.13 VIOLATION OF STANDARD ASSUMPTION OF UNCORRELATED ERRORS

In the past decade there has been widespread use of automatic digital data acquisition equipment in connection with dynamic experiments. Transient temperatures have been measured, for example, by using such equipment to digitize the response of thermocouples. However, measurement errors tend to become correlated as the high sampling rate capability is used. In such cases the standard assumption of independent observation errors is not valid.

One might also obtain correlated measurements by testing the *same* specimen using the same sensors for different ranges of the independent variable X_i. Examples are measurements for a *particular* steel specimen at different temperatures for a property such as thermal conductivity, electric resistance, or hardness.

The standard assumptions of zero mean and uncorrelated measurement errors given by (5.1.7) and (5.1.9) result in

$$E(\varepsilon_i\varepsilon_k)=0 \qquad \text{for} \quad i\neq k \tag{5.13.1}$$

When this equation is not true many descriptive terms have been used; these terms include colored, correlated, not independent, and dependent errors. Some specific types of correlated errors are called *autoregressive* (AR), *moving average* (MA), and *autoregressive-moving average* (ARMA). Only AR errors are considered in this section. For further discussion see Chapter 6.

Let us consider a case with additive, zero mean, autoregressive errors in Y_i. There are no errors in the X_i's. We can then write

$$Y_i=\eta_i+\varepsilon_i, \qquad E(Y_i|\boldsymbol{\beta})=\eta_i \tag{5.13.2}$$

The measurements errors are described by the model

$$\varepsilon_i=\rho_i\varepsilon_{i-1}+u_i, \quad E(u_i)=0, \quad E(u_iu_j) \begin{array}{ll} =\sigma_i^2 & \text{for} \quad i=j \\ =0 & \text{for} \quad i\neq j \end{array} \tag{5.13.3}$$

which is called first-order autoregressive since the error ε_i depends on the error ε_{i-1} which is for the preceding time. (*Second*-order errors would depend on two preceding times, etc.) In the following analysis the ρ_i and σ_i^2 values are assumed to be known. There is no prior information. The associated assumptions are designated 1102-111.

Rather than using the direct matrix maximum likelihood approach of Chapter 6, we shall attempt to construct some sums of squares of terms that are uncorrelated and have constant variance. In other words a transformation is to be used to obtain modified measurements for which the assumptions 1111-111 are valid. Then write (5.13.3) at time i and $i-1$ as

$$Y_i=\eta_i+\rho_i\varepsilon_{i-1}+u_i \tag{5.13.4a}$$

$$Y_{i-1}=\eta_{i-1}+\varepsilon_{i-1} \tag{5.13.4b}$$

Multiply (5.13.4b) by ρ_i and subtract from (5.13.4a) to get

$$Y_i - \rho_i Y_{i-1} = (\eta_i - \rho_i \eta_{i-1}) + u_i \tag{5.13.5}$$

Define the transformed observation F_i and model H_i as

$$F_i = Y_i - \rho_i Y_{i-1}, \qquad H_i = \eta_i - \rho_i \eta_{i-1} \tag{5.13.6a, b}$$

Then analogous to (5.13.2) a transformed model is

$$F_i = H_i + u_i \tag{5.13.7}$$

where the model value F_i is now independent from other F_j $(j \neq i)$ values. Notice that the term u_i divided by σ_i has a variance of unity for all i's. This suggests that a sum of squares of independent, constant variance terms can be constructed from the u_i/σ_i values, or

$$S = \sum_{i=1}^{n} \left(\frac{u_i}{\sigma_i} \right)^2 = \sum_{i=1}^{n} (F_i - H_i)^2 \sigma_i^{-2} \tag{5.13.8}$$

where F_i and H_i are given by (5.13.6) provided

$$Y_0 \equiv 0, \qquad \eta_0 \equiv 0 \tag{5.13.9}$$

It is important to note that (5.13.8) has been derived without restricting the problem to cases for which η is linear in the parameters; hence it can be used for linear and nonlinear cases.

In Chapter 6 it is shown that the function given by (5.13.8) must be minimized for ML estimation if, in addition to the assumptions given above, the errors u_i are normal.

5.14 ERRORS IN INDEPENDENT AND DEPENDENT VARIABLES

Another violation of the standard assumptions is that of the independent variables, designated X_{ij} in this chapter, being stochastic as well as Y_i. In order to present a method of solution that can be generalized to complex situations the method of Lagrange multipliers is introduced in this section. For the simple example to be given it is not required, but this method of solution is illustrated. Before giving the example, the method of Lagrange multipliers is presented.

5.14.1 Method of Lagrange Multipliers

We consider the problem of finding a *stationary* (a relative maximum or minimum) value of the continuously differentiable function $f(a_1, a_2, \ldots, a_m)$

that is subject to n equality *constraints*,

$$\phi_i(a_1,a_2,\ldots,a_m)=0, \qquad i=1,2,\ldots,n \tag{5.14.1}$$

where $m>n$ and the ϕ_i are differentiable functions. Since the m variables $a_1,a_2,\ldots,a_m$ must satisfy n constraints, there are in effect only $m-n$ independent variables. A stationary value of $f(a_1,\ldots,a_m)$ requires that

$$df \equiv \frac{\partial f}{\partial a_1}\,da_1+\cdots+\frac{\partial f}{\partial a_m}\,da_m=0 \tag{5.14.2}$$

but the differentials da_i are not independent. The constraints (5.14.1) imply the n differential relations

$$\begin{aligned}
\frac{\partial \phi_1}{\partial a_1}da_1+\frac{\partial \phi_1}{\partial a_2}da_2+\cdots+\frac{\partial \phi_1}{\partial a_m}da_m&=0\\
\vdots \qquad\qquad \vdots \qquad\qquad\qquad \vdots\quad&\\
\frac{\partial \phi_n}{\partial a_1}da_1+\frac{\partial \phi_n}{\partial a_2}da_2+\cdots+\frac{\partial \phi_n}{\partial a_m}da_m&=0
\end{aligned} \tag{5.14.3}$$

A direct method of solution can be illustrated by a simple case. Suppose $m=3$ so that (5.14.2) becomes

$$df \equiv \frac{\partial f}{\partial a_1}\,da_1+\frac{\partial f}{\partial a_2}\,da_2+\frac{\partial f}{\partial a_3}\,da_3=0 \tag{5.14.4a}$$

Let there be only one constraint so that $n=1$ and then (5.14.3) gives

$$\frac{\partial \phi_1}{\partial a_1}\,da_1+\frac{\partial \phi_1}{\partial a_2}\,da_2+\frac{\partial \phi_1}{\partial a_3}\,da_3=0 \tag{5.14.4b}$$

which could be solved for da_3, say. This expression substituted for da_3 in (5.14.4a) then would give

$$df \equiv (\ldots)\,da_1+(\ldots)\,da_2 \tag{5.14.4c}$$

where the two different expressions in the parentheses are set equal to zero because the da_1 and da_2 terms can now be arbitrarily assigned. These two equations coming from the parentheses in (5.14.4c) plus $\phi_1=0$ would provide three equations for the three unknowns, a_1, a_2, and a_3.

An alternative procedure is called the Lagrange multiplier method. This method is introduced using the same example of $m=3$ and one constraint.

Multiply (5.14.4b) by λ_1 and add the results to (5.14.4a). Since the right-hand members are zeros, there follows

$$\left(\frac{\partial f}{\partial a_1}+\lambda_1\frac{\partial \phi_1}{\partial a_1}\right)da_1+\left(\frac{\partial f}{\partial a_2}+\lambda_1\frac{\partial \phi_1}{\partial a_2}\right)da_2+\left(\frac{\partial f}{\partial a_3}+\lambda_1\frac{\partial \phi_1}{\partial a_3}\right)da_3=0 \quad (5.14.5)$$

for an *arbitrary value of* λ_1. Now let λ_1 be determined so that one of the parentheses in (5.14.5) vanishes. Then the two differentials multiplying the remaining parentheses can be arbitrarily assigned and hence these two parentheses must also vanish. Consequently we must have

$$\frac{\partial f}{\partial a_1}+\lambda_1\frac{\partial \phi_1}{\partial a_1}=0 \quad (5.14.6a)$$

$$\frac{\partial f}{\partial a_2}+\lambda_1\frac{\partial \phi_1}{\partial a_2}=0 \quad (5.14.6b)$$

$$\frac{\partial f}{\partial a_3}+\lambda_1\frac{\partial \phi_1}{\partial a_3}=0 \quad (5.14.6c)$$

Then these three equations, (5.14.6a, b, c) plus the constraint $\phi_1=0$ comprise four equations for solving for the four unknowns a_1, a_2, a_3, and λ_1. The quantity λ_1 is known as a *Lagrange multiplier*. The introduction of these multipliers frequently simplifies and organizes the relevant algebra in minimization problems with equality constraints. It is important to note that the conditions given by (5.14.6) are equivalent to requiring that $f+\lambda_1\phi_1$ be stationary without any further constraints being imposed.

Applying this observation to the more general problem given above suggests that

$$f+\lambda_1\phi_1+\lambda_2\phi_2+\cdots+\lambda_n\phi_n$$

be extremized with respect to $a_1,a_2,\ldots,a_m$. Hence the following m equations must be satisfied,

$$\frac{\partial f}{\partial a_i}+\sum_{j=1}^{n}\lambda_j\frac{\partial \phi_j}{\partial a_i}=0, \qquad i=1,2,\ldots,m \quad (5.14.7)$$

along with the n constraints given by (5.14.1). Thus (5.14.1) and (5.14.7) constitute a set of $m+n$ equations for the $m+n$ unknowns

$$a_1,a_2,\ldots,a_m, \qquad \lambda_1,\lambda_2,\ldots,\lambda_n$$

5.14.2 Problem of Errors in the Independent and Dependent Variables

A problem which is nonlinear even though the model is linear in the parameters is the estimation of the parameters in the presence of errors in the independent variables as well as the dependent variables. The problem is formulated in this subsection and the solution of a simple case is considered in the next.

Consider first the dependent variable Y_i which is related to the model by

$$Y_i = \eta_i + \varepsilon_{Y_i} \tag{5.14.8}$$

and thus the error ε_{Y_i} is additive. Also let ε_{Y_i} have a zero mean, be independent from ε_{Y_j} for $i \neq j$, have a normal probability density with known variance terms, or

$$E(\varepsilon_{Y_i}) = 0, \qquad E(\varepsilon_{Y_i}\varepsilon_{Y_j}) = 0 \qquad \text{for} \quad i \neq j,$$

$$E\left(\varepsilon_{Y_i}^2\right) = \sigma_{Y_i}^2 \qquad \text{and} \qquad \varepsilon_{Y_i} \qquad \text{is normal} \tag{5.14.9}$$

These assumptions are designated 110111--. With this information the probability density of $\varepsilon_{Y_1}, \varepsilon_{Y_2}, \ldots, \varepsilon_{Y_n}$ is

$$f(\varepsilon_{Y_1}, \ldots, \varepsilon_{Y_n}) = \frac{1}{(2\pi)^{n/2}\sigma_{Y_1} \cdots \sigma_{Y_n}} \exp\left[-\frac{1}{2}\sum_{j=1}^{n} \varepsilon_{Y_j}^2 \sigma_{Y_j}^{-2}\right] \tag{5.14.10}$$

There are also errors in the independent variables X_{ij} which are described by

$$X_{ij} = \xi_{ij} + \varepsilon_{X_{ij}}, \qquad E(\varepsilon_{X_{ij}}) = 0,$$

$$E(\varepsilon_{X_{ij}}\varepsilon_{X_{kl}}) = 0 \quad \text{except when } i = k \quad \text{and} \quad j = l, \qquad E\left(\varepsilon_{X_{ij}}^2\right) = \sigma_{X_{ij}}^2 \tag{5.14.11}$$

and $\varepsilon_{X_{ij}}$ has a normal density. The $\sigma_{X_{ij}}^2$ values are assumed to be known. The value X_{ij} is measured and ξ_{ij} is the true value of X_{ij}. The errors ε_{Y_i} and $\varepsilon_{X_{kj}}$ are considered to be independent for all values of i, j, and k. Analogous to (5.14.10) we can write

$$f(\varepsilon_{X_{11}}, \ldots, \varepsilon_{X_{np}}) = \frac{1}{(2\pi)^{np/2}\sigma_{X_{11}} \cdots \sigma_{X_{np}}} \exp\left[-\frac{1}{2}\sum_{j=1}^{p}\sum_{i=1}^{n} \varepsilon_{X_{ij}}^2 \sigma_{X_{ij}}^{-2}\right] \tag{5.14.12}$$

Owing to the independence of the ε_{Y_i} and $\varepsilon_{X_{kj}}$ errors, the maximum likelihood method of estimating the parameters requires that the product of (5.14.10) and (5.14.12) be maximized with respect to the parameters $\beta_1,\beta_2,\ldots,\beta_p$ and the $\eta_1,\ldots,\eta_n,\xi_{11},\ldots,\xi_{np}$ values. This is equivalent to minimizing

$$S(\eta,\xi)=\sum_{i=1}^{n}(Y_i-\eta_i)^2\sigma_{Y_i}^{-2}+\sum_{j=1}^{p}\sum_{i=1}^{n}(X_{ij}-\xi_{ij})^2\sigma_{X_{ij}}^{-2} \tag{5.14.13}$$

with respect to $\beta_1,\ldots,\xi_{np}$ or a total of $(n+p+np)$ parameters. This will produce the estimates $b_1,b_2,\ldots,b_p,\hat{Y}_1,\ldots,\hat{Y}_n,\hat{X}_{11},\ldots,\hat{X}_{np}$. The η_i, β_k, and ξ_{ij} values are not independent, however, and must be related through the model for η_i which can be written as the equality constraint

$$g_i(\eta,\beta,\xi)\equiv\eta_i-\beta_1\xi_{i1}-\beta_2\xi_{i2}-\cdots-\beta_p\xi_{ip}=0 \tag{5.14.14}$$

which applies for $i=1,2,\ldots,n$.

The method of Lagrange multipliers involves minimizing the function

$$L=\frac{1}{2}S(\eta,\xi)+\sum_{j=1}^{n}\lambda_j g_j(\eta,\beta,\xi) \tag{5.14.15}$$

with respect to parameters $\beta_1,\beta_2,\ldots,\beta_p$. Necessary conditions for a minimum are

$$\frac{\partial L}{\partial\beta_k}=\frac{1}{2}\frac{\partial S}{\partial\beta_k}+\sum_{j=1}^{n}\lambda_j\frac{\partial g_j}{\partial\beta_k}=0,\qquad k=1,2,\ldots,p \tag{5.14.16a}$$

$$\frac{\partial L}{\partial\eta_i}=\frac{1}{2}\frac{\partial S}{\partial\eta_i}+\sum_{j=1}^{n}\lambda_j\frac{\partial g_j}{\partial\eta_i}=0,\qquad i=1,2,\ldots,n \tag{5.14.16b}$$

$$\frac{\partial L}{\partial\xi_{ik}}=\frac{1}{2}\frac{\partial S}{\partial\xi_{ik}}+\sum_{j=1}^{n}\lambda_j\frac{\partial g_j}{\partial\xi_{ik}}=0,\qquad i=1,\ldots,n;k=1,\ldots,p \tag{5.14.16c}$$

The expressions in (5.14.16) are all evaluated at $\beta_1=b_1,\ldots,\beta_p=b_p$, $\eta_1=\hat{Y}_1,\ldots,\eta_n=\hat{Y}_n$, $\xi_{11}=\hat{X}_{11},\ldots,\xi_{np}=\hat{X}_{np}$. It is important to note that

$$S=S(\eta_1,\ldots,\eta_n,\xi_{11},\ldots,\xi_{np}) \tag{5.14.17a}$$

$$g_j=g_j(\eta_j,\beta_1,\beta_2,\ldots,\beta_p,\xi_{j1},\xi_{j2},\ldots,\xi_{jp}) \tag{5.14.17b}$$

Thus S is not an explicit function of the parameters $\beta_1,\ldots,\beta_p$. Then

(5.14.16) for any model, linear or nonlinear, can be written as

$$\sum_{j=1}^{n} \lambda_j \frac{\partial g_j}{\partial \beta_k}\bigg|_{b_1,\ldots,\hat{X}_{jp}} = 0, \qquad k = 1,\ldots,p \tag{5.14.18a}$$

$$-(Y_i - \hat{Y}_i)\sigma_{Y_i}^{-2} + \lambda_i = 0, \qquad i = 1,\ldots,n \tag{5.14.18b}$$

$$-(X_{ik} - \hat{X}_{ik})\sigma_{X_{ik}}^{-2} - \lambda_i \frac{\partial g_i}{\partial \xi_{ik}}\bigg|_{\hat{Y}_i, b_1,\ldots,\hat{X}_{ik}} = 0 \tag{5.14.18c}$$

where (5.14.18c) applies for $i = 1,\ldots,n$; $k = 1,\ldots,p$. In addition to the equations given by (5.14.18) there are the constraints $g_i = 0$ which, for the linear model considered in this section, are equivalent to

$$\hat{Y}_i = b_1\hat{X}_{i1} + b_2\hat{X}_{i2} + \cdots + b_p\hat{X}_{ip}, \qquad i = 1,2,\ldots,n \tag{5.14.19}$$

Then (5.14.18) and (5.14.19) provide $p + 2n + np$ equations for the same number of unknowns which are $b_1,\ldots,b_p$, $\hat{Y}_1,\ldots,\hat{Y}_n$, $\lambda_1,\ldots,\lambda_n$, $\hat{X}_{11},\ldots,\hat{X}_{np}$.

Consider first (5.14.18) without introducing the assumption of a model linear in the parameters such as (5.14.19). Then in general (5.14.18b) yields

$$\lambda_i = (Y_i - \hat{Y}_i)\sigma_{Y_i}^{-2} \tag{5.14.20}$$

Thus the Lagrange multipliers are weighted residuals. Introducing (5.14.20) into (5.14.18a, c) eliminates λ_i and gives

$$\sum_{j=1}^{n} (Y_j - \hat{Y}_j)\sigma_{Y_j}^{-2} \frac{\partial g_j}{\partial \beta_k}\bigg| = 0, \qquad k = 1,\ldots,p \tag{5.14.21a}$$

$$(X_{ik} - \hat{X}_{ik})\sigma_{X_{ik}}^{-2} + (Y_i - \hat{Y}_i)\sigma_{Y_i}^{-2} \frac{\partial g_i}{\partial \xi_{ik}}\bigg| = 0, \qquad i = 1,\ldots,n;\, k = 1,\ldots,p \tag{5.14.21b}$$

Hence, for the general nonlinear case, (5.14.21) and a set of constraint equations, $g_i = 0$, can be solved for the $p + n + np$ unknowns of $b_1,\ldots,\hat{Y}_1,\ldots,\hat{X}_{11},\ldots$

Let now the linear model and its constraint, (5.14.19), be used. Then

(5.14.21) can be given as

$$\sum_{j=1}^{n}\left(Y_j-\sum_{i=1}^{p} b_i\hat{X}_{ji}\right)\sigma_{Y_j}^{-2}\hat{X}_{jk}=0, \qquad k=1,2,\ldots,p \tag{5.14.22a}$$

$$(X_{ik}-\hat{X}_{ik})\sigma_{X_{ik}}^{-2}+\left(Y_i-\sum_{j=1}^{p} b_j\hat{X}_{ij}\right)b_k\sigma_{Y_i}^{-2}=0, \qquad i=1,\ldots,n;k=1,\ldots,p \tag{5.14.22b}$$

which comprise $p+np$ equations for the unknowns $b_1,\ldots,b_p$, $\hat{X}_{11},\ldots,\hat{X}_{np}$. Notice that, even though the model is linear in the parameters $\beta_1,\ldots,\beta_p$, the solution of (5.14.22) is nonlinear and thus is not straightforward. One way to start is to note that (5.14.22b) for fixed i provides a set of linear equations for $\hat{X}_{i1},\ldots,\hat{X}_{ip}$ which can be solved in terms of the $b_1,\ldots,b_p$ values. When the $\hat{X}_{ji}$ values are substituted into (5.14.22a), a set of p nonlinear equations results for the unknowns $b_1,\ldots,b_p$. The simplest case is for $p=1$, which is considered next.

5.14.3 Model 2 ($\eta_i=\beta\xi_i$) Example with Errors in both η_i and ξ_i

As an example of the above procedure consider a case involving model 2, $\eta_i=\beta\xi_i$, where there are errors in both the dependent variable η_i and the independent variable ξ_i

$$Y_i=\eta_i+\varepsilon_{Y_i} \tag{5.14.23a}$$

$$X_i=\xi_i+\varepsilon_{X_i} \tag{5.14.23b}$$

Let the assumptions given above for ε_{Y_i} and ε_{X_i} apply except let $\sigma_{Y_i}^2$ and $\sigma_{X_i}^2$ be the constants σ_Y^2 and σ_X^2, respectively.

We can obtain the solution for b, an estimate of β, through the use of (5.14.22a, b). Using first (5.14.22b) gives

$$(X_i-\hat{X}_i)\sigma_X^{-2}+(Y_i-b\hat{X}_i)b\sigma_Y^{-2}=0 \tag{5.14.24}$$

which can be solved for $\hat{X}_i$ to obtain

$$\hat{X}_i=\frac{[X_i+Y_ib\alpha]}{[1+b^2\alpha]}, \qquad i=1,\ldots,n \tag{5.14.25}$$

where $\alpha\equiv\sigma_X^2/\sigma_Y^2$. Note that $\hat{X}_i$ is a nonlinear function of b. Introducing

(5.14.25) into (5.14.22a) gives the nonlinear equation,

$$\Sigma\left[Y_j - b\frac{X_j + Y_j b\alpha}{1+b^2\alpha}\right]\sigma_Y^{-2}\frac{X_j + Y_j b\alpha}{1+b^2\alpha} = 0 \tag{5.14.26}$$

For convenience let

$$S_{YY} = \Sigma Y_j^2, \qquad S_{XY} = \Sigma X_j Y_j, \tag{5.14.27a, b}$$

$$S_{XX} = \Sigma X_j^2 \tag{5.14.27c}$$

and then (5.14.26) can be expanded to

$$S_{XY} + b\alpha S_{YY} + b^2\alpha S_{XY} + b^3\alpha^2 S_{YY} = bS_{XX} + 2b^2\alpha S_{XY} + b^3\alpha^2 S_{YY}$$

which can be simplified to

$$(\alpha S_{XY})b^2 + (S_{XX} - \alpha S_{YY})b - S_{XY} = 0 \tag{5.14.28}$$

which in turn can be solved for b,

$$b = \frac{\alpha S_{YY} - S_{XX} \pm \left[(\alpha S_{YY} - S_{XX})^2 + 4\alpha S_{XY}^2\right]^{1/2}}{2\alpha S_{XY}} \tag{5.14.29}$$

The *positive* sign is chosen in the $\pm$ sign in (5.14.29) because then the estimate will converge to the correct value of S_{XY}/S_{XX} when $\alpha = \sigma_X^2/\sigma_Y^2 \to 0$. If $\alpha \to \infty$, b approaches S_{YY}/S_{XY}. Equation 5.14.29 also gives $b = S_{XY}/S_{XX}$ for all values of α if it happens that S_{XY}/S_{XX} is equal to S_{YY}/S_{XY} or in other terms, $S_{XX}S_{YY} - S_{XY}^2 = 0$. In ordinary least squares estimation involving Model 2, we do not permit S_{XX} to be equal to zero. If S_{XY} is equal to zero, (5.14.28) gives $b = 0$.

After b is calculated using (5.14.29), the estimated values $\hat{X}_i$ can be obtained from (5.14.25). Observe that a different $\hat{X}_i$ is calculated from (5.14.25) for each i value if the Y_i values are different even if the X_i values are actually the same. Physically a given X_i value may not be known precisely, but it may be known that it is constant for several measurements. However, if this is the case the assumption of independent errors in each X_i is violated. Hence another analysis is required for this special case of repeated Y_i values at precisely the same X_i value.

Example 5.14.1

Consider a case involving Model 2, with errors in either Y_i or X_i or both, that satisfies the assumptions given above in Sections 5.14.2 and 5.14.3. The data are given below.

i	X_i	Y_i
1	-1	$1-\delta$
2	1	1

Let δ be a positive value. Also investigate the case for $\delta \to 0$.

(*a*) Find b, $\hat{X}_i$, and $\hat{Y}_i$ for $\sigma_X = \sigma_Y$.

(*b*) Find b and $\hat{Y}_i$ for $\sigma_X = 0$.

(*c*) Find b and $\hat{X}_i$ for $\sigma_Y = 0$.

Solution

To find the b values (5.14.29) can be used. Hence find S_{XX}, S_{YY}, and S_{XY} from (5.14.27) to be $S_{XX} = 2$, $S_{YY} = 2 - 2\delta + \delta^2$, and $S_{XY} = \delta$.

(*a*) In this case $\alpha = 1$ and (5.14.29) gives

$$b = \frac{\left[-2+\delta+(8-4\delta+\delta^2)^{1/2}\right]}{2}$$

If $\delta \to 0$, $b \to -1 + 2^{1/2} = 0.4142136$.

The $\hat{X}_i$ values are found from (5.14.25),

$$\hat{X}_i = \frac{X_i + Y_i b}{1+b^2}$$

and $\hat{Y}_i$ is equal to $b\hat{X}_i$. For $\delta \to 0$ we obtain $\hat{X}_1 = -0.5$, $\hat{Y}_1 = -0.2071067$ and $\hat{X}_2 = 1.2071067$, $\hat{Y}_2 = 0.5$. For $\sigma_X = \sigma_Y = 1$ the sum S given by (5.14.13) is precisely 2.

(*b*) This is the usual least squares case and $b = S_{XY}/S_{XX}$ is equal to $\delta/2$. The $\hat{Y}_i$ values are $\hat{Y}_1 = -\delta/2$ and $\hat{Y}_2 = \delta/2$. For $\delta = 0$, the values are zero; hence the predicted line is $\hat{Y}_i = 0$. Again the minimum S for $\delta = 0$ is 2.

(*c*) For this case $b = S_{YY}/S_{XY} = (2 - 2\delta + \delta^2)/\delta$. The $\hat{X}_i$ values are found from

$$\hat{X}_i = \frac{Y_i}{b} = \frac{Y_1\delta}{2-2\delta+\delta^2}$$

For $\delta \to 0$, $b \to \infty$ and $\hat{X}_1 = \hat{X}_2 = 0$. Unlike part (*b*) the predicted line is now the vertical axis for $\hat{X}_i = 0$ for all i. The minimum S is again 2.

It is instructive to examine the predicted lines for each of the cases above. See Fig. 5.8 for $\delta \to 0^+$. Notice that the usual least squares case ($\alpha = 0$) has the predicted line of $Y = 0$; the $Y_1 = 1$, $X_1 = -1$ observation is replaced with $\hat{Y}_1 = 0$ and $X_1 = -1$, and $Y_2 - 1$, $X_2 - 1$ is replaced with $\hat{Y}_2 - 0$, $X_2 - 1$. The case for $\alpha \to \infty$ has the vertical predicted line of $X = 0$; the two observations are replaced by the single point $Y_1 = Y_2 = 1$ with $X = 0$. For the $\alpha = 1$ case, the predicted line is inclined as shown. It is thus clear that the three α values can yield quite different predicted values. In other words it can make a large difference in the predicted line whether the errors are in Y or X or both. This case shown in Fig. 5.8 is an extreme one, however, because many times the predicted lines are quite close.

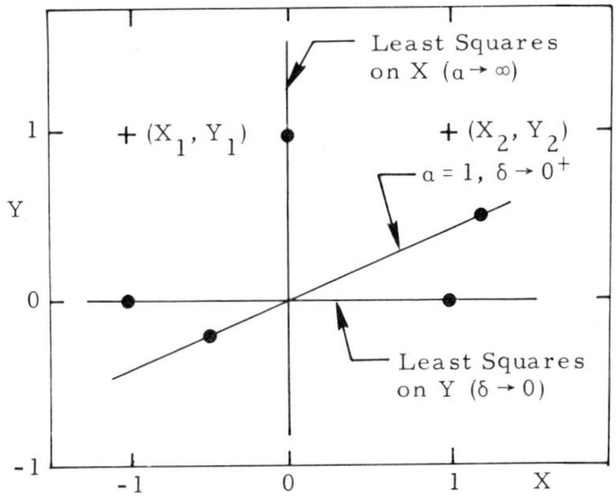

Figure 5.8 Predicted lines for errors in dependent and independent variables for data points $(-1,1)$ and $(1,1)$ for Example 5.14.1.

A case for which the predicted lines are much closer together is when $\delta=1$ causing $Y_1=0$ with $X_1=-1$ as shown in Fig. 5.9. If δ were equal to 2 so that $Y_1=-1$, then for any α such that $0 \leqslant \alpha \leqslant 1$ the predicted lines are all the same.

Example 5.14.2

Near a wall over which a turbulent fluid is flowing, the velocity is a linear function of position. Let the velocity (in cm/sec) be designated u and the distance from the wall (in cm) be designated x. The below data were taken from Fig. 6.20 of Kreith

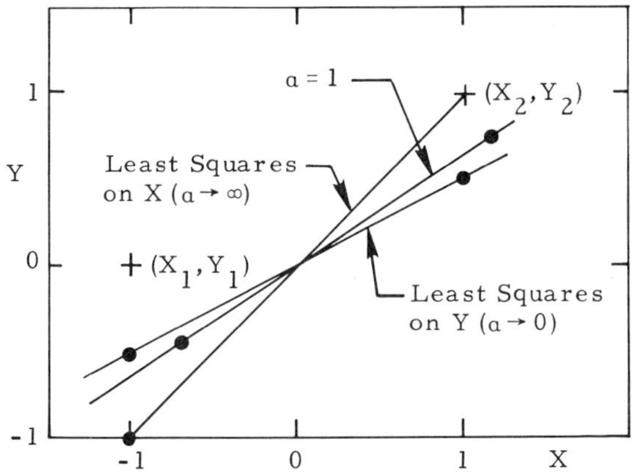

Figure 5.9 Predicted lines for errors in dependent and independent variables for data points $(-1,0)$ and $1,1)$ for Example 5.14.1.

[5]. Estimated values of σ_X and σ_U are also given.

i	X (cm)	σ_X (cm)	U (cm/sec)	σ_U (cm/sec)
1	0.0112	0.0003	80	3
2	0.0162	0.0003	125	3
3	0.0215	0.0003	165	3
4	0.0310	0.0003	235	3

The models for the true velocity u and the true distance x are

$$U = u + \varepsilon_U, \qquad u = \beta x$$

$$X = x + \varepsilon_X$$

where U and X are measured values and β is a parameter which is proportional tc the shear stress at the wall. Estimates for β are to be obtained using

(a) the above information;

(b) $\sigma_X = 0$ and σ_U is unknown; and

(c) σ_X is unknown and $\sigma_U = 0$. Also calculate the $\hat{U}$ and $\hat{X}$ values for each case. The assumptions indicated in Section 5.14.2 are valid.

Solution

In each of the cases u and U are analogous to η and Y, and x and X to ξ and X in the notation given in this section. With this in mind let us then evaluate S_{YY}, S_{XX}, S_{XY}, and α in (5.14.29).

$$S_{YY} = \sum_{i=1}^{4} U_i^2 = (80)^2 + (125)^2 + (165)^2 + (235)^2 = 104475$$

$$S_{XX} = \sum_{i=1}^{4} X_i^2 = (0.0112)^2 + (0.0162)^2 + (0.0215)^2 + (0.031)^2$$

$$= 0.00181113$$

$$S_{XY} = \sum_{i=1}^{4} X_i U_i = 0.0112(80) + 0.0162(125) + 0.0215(165) + 0.031(235)$$

$$= 13.7535$$

$$\alpha = \frac{\sigma_X^2}{\sigma_U^2} = \left(\frac{0.0003}{3}\right)^2 = 10^{-8}$$

(*a*) For the above values the parameter β is estimated using (5.14.29) to be $b = 7594.7438/\text{sec}$. The values of $\hat{X}_i$ are obtained from (5.14.25) to be

$$\hat{X}_i = \frac{X_i + U_i b\alpha}{1 + b^2\alpha} = \frac{X_i + 7.5947438 \times 10^{-5} U_i}{1.5768013}$$

After values are calculated for $\hat{X}_i$, the values for $\hat{U}_i$ are found using

$$\hat{U}_i = b\hat{X}_i = 7594.7438\,\hat{X}_i$$

The resulting values are given below.

i	X_i	U_i
1	0.01096	83.21
2	0.01629	123.75
3	0.02158	163.91
4	0.03098	235.28

(*b*) This is the usual least squares analysis for which $b = S_{XY}/S_{XX} = 7593.8778$. The predicted or regression line is now $\hat{U}_i = bX_i$. The values for X_i and $\hat{U}_i$ are tabulated next.

i	X_i	$\hat{U}_i$
1	0.0112	85.05
2	0.0162	123.02
3	0.0215	163.27
4	0.0310	235.41

(*c*) In this case the role of X and Y are interchanged in the least squares analysis. Here $b = S_{YY}/S_{XY} = 7596.2482$; $\hat{X}_i$ is obtained from $\hat{X}_i = U_i/b$; and U_i is the measured value. The results of the calculations are as follows:

i	X_i	U_i
1	0.01053	80
2	0.01646	125
3	0.02172	165
4	0.03094	235

A comparison of the b values in this example reveals that there are some differences but they are very small–the largest difference in the b values is 0.03%. This case is more common than that shown in Figs. 5.8 and 5.9 where the predicted lines are quite different. Because the curves are so similar in this example, only the lower two points are shown in Fig. 5.10. There are negligible differences in the curves that can be drawn between the three sets of predicted points.

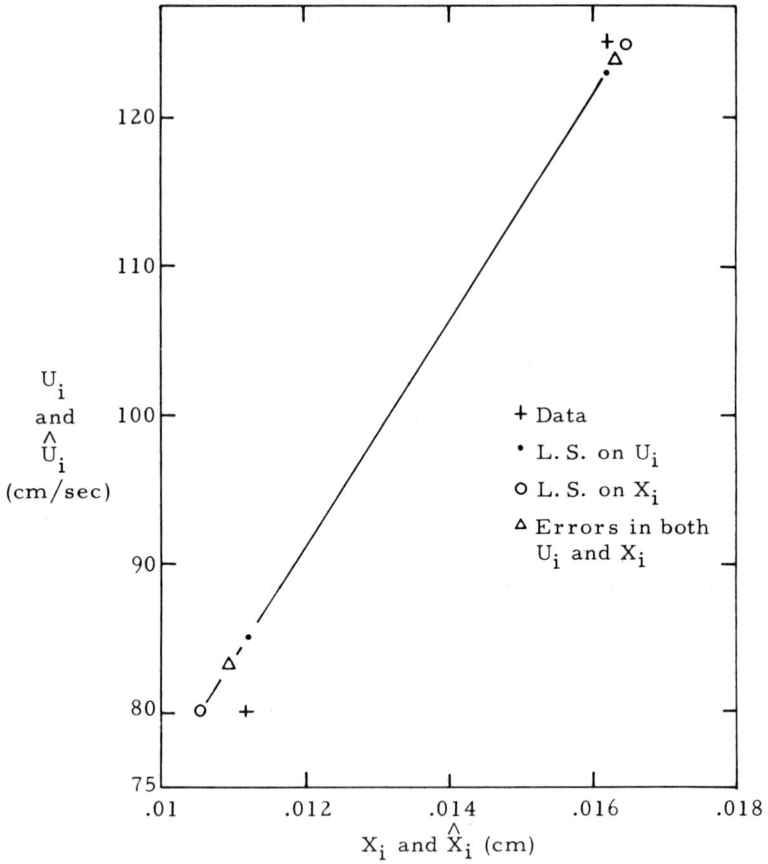

Figure 5.10 Predicted line for errors in dependent and independent variables for Example 5.14.2.

REFERENCES

1. Draper, N. R. and Smith H., *Applied Regression Analysis*, John Wiley and Sons, Inc., New York, 1966.
2. Burington, R. S. and May, D. C., *Handbook of Probability and Statistics with Tables*, 2nd ed., McGraw-Hill Book Company, New York, 1970.
3. Box, G. E. P. and Tiao, G. C., *Bayesian Inference in Statistical Analysis*, Addison-Wesley Publishing Co., Reading, Mass., 1973.
4. Brownlee, K. A., *Statistical Theory and Methodology in Science and Engineering*, 2nd ed., John Wiley and Sons, Inc., New York, 1965.
5. Kreith, F., *Principles of Heat Transfer*, 3rd ed., Intext Educational Publishers, New York, 1973.

PROBLEMS

5.1 Prove using the standard assumptions that

$$V(Y_i)=\sigma^2, \qquad E(Y_iY_j)=E(Y_i)E(Y_j) \quad \text{for } i\neq j$$

and indicate which assumptions are used.

5.2 Show that

$$\sum_i (X_i-\bar{X})Y_i=\sum_i X_iY_i-\frac{1}{n}\left(\sum_j X_j\right)\sum_i Y_i$$

and use to show that (5.2.37) follows from (5.2.36).

5.3 What is the expected value of e_i for OLS estimation when the following assumptions apply?

(*a*) $Y_i=E(Y_i|\beta)+\epsilon_i=\eta_i+\epsilon_i=\beta X_i+\epsilon_i$
(*b*) $E(\epsilon_i)=a\neq 0$
(*c*) $V(X_i)=0$
(*d*) $V(\beta)\doteq 0$

5.4 Prove (5.2.43) for $Y_{ij}=\beta_0+\beta_1X_i+\beta_2X_i^2+\beta_3X_i^3+\epsilon_i$ when using OLS.

5.5 For Model 5 what weighting functions for maximum likelihood estimation would cause the sum of the residuals to be equal to zero? Assume that the assumptions designated 11011111 apply.

5.6 Show that the minimum value of S for $Y_i=\beta_0+\beta_1X_i+\epsilon_i$ is

$$S_{\min}\equiv\sum_i e_i^2=\sum_i e_i\epsilon_i$$

5.7 The following data are given

i	1	2	3	4	5
X_i	1	2	3	4	5
Y_i	2	1	3	3	6

Assume that the standard assumptions apply.

(*a*) Find estimates of the parameters in $Y_i=\beta_0+\beta_1X_i+\epsilon_i$.

Answer. 0, 1.

(*b*) Find estimates of the parameters in $Y_i=\beta_0'+\beta_1(X_i-\bar{X})+\epsilon_i$.

Answer. 3, 1.

(*c*) Give the residuals e_i. (Do they add up to zero?)
(*d*) Estimate the variance of ϵ_i.

Answer. 1.333.

(e) Give the estimated standard error of b_0.

Answer. 1.211.

(f) Give the estimated standard error of b_0'.

Answer. 0.516.

(g) Give the estimated standard error of b_1.

Answer. 0.365.

(h) Give the estimated covariance of b_0 and b_1.

Answer. -0.4.

(i) Give the estimated covariance of b_0' and b_1.

Answer. 0.

5.8 The following data are given

i	1	2	3	4	5	6	7	8	9	10
X_i	0	0	0	0	0	10	10	10	10	10
Y_i	110	95	90	100	105	40	50	55	45	60

Assume that the standard assumptions apply. Answer the same questions as in Problem 5.7.

5.9 The following values have been reported for a certain set of experiments.

i	1	2	3	4	5	6	7
X_i	40	50	60	70	80	90	100
Y_i	0.325	0.332	0.340	0.347	0.353	0.359	0.364

Assume that the standard assumptions apply.

(a) Estimate β_0 and β_1 for the model $Y_i = \beta_0 + \beta_1 X_i + \epsilon_i$.

Answer. 0.2997, 0.000657.

(b) Estimate variances for b_0 and b_1.

Answer. 2.38×10^{-6}, 4.49×10^{-10}.

(c) Calculate e_i and plot.
(d) Are the residuals correlated?
(e) Based on the conclusions of (d), are the estimates given in (b) valid?
(f) How could the model be improved?

5.10 A study was made on the effect of temperature on the yield of a chemical process. The following data were collected with X linearly related to temperature and Y to the yield:

X_i	−5	−4	−3	−2	−1	0	1	2	3	4	5
Y_i	0	4	3	6	9	7	8	12	13	12	17

Assume that the standard assumptions apply.

(*a*) For the model $Y_i = \beta_0 + \beta_1 X_i + \epsilon_1$ estimate β_0 and β_1. What is the prediction equation?
(*b*) Construct an analysis of variance table. Let the null hypothesis be that $\beta_1 = 0$ with a risk 0.05.
(*c*) What are the 95% confidence limits for β_1?
(*d*) What are the confidence limits about η_i at $X = 3$?
(*e*) Are there any indications that another model should be tried?

5.11 Consider the model

$$Y_i = \beta + \epsilon_i$$

where the standard assumptions apply for ϵ_i.

(*a*) Derive an unbiased, minimum variance estimator for β.
(*b*) Give an unbiased estimate of the variance of ϵ_i (σ^2 is unknown).

5.12 Repeat Problem 5.11 for the model

$$Y_i = \beta X_i^2 + \epsilon_i$$

5.13 Repeat Problem 5.11 for the model

$$Y_i = \beta \sin X_i + \epsilon_i$$

5.14 Use the ϵ_i column of Table 5.1 as data (that is, $Y_1 = -0.742$, $Y_2 = -0.034$, etc.) and use the model of Problem 5.11.

(*a*) Estimate β.
(*b*) Estimate σ.

5.15 Repeat Example 5.2.4 with the ϵ_i values replaced with nine consecutive values of a column of Table XXIII of reference 2. The column is to be the one corresponding to your birth date and the first value used in the column is to correspond to the birthday month. For example, if your birthday is March 14, then pick the fourteenth column and start with the third entry since March is the third month.

5.16 The temperature of a fluid flowing over a plate is nearly linear near the plate. Let Y be proportional to the temperature and X be the distance from the wall. The following results are obtained:

$$\bar{X} = 0.05, \qquad \Sigma(X_i - \bar{X})^2 = 0.016, \qquad \bar{Y} = 300,$$

$$\Sigma(X_i - \bar{X})(Y_i - \bar{Y}) = 80, \qquad \Sigma(Y_i - \bar{Y})^2 = 8320, \qquad \Sigma(\hat{Y}_i - \bar{Y})^2 = 8100$$

Assume that the model $Y_i = \beta_0 + \beta_1 X_i + \epsilon_i$ and the standard assumptions apply.

(a) Estimate β_0 and β_1.
(b) Prepare the analysis of variance table.

5.17 Show that $\Sigma_{i=1}^{n}(X_i - \bar{X})^2$ is maximized for n even with $-R < X_i < R$ by choosing one-half of the X_i to be $-R$ and the other half to be at R.

5.18 Derive an expression for $\text{cov}(Y_i, \hat{Y}_i)$, thus proving (5.2.29).

5.19 Derive the expression for $V(b_1 - \beta_1)$ given in (5.4.12).

5.20 Modify the analysis given in Section 5.5 to obtain estimates for β_1 and β_2 in Model 5 when maximum likelihood estimation is used for

$$E(\epsilon_{ij}^2) = \sigma_{ij}^2$$

$$E(\epsilon_{ij}\epsilon_{kl}) = 0 \qquad \text{for} \quad i \neq k \text{ or } j \neq l$$

and whatever other standard assumptions are needed.

5.21 Consider MAP estimation for a random parameter for Model 5. Let the standard assumptions implied by 11011110 be valid. All the measurements are taken from the same batch. The random parameters β_1 and β_2 have the joint density

$$f(\beta_1, \beta_2) = \left[2\pi(\Delta_\beta)^{1/2}\right]^{-1} \exp\left(-\frac{1}{2} S_\beta\right), \qquad \Delta_\beta \equiv V_1 V_2 - V_{12}^2$$

$$S_\beta = \frac{V_2(\beta_1 - \mu_1)^2 + V_1(\beta_2 - \mu_2)^2 - 2V_{12}(\beta_1 - \mu_1)(\beta_2 - \mu_2)}{\Delta_\beta}$$

The quantities V_1, V_2, and Δ_β must be greater than zero.

(a) Derive for $i = 1$ and $j = 2$ or $i = 2$ and $j = 1$,

$$b_i = \mu_i + \frac{D_i[jj] - D_j[12]}{\Delta}; \qquad \Delta \equiv [11][22] - [12]^2$$

$$[11] \equiv C_{11} + V_2 \Delta_\beta^{-1}, \qquad [22] \equiv C_{22} + V_1 \Delta_\beta^{-1}, \qquad [12] \equiv C_{12} - V_{12} \Delta_\beta^{-1}$$

$$C_{rs} \equiv \sum_t Z_{tr} Z_{ts}, \qquad Z_{tr} \equiv \frac{X_{tr}}{\sigma_t}$$

$$D_r \equiv \sum_t (F_t - \mu_1 Z_{t1} - \mu_2 Z_{t2}) Z_{tr}, \qquad F_t \equiv \frac{Y_t}{\sigma_t}$$

(b) It can be shown that

$$V(b_1 - \beta_1) = \frac{[22]}{\Delta}, \qquad V(b_2 - \beta_2) = \frac{[11]}{\Delta}$$

and

$$\operatorname{cov}(b_1-\beta_1, b_2-\beta_2) = -\frac{[12]}{\Delta}$$

Show that $b_1-\beta_1$ can be put in the form

$$b_1-\beta_1 = \frac{\beta_1 A_1 + \beta_2 A_2 + \Sigma \varepsilon_t B_t}{\Delta} + c$$

where

$$A_1 = \frac{V_2[22] + V_{12}[12]}{\Delta_\beta}, \qquad A_2 = \frac{V_{12}[22] + V_1[12]}{\Delta_\beta}$$

$$B_t = (Z_{t1}[22] - Z_{t2}[12])\sigma_t^{-1}, c \text{ is not a random variable}$$

Derive the given expression for $V(b_1-\beta_1)$.

(*c*) The expressions given in (*a*) and (*b*) can also be applied for the case of subjective prior information. Reinterpret the meaning of β_1, β_2, V_1, V_2, V_{12}, b_1, b_2, $V(b_1-\beta_1)$, $V(b_2-\beta_2)$, and $\operatorname{cov}(b_1-\beta_1, b_2-\beta_2)$ for this case.

5.22 Before measuring the thermal conductivity of a particular steel alloy, a research engineer has developed from experience knowledge relative to values for steel alloys in general. The thermal conductivity over a limited range of temperature can be described by the regression model $\eta_i = \beta_1 + \beta_2 X_i$ where X_i is temperature in °C. This prior information regarding β_1 and β_2 can be described by $f(\beta_1, \beta_2)$ given by that in Problem 5.21 with $\mu_1 = 38$, $\mu_2 = -0.01$, $V_1 = 2$, $V_2 = 10^{-6}$, and $V_{12} = -0.001$. Assume that the standard assumptions designated 11011113 apply. Using the results of Problem 5.21 find b_1, b_2, $V(b_1-\beta_1)$, $V(b_2-\beta_2)$, and $\operatorname{cov}(b_1-\beta_1, b_2-\beta_2)$ for the following data:

i	X_i (°C)	Y_i (W/m-°C)	σ_i
1	100	36.3	0.2
2	200	36.3	0.3
3	300	34.6	0.5
4	400	32.9	0.7
5	600	31.2	1.0

5.23 Utilizing (5.13.8) derive for Model 2 ($\eta_i = \beta_1 X_i$) the following estimator for first-order autoregressive errors

$$b_1 = \frac{\Sigma Z_i F_i}{\Delta}, \qquad \Delta = \Sigma Z_j^2$$

$$Z_i = (X_i - \rho_i X_{i-1})\sigma_i^{-1}, \qquad i = 1, 2, \ldots, n$$

$$F_i = (Y_i - \rho_i Y_{i-1})\sigma_i^{-1}, \qquad i = 1, 2, \ldots, n$$

and where $X_0 \equiv 0$ and $Y_0 \equiv 0$.

5.24 Simplify the results of Problem 5.23 for the case of Model 1. Show that

$$b_0 = \frac{\sigma_1^{-2} Y_1 + \sum_2^n (1-\rho_i)(Y_i - \rho_i Y_{i-1})\sigma_i^{-2}}{\Delta}$$

$$V(b_0) = \Delta^{-1}, \qquad \Delta \equiv \sigma_1^{-2} + \sum_2^n (1-\rho_i)^2 \sigma_i^{-2}$$

5.25 For the case of first-order autoregressive errors show that the variance of ε_i given by (5.13.3) is the *constant* value

$$V(\varepsilon_i) = \sigma_\varepsilon^2 = \sigma_u^2 (1-\rho^2)^{-1} \qquad \text{for} \quad i \geq 1$$

when

$$\rho_i = \rho, \qquad \rho^2 < 1$$

$$\sigma_1^2 = \sigma_u^2 (1-\rho^2)^{-1}, \qquad \sigma_i^2 = \sigma_u^2 \qquad \text{for} \quad i = 2, \ldots, n$$

5.26 (*a*) Using the results of Problem 5.25 show that Δ^{-1} of Problem 5.24 can be written as

$$V(b_0) = \sigma_\varepsilon^2 \left[\frac{1+\rho}{2\rho + n(1-\rho)} \right]$$

(*b*) Suppose that as n becomes larger the measurements become more correlated as indicated by the expression $\rho = \exp(-a/n)$ where a is some positive constant characteristic of the errors. Show that

$$\lim_{n \to \infty} V(b_0) = \sigma_\varepsilon^2 \frac{2}{a+2}$$

for fixed σ_ε^2. What is the physical significance of this result?

(*c*) Modify the result of part (*a*) for fixed σ_ε^2 and ρ and large n. What is the physical significance of this result?

5.27 The following are actual data obtained for the thermal conductivity k of Pyrex. The temperature T (in K) is related to the voltage (in mVs) by $T = 301.6 + 18.24$ V.

Test	V_i (mV)	k_i (W/m-K)	Test	V_i	k_i
1	8.81	1.178	11	6.11	1.129
2	8.35	1.133	12	5.96	1.133
3	7.97	1.148	13	5.75	1.101
4	7.66	1.159	14	5.51	1.101
5	7.38	1.148	15	5.34	1.091
6	7.10	1.136	16	5.10	1.087
7	6.86	1.144	17	4.77	1.084
8	6.64	1.136	18	4.52	1.087
9	6.44	1.133	19	4.19	1.080
10	6.27	1.136	20	3.87	1.058

For the model $k_i = \beta_0 + \beta_1 T_i$ find the estimates b_0 and b_1. Also find est. s.e.(b_0), est. s.e.(b_1), s, and e_i. Let the standard assumptions be valid.

5.28 Modify the program for Problem 5.27 for variable σ_i^2. Let σ_i^2 be a table of input values. In particular, let $\sigma_i^2 = .01\, V_i$ and obtain new b_0 and b_1 values for the data of Problem 5.27.

5.29 The Moody chart provides the following data for the friction factor f_{DW} as a function of the Reynolds number Re for a roughness ratio $\varepsilon/D = 0.0001$. Fit these data to an equation of the form $f_{DW} = c + d(\text{Re})^m$ using the linear OLS method with c set equal to 0.0118. (Notice that f_{DW} is approaching a constant for large values of Re.)

Re	f_{DW}	Re	f_{DW}
5×10^3	0.0370	5×10^6	0.0123
1×10^4	0.0310	1×10^7	0.0121
5×10^4	0.0214	5×10^7	0.0120
1×10^5	0.0180	1×10^8	0.0120
5×10^5	0.0145		
1×10^6	0.0135		

Let $\log(f_{DW} - c)$ be the dependent variable and $\log \text{Re}$ be the independent variable. Calculate also the residuals in terms of $f_{DW} - \hat{f}_{DW}$ and the relative residuals, $(f_{DW} - \hat{f}_{DW})/\hat{f}_{DW}$.

5.30 The United States draft lottery issued in March 1975 gave the call order for the standby draft for men born in 1956. Results for birthday months of April and September are given below.

April

1-170	9-025	17-264	25-110
2-228	10-147	18-134	26-053
3-008	11-031	19-036	27-277
4-340	12-133	20-359	28-050
5-005	13-205	21-183	29-105
6-092	14-047	22-101	30-343
7-303	15-093	23-280	
8-180	16-131	24-080	

September

1-175	9-349	17-307	25-209
2-263	10-347	18-019	26-231
3-087	11-173	19-041	27-022
4-199	12-161	20-230	28-102
5-236	13-325	21-086	29-089
6-221	14-343	22-128	30-064
7-322	15-135	23-156	
8-341	16-117	24-227	

Use the model $\eta = \beta_0$.

(*a*) Which standard assumptions are valid?
(*b*) Estimate β_0 using OLS using the April data.
(*c*) Estimate σ^2 using the April data.

5.31 Repeat Problem 5.30*b* and *c* using the September data.

5.32 Using the April data in Problem 5.30, estimate β_0 and β_1 in the model $\eta_i = \beta_0 + \beta_1 X_i$ using OLS. Also estimate their standard errors.

CHAPTER 6

MATRIX ANALYSIS FOR LINEAR PARAMETER ESTIMATION

6.1 INTRODUCTION TO MATRIX NOTATION AND OPERATIONS

The extension of parameter estimation to more than two parameters is effectively accomplished through the use of matrices. The notation becomes more compact, facilitating manipulations, encouraging further insights, and permitting greater generality. This chapter develops matrix methods for linear parameter estimation and Chapter 7 considers the nonlinear case.

Linear estimation requires that the model be linear in the parameters. For linear maximum likelihood estimation, it is also necessary that the independent variables be errorless and that the covariances of the measurement errors be known to within the same multiplicative constant.

Before discussing various estimation procedures, this section presents various properties of matrices and matrix calculus that are used in both linear and nonlinear parameter estimation.

6.1.1 Elementary Matrix Operations

A *matrix* **Y** consisting of a single column is called a column *vector*. We use boldface letters to designate matrices. In display form $\mathbf{Y}$ is written as

$$\mathbf{Y}=\begin{bmatrix} Y_1 \\ Y_2 \\ \vdots \\ Y_n \end{bmatrix} \tag{6.1.1}$$

Square brackets are used to enclose the elements of a matrix. This matrix has n *elements* and is sometimes called an $n\times 1$ matrix, that is, n rows and one column.

An $m\times n$ *rectangular matrix* $\mathbf{A}$ is given by

$$\mathbf{A}=[a_{ij}]=\begin{bmatrix} a_{11} & a_{12} & \cdots & a_{1n} \\ a_{21} & a_{22} & \cdots & a_{2n} \\ \vdots & \vdots & & \vdots \\ a_{m1} & a_{m2} & \cdots & a_{mn} \end{bmatrix} \tag{6.1.2}$$

In the notation for an element, a_{ij}, the first subscript refers to the row and the second to the column. You may may find it helpful in memorizing this order by mentally visualizing $\downarrow\rightarrow$ with subscripts ij.

If $m=n$ in (6.1.2), the matrix $\mathbf{A}$ is said to be *square*. If the matrix is square and also $a_{ij}=a_{ji}$ for all i and j, the matrix is termed *symmetric*.

6.1.1.1 Product of Matrices

The product of $\mathbf{A}$ times $\mathbf{B}$, where $\mathbf{A}$ is an $m\times n$ matrix and $\mathbf{B}$ is $n\times s$, is an $m\times s$ matrix $\mathbf{C}$,

$$\mathbf{AB}=[a_{ik}][b_{kj}]=\left[\sum_{k=1}^{n} a_{ik}b_{kj}\right]=[c_{ij}]=\mathbf{C} \tag{6.1.3a}$$

Note that it is meaningful to form the product $\mathbf{A}$ times $\mathbf{B}$ only when the number of columns in $\mathbf{A}$ is equal to the number of rows in $\mathbf{B}$. A square matrix $\mathbf{A}$ is said to be *idempotent* if $\mathbf{AA}=\mathbf{A}$.

The triple matrix product of $\mathbf{ABC}$, where $\mathbf{A}$ is an $m\times n$ matrix, $\mathbf{B}$ is $n\times s$, and **C** is $s \times r$, is found using (6.1.3a) to be

$$\mathbf{ABC} = [a_{ik}][b_{kj}][c_{jl}] = [d_{il}] = \left[\sum_{k=1}^{n} \sum_{j=1}^{s} a_{ik} b_{kj} c_{jl} \right] = \mathbf{D} \qquad (6.1.3b)$$

where **D** is an $m \times r$ matrix.

Example 6.1.1

For the matrix **A** being 3×2 and **B** being 2×2, find the product **AB**.

Solution

Using (6.1.3a) we get

$$\mathbf{AB} = \begin{bmatrix} a_{11} & a_{12} \\ a_{21} & a_{22} \\ a_{31} & a_{32} \end{bmatrix} \begin{bmatrix} b_{11} & b_{12} \\ b_{21} & b_{22} \end{bmatrix} = \begin{bmatrix} a_{11}b_{11} + a_{12}b_{21} & a_{11}b_{12} + a_{12}b_{22} \\ a_{21}b_{11} + a_{22}b_{21} & a_{21}b_{12} + a_{22}b_{22} \\ a_{31}b_{11} + a_{32}b_{21} & a_{31}b_{12} + a_{32}b_{22} \end{bmatrix}$$

6.1.1.2 Transpose of Matrix

The transpose of an $m \times n$ matrix **A** is given by

$$\mathbf{A}^T = \begin{bmatrix} a_{11} & a_{21} & \cdots & a_{m1} \\ \vdots & \vdots & & \vdots \\ a_{1n} & a_{2n} & \cdots & a_{mn} \end{bmatrix} \qquad (6.1.4)$$

If **A** is a symmetric matrix, $\mathbf{A}^T = \mathbf{A}$. The transpose of the product **AB** is equal to

$$(\mathbf{AB})^T = \mathbf{B}^T \mathbf{A}^T \qquad (6.1.5a)$$

which can be extended to the product of any finite number of matrices,

$$(\mathbf{A} \cdots \mathbf{K})^T = \mathbf{K}^T \cdots \mathbf{A}^T \qquad (6.1.5b)$$

6.1.1.3 Inverse, Determinant, and Eigenvalues

For a square matrix, **A**, the notation $\mathbf{A}^{-1}$ indicates the *inverse* of **A**. A *nonsingular* matrix is a *square* matrix whose *determinant* is *not zero*.

(Examples of determinants are given below.) It can be shown that if and only if the matrix **A** is nonsingular, it possesses an inverse $\mathbf{A}^{-1}$, that is

$$\mathbf{A}^{-1}\mathbf{A}=\mathbf{A}\mathbf{A}^{-1}=\mathbf{I} \tag{6.1.6}$$

where **I** is a *diagonal* matrix and is called the *unit* or *identity* matrix. **I** is given by

$$\mathbf{I}=\begin{bmatrix} 1 & 0 & 0 & & \\ 0 & 1 & 0 & & \\ 0 & 0 & 1 & & \\ & & & \ddots & \\ & & & & 1 \end{bmatrix}=\operatorname{diag}[1\ 1\ \cdots\ 1] \tag{6.1.7}$$

Notice the notation "diag" in (6.1.7), which indicates the elements of a diagonal matrix.

Let $\boldsymbol{\phi}$ be a diagonal matrix,

$$\boldsymbol{\phi}=\operatorname{diag}[\phi_{11}\ \phi_{22}\ \cdots\ \phi_{nn}] \tag{6.1.8}$$

Its inverse and determinant are given by

$$\boldsymbol{\phi}^{-1}=\operatorname{diag}[\phi_{11}^{-1}\ \phi_{22}^{-1}\ \cdots\ \phi_{nn}^{-1}] \tag{6.1.9a}$$

$$|\boldsymbol{\phi}|=\phi_{11}\phi_{22}\cdots\phi_{nn} \tag{6.1.9b}$$

Evidently diagonal matrices have some very convenient mathematical properties.

A 2×2 matrix **A** and its inverse are

$$\mathbf{A}=\begin{bmatrix} a_{11} & a_{12} \\ a_{21} & a_{22} \end{bmatrix}; \qquad \mathbf{A}^{-1}=\frac{1}{|\mathbf{A}|}\begin{bmatrix} a_{22} & -a_{12} \\ -a_{21} & a_{11} \end{bmatrix} \tag{6.1.10a}$$

where $|\mathbf{A}|$ is the determinant of **A**,

$$|\mathbf{A}|=a_{11}a_{22}-a_{12}a_{21} \tag{6.1.10b}$$

The inverse of the *symmetric* 3×3 matrix **A** is

$$\mathbf{A}^{-1} = \begin{bmatrix} a_{11} & a_{12} & a_{13} \\ a_{12} & a_{22} & a_{23} \\ a_{13} & a_{23} & a_{33} \end{bmatrix}^{-1}$$

$$= \frac{1}{|\mathbf{A}|} \begin{bmatrix} a_{22}a_{33} - a_{23}^2 & a_{13}a_{23} - a_{12}a_{33} & a_{12}a_{23} - a_{13}a_{22} \\ & a_{11}a_{33} - a_{13}^2 & a_{12}a_{13} - a_{11}a_{23} \\ \text{symmetric} & & a_{11}a_{22} - a_{12}^2 \end{bmatrix} \quad (6.1.11a)$$

$$|\mathbf{A}| = a_{11}a_{22}a_{33} + 2a_{12}a_{13}a_{23} - \left(a_{11}a_{23}^2 + a_{22}a_{13}^2 + a_{33}a_{12}^2\right) \quad (6.1.11b)$$

Clearly nondiagonal matrices are much more difficult to invert than diagonal matrices. For the inverses to exist the determinants given by (6.1.9b), (6.1.10b), and (6.1.11b) cannot be equal to zero. For a method for evaluating higher-order determinants, see Hildebrand [1]. See Problem 6.4 for a method for finding inverses of larger matrices.

The determinant of the product of two square matrices is

$$|\mathbf{AB}| = |\mathbf{A}|\,|\mathbf{B}| \quad (6.1.12a)$$

If (6.1.12a) is applied to (6.1.6), the reciprocal of $|\mathbf{A}^{-1}|$ is found to be equal to $|\mathbf{A}|$,

$$\frac{1}{|\mathbf{A}^{-1}|} = |\mathbf{A}| \quad (6.1.12b)$$

Another convenient relation involving products of two square $n \times n$ matrices **A** and **B** is

$$(\mathbf{AB})^{-1} = \mathbf{B}^{-1}\mathbf{A}^{-1} \quad (6.1.13)$$

which is a relation similar to (6.1.5) for transposes.

For any square matrix **A** it is also true that

$$(\mathbf{A}^{-1})^T = (\mathbf{A}^T)^{-1} \quad (6.1.14)$$

or in words, the transpose of an inverse is equal to the inverse of the transpose.

An eigenvalue (also called characteristic value or latent root) λ_i of a square matrix $\mathbf{A}$ is obtained by solving the equation

$$|\mathbf{A}-\lambda_i\mathbf{I}|=0 \tag{6.1.15}$$

There are n eigenvalues if $\mathbf{A}$ is $n\times n$ (counting repeated values).

6.1.1.4 Partitioned Matrix

Let $\mathbf{B}$ be an $n\times n$ matrix that is *partitioned* as follows

$$\mathbf{B}=\begin{bmatrix}\mathbf{B}_{11} & \mathbf{B}_{12}\\ \mathbf{B}_{21} & \mathbf{B}_{22}\end{bmatrix} \tag{6.1.16a}$$

where $\mathbf{B}_{ij}$ has size $n_i\times n_j, i,j=1,2$, and where $n_1+n_2=n$. Suppose that $|\mathbf{B}|\neq 0$, $|\mathbf{B}_{11}|\neq 0$, and $|\mathbf{B}_{22}|\neq 0$. Set $\mathbf{A}=\mathbf{B}^{-1}$ and partition $\mathbf{A}$ as

$$\mathbf{B}^{-1}=\mathbf{A}=\begin{bmatrix}\mathbf{A}_{11} & \mathbf{A}_{12}\\ \mathbf{A}_{21} & \mathbf{A}_{22}\end{bmatrix} \tag{6.1.16b}$$

where $\mathbf{A}_{ij}$ has size $n_i\times n_j$ for $i,j=1,2$. The components $\mathbf{A}_{ij}$ are

$$\mathbf{A}_{11}=\left[\mathbf{B}_{11}-\mathbf{B}_{12}\mathbf{B}_{22}^{-1}\mathbf{B}_{21}\right]^{-1}=\mathbf{B}_{11}^{-1}+\mathbf{B}_{11}^{-1}\mathbf{B}_{12}\mathbf{A}_{22}\mathbf{B}_{21}\mathbf{B}_{11}^{-1} \tag{6.1.17a}$$

$$\mathbf{A}_{12}=-\mathbf{B}_{11}^{-1}\mathbf{B}_{12}\left[\mathbf{B}_{22}-\mathbf{B}_{21}\mathbf{B}_{11}^{-1}\mathbf{B}_{12}\right]^{-1}=-\mathbf{B}_{11}^{-1}\mathbf{B}_{12}\mathbf{A}_{22} \tag{6.1.17b}$$

$$\mathbf{A}_{22}=\left[\mathbf{B}_{22}-\mathbf{B}_{21}\mathbf{B}_{11}^{-1}\mathbf{B}_{12}\right]^{-1}=\mathbf{B}_{22}^{-1}+\mathbf{B}_{22}^{-1}\mathbf{B}_{21}\mathbf{A}_{11}\mathbf{B}_{12}\mathbf{B}_{22}^{-1} \tag{6.1.17c}$$

$$\mathbf{A}_{21}=-\mathbf{B}_{22}^{-1}\mathbf{B}_{21}\left[\mathbf{B}_{11}-\mathbf{B}_{12}\mathbf{B}_{22}^{-1}\mathbf{B}_{21}\right]^{-1}=-\mathbf{B}_{22}^{-1}\mathbf{B}_{21}\mathbf{A}_{11} \tag{6.1.17d}$$

The determinant of $\mathbf{B}$ is

$$|\mathbf{B}|=|\mathbf{B}_{22}|\,|\mathbf{B}_{11}-\mathbf{B}_{12}\mathbf{B}_{22}^{-1}\mathbf{B}_{21}|=\frac{|\mathbf{B}_{22}|}{|\mathbf{A}_{11}|} \tag{6.1.17e}$$

If $\mathbf{B}_{12}$ and $\mathbf{B}_{21}$ are *null* matrices (only zero components), then the inverse of $\mathbf{B}$ is more simply given by

$$\mathbf{B}^{-1}=\operatorname{diag}\left[\mathbf{B}_{11}^{-1}\mathbf{B}_{22}^{-1}\right] \tag{6.1.17f}$$

6.1.1.5 Positive Definite Matrices

Necessary and sufficient conditions for the square symmetric matrix $\mathbf{A}$ to be *positive definite* are that the inequalities,

$$|a_{11}|>0, \qquad \begin{vmatrix} a_{11} & a_{12} \\ a_{12} & a_{22} \end{vmatrix}>0, \qquad \begin{vmatrix} a_{11} & a_{12} & a_{13} \\ a_{12} & a_{22} & a_{23} \\ a_{13} & a_{23} & a_{33} \end{vmatrix}>0, \qquad \text{etc.} \tag{6.1.18}$$

be satisfied. Also A is positive definite (semidefinite) if and only if all of its eigenvalues are positive (nonnegative, i.e., $\lambda_1 \geqslant 0$ with at least one equal to zero). If $-A$ is positive definite (semidefinite), A is negative definite (semidefinite).

6.1.1.6 Trace

The *trace* of a square matrix $\mathbf{A}$ is the sum of the diagonal elements and is also equal to the sum of the eigenvalues λ_i,

$$\text{tr}(\mathbf{A}) = \sum_{i=1}^{n} a_{ii} = \sum_{i=1}^{n} \lambda_i \tag{6.1.19}$$

It can also be shown that the determinant of $\mathbf{A}$ is equal to the product of the eigenvalues,

$$|\mathbf{A}| = \prod_{i=1}^{n} \lambda_i \tag{6.1.20}$$

6.1.2 Matrix Calculus

In parameter estimation the first derivative with respect to the elements of a p vector of parameters, $\boldsymbol{\beta}$, is frequently needed. Let us define the operator $\boldsymbol{\nabla}_\beta$ by

$$\boldsymbol{\nabla}_\beta \equiv \begin{bmatrix} \dfrac{\partial}{\partial \beta_1} \\ \vdots \\ \dfrac{\partial}{\partial \beta_p} \end{bmatrix} \tag{6.1.21}$$

Because $\boldsymbol{\nabla}_\beta$ is $p \times 1$ it must be applied to the transpose of a column vector or to a row vector. Let $\mathbf{C}$ be an $n \times 1$ column vector which is a function of $\boldsymbol{\beta}$.

Then we obtain

$$\nabla_\beta \mathbf{C}^T = \begin{bmatrix} \dfrac{\partial}{\partial \beta_1} \\ \vdots \\ \dfrac{\partial}{\partial \beta_p} \end{bmatrix} \begin{bmatrix} c_1 \cdots c_n \end{bmatrix} = \begin{bmatrix} \dfrac{\partial c_1}{\partial \beta_1} & \dfrac{\partial c_2}{\partial \beta_1} & \cdots & \dfrac{\partial c_n}{\partial \beta_1} \\ \vdots & \vdots & & \vdots \\ \dfrac{\partial c_1}{\partial \beta_p} & \dfrac{\partial c_2}{\partial \beta_p} & \cdots & \dfrac{\partial c_n}{\partial \beta_p} \end{bmatrix} \tag{6.1.22}$$

which is a $p \times n$ matrix and whose determinant is called the *Jacobian*.

Many times the matrix derivative is applied to a scalar. Consider ∇_β operating upon the matrix product $\mathbf{AB}$ where $\mathbf{A}$ is $1 \times n$ and $\mathbf{B}$ is $n \times 1$; then

$$\nabla_\beta^{[p\times 1]}(\mathbf{AB})^{[1\times 1]} = \left[\frac{\partial}{\partial \beta_k}\right]\left(\sum_{i=1}^{n} a_i b_i\right) = \left[\sum_{i=1}^{n}\left(\frac{\partial a_i}{\partial \beta_k} b_i + a_i \frac{\partial b_i}{\partial \beta_k}\right)\right]$$

$$= \left(\nabla_\beta^{[p\times 1]}\mathbf{A}^{[1\times n]}\right)\mathbf{B}^{[n\times 1]} + \left(\nabla_\beta^{[p\times 1]}\mathbf{B}^{T[1\times n]}\right)\mathbf{A}^{T[n\times 1]} \tag{6.1.23}$$

The sizes of the matrices are indicated by the superscripts in brackets. One application of (6.1.23) is for $\mathbf{A}=\mathbf{C}^T$, $\mathbf{B}=\boldsymbol{\beta}$, and $\mathbf{C}$ a $p \times 1$ column vector whose elements are not functions of the β's. Using (6.1.23) gives

$$\boxed{\nabla_\beta^{[p\times 1]}\mathbf{C}^{T[1\times p]}\boldsymbol{\beta}^{[p\times 1]} = \left(\nabla_\beta \boldsymbol{\beta}^T\right)^{[p\times p]}\mathbf{C}^{[p\times 1]} = \mathbf{C}^{[p\times 1]}} \tag{6.1.24}$$

since

$$\nabla_\beta \boldsymbol{\beta}^T = \mathbf{I} \tag{6.1.25}$$

Another important application of the matrix derivative is ∇_β operating on the product $\mathbf{AB}$ where $\mathbf{A}$ is a $1 \times n$ matrix that is a function of $\boldsymbol{\beta}$ and $\mathbf{B}$ is an $n \times m$ matrix which is not a function of $\boldsymbol{\beta}$. In the same manner that (6.1.23) was derived we obtain

$$\nabla_\beta^{[p\times 1]}\mathbf{A}^{[1\times n]}\mathbf{B}^{[n\times m]} = \left(\nabla_\beta \mathbf{A}\right)^{[p\times n]}\mathbf{B}^{[n\times m]} \tag{6.1.26a}$$

If $\mathbf{A}=\boldsymbol{\beta}^T$ and $\mathbf{B}$ has the size $p \times m$, (6.1.26a) then yields

$$\boxed{\nabla_\beta^{[p\times 1]}\boldsymbol{\beta}^{T[1\times p]}\mathbf{B}^{[p\times m]} = \mathbf{B}^{[p\times m]}} \tag{6.1.26b}$$

6.1.3 Quadratic Form

In estimation sum of squares functions are frequently encountered. These functions can be considered to be *quadratic forms* Q,

$$Q = \mathbf{A}^{T[1\times n]}\mathbf{\Phi}^{[n\times n]}\mathbf{A}^{[n\times 1]} = \sum_{i=1}^{n}\sum_{j=1}^{n} a_i\phi_{ij}a_j \tag{6.1.27a}$$

$$= a_1^2\phi_{11} + \cdots + a_n^2\phi_{nn} + 2a_1a_2\phi_{12} + 2a_1a_3\phi_{13} + \cdots \tag{6.1.27b}$$

where $\mathbf{\Phi}$ is a symmetric positive semidefinite matrix. Q is termed positive definite if the conditions given in (6.1.18) are satisfied by $\mathbf{\Phi}$ [1, pp. 48–52]; for these conditions Q is always nonnegative for all real values of the variable a_i. In some analytical considerations the equivalent expression for Q,

$$Q = \operatorname{tr}(\mathbf{\Phi}\mathbf{A}\mathbf{A}^T) \tag{6.1.28}$$

is more convenient than (6.1.27).

Applying the operator ∇_β to Q defined by (6.1.27), assuming that $\mathbf{A}$ is a function of $\boldsymbol{\beta}$ and $\mathbf{\Phi}$ is not, gives

$$\nabla_\beta Q = \nabla_\beta(\mathbf{A}^T\mathbf{\Phi}\mathbf{A}) = (\nabla_\beta\mathbf{A}^T)\mathbf{\Phi}\mathbf{A} + \left[\nabla_\beta(\mathbf{A}^T\mathbf{\Phi})\right]\mathbf{A} \tag{6.1.29}$$

$$\boxed{\nabla_\beta Q = 2(\nabla_\beta\mathbf{A}^T)\mathbf{\Phi}\mathbf{A}} \tag{6.1.30}$$

where (6.1.23) and (6.1.26a) are used.

In linear estimation, $\mathbf{A}$ in the quadratic form is given by

$$\mathbf{A}^{[n\times 1]} = \mathbf{X}^{[n\times p]}\boldsymbol{\beta}^{[p\times 1]} \tag{6.1.31}$$

where $\mathbf{X}$ is independent of $\boldsymbol{\beta}$. Then substituting (6.1.31) into (6.1.30) gives

$$\nabla_\beta Q = 2(\nabla_\beta\boldsymbol{\beta}^T\mathbf{X}^T)\mathbf{\Phi}\mathbf{X}\boldsymbol{\beta} \tag{6.1.32}$$

$$\boxed{\nabla_\beta Q = 2\mathbf{X}^T\mathbf{\Phi}\mathbf{X}\boldsymbol{\beta}}\,; \qquad Q \equiv (\mathbf{X}\boldsymbol{\beta})^T\mathbf{\Phi}\mathbf{X}\boldsymbol{\beta} \tag{6.1.33}$$

where (6.1.26b) is used.

6.1.4 Expected Value of Matrix and Variance–Covariance Matrix

6.1.4.1 Expected Value Matrix

The integral of a matrix is given by

$$\int \mathbf{A}^{[m\times n]}(t)\,dt = \begin{bmatrix} \int a_{11}(t)\,dt & \cdots & \int a_{1n}(t)\,dt \\ \vdots & & \vdots \\ \int a_{m1}(t)\,dt & \cdots & \int a_{mn}(t)\,dt \end{bmatrix} \tag{6.1.34}$$

This result can be used to find the *expected value* of a column vector $\mathbf{Y}^{[n\times 1]}$,

$$E(\mathbf{Y}^{[n\times 1]}) = \begin{bmatrix} E(Y_1) \\ \vdots \\ E(Y_n) \end{bmatrix} = \begin{bmatrix} \int Y_1 f(Y_1)\,dY_1 \\ \vdots \\ \int Y_n f(Y_n)\,dY_n \end{bmatrix} \tag{6.1.35}$$

where $f(Y_i)$ is the probability density of Y_i.

6.1.4.2 Variance–Covariance Matrix

An important application of the expected value of a matrix is the *variance–covariance matrix* of a random column vector $\mathbf{Y}$ which is

$$\operatorname{cov}(\mathbf{Y}^{[n\times 1]}) = E\left\{[\mathbf{Y}-E(\mathbf{Y})][\mathbf{Y}-E(\mathbf{Y})]^T\right\}$$

$$= E\left\{\begin{bmatrix} Y_1 - E(Y_1) \\ \vdots \\ Y_n - E(Y_n) \end{bmatrix} [Y_1 - E(Y_1) \ \cdots \ Y_n - E(Y_n)]\right\}$$

$$= \begin{bmatrix} \operatorname{cov}(Y_1,Y_1) & \operatorname{cov}(Y_1,Y_2) & \cdots & \operatorname{cov}(Y_1,Y_n) \\ \operatorname{cov}(Y_1,Y_2) & \operatorname{cov}(Y_2,Y_2) & \cdots & \operatorname{cov}(Y_2,Y_n) \\ \vdots & \vdots & & \vdots \\ \operatorname{cov}(Y_1,Y_n) & \operatorname{cov}(Y_2,Y_n) & \cdots & \operatorname{cov}(Y_n,Y_n) \end{bmatrix} \tag{6.1.36}$$

Notice that this is a symmetric $n\times n$ matrix.

An important special case in connection with the covariance matrix (sometimes this term is used rather than variance–covariance) is for $\mathbf{Y}$ containing additive zero mean errors $\boldsymbol{\varepsilon}$, or

$$\mathbf{Y} = \boldsymbol{\eta} + \boldsymbol{\varepsilon}; \qquad E(\boldsymbol{\varepsilon}) = \mathbf{0} \quad \text{and} \quad \text{cov}(\boldsymbol{\eta}) = \mathbf{0} \tag{6.1.37}$$

Let $\boldsymbol{\psi}$ be defined as the covariance matrix of $\mathbf{Y}$. For the conditions given in (6.1.37), $\boldsymbol{\psi}$ is also equal to cov($\boldsymbol{\varepsilon}$) or

$$\boldsymbol{\psi} \equiv \text{cov}(\mathbf{Y}) = \text{cov}(\boldsymbol{\varepsilon}) \tag{6.1.38a}$$

where $\boldsymbol{\psi}$ is in detail

$$\boldsymbol{\psi} = \begin{bmatrix} E(\varepsilon_1^2) & E(\varepsilon_1\varepsilon_2) & \cdots & E(\varepsilon_1\varepsilon_n) \\ \vdots & \vdots & & \vdots \\ E(\varepsilon_1\varepsilon_n) & E(\varepsilon_2\varepsilon_n) & \cdots & E(\varepsilon_n^2) \end{bmatrix}$$

$$= \begin{bmatrix} \sigma_1^2 & E(\varepsilon_1\varepsilon_2) & \cdots & E(\varepsilon_1\varepsilon_n) \\ & & & \vdots \\ \text{symmetric} & & & \sigma_n^2 \end{bmatrix} \tag{6.1.38b}$$

since $\sigma_i^2 = V(\varepsilon_i) = E(\varepsilon_i^2)$. In many estimation methods, $\boldsymbol{\psi}$ given by (6.1.38b) plays an important role.

6.1.4.3 *Covariance of Linear Combination of Vector Random Variables*

Let $\mathbf{z}^{[m\times 1]}$ be linear in the random vector $\boldsymbol{\varepsilon}^{[n\times 1]}$ or

$$\mathbf{z}^{[m\times 1]} = \mathbf{G}^{[m\times n]}\boldsymbol{\varepsilon}^{[n\times 1]} \tag{6.1.39}$$

The matrix $\mathbf{G}$ is *not* a random matrix. It can be proved that the covariance matrix of $\mathbf{z}$ is given by

$$\text{cov}(\mathbf{z}) = E(\mathbf{z}\mathbf{z}^T) = E[\mathbf{G}\boldsymbol{\varepsilon}\boldsymbol{\varepsilon}^T\mathbf{G}^T] = \mathbf{G}\boldsymbol{\psi}\mathbf{G}^T \tag{6.1.40}$$

where $\boldsymbol{\psi} = \text{cov}(\boldsymbol{\varepsilon})$.

A derivation of (6.1.40) is given next. Let

$$\mathbf{H}^{[n\times n]} = \boldsymbol{\varepsilon}\boldsymbol{\varepsilon}^T \tag{6.1.41}$$

and then $E(\mathbf{zz}^T)$ can be written as

$$E(\mathbf{zz}^T) = E(\mathbf{GHG}^T) = E\left\{\left[\sum_{j=1}^{m}\sum_{k=1}^{m} g_{ij}h_{jk}g_{lk}\right]\right\} \tag{6.1.42}$$

where $i, l = 1, 2, \ldots, n$ and where we used

$$\mathbf{G} = [\, g_{ij}\,], \qquad \mathbf{H} = [\, h_{ij}\,], \qquad \mathbf{G}^T = [\, g_{ji}\,] \tag{6.1.43}$$

and (6.1.3b). Using the linearity property of the expected value operator,

$$E(a\varepsilon_1 + b\varepsilon_2) = aE(\varepsilon_1) + bE(\varepsilon_2) \tag{6.1.44}$$

permits (6.1.42) to be written as

$$E(\mathbf{zz}^T) = \left[\sum\sum g_{ij}E(h_{jk})\, g_{lk}\right] = \mathbf{G}E(\mathbf{H})\mathbf{G}^T \tag{6.1.45}$$

Since $\boldsymbol{\psi} = E(\mathbf{H})$, we have proved (6.1.40).

A more general form of (6.1.40) which can also be proved is

$$\operatorname{cov}(\mathbf{z}) = \mathbf{G}\operatorname{cov}(\boldsymbol{\varepsilon})\mathbf{G}^T \tag{6.1.46}$$

where the expected values of the components of $\boldsymbol{\varepsilon}$ need not be zero [2].

6.1.4.4 Expected Value of a Quadratic Form

The quadratic form given by (6.1.27) and (6.1.28) is

$$Q = \mathbf{A}^T\boldsymbol{\Phi}\mathbf{A} = \operatorname{tr}(\boldsymbol{\Phi}\mathbf{AA}^T) \tag{6.1.47}$$

Let $\mathbf{A}$ be an $[n \times 1]$ random vector but $\boldsymbol{\Phi}$ be not a random matrix. Using the expanded form of a matrix product given by (6.1.3b) and the linearity property of the expected value operator indicated by (6.1.44), it can be shown that

$$E(Q) = \operatorname{tr}\{\boldsymbol{\Phi}E(\mathbf{AA}^T)\} \tag{6.1.48}$$

In general the covariance matrix of $\mathbf{A}$ is equal to

$$\operatorname{cov}(\mathbf{A}) = E(\mathbf{AA}^T) - E(\mathbf{A})E(\mathbf{A}^T) \tag{6.1.49}$$

Using this equation in (6.1.48) and also using (6.1.47) yields

$$E(Q) = E(\mathbf{A}^T)\boldsymbol{\Phi}E(\mathbf{A}) + \operatorname{tr}\{\boldsymbol{\Phi}\operatorname{cov}(\mathbf{A})\} \tag{6.1.50}$$

This expression can be used to obtain an estimator for σ^2 for certain cases.

6.1.5 Model in Matrix Terms

Certain conditions regarding the model and assumptions can be expressed more generally using matrix rather than algebraic notation. Let the dependent variable $\boldsymbol{\eta}$ be given by the linear-in-parameters model

$$\boldsymbol{\eta} = \mathbf{X}\boldsymbol{\beta} \tag{6.1.51}$$

where for the same dependent variable measured n times, the dimensions of $\boldsymbol{\eta}$ are $[n \times 1]$, and those of $\mathbf{X}$ are $[n \times p]$. We term $\mathbf{X}$ the sensitivity matrix; it is a matrix of derivatives with respect to the parameters. The parameter vector $\boldsymbol{\beta}$ is $[p \times 1]$. Thus we have

$$\boldsymbol{\eta} = \begin{bmatrix} \eta_1 \\ \eta_2 \\ \vdots \\ \eta_n \end{bmatrix}, \quad \mathbf{X} = \begin{bmatrix} X_{11} & X_{12} & \cdots & X_{1p} \\ X_{21} & X_{22} & \cdots & X_{2p} \\ \vdots & \vdots & & \vdots \\ X_{n1} & X_{n2} & \cdots & X_{np} \end{bmatrix}, \quad \boldsymbol{\beta} = \begin{bmatrix} \beta_1 \\ \beta_2 \\ \vdots \\ \beta_p \end{bmatrix} \tag{6.1.52}$$

Equation 6.1.51 includes all the models given in Chapter 5. For example, for Model 5 ($\eta_i = \beta_1 X_{i1} + \beta_2 X_{i2}$) the $\mathbf{X}$ matrix is

$$\mathbf{X} = \begin{bmatrix} X_{11} & X_{12} \\ \vdots & \vdots \\ X_{n1} & X_{n2} \end{bmatrix} \tag{6.1.53}$$

It is possible that the terms X_{ij} could represent different functions of the same variable such as t_i. Some examples are

$$X_{i1} = 1, \qquad X_{i2} = t_i, \qquad X_{i3} = t_i^2, \qquad X_{i4} = t_i^3$$

$$X_{i1} = t_i, \qquad X_{i2} = e^{-5t_i}, \qquad X_{i3} = \sin 6t_i$$

In other cases X_{ij} might be composed of different coordinates such as time t_i and position z_i,

$$X_{i1} = t_i, \qquad X_{i2} = z_i$$

Similar to the last case $\boldsymbol{\eta}$ might represent the output of a chemical process which might be a linear function of air flow rate q_i, cooling water inlet temperature T_i, and acid concentration C_i; here if there is a constant term we have

$$X_{i1}=1, \qquad X_{i2}=q_i, \qquad X_{i3}=T_i, \qquad \text{and} \quad X_{i4}=C_i$$

This same example involving these independent variables is given in Section 13.12 of Brownlee [3], Chapter 6 of Draper and Smith [4], and Chapter 5 of Daniel and Wood [5].

Another situation that can be represented by $\boldsymbol{\eta}$ in (6.1.51) is called the multiresponse case in the statistical literature. In this case $\boldsymbol{\eta}$ and $\mathbf{X}$ may be given by

$$\boldsymbol{\eta}=\begin{bmatrix}\boldsymbol{\eta}(1)\\ \boldsymbol{\eta}(2)\\ \vdots\\ \boldsymbol{\eta}(n)\end{bmatrix} \qquad \text{where } \boldsymbol{\eta}(i)=\begin{bmatrix}\eta_1(i)\\ \eta_2(i)\\ \vdots\\ \eta_m(i)\end{bmatrix} \tag{6.1.54}$$

$$\mathbf{X}=\begin{bmatrix}\mathbf{X}(1)\\ \mathbf{X}(2)\\ \vdots\\ \mathbf{X}(n)\end{bmatrix} \qquad \text{where } \mathbf{X}(i)=\begin{bmatrix}X_{11}(i) & \cdots & X_{1p}(i)\\ X_{21}(i) & \cdots & X_{2p}(i)\\ \vdots & & \vdots\\ X_{m1}(i) & \cdots & X_{mp}(i)\end{bmatrix} \tag{6.1.55}$$

Note that the variable $X_{jk}(i)$, which is called a sensitivity coefficient, is for the jth dependent variable in η, for the kth parameter, and at the ith time. Also observe that

$$X_{jk}(i)=\frac{\partial\eta_j(i)}{\partial\beta_k}; \qquad i=1,\ldots,n; \quad j=1,\ldots,m \tag{6.1.56}$$

In the above definitions of $\boldsymbol{\eta}$ and $\mathbf{X}$ in (6.1.54) and (6.1.55) there is a total of n "times" and m dependent variables resulting in $\boldsymbol{\eta}$ being a $[mn\times 1]$ vector and $\mathbf{X}$ having dimensions of $[mn\times p]$. If $m=1$, then the j subscripts in (6.1.54)–(6.1.56) are dropped and replaced by i, or $X_{1k}(i)=X_{ik}$ and $\eta_1(i)=\eta_i$.

Multiresponse cases commonly arise in parameter estimation problems involving ordinary and partial differential equations. An example involving ordinary differential equations occurs in chemical engineering when the concentrations of several components present in a chemical reaction are measured as a function of time. To be more precise, let the concentrations of components A and B be designated $C_A(t_i)$ and $C_B(t_i)$ and then we would set

$$\eta_1(i)=C_A(t_i), \qquad \eta_2(i)=C_B(t_i)$$

Unfortunately in most such cases the model is nonlinear in terms of the parameters which are usually rate constants in the reaction problem just mentioned. For didactic purposes, however, let $C_A(t_i)$ and $C_B(t_i)$ be given by

$$C_A(t_i)=\beta_1 f_1(t_i)+\beta_2 f_2(t_i)$$

$$C_B(t_i)=\beta_1 f_3(t_i)+\beta_3 f_4(t_i)$$

where the $f_j(t_i)$ are known functions. Then the sensitivity coefficients are

$$\mathbf{X}(i)=\begin{bmatrix} f_1(t_i) & f_2(t_i) & 0 \\ f_3(t_i) & 0 & f_4(t_i) \end{bmatrix}$$

An example involving the partial differential equation describing heat conduction is heat transfer in a plate of thickness L with temperatures measured at several locations as a function of time. Suppose that the observations at m positions x_j have been made at times $t_1, t_2, \ldots, t_n$. The temperatures, which would be the dependent variables, are designated $T(x_j, t_i)$ and then

$$\eta_j(i)=T(x_j,t_i) \qquad \text{for } i=1,\ldots,n; \quad j=1,\ldots,m$$

For the boundary conditions of a known constant heat flux of q_0 at $x=0$ and no heat flow at $x=L$, the temperature distribution for large times is given by

$$T(x_j,t_i)=T_0+\beta_1 f_1(i)+\beta_2 f_2(j)$$

where

$$f_1(i)=\frac{q_0 t_i}{L}, \qquad f_2(j)=q_0 L\left[\frac{1}{3}-\frac{x_j}{L}+\frac{1}{2}\left(\frac{x_j}{L}\right)^2\right]$$

$$\beta_1=(\rho c)^{-1}, \qquad \beta_2=k^{-1}$$

The quantity ρ is density, c is specific heat, and k is thermal conductivity. The initial value (i.e., at $t=0$) of T is T_0, which is known. If there are measurements at two locations (such as at $x=0$ and $x=L$), $m=2$ and the $\eta_j(i)$ values are

$$\eta_j(i)=T_0+\beta_1 f_1(i)+\beta_2 f_2(j), \qquad j=1,2$$

and the sensitivity coefficients for $i=1,\ldots,n$ are

$$X_{11}(i)=f_1(i), \qquad X_{12}(i)=f_2(1)$$

$$X_{21}(i)=f_1(i), \qquad X_{22}(i)=f_2(2)$$

Actually this case could also be considered a single response but multiple sensors are required for $m>1$. Many other examples can be generated.

6.1.5.1 *Identifiability Condition*

In order to *uniquely* estimate *all* p parameters in the $\boldsymbol{\beta}$ vector it is necessary that

$$|\mathbf{X}^T\mathbf{X}|\neq 0 \qquad (6.1.57)$$

for estimation problems involving least squares or maximum likelihood. This criterion is analogous to those given by (5.1.2) for some simple models. When maximum a posteriori estimation is used, it may not be necessary that (6.1.57) be true. See Appendix A.

6.1.5.2 *Assumptions*

The *standard assumptions* given at the end of Section 5.1 are repeated and amplified below. In making this list all cases considered in this text are included but doubtless many other possibilities exist that are not explicitly covered below.

1. $\mathbf{Y}=E(\mathbf{Y}|\boldsymbol{\beta})+\boldsymbol{\varepsilon}=\boldsymbol{\eta}(\mathbf{X},\boldsymbol{\beta})+\boldsymbol{\varepsilon}$ (additive errors in measurements) (6.1.58)
- 0 No, *measurement* errors are not additive.
- 1 Yes, measurement errors are additive. Also the regression function $\boldsymbol{\eta}$ is correct and does not involve any random variables.

2. $E(\boldsymbol{\varepsilon})=\mathbf{0}$ (zero mean measurement errors) (6.1.59)
- 0 No, the errors in $\mathbf{Y}$ do not have zero means. In other words, the errors are biased.
- 1 Yes, the errors in $\mathbf{Y}$ have zero mean.

3. $V(Y_i|\boldsymbol{\beta})=\sigma^2$ (constant variance errors) (6.1.60)
- 0 No, $V(Y_i|\boldsymbol{\beta})\neq$ constant; that is, $V(Y_i|\boldsymbol{\beta})=\sigma_i^2$.
- 1 Yes, the errors ε_i, $i=1,\ldots,n$ have a common variance σ^2.

4. $E\{[\varepsilon_i-E(\varepsilon_i)][\varepsilon_j-E(\varepsilon_j)]\}=0$ for $i\neq j$ (uncorrelated errors) (6.1.61)
- 0 No, the errors ε_i are correlated.
- 1 Yes, the errors ε_i are uncorrelated.

2 No, the errors are described by an autoregressive process. See Section 6.9

5. $\boldsymbol{\varepsilon}$ has a normal distribution (normality) (6.1.62)

0 No, the distribution is not normal.

1 Yes, the distribution is normal.

6. Known statistical parameters describing $\boldsymbol{\varepsilon}$ (6.1.63)

0 No, $\boldsymbol{\psi} \equiv \text{cov}(\boldsymbol{\varepsilon})$ is known only to within an arbitrary multiplicative constant, or $\boldsymbol{\psi} = \sigma^2\boldsymbol{\Omega}$ where σ^2 is unknown and $\boldsymbol{\Omega}$ is known.

1 Yes, $\boldsymbol{\psi} \equiv \text{cov}(\boldsymbol{\varepsilon})$ is completely known.

2 No, but $\boldsymbol{\varepsilon}$ is known to be described by an autoregressive process although the associated parameters are unknown. See Section 6.9.

3 No, $\boldsymbol{\psi} \equiv \text{cov}(\boldsymbol{\varepsilon})$ is completely unknown.

7. Errorless independent variables [$V(X_{ij}) = 0$ for linear models] (6.1.64)

0 No, there are errors in the independent variables.

1 Yes, there are no errors in the independent variables.

8. $\boldsymbol{\beta}$ is a *constant* parameter vector and there is no prior information (nature of parameters) (6.1.65)

0 No, $\boldsymbol{\beta}$ is a random parameter vector and the statistics of $\boldsymbol{\beta}$ are unknown.

1 Yes, $\boldsymbol{\beta}$ is a constant parameter vector and there is no prior information.

2 $\boldsymbol{\beta}$ is a random parameter vector with a known mean of $\boldsymbol{\mu}$ and known covariance matrix $\mathbf{V}_\beta$. Also $\boldsymbol{\beta}$ is normal. The random vectors $\boldsymbol{\beta}$ and $\mathbf{Y}$ are uncorrelated. All the measurements are considered to be from the same "batch."

3 $\boldsymbol{\beta}$ is a constant but unknown vector but there is subjective prior information. Since $\boldsymbol{\beta}$ is unknown, the prior information regarding $\boldsymbol{\beta}$ is summarized by $\boldsymbol{\beta}$ being $N(\boldsymbol{\mu}_\beta, \mathbf{V}_\beta)$ and by $\text{cov}(\beta_i, Y_j) = 0$.

As indicated in Chapter 5 the eight standard assumptions are denoted 11111111.

The estimation problems are generally less difficult (if the "best" estimates and their variances are of interest) when the standard assumptions are valid. If, for example, the covariance matrix $\boldsymbol{\psi}$ of the errors is completely unknown and there are no repeated observations, the complete $\boldsymbol{\psi}$ matrix cannot be estimated along with the parameters. From repeated observations, investigations of the measuring devices, and familiarity with estimation in such cases, the $\boldsymbol{\psi}$ matrix gradually becomes better known, however. Then estimators other than OLS can be used although a nondiagonal $\boldsymbol{\psi}$ does introduce complications; see Sections 5.13 and 6.9. If $\boldsymbol{\psi}$ is completely unknown in a multiresponse case, Box and Draper [6] recommend the use of a different method than OLS; see (6.1.73a).

6.1.6 Maximum Likelihood Sum of Squares Functions

Suppose that the measurement errors are additive, and have zero mean, normal distribution, known statistical parameters, and errorless independent variables. Also there is no prior information. This case is designated 11--1111.

The probability density function for the above assumptions is

$$f(\mathbf{Y}|\boldsymbol{\beta})=(2\pi)^{-N/2}|\boldsymbol{\psi}|^{-1/2}\exp\left[-(\mathbf{Y}-\boldsymbol{\eta})^T\boldsymbol{\psi}^{-1}(\mathbf{Y}-\boldsymbol{\eta})/2\right] \quad (6.1.66)$$

where $\boldsymbol{\psi}^{-1}$ is the inverse of $\boldsymbol{\psi}$ given by (6.1.38) and N is the total number of observations. Before the data are available $f(\mathbf{Y}|\boldsymbol{\beta})$, the probability density of $\mathbf{Y}$ given $\boldsymbol{\beta}$, associates a probability density with each different outcome $\mathbf{Y}$ of the experiment for a fixed parameter vector $\boldsymbol{\beta}$. After the data are available, we wish to find the various values in $\boldsymbol{\beta}$ which might have led to the set of measurement $\mathbf{Y}$ actually obtained. In the maximum likelihood method the *likelihood function* $L(\boldsymbol{\beta}|\mathbf{Y})$ is maximized; $L(\boldsymbol{\beta}|\mathbf{Y})$ has the same form as $f(\mathbf{Y}|\boldsymbol{\beta})$ but now $\boldsymbol{\beta}$ is considered variable and $\mathbf{Y}$ is fixed.

The procedure for forming estimators from $L(\boldsymbol{\beta}|Y)$ usually first involves taking the natural logarithm of (6.1.66) to get

$$l(\boldsymbol{\beta}|\mathbf{Y})=\ln L(\boldsymbol{\beta}|\mathbf{Y})=-\frac{1}{2}\left[N\ln(2\pi)+\ln|\boldsymbol{\psi}|+S_{\mathrm{ML}}\right] \quad (6.1.67a)$$

where

$$S_{\mathrm{ML}}\equiv(\mathbf{Y}-\boldsymbol{\eta})^T\boldsymbol{\psi}^{-1}(\mathbf{Y}-\boldsymbol{\eta}) \quad (6.1.67b)$$

When $\boldsymbol{\psi}$ is known as assumed above, maximizing $L(\boldsymbol{\beta}|\mathbf{Y})$ is equivalent to minimizing S_{ML}. Several possible algebraic forms of S_{ML} are given next.

6.1.6.1 Single Dependent Variable (Single Response) Case

Let η represent a single dependent variable that is measured at $N=n$ different times or conditions. For correlated errors, $\boldsymbol{\psi}^{-1}$ is not diagonal. Let the inverse of $\boldsymbol{\psi}$ be given by the symmetric matrix $\mathbf{W}$ with components W_{ij}. For this case S_{ML} is given in algebraic form by

$$S_{\mathrm{ML}}=\sum_{i=1}^{n}\sum_{j=1}^{n}(Y_i-\eta_i)(Y_j-\eta_j)W_{ij} \quad (6.1.68a)$$

If $\boldsymbol{\psi}$ is diagonal with components σ_i^2, then

$$S_{\mathrm{ML}}=\sum_{i=1}^{n}(Y_i-\eta_i)^2\sigma_i^{-2} \quad (6.1.68b)$$

For further discussion of S_{ML} for single response cases, see Section 6.5.

6.1.6.2 Several Dependent Variables (Multiresponse) Case

Suppose that m different dependent variables are measured at n different discrete instants. For example, temperature and concentration might be recorded with time. The formulation below is also appropriate if the same physical quantity, such as temperature, is dynamically measured utilizing several thermocouples each positioned at a different location.

In multiresponse cases, the notation given by (6.1.54) is used for $\boldsymbol{\eta}$ and a similar expression is used for the mn observation column vector, $\mathbf{Y}$. Let the inverse of $\boldsymbol{\psi}$ be $\mathbf{W}$ which is partitioned into a symmetric $n \times n$ matrix with components being $m \times m$ matrices,

$$\mathbf{W} = \begin{bmatrix} \mathbf{W}(1,1) & \mathbf{W}(1,2) & \cdots & \mathbf{W}(1,n) \\ \vdots & \vdots & & \vdots \\ \mathbf{W}(1,n) & \mathbf{W}(2,n) & \cdots & \mathbf{W}(n,n) \end{bmatrix} \tag{6.1.69}$$

Then the maximum likelihood sum of squares function is

$$S_{\mathrm{ML}} = \sum_{i=1}^{n} \sum_{j=1}^{n} \left[\mathbf{Y}(i) - \boldsymbol{\eta}(i)\right]^T \mathbf{W}(i,j)\left[\mathbf{Y}(j) - \boldsymbol{\eta}(j)\right] \tag{6.1.70}$$

which can be simplified further for independent errors. For example, let the errors be zero mean and independent in "time," that is, $E[\boldsymbol{\varepsilon}(i)\boldsymbol{\varepsilon}(j)^T] = \mathbf{0}$ for $i \neq j$. Then (6.1.70) reduces to

$$S_{\mathrm{ML}} = \sum_{i=1}^{n} \left[\mathbf{Y}(i) - \boldsymbol{\eta}(i)\right]^T \mathbf{W}(i,i)\left[\mathbf{Y}(i) - \boldsymbol{\eta}(i)\right] \tag{6.1.71}$$

If, in addition, the $\varepsilon_j(i)$ values are independent for a given i, that is, $E[\varepsilon_j(i)\varepsilon_k(i)] = 0$ for $j \neq k$, S_{ML} further simplifies to

$$S_{\mathrm{ML}} = \sum_{j=1}^{m} \sum_{i=1}^{n} \left[Y_j(i) - \eta_j(i)\right]^2 \sigma_{jj}^{-2}(i) \tag{6.1.72}$$

since $W_{jj}(i) = \sigma_{jj}^{-2}(i)$.

In this section, S_{ML} is given with $\boldsymbol{\psi}$ and its inverse assumed known. Exactly the same estimates for the physical parameters ($\boldsymbol{\beta}$) are obtained if $\boldsymbol{\psi}$ is known only to within a multiplicative constant, that is, $\boldsymbol{\psi} = \sigma^2 \boldsymbol{\Omega}$ where σ^2 is unknown but $\boldsymbol{\Omega}$ is known. This case is designated 11--1011. Should the $\boldsymbol{\psi}$ matrix contain more unknown parameters than σ^2, the ML procedure becomes more complicated. See Section 6.9.5.

When the ψ matrix is completely unknown, Box and Draper [6] recommend for the multiresponse case that the following determinant be minimized,

$$S_{\mathrm{BD}} = \begin{vmatrix} S_{11} & S_{12} & \cdots & S_{1m} \\ \vdots & \vdots & & \vdots \\ S_{1m} & S_{2m} & \cdots & S_{mm} \end{vmatrix} \tag{6.1.73a}$$

where S_{ij} is given by

$$S_{ij} = (\mathbf{Y}_i - \boldsymbol{\eta}_i)^T (\mathbf{Y}_j - \boldsymbol{\eta}_j) \tag{6.1.73b}$$

For $m = 1$, this reduces to ordinary least squares. For $m \geqslant 2$ the minimization of (6.1.73a) must be accomplished by the solution of a set of nonlinear equations even when the model is linear in the parameters. For other discussion of multiresponse cases, see Sections 6.7.5 and 7.8 and the Hunter paper [7].

6.1.7 Gauss–Markov Theorem

An important result in estimation is the Gauss–Markov theorem: if $\boldsymbol{\eta} = \mathbf{X}\boldsymbol{\beta}$ is a linear regression function, if errors are additive and have zero mean, if the covariance matrix of $\boldsymbol{\varepsilon}, \psi$, is known to within a multiplicative constant ($\psi = \boldsymbol{\Omega}\sigma^2$), and is positive definite, if there is no error in the independent variable, and if the estimation procedure does not use prior information (assumptions 11---011), then of all unbiased estimators of any component β_i of $\boldsymbol{\beta}$ which are unbiased and which are linear combinations of the observations, the component $b_{i\mathrm{MV}}$ of $\mathbf{b}_{\mathrm{MV}}$ has minimum variance where

$$\mathbf{b}_{\mathrm{MV}} = (\mathbf{X}^T \boldsymbol{\Omega}^{-1} \mathbf{X})^{-1} \mathbf{X}^T \boldsymbol{\Omega}^{-1} \mathbf{Y} \tag{6.1.74}$$

The covariance matrix of $\mathbf{b}_{\mathrm{MV}}$ is

$$\operatorname{cov}(\mathbf{b}_{\mathrm{MV}}) = (\mathbf{X}^T \boldsymbol{\Omega}^{-1} \mathbf{X})^{-1} \sigma^2 \tag{6.1.75}$$

Since we are usually more interested in the use of our estimates of $\boldsymbol{\beta}$ in estimating expected values of Y or other linear combinations of the components of $\boldsymbol{\beta}$, we shall prove a stronger theorem which includes the one just stated, namely: under the conditions stated above, of all unbiased estimators of any particular linear combination of the components of $\boldsymbol{\beta}$,

$\ell^T\boldsymbol{\beta}$, say, which can be expressed as a linear combination of the observations, the one with minimum variance is $\ell^T\mathbf{b}_{\mathrm{MV}}$.

To start the proof, we note that $\mathbf{b}_{\mathrm{MV}}$ is simply expressible in terms of $\boldsymbol{\beta}$ and $\boldsymbol{\varepsilon}$

$$\mathbf{b}_{\mathrm{MV}}=(\mathbf{X}^T\boldsymbol{\Omega}^{-1}\mathbf{X})^{-1}\mathbf{X}^T\boldsymbol{\Omega}^{-1}(\mathbf{X}\boldsymbol{\beta}+\boldsymbol{\varepsilon})=\boldsymbol{\beta}+\mathbf{A}\boldsymbol{\varepsilon} \tag{6.1.76}$$

where

$$\mathbf{A}\equiv(\mathbf{X}^T\boldsymbol{\Omega}^{-1}\mathbf{X})^{-1}\mathbf{X}^T\boldsymbol{\Omega}^{-1} \tag{6.1.77}$$

so that

$$E(\mathbf{b}_{\mathrm{MV}})=\boldsymbol{\beta}+\mathbf{A}E(\boldsymbol{\varepsilon})=\boldsymbol{\beta} \tag{6.1.78}$$

or in words, $\mathbf{b}_{\mathrm{MV}}$ is an unbiased estimator of $\boldsymbol{\beta}$. The covariance matrix of $\mathbf{b}_{\mathrm{MV}}$ is

$$\operatorname{cov}(\mathbf{b}_{MV})=E\left[(\mathbf{A}\boldsymbol{\varepsilon})(\mathbf{A}\boldsymbol{\varepsilon})^T\right]=E\left[\mathbf{A}\boldsymbol{\varepsilon}\boldsymbol{\varepsilon}^T\mathbf{A}^T\right]=\mathbf{A}\boldsymbol{\Omega}\mathbf{A}^T\sigma^2=(\mathbf{X}^T\boldsymbol{\Omega}^{-1}\mathbf{X})^{-1}\sigma^2 \tag{6.1.79}$$

Consider now the scalar $\ell^T\mathbf{b}_{\mathrm{MV}}$. It follows from (6.1.78) and (6.1.79) that

$$E(\ell^T\mathbf{b}_{\mathrm{MV}})=\ell^T\boldsymbol{\beta} \tag{6.1.80}$$

and

$$V(\ell^T\mathbf{b}_{\mathrm{MV}})=\ell^T(\mathbf{X}^T\boldsymbol{\Omega}^{-1}\mathbf{X})^{-1}\ell\sigma^2 \tag{6.1.81}$$

Next we consider any unbiased linear estimator of $\ell^T\boldsymbol{\beta}$. It can be written as

$$\ell^T\mathbf{b}_{\mathrm{MV}}+\mathbf{C}^T\mathbf{Y} \tag{6.1.82}$$

where $\mathbf{C}$ is a vector of constants.

Since $\ell^T\mathbf{b}_{\mathrm{MV}}+\mathbf{C}^T\mathbf{Y}$ is unbiased and since

$$E(\ell^T\mathbf{b}_{\mathrm{MV}}+\mathbf{C}^T\mathbf{Y})=\ell^T\boldsymbol{\beta}+\mathbf{C}^T\mathbf{X}\boldsymbol{\beta} \tag{6.1.83}$$

we see that $\mathbf{C}^T\mathbf{X}\boldsymbol{\beta}$ is simply zero. Since, by definition of unbiased estimator, this condition cannot depend on the value of $\boldsymbol{\beta}$, $\mathbf{C}^T\mathbf{X}$ must be a null vector. The variance of $\ell^T \mathbf{b}_{\text{MV}} + \mathbf{C}^T\mathbf{Y}$ is

$$V\left(\ell^T \mathbf{b}_{\text{MV}} + \mathbf{C}^T\mathbf{Y}\right) = E\left(\ell^T(\mathbf{X}^T\boldsymbol{\Omega}^{-1}\mathbf{X})^{-1}\mathbf{X}^T\boldsymbol{\Omega}^{-1}\boldsymbol{\varepsilon} + \mathbf{C}^T\boldsymbol{\varepsilon}\right)^2$$

$$= E\left(\ell^T(\mathbf{X}^T\boldsymbol{\Omega}^{-1}\mathbf{X})^{-1}\mathbf{X}^T\boldsymbol{\Omega}^{-1}\boldsymbol{\varepsilon}\boldsymbol{\varepsilon}^T\boldsymbol{\Omega}^{-1}\mathbf{X}(\mathbf{X}^T\boldsymbol{\Omega}^{-1}\mathbf{X})^{-1}\ell\right)$$

$$+ E(\mathbf{C}^T\boldsymbol{\varepsilon}\boldsymbol{\varepsilon}^T\mathbf{C}) + 2E\left(\mathbf{C}^T\boldsymbol{\varepsilon}\boldsymbol{\varepsilon}^T\boldsymbol{\Omega}^{-1}\mathbf{X}(\mathbf{X}^T\boldsymbol{\Omega}^{-1}\mathbf{X})^{-1}\ell\right)$$

$$= \ell^T(\mathbf{X}^T\boldsymbol{\Omega}^{-1}\mathbf{X})^{-1}\ell\sigma^2 + \mathbf{C}^T\boldsymbol{\Omega}\mathbf{C}\sigma^2 + 2\mathbf{C}^T\mathbf{X}(\mathbf{X}^T\boldsymbol{\Omega}^{-1}\mathbf{X})^{-1}\ell\sigma^2 \quad (6.1.84)$$

Since $\mathbf{C}^T\mathbf{X} = \mathbf{0}$, the third term is 0. The first term is unaffected by a choice of $\mathbf{C}$. Since $\boldsymbol{\Omega}$ is positive definite $\mathbf{C}^T\boldsymbol{\Omega}\mathbf{C}$ is positive unless $\mathbf{C}$ is the null vector. Thus no unbiased linear estimator of $\ell^T\boldsymbol{\beta}$ has a variance less than the variance of $\ell^T\mathbf{b}_{\text{MV}}$. Hence $\ell^T\mathbf{b}_{\text{MV}}$ is a minimum variance unbiased linear estimator of $\ell^T\boldsymbol{\beta}$ and in particular $b_{i\text{MV}}$ is a minimum variance unbiased linear estimator of $\beta_{i\text{MV}}$.

6.2 LEAST SQUARES ESTIMATION

In order to obtain parameter estimates using ordinary least squares (OLS), none of the standard assumptions need be valid. Instead a prescribed sum of squares is always minimized. When nothing is known regarding the measurement errors, OLS is recommended. The reader should be made aware, however, that he may know more about the measurement errors than he first thinks. After analyzing similar sets of data, for example, he can learn much regarding the errors.

When information regarding the measurement errors is present and σ_i^2 is far from constant, some estimation procedure other than OLS is recommended.

6.2.1 Ordinary Least Squares Estimator (OLS)

The sum of squares function used for ordinary least squares with the linear model $\eta = \mathbf{X}\boldsymbol{\beta}$ is

$$S_{\text{LS}} = (\mathbf{Y} - \mathbf{X}\boldsymbol{\beta})^T(\mathbf{Y} - \mathbf{X}\boldsymbol{\beta}) \quad (6.2.1)$$

which is a quadratic form. The principle of least squares asserts that a set of estimates of parameters can be obtained by minimizing S_{LS}. This is accomplished by setting the matrix derivatives of S_{LS} with respect to $\boldsymbol{\beta}$

equal to zero. Using (6.1.30) and (6.1.26b) we get

$$\nabla_\beta S_{LS} = 2\left[\nabla_\beta (\mathbf{Y}-\mathbf{X}\boldsymbol{\beta})^T\right]\left[\mathbf{Y}-\mathbf{X}\boldsymbol{\beta}\right] \tag{6.2.2}$$

$$\nabla_\beta (\mathbf{Y}-\mathbf{X}\boldsymbol{\beta})^T = -\nabla_\beta \boldsymbol{\beta}^T\mathbf{X}^T = -\mathbf{X}^T \tag{6.2.3}$$

and thus (6.2.2) equated to zero at $\boldsymbol{\beta}=\mathbf{b}_{LS}$ produces

$$-\mathbf{X}^T\mathbf{Y}+\mathbf{X}^T\mathbf{X}\mathbf{b}_{LS}=\mathbf{0} \tag{6.2.4}$$

Premultiplying by the inverse $(\mathbf{X}^T\mathbf{X})^{-1}$ results in the ordinary least squares estimator,

$$\boxed{\mathbf{b}_{LS} = (\mathbf{X}^T\mathbf{X})^{-1}\mathbf{X}^T\mathbf{Y}} \tag{6.2.5}$$

This estimator requires for unique estimation of all the p parameters that the $p\times p$ matrix $\mathbf{X}^T\mathbf{X}$ be nonsingular or $|\mathbf{X}^T\mathbf{X}|\neq 0$. This means that any one column in $\mathbf{X}$ *cannot* be proportional to any other column or any linear combination of other columns because if such a proportionality (i.e., linear dependence) exists, $|\mathbf{X}^T\mathbf{X}|=0$. (See Appendix A at the end of the text.) The condition $|\mathbf{X}^T\mathbf{X}|\neq 0$ also requires that n, the number of measurements of Y_i, be equal to or greater than the number of parameters p. If the predicted curve $\hat{Y}_i$ is *not* to pass through each observation it is further necessary that $n \geqslant p+1$.

Estimators obtained from (6.2.5) for Model 2, $\eta_i = \beta X_i$, and Model 5, $\eta_i = \beta_1 X_{i1} + \beta_2 X_{i2}$, are those obtained in Chapter 5; in particular see Table 5.3. For the single response case and using $\mathbf{X}$ given by (6.1.52), $\mathbf{X}^T\mathbf{X}$ and $\mathbf{X}^T\mathbf{Y}$ are given by

$$\mathbf{X}^T\mathbf{X} = \begin{bmatrix} \sum X_{i1}^2 & \sum X_{i1}X_{i2} & \cdots & \sum X_{i1}X_{ip} \\ & \sum X_{i2}^2 & \cdots & \sum X_{i2}X_{ip} \\ & & & \vdots \\ \text{symmetric} & & & \sum X_{ip}^2 \end{bmatrix} \tag{6.2.6}$$

$$\mathbf{X}^T\mathbf{Y} = \begin{bmatrix} \Sigma X_{i1} Y_i \\ \vdots \\ \Sigma X_{ip} Y_i \end{bmatrix} \tag{6.2.7}$$

where the summations are from 1 to n.

There are many possible applications of OLS. One is to find a curve which is "best" fit to some actual data. Another is to find parameters having physical significance. One can also use OLS to pass an approximate curve though some points that have been analytically derived.

Example 6.2.1

Give the components of the $\mathbf{X}^T\mathbf{X}$ and $\mathbf{X}^T\mathbf{Y}$ matrices for the model

$$\eta_i = \beta_1 + \beta_2 t_i + \beta_3 t_i^2$$

Solution

In this example $\mathbf{X}$ is

$$\mathbf{X} = \begin{bmatrix} 1 & t_1 & t_1^2 \\ 1 & t_2 & t_2^2 \\ \vdots & \vdots & \vdots \\ 1 & t_n & t_n^2 \end{bmatrix}$$

and then

$$\mathbf{X}^T\mathbf{X} = \begin{bmatrix} n & \Sigma t_i & \Sigma t_i^2 \\ & \Sigma t_i^2 & \Sigma t_i^3 \\ \text{symmetric} & & \Sigma t_i^4 \end{bmatrix}, \qquad \mathbf{X}^T\mathbf{Y} = \begin{bmatrix} \Sigma Y_i \\ \Sigma t_i Y_i \\ \Sigma t_i^2 Y_i \end{bmatrix}$$

Example 6.2.2

Many experimentally based heat transfer correlations are given by the Nusselt number (Nu) as a function of the Reynolds (Re) and Prandtl (Pr) numbers. One such correlation is for flow normal to heated cylinders and wires. Even though Nu may vary several orders of magnitude, typically the variation in the measurements for Nu is $\pm 30\%$ irrespective of the magnitude of Nu. On a log–log plot of Nu versus Re for the geometry mentioned, the data appear to be described by a second degree curve and the variance of log Nu is nearly constant. For the below values given by Welty [8, p. 268], use OLS to find the parameters in the linear-in-the-parameters model

$$\log \mathrm{Nu} = \beta_1 + \beta_2 \log \mathrm{Re} + \beta_3 (\log \mathrm{Re})^2 + \varepsilon$$

Also give $\hat{\mathrm{Nu}}$ as a function of Re and compare with the below given values.

Re	0.1	1	10	10^2	10^3	10^4	10^5
Nu	0.45	0.84	1.83	5.1	15.7	56.5	245

Solution

The model can be written in the more familiar form

$$\eta = \beta_1 + \beta_2 t + \beta_3 t^2, \qquad Y = \eta + \varepsilon$$

where $Y = \log \mathrm{Nu}$, $\eta = E(Y)$, and $t = \log \mathrm{Re}$. In terms of Y and t, the data are

i	1	2	3	4	5	6	7
t_i	−1	0	1	2	3	4	5
Y_i	−0.347	−0.076	0.263	0.708	1.20	1.75	2.39

which is equally spaced in t. The $\mathbf{X}^T$ matrix is given by

$$\mathbf{X}^T = \begin{bmatrix} 1 & 1 & 1 & 1 & 1 & 1 & 1 \\ -1 & 0 & 1 & 2 & 3 & 4 & 5 \\ 1 & 0 & 1 & 4 & 9 & 16 & 25 \end{bmatrix}$$

Note that the rows in $\mathbf{X}^T$ (or columns of $\mathbf{X}$) are not proportional or even nearly so. This indicates that little difficulty will be encountered due to $\mathbf{X}^T\mathbf{X}$ being nearly singular. Using the results of Example 6.2.1, the $\mathbf{X}^T\mathbf{X}$ and $\mathbf{X}^T\mathbf{Y}$ matrices are found to be

$$\mathbf{X}^T\mathbf{X} = \begin{bmatrix} 7 & 14 & 56 \\ 14 & 56 & 224 \\ 56 & 224 & 980 \end{bmatrix}, \qquad \mathbf{X}^T\mathbf{Y} = \begin{bmatrix} 5.885 \\ 24.57 \\ 101.3 \end{bmatrix}$$

The inverse of the symmetric $\mathbf{X}^T\mathbf{X}$ matrix is the symmetric matrix [see (6.1.11)],

$$(\mathbf{X}^T\mathbf{X})^{-1} = \frac{1}{16464}\begin{bmatrix} 4704 & -1176 & 0 \\ -1176 & 3724 & -784 \\ 0 & -784 & 196 \end{bmatrix}$$

Then from $\mathbf{b}_{\mathrm{LS}} = (\mathbf{X}^T\mathbf{X})^{-1}\mathbf{X}^T\mathbf{Y}$ we find the estimated parameters to be

$$\mathbf{b}_{\mathrm{LS}}^T = [-0.0734 \quad 0.314 \quad 0.0358]$$

In order to express the model in terms of Nu, let $\beta_1 = \log B$ which produces $\hat{B} = 0.844$. With B in the $\log \mathrm{Nu}$ expression we can write

$$\hat{\mathrm{N}}\mathrm{u} = \hat{\mathrm{B}}\,\mathrm{Re}^{b_2 + b_3 \log \mathrm{Re}} = 0.844\,\mathrm{Re}^{0.314 + 0.0358 \log \mathrm{Re}}$$

Values obtained using this expression are given in the *third* column of Table 6.1. A comparison of the recommended and $\hat{\mathrm{N}}\mathrm{u}$ values shows an agreement within ±3%. Note that the largest residual ($\mathrm{Nu} - \hat{\mathrm{N}}\mathrm{u}$) occurs at $\mathrm{Re} = 10^5$ and is 0.9, which is much larger than those for small Re. If an OLS analysis on Nu (and not $\log \mathrm{Nu}$) had been used, the magnitude of the residuals would have been more uniform over the Nu range, causing the relative differences in columns two and three of Table 6.1 to be much larger for the small Nu values than the large values. Hence in this case in which the relative differences in the given and predicted values of Nu were to be nearly constant, OLS estimation on $\log \mathrm{Nu}$ produced the desired result whereas OLS on Nu would not. Furthermore the transformation from Nu and Re to log Nu and log Re, respectively, results in a linear estimation problem while the problem in terms of Nu and Re is nonlinear. These data are considered further in Example 6.3.1.

Table 6.1 Given and OLS Values for Nu_i for Example 6.2.2

	Nu_i	
Re	Given Values	OLS Values
0.1	0.45	0.4452
1	0.84	0.8445
10	1.83	1.889
100	5.1	4.983
10^3	15.7	15.50
10^4	56.5	56.85
10^5	245	245.9

6.2.2 Mean of the OLS Estimator

In order to obtain the parameter vector $\mathbf{b}_{\mathrm{LS}}$ it is not necessary to invoke any of the standard assumptions. However, if some statistical statements are to be made regarding $\mathbf{b}_{\mathrm{LS}}$, we must know several facts regarding the errors. Let there be additive, zero mean errors in $\mathbf{Y}$ and let $\mathbf{X}$ and $\boldsymbol{\beta}$ be nonstochastic (11----11). Then the expected value of $\mathbf{b}_{\mathrm{LS}}$ is

$$E(\mathbf{b}_{\mathrm{LS}}) = E\left[(\mathbf{X}^T\mathbf{X})^{-1}\mathbf{X}^T\mathbf{Y}\right] = (\mathbf{X}^T\mathbf{X})^{-1}\mathbf{X}^T E(\mathbf{Y})$$

$$= (\mathbf{X}^T\mathbf{X})^{-1}\mathbf{X}^T\mathbf{X}\boldsymbol{\beta} = \boldsymbol{\beta} \tag{6.2.8}$$

With the four assumptions given above, the least squares estimator is thus shown to be *unbiased*.

6.2.3 Variance–Covariance Matrix of $\mathbf{b}_{\mathrm{LS}}$

With the use of the four assumptions denoted (11----11) and utilizing (6.2.8) we can find the covariance of $\mathbf{b}_{\mathrm{LS}}$. Let

$$\mathbf{b}_{\mathrm{LS}} = \mathbf{A}(\mathbf{X}\boldsymbol{\beta} + \boldsymbol{\varepsilon}) = \boldsymbol{\beta} + \mathbf{A}\boldsymbol{\varepsilon} \qquad \text{where } \mathbf{A} \equiv (\mathbf{X}^T\mathbf{X})^{-1}\mathbf{X}^T \tag{6.2.9}$$

since $\mathbf{AX}=\mathbf{I}$. Then $\text{cov}(\mathbf{b}_{\text{LS}})$ is

$$\text{cov}(\mathbf{b}_{\text{LS}})=E\left[(\mathbf{b}_{\text{LS}}-\boldsymbol{\beta})(\mathbf{b}_{\text{LS}}-\boldsymbol{\beta})^T\right]$$

$$=E\left[(\boldsymbol{\beta}+\mathbf{A}\boldsymbol{\varepsilon}-\boldsymbol{\beta})(\boldsymbol{\beta}+\mathbf{A}\boldsymbol{\varepsilon}-\boldsymbol{\beta})^T\right]=\mathbf{A}\boldsymbol{\psi}\mathbf{A}^T \tag{6.2.10}$$

since $\boldsymbol{\psi}=E(\boldsymbol{\varepsilon}\boldsymbol{\varepsilon}^T)$. Utilizing the definition of $\mathbf{A}$ then gives

$$\boxed{\text{cov}(\mathbf{b}_{\text{LS}})=(\mathbf{X}^T\mathbf{X})^{-1}\mathbf{X}^T\boldsymbol{\psi}\mathbf{X}(\mathbf{X}^T\mathbf{X})^{-1}=\mathbf{P}_{\text{LS}}} \tag{6.2.11}$$

in which we also introduce the symbol $\mathbf{P}_{\text{LS}}$.

Without the additional standard assumptions of uncorrelated and constant variance measurement errors (1111--11), the OLS estimator does not provide the minimum variance estimator. For this reason, $\mathbf{b}_{\text{LS}}$ is said to be not efficient. Suppose that the standard assumptions of 1111--11 are valid, then $\boldsymbol{\psi}$ is given by

$$\boldsymbol{\psi}=E(\boldsymbol{\varepsilon}\boldsymbol{\varepsilon}^T)=\sigma^2\mathbf{I}$$

and we have

$$\text{cov}(\mathbf{b}_{\text{LS}})=(\mathbf{X}^T\mathbf{X})^{-1}\sigma^2$$

which is the *minimum* covariance matrix of $\mathbf{b}_{\text{LS}}$ and thus for these assumptions, which include additive, zero mean, uncorrelated, and constant variance errors, $\mathbf{b}_{\text{LS}}$ does provide an efficient estimator. (See Section 6.1.7 on the Gauss–Markov theorem.)

In addition to needing the covariance of the parameters, one may desire the covariance matrix of a collection of predicted *points* on the regression line or surface. Using (6.1.46) with the assumptions of additive, zero mean errors and nonstochastic $\mathbf{X}$ and $\boldsymbol{\beta}$ (11----11) we find for the set of points represented by $\hat{\mathbf{Y}}=\mathbf{X}_1\mathbf{b}_{\text{LS}}$,

$$\text{cov}(\hat{\mathbf{Y}})=\text{cov}(\mathbf{X}_1\mathbf{b}_{\text{LS}})=\mathbf{X}_1\,\text{cov}(\mathbf{b}_{\text{LS}})\mathbf{X}_1^T$$

$$=\mathbf{X}_1(\mathbf{X}^T\mathbf{X})^{-1}\mathbf{X}^T\boldsymbol{\psi}\mathbf{X}(\mathbf{X}^T\mathbf{X})^{-1}\mathbf{X}_1^T \tag{6.2.12a}$$

See Fig. 5.7. For independent, constant variance errors this expression reduces to

$$\text{cov}(\hat{\mathbf{Y}})=\mathbf{X}_1(\mathbf{X}^T\mathbf{X})^{-1}\mathbf{X}_1^T\sigma^2 \tag{6.2.12b}$$

The diagonal elements of this square matrix give the variances of $\hat{Y}_i$. For a typical element of (6.2.12b) see (5.2.41a), which is for the simple model $\eta = \beta_0 + \beta_1 X$.

6.2.4 Relations Involving the Sum of Squares of Residuals

For ordinary least squares with the linear model $\boldsymbol{\eta} = \mathbf{X}\boldsymbol{\beta}$, the minimum sum of squares is

$$R_{\text{LS}} = (\mathbf{Y} - \mathbf{X}\mathbf{b}_{\text{LS}})^T (\mathbf{Y} - \mathbf{X}\mathbf{b}_{\text{LS}}) \tag{6.2.13}$$

which is the sum of squares of residuals. (It is also called error sum of squares and designated SSE.) By expanding R_{LS} and employing the $\mathbf{b}_{\text{LS}}$ expression given by (6.2.5), we can also write

$$R_{\text{LS}} = \mathbf{Y}^T\mathbf{Y} - \mathbf{b}_{\text{LS}}^T\mathbf{X}^T\mathbf{Y} = \mathbf{Y}^T\mathbf{Y} - \mathbf{b}_{\text{LS}}^T\mathbf{X}^T\mathbf{X}\mathbf{b}_{\text{LS}} \tag{6.2.14a, b}$$

since (6.2.4) and the scalar nature of R_{LS} can be used to obtain

$$\mathbf{Y}^T\mathbf{X}\mathbf{b}_{\text{LS}} = \mathbf{b}_{\text{LS}}^T\mathbf{X}^T\mathbf{Y} = \mathbf{b}_{\text{LS}}^T\mathbf{X}^T\mathbf{X}\mathbf{b}_{\text{LS}} \tag{6.2.15}$$

The expression given by (6.2.14a) can be convenient to use for numerically evaluating R_{LS} because the individual residuals need not be calculated and also because $\mathbf{X}^T\mathbf{Y}$ is known from the $\mathbf{b}_{\text{LS}}$ calculation as indicated by (6.2.5) and (6.2.7).

For the standard assumptions of additive, zero mean, constant variance, uncorrelated errors; errorless $\mathbf{X}$ matrix; and nonrandom parameters (1111--11), the expected value of R_{LS} given by (6.2.13) is obtained next. Substituting (6.2.5) into (6.2.13) gives

$$\begin{aligned} R_{\text{LS}} &= \left(\mathbf{Y} - \mathbf{X}(\mathbf{X}^T\mathbf{X})^{-1}\mathbf{X}^T\mathbf{Y}\right)^T \left(\mathbf{Y} - \mathbf{X}(\mathbf{X}^T\mathbf{X})^{-1}\mathbf{X}^T\mathbf{Y}\right) \\ &= \mathbf{Y}^T\left(\mathbf{I} - \mathbf{X}(\mathbf{X}^T\mathbf{X})^{-1}\mathbf{X}^T\right)^T \left(\mathbf{I} - \mathbf{X}(\mathbf{X}^T\mathbf{X})^{-1}\mathbf{X}^T\right)\mathbf{Y} \\ &= \mathbf{Y}^T\left(\mathbf{I} - \mathbf{X}(\mathbf{X}^T\mathbf{X})^{-1}\mathbf{X}^T\right)\mathbf{Y} \end{aligned} \tag{6.2.16}$$

since $\mathbf{I} - \mathbf{X}(\mathbf{X}^T\mathbf{X})^{-1}\mathbf{X}^T$ is symmetric and idempotent. Let us substitute $\mathbf{X}\boldsymbol{\beta} + \boldsymbol{\varepsilon}$ for $\mathbf{Y}$ in (6.2.16) to find

$$R_{\text{LS}} = \boldsymbol{\varepsilon}^T\left(\mathbf{I} - \mathbf{X}(\mathbf{X}^T\mathbf{X})^{-1}\mathbf{X}^T\right)\boldsymbol{\varepsilon} \tag{6.2.17}$$

since $\mathbf{X}^T(\mathbf{I} - \mathbf{X}(\mathbf{X}^T\mathbf{X})^{-1}\mathbf{X}^T) = \mathbf{X}^T - \mathbf{X}^T = \mathbf{0}$. Note that R_{LS} given by (6.2.17) is a quadratic form in terms of $\boldsymbol{\varepsilon}$. Then using (6.1.50) with $E(\boldsymbol{\varepsilon})=\mathbf{0}$ and $\operatorname{cov}(\boldsymbol{\varepsilon})=\sigma^2\mathbf{I}$, the expected value of R_{LS} is

$$E(R_{\mathrm{LS}})=\operatorname{tr}\left[\mathbf{I}-\mathbf{X}(\mathbf{X}^T\mathbf{X})^{-1}\mathbf{X}^T\right]\sigma^2$$

$$=\operatorname{tr}(\mathbf{I})\sigma^2-\operatorname{tr}\left[(\mathbf{X}^T\mathbf{X})^{-1}(\mathbf{X}^T\mathbf{X})\right]\sigma^2=(n-p)\sigma^2 \qquad (6.2.18)$$

It follows that an unbiased estimate of σ^2 (provided the assumptions 1111--11 are valid) is

$$s^2=\hat{\sigma}^2=\frac{R_{\mathrm{LS}}}{n-p} \qquad (6.2.19)$$

This is a generalization of the results in Chapter 5 where $p=1$ and 2.

A summary of the basic equations for OLS estimation is given in Appendix B.

6.2.5 Distribution of R_{LS} and $\mathbf{b}_{\mathrm{LS}}$

If in addition to the assumptions used above, the errors ε_i, $i=1,\ldots,n$ have a normal probability density, several theorems can be stated regarding R_{LS} and b_{LS}. They will be presented without proof.

Theorem 6.2.1

The quantity R_{LS}/σ^2 has the $\chi^2(n-p)$ distribution.

Theorem 6.2.2

The vector $\mathbf{b}_{\mathrm{LS}}-\boldsymbol{\beta}$ is distributed as $N(\mathbf{0},\sigma^2(\mathbf{X}^T\mathbf{X})^{-1})$.

Theorem 6.2.3

Let $\hat{\sigma}_{b_i}$ be the ith diagonal element of $(\mathbf{X}^T\mathbf{X})^{-1}s^2$. Then the quantity $(b_{i,\mathrm{LS}}-\beta_i)/\hat{\sigma}_{b_i}$ has the $t(n-p)$ distribution.

Another theorem is concerned with testing whether certain subsets of the parameters are needed in the model. Let the parameter vector $\boldsymbol{\beta}$ be partitioned so that $\boldsymbol{\beta}^T=[\boldsymbol{\beta}_1^T\ \boldsymbol{\beta}_2^T]$ and let $\mathbf{X}$ be partitioned conformably so that $\mathbf{X}=[\mathbf{X}_1\ \mathbf{X}_2]$. Then $\mathbf{Y}$ can be written as

$$\mathbf{Y}=\mathbf{X}_1\boldsymbol{\beta}_1+\mathbf{X}_2\boldsymbol{\beta}_2+\boldsymbol{\varepsilon} \qquad (6.2.20)$$

where $\boldsymbol{\beta}_1$ contains $p-q$ elements, say, and $\boldsymbol{\beta}_2$ contains q elements. Assume that the hypothesis to be tested, simultaneously specifies all the components of $\boldsymbol{\beta}_2$. One possible hypothesis is that $\boldsymbol{\beta}_2=\boldsymbol{\beta}_2^*$ where $\boldsymbol{\beta}_2^*$ could be a zero vector. The test is based upon the relative reduction in the residual sum of squares when all p parameters are estimated as compared with the residual sum of squares when $\boldsymbol{\beta}_1$ is estimated with $\boldsymbol{\beta}_2=\boldsymbol{\beta}_2^*$. If the reduction is large, the hypothesis that $\boldsymbol{\beta}_2=\boldsymbol{\beta}_2^*$ is untenable.

The least squares estimate for all the parameters is the usual expression

$$\hat{\boldsymbol{\beta}}_{\mathrm{LS}}=\begin{bmatrix}\mathbf{b}_{1,\mathrm{LS}}\\ \mathbf{b}_{2,\mathrm{LS}}\end{bmatrix}=(\mathbf{X}^T\mathbf{X})^{-1}\mathbf{X}^T\mathbf{Y} \tag{6.2.21}$$

which has the associated residual sum of squares corresponding to (6.2.17) of

$$R(\mathbf{b}_{\mathrm{LS}})=R(\mathbf{b}_{1,\mathrm{LS}},\mathbf{b}_{2,\mathrm{LS}})=\boldsymbol{\varepsilon}^T\left(\mathbf{I}-\mathbf{X}(\mathbf{X}^T\mathbf{X})^{-1}\mathbf{X}^T\right)\boldsymbol{\varepsilon} \tag{6.2.22}$$

Now suppose that $\boldsymbol{\beta}_2$ were set equal to $\boldsymbol{\beta}_2^*$ and estimates of $\boldsymbol{\beta}_1$, denoted by $\mathbf{b}_{1,\mathrm{LS}}(\boldsymbol{\beta}_2^*)$ are obtained,

$$\mathbf{b}_{1,\mathrm{LS}}(\boldsymbol{\beta}_2^*)=(\mathbf{X}_1^T\mathbf{X}_1)^{-1}\mathbf{X}_1^T(\mathbf{Y}-\mathbf{X}_2\boldsymbol{\beta}_2^*) \tag{6.2.23a}$$

which produces the R expression,

$$R(\mathbf{b}_{1,\mathrm{LS}}(\boldsymbol{\beta}_2^*),\boldsymbol{\beta}_2^*)=\boldsymbol{\varepsilon}^T\left(\mathbf{I}-\mathbf{X}_1(\mathbf{X}_1^T\mathbf{X}_1)^{-1}\mathbf{X}_1^T\right)\boldsymbol{\varepsilon} \tag{6.2.23b}$$

when $\boldsymbol{\beta}_2=\boldsymbol{\beta}_2^*$. Now we give the theorem.

Theorem 6.2.4

The statistic

$$F=\frac{\left\{R\left[\mathbf{b}_{1,\mathrm{LS}}(\boldsymbol{\beta}_2^*),\boldsymbol{\beta}_2^*\right]-R(\mathbf{b}_{1,\mathrm{LS}},\mathbf{b}_{2,\mathrm{LS}})\right\}/q}{R(\mathbf{b}_{1,\mathrm{LS}},\mathbf{b}_{2,\mathrm{LS}})/(n-p)}=\frac{\Delta R/q}{R(\mathbf{b}_{1,\mathrm{LS}},\mathbf{b}_{2,\mathrm{LS}})/(n-p)} \tag{6.2.24}$$

has the $F_{1-\alpha}(q,n-p)$ distribution. See Section 2.8.10 for an F table. Also see Section 5.7 for other discussion.

For a proof, see Goldfeld and Quandt [9, p. 46].

In using Theorem 6.2.4, the hypothesis that $\boldsymbol{\beta}_2=\boldsymbol{\beta}_2^*$ is tested at the α level of significance by comparing F of Theorem 6.2.4 with the critical value $F_{1-\alpha}(q,n-p)$. If this critical value is exceeded, the hypothesis that $\boldsymbol{\beta}_2=\boldsymbol{\beta}_2^*$ is rejected. Note that the particular null vector of $\boldsymbol{\beta}_2^*=\mathbf{0}$ can be used to investigate whether the model should include the $\mathbf{X}_2\boldsymbol{\beta}_2$ terms; in this case if the ratio in Theorem 6.2.4 is greater than $F_{1-\alpha}(q,n-p)$, then we have an indication that the $\mathbf{X}_2\boldsymbol{\beta}_2$ terms may be needed.

Example 6.2.3

A solid copper billet 1.82 in. (0.0462 m) long and 1 in. (0.0254 m) in diameter was heated in a furnace and then removed. Two thermocouples were attached to the billet. The temperatures, Y_i, given by one thermocouple are given in Table 6.2 as a function of time. See also the plot of Y_i versus time shown in Fig. 6.1. The other thermocouple gave values about 0.2 F° (0.11 K) larger for smaller times and decreased gradually to about 0.14 F° (0.08 K) larger at 1536 seconds. For a constant temperature environment the thermocouple temperature readings have an estimated standard deviation of 0.03 F° (0.017 K).

Table 6.2 Data for Example 6.2.3

Observation No.	Time (sec)	Y_i, Temperature of Billet (°F)
1	0	279.59
2	96	264.87
3	192	251.53
4	288	239.30
5	384	228.18
6	480	217.24
7	576	207.86
8	672	199.36
9	768	191.65
10	864	184.44
11	960	177.64
12	1056	171.41
13	1152	165.04
14	1248	159.89
15	1344	155.19
16	1440	150.78
17	1536	146.68

Find a satisfactory power series approximation to the temperature history of the billet given in Table 6.2. Assume that all the standard assumptions except known σ^2 are valid. Use the F test at the 5% level of significance.

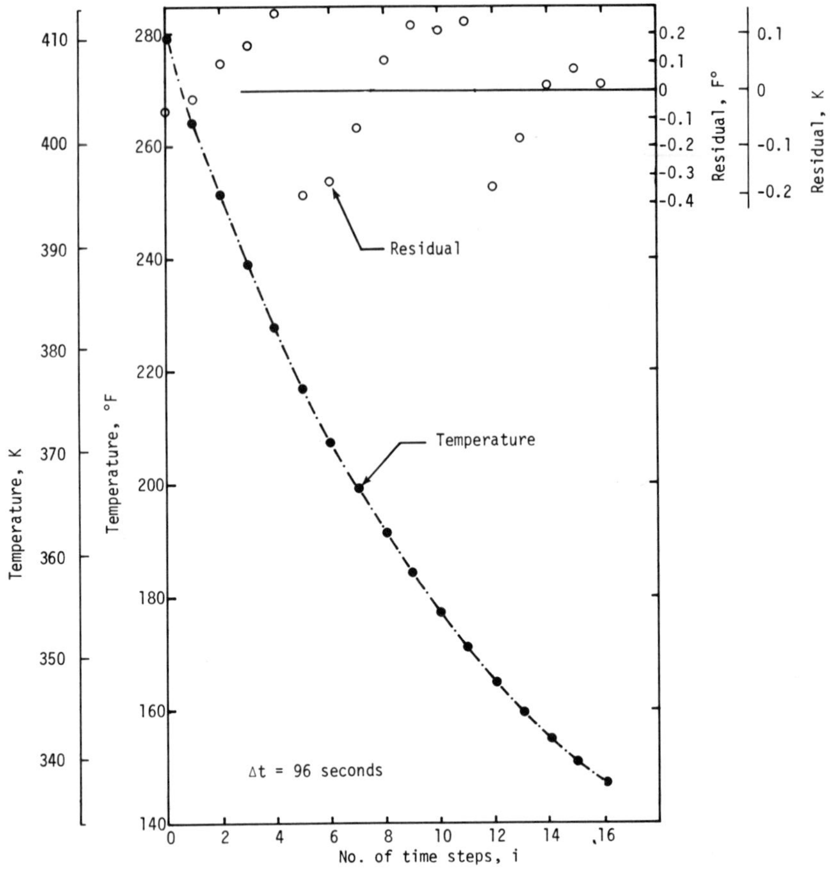

Figure 6.1 Temperatures and residuals for cooling billet example (Example 6.2.3)

Solution

The suggested model for the temperature in this case has the form

$$Y = \beta_1 + \beta_2 t + \beta_3 t^2 + \cdots + \beta_p t^{p-1} + \varepsilon$$

If the actual values of t as given in Table 6.2 are used, the components of the $\mathbf{X}^T\mathbf{X}$ matrix would have disparate values. For example, the first diagonal term is 17 and the fifth is 2.07×10^{13}. For this reason, the model is written as

$$Y = \beta_1 + \beta_2 \left(\frac{t}{\Delta t}\right) + \cdots + \beta_p \left(\frac{t}{\Delta t}\right)^{p-1} + \varepsilon \qquad (a)$$

where Δt is 96 sec.

There are many ways of determining the "best" model. One way for this model is to calculate the parameters and residual sum of squares, R, for p and $p-1$ parameters for $p=2,3,4,\dots$. Then an F test based on Theorem 6.2.4 is used. In each case we have $q=1$ and the hypothesis is $\beta_p^*=0$. The sum of squares, R_{LS}, for $p=1$ to 6 are listed in Table 6.3. The mean square (R_{LS} divided by the number of degrees of freedom) is also given. The F ratio is the ΔR value in the table, $R(b_1,\dots,b_p)-R(b_1,\dots,b_{p-1})$, divided by the mean square, s^2. This ratio is compared with $F_{.95}(1,17-p)$, which is 4.75 and 4.84 for $p=5$ and 6, respectively. Then at the 5% level of significance, the $p=5$ model is selected. (This decision is based on the standard assumptions, 11111011, being valid.) The parameters corresponding to equation a in English units are 279.660, −15.451, 0.71350, −0.023024, and 0.00039479 for $b_1,\dots,b_5$, respectively.

Table 6.3 Sum of Squares and F ratio for Example 6.2.3

p	Degrees of Freedom (d.f.)	$R(\hat{\beta}_{LS})$, Residual Sum of Squares	Mean Square, s^2 ($s^2=R/\text{d.f.}$)	ΔR	$F=\Delta R/s^2$
1	16	27624.667			
2	15	894.491	59.633	26730.176	448.3
3	14	16.1613	1.1544	878.33	760.9
4	13	1.0959	0.0843	15.0654	178.8
5	12	0.7186	0.0599	0.3773	6.30
6	11	0.5781	0.0526	0.1405	2.67

The residuals for the $p=5$ model are shown in upper portion of Fig. 6.1. They seem to be somewhat correlated and are even more so for smaller time steps. (See Fig. 6.6.) The assumption of uncorrelated measurements may not be valid. Hence the selection of a five-parameter model based on the uncorrelated errors assumption may not be correct. The $p=4$ model (cubic in t) might be actually more appropriate.

The inverse of the $\mathbf{X}^T\mathbf{X}$ matrix is

$$(\mathbf{X}^T\mathbf{X})^{-1}=\begin{bmatrix} 0.7853 & -0.5581 & 0.1174 & -0.00945 & 2.58\times10^{-4} \\ & 0.6697 & -0.1709 & 0.01525 & -4.43\times10^{-4} \\ & & 0.04739 & -0.00044 & 1.34\times10^{-4} \\ & & & 4.305\times10^{-4} & -1.32\times10^{-5} \\ \text{symmetric} & & & & 4.13\times10^{-7} \end{bmatrix}$$

If the assumptions 11111011 are valid, the estimated covariance matrix of $\mathbf{b}_{LS}$ is $(\mathbf{X}^T\mathbf{X})^{-1}$, given above, multiplied by s^2, which is 0.0599 (the mean square value in Table 6.3 for $p=5$). From these values we can find, for example, that the estimated standard error of b_1 is $[0.7853\ (0.0599)]^{1/2}=0.217$ and that of b_5 is 0.000157. A comparison of these values with b_1 and b_5 shows that the relative uncertainty in b_5 is much greater than for b_1.

The value of $s=(0.0599)^{1/2}=0.245$ F° is worthy of comment. First, for $p=6$, 7, and 8, its value would not decrease greatly. Next, this value of 0.245 is considerably larger than 0.03 found for a constant, low temperature environment. It is, however, close to the difference in temperature between the two thermocouples. At the higher temperatures the specimen's temperature may be more sensitive to air currents, etc. Also the calibration error over the whole temperature range is greater than 0.03 F°. For these reasons additive errors with as large a standard error as 0.24 are quite possible.

Example 6.2.4

Using the model for temperature found in the preceding example, estimate h in the differential equation,

$$\rho c V \frac{dT}{dt} = hA(T_\infty - T)$$

This equation describes the temperature T of the billet, assuming that there is negligible internal resistance to heat flow. The density, ρ, of copper is 555 lb_m/ft^3 (8890 kg/m^3); c is the specific heat and has the value of 0.092 Btu/lb_m-°F (0.385 kJ/kg-K); V is volume; A is heated surface area; h is the heat transfer coefficient; and T_∞ is the air temperature.

Solution

Solving the above equation for h gives

$$\hat{h} = \frac{\rho c V}{A(T_\infty - \hat{Y})} \frac{d\hat{Y}}{dt}$$

where the true temperature T has been replaced by the estimated value $\hat{Y}$. The ratio V/A is found to be 0.01634 ft (0.00498 m). The temperature $\hat{Y}$ and the derivative is obtained from the power series model with $p=5$. The derivatives are calculated using

$$\frac{d\hat{Y}}{dt} = \frac{b_2 + 2b_3 \frac{t}{\Delta t} + 3b_4\left(\frac{t}{\Delta t}\right)^2 + 4b_5\left(\frac{t}{\Delta t}\right)^3}{\Delta t}$$

where the parameters are given in the preceding example and $\Delta t = 96$ sec or 0.02667 hr. The temperature T_∞ is the room temperature which is nearly constant at 81.5 °F (300.6 K).

The h_i estimates are depicted in Fig. 6.2. From knowledge of the heat transfer process we know that the magnitude of the h_i values are reasonable and that they should drop with the temperature difference, $T - T_\infty$.

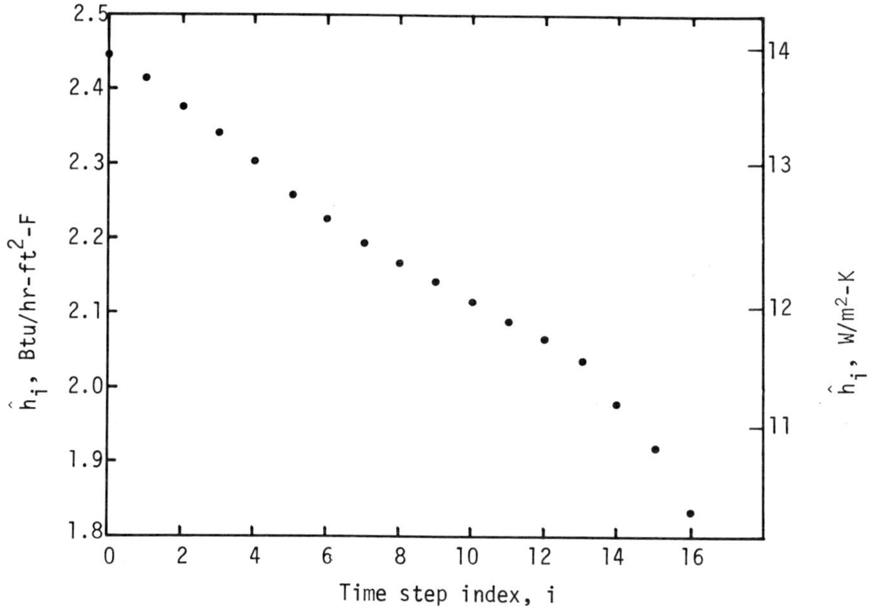

Figure 6.2 Heat transfer coefficient as a function of time for Example 6.2.4. ($i = t_i/\Delta t$).

One advantage of this method of analysis for this problem is that it is not necessary to specify a functional form for h_i versus time. Another is that the estimation for h_i is a linear problem. Some disadvantages are that (1) a functional form for T is needed; (2) sometimes this method yields h_i values that are more variable than are expected; and (3) the method is not "parsimonious" in its use of parameters. The method that is usually recommended for this problem involves the solution of the differential equation given above. This would give T as a nonlinear function of time and h and thus nonlinear estimation would be needed. See Section 7.5.2.

6.2.6 Weighted Least Squares (WLS)

There are cases when the covariance matrix of the errors (ψ) is not known but yet the experimenter wishes to include some general information regarding the errors. He might know, for example, that as Y_i increases in magnitude the variance of Y_i also increases. This was the case discussed with the heat transfer example in Example 6.2.2 in which σ_i / Y_i was about constant. It might be appropriate to assume that some symmetric weighting matrix ω is given. This matrix may or may not be equal to ψ^{-1}, the inverse of the error covariance matrix.

In this estimation problem, the function to minimize for the linear model of $\boldsymbol{\eta}=\mathbf{X}\boldsymbol{\beta}$ is

$$S_{\mathrm{WLS}}=(\mathbf{Y}-\mathbf{X}\boldsymbol{\beta})^T\omega(\mathbf{Y}-\mathbf{X}\boldsymbol{\beta}) \tag{6.2.25}$$

which yields the estimator

$$\mathbf{b}_{\mathrm{WLS}}=(\mathbf{X}^T\omega\mathbf{X})^{-1}\mathbf{X}^T\omega\mathbf{Y} \tag{6.2.26}$$

Using the standard assumptions of additive, zero mean errors and errorless $\mathbf{X}$ and $\boldsymbol{\beta}$ (11----11), the covariance matrix of $\mathbf{b}_{\mathrm{WLS}}$ can be shown using (6.2.10) to be

$$\operatorname{cov}(\mathbf{b}_{\mathrm{WLS}})=(\mathbf{X}^T\omega\mathbf{X})^{-1}\mathbf{X}^T\omega\psi\omega\mathbf{X}(\mathbf{X}^T\omega\mathbf{X})^{-1} \tag{6.2.27}$$

6.3 ORTHOGONAL POLYNOMIALS IN OLS ESTIMATION

The polynomial regression model with evenly spaced (in t) observations can be analyzed with certain computational simplifications through the use of *orthogonal polynomials*. There are many different sets of orthogonal polynomials that could be employed. Some are appropriate for use with uniform spacing in the variable t. Others have been developed for other particular spacings and for weighted least squares, Himmelblau [10, p. 152]. A set of orthogonal polynomials is given below for OLS and uniform spacing.

Suppose that the model is given by the polynomial which is a function of a single independent variable,

$$Y_i=\beta_0+\beta_1 t_i+\beta_2 t_i^2+\cdots+\beta_r t_i^r+\varepsilon_i \tag{6.3.1}$$

and that Y_i has been measured at n values of t_i which are evenly spaced or $t_{i+1}=t_i+\Delta t$. One can rewrite this model as

$$Y_i=\alpha_0 p_0(t_i)+\alpha_1 p_1(t_i)+\alpha_2 p_2(t_i)+\cdots+\alpha_r p_r(t_i)+\varepsilon_i \tag{6.3.2}$$

where $p_j(t)$ is a polynomial in t of order j. The $p_j(t)$ functions are chosen so that they are *orthogonal*, that is, for j and $k=0,1,\ldots,r$,

$$\sum_{i=1}^{n} p_j(t_i)\,p_k(t_i)=0 \qquad \text{for } j\neq k$$

$$\neq 0 \qquad \text{for } j=k \tag{6.3.3}$$

For these orthogonal polynomials, OLS yields

$$\hat{\alpha}_m = \sum_i Y_i p_m(t_i) \left[\sum_j p_m^2(t_j) \right]^{-1} \tag{6.3.4}$$

The first few orthogonal polynomials satisfying the above conditions are

$$\begin{aligned} &p_0(t_i) = 1, \qquad p_1(t_i) = \frac{t_i - \bar{t}}{\Delta t} = z_i = i - \frac{n+1}{2} \\ &p_2(t_i) = z_i^2 - \frac{n^2-1}{12}, \qquad p_3(t_i) = z_i\left(z_i^2 - \frac{3n^2-7}{20}\right) \end{aligned} \tag{6.3.5}$$

where $\bar{t} = \sum t_i / n$. Notice that the $p_j(t_i)$ values are dimensionless and are functions only of n (for given j and i values). A recursive scheme for obtaining $p_{j+1}(t_i)$ in terms of $p_j(t_i)$ is

$$p_{j+1}(t_i) = p_j(t_i)p_1(t_i) - \frac{j^2(n^2-j^2)}{4(4j^2-1)} p_{j-1}(t_i),$$

$$j = 1, 2, \ldots, i-1; \quad i = 1, 2, \ldots, n \tag{6.3.6}$$

The starting conditions on j are $p_0(t_i) = 1$ and $p_1(t_i) = i - (n+1)/2$.

If desired, after the parameters $\alpha_0, \alpha_1, \ldots, \alpha_r$ have been estimated, one can find the $\hat{\beta}_0, \ldots, \hat{\beta}_r$ from the $\hat{\alpha}_j$ values. For example, for $r = 0$, 1, and 2 we have [since (6.3.1) and (6.3.2) are identical polynomials in t]

$$\begin{aligned} r=0: &\quad \hat{\beta}_0 = \hat{\alpha}_0, \\ r=1: &\quad \hat{\beta}_0 = \hat{\alpha}_0 - \hat{\alpha}_1 \frac{\bar{t}}{\Delta t}, \qquad \hat{\beta}_1 = \frac{\hat{\alpha}_1}{\Delta t} \\ r=2: &\quad \hat{\beta}_0 = \hat{\alpha}_0 - \hat{\alpha}_1 \frac{\bar{t}}{\Delta t} + \hat{\alpha}_2 \left[\left(\frac{\bar{t}}{\Delta t} \right)^2 - \frac{n^2-1}{12} \right], \\ &\quad \hat{\beta}_1 = \left(\hat{\alpha}_1 - \hat{\alpha}_2 \frac{2\bar{t}}{\Delta t} \right) \frac{1}{\Delta t}, \qquad \hat{\beta}_2 = \frac{\hat{\alpha}_2}{(\Delta t)^2} \end{aligned} \tag{6.3.7}$$

Example 6.3.1

Using the evenly spaced in log Re data of Example 6.2.2, estimate $\alpha_0, \ldots, \alpha_r$ and compare the results with those in Example 6.2.2. Let $r = 3$ which causes η to be a cubic in t.

Solution

Using the orthogonal polynomials given by (6.3.5), we can find for the given data (the i, t_i, Y_i table of Example 6.2.2) that $n=7$, $\Delta t=1$, and $\mathbf{X}$ is given by

$$\mathbf{X}=\begin{bmatrix} 1 & -3 & 5 & -6 \\ 1 & -2 & 0 & 6 \\ 1 & -1 & -3 & 6 \\ 1 & 0 & -4 & 0 \\ 1 & 1 & -3 & -6 \\ 1 & 2 & 0 & -6 \\ 1 & 3 & 5 & 6 \end{bmatrix}$$

Note that $\mathbf{X}$ is composed of orthogonal vectors which results in $\mathbf{X}^T\mathbf{X}$ reducing to the diagonal matrix,

$$\mathbf{X}^T\mathbf{X}=\begin{bmatrix} 7 & 0 & 0 & 0 \\ 0 & 28 & 0 & 0 \\ 0 & 0 & 84 & 0 \\ 0 & 0 & 0 & 216 \end{bmatrix}$$

which has the simple inverse

$$(\mathbf{X}^T\mathbf{X})^{-1}=\text{diag}[7^{-1} \quad 28^{-1} \quad 84^{-1} \quad 216^{-1}]$$

These are the $\mathbf{X}$, $\mathbf{X}^T\mathbf{X}$, and $(\mathbf{X}^T\mathbf{X})^{-1}$ matrices for $r=3$ and for any set of seven equally spaced values of t. The $\mathbf{X}^T\mathbf{Y}$ vector becomes

$$\mathbf{X}^T\mathbf{Y}=\begin{bmatrix} 5.8846 \\ 12.7968 \\ 3.0066 \\ -0.1516 \end{bmatrix}$$

Then from either $\hat{\boldsymbol{\alpha}}=(\mathbf{X}^T\mathbf{X})^{-1}\mathbf{X}^T\mathbf{Y}$ or from (6.3.4) we find

$$\hat{\boldsymbol{\alpha}}=\begin{bmatrix} 0.84066 \\ 0.45703 \\ 0.03579 \\ -0.00070 \end{bmatrix}$$

Since the polynomials $p_j(t)$ are orthogonal, the $\hat{\alpha}_j$'s are independent. That is, regardless of the value of r in (6.3.2) we find $\hat{\alpha}_0=0.84$.

The residual sum of squares as r increases can be easily calculated from the above information. The total sum of squares, $\mathbf{Y}^T\mathbf{Y}$, is given by

$$\mathbf{Y}^T\mathbf{Y}=\Sigma Y_i^2=10.90349564$$

The residual sum of squares can be found using $R_{\text{LS}}=\mathbf{Y}^T\mathbf{Y}-\boldsymbol{\alpha}^T\mathbf{X}^T\mathbf{Y}$.

The models for $r=0$, 1, and 2 in terms of $p_j(t_i)$ are

$$r=0: \quad \hat{Y}_i=0.84066$$

$$r=1: \quad \hat{Y}_i=0.84066+0.45703p_1(t_i)$$

$$r=2: \quad \hat{Y}_i=0.84066+0.45703p_1(t_i)+0.03579p_2(t_i)$$

where we note that each parameter $\hat{\alpha}_i$ is unchanged as r is increased. In the standard form given by (6.3.1) this is not true; using (6.3.7) yields

$$r=0: \quad \hat{Y}_i=0.84066$$

$$r=1: \quad \hat{Y}_i=-0.07340+0.45703t_i$$

$$r=2: \quad \hat{Y}_i=-0.07340+0.31386t_i+0.03579t_i^2$$

where the parameter estimates change in most cases as r is increased.

In the above example the $\mathbf{X}$ matrix for seven equally spaced observations contained only whole numbers. If another column were added, fractions would enter. To avoid fractions, tables have been prepared [11, 12] which contain factors λ_i and elements ϕ_{ji}. For the $n=7$ case above, a table frequently has the information as given in Table 6.4. The component X_{ij} of $\mathbf{X}$ is related to ϕ_{ij} by

$$p_j(t_i)=X_{ij}=\frac{\phi_{ij}}{\lambda_j}; \qquad j=0,1,\ldots,r; \quad i=1,2,\ldots,n \tag{6.3.8}$$

Table 6.4 Orthogonal Polynomials for $n=7$

i	ϕ_{i0}	ϕ_{i1}	ϕ_{i2}	ϕ_{i3}	ϕ_{i4}
1	1	−3	5	−1	3
2	1	−2	0	1	−7
3	1	−1	−3	1	1
4	1	0	−4	0	6
5	1	1	−3	−1	1
6	1	2	0	−1	−7
7	1	3	5	1	3
D_j	7	28	84	6	154
λ_j	1	1	1	$\frac{1}{6}$	$\frac{7}{12}$

and D_j is related to the ith diagonal term of $\mathbf{X}^T\mathbf{X}$ by

$$\sum_{i=1}^{n} X_{ij}^2 = \frac{D_j}{\lambda_j^2} \tag{6.3.9}$$

Using these relations in (6.3.4) gives

$$\hat{\alpha}_m = \left(\sum_{i=1}^{n} Y_i \phi_{im} \right) \left(\frac{\lambda_m}{D_m} \right); \qquad m = 0, 1, \ldots, \quad \lambda_0 \equiv 1 \tag{6.3.10}$$

Some of the advantages of the use of orthogonal polynomials are the following. First, no difficulty can occur in finding the inverse of the $\mathbf{X}^T\mathbf{X}$ matrix because it is a diagonal matrix with nonzero diagonal components. If r is increased by one, $(\mathbf{X}^T\mathbf{X})^{-1}$ is changed merely by the addition of a diagonal element and the sum of squares R_{LS} is reduced by

$$\frac{\left(\sum Y_i X_{ir} \right)^2}{\sum X_{jr}^2} = \frac{\left(\sum Y_i \phi_{ir} \right)^2}{D_r}$$

Second, for a number of cases, the terms of $\mathbf{X}^T\mathbf{X}$ are tabulated, thereby reducing the number of calculations. Third, the parameters $\hat{\alpha}_0, \hat{\alpha}_1, \ldots$ are unchanged as additional values are estimated. Thus one can easily add additional parameters by increasing the degree of the orthogonal polynomials. Fourth, when orthogonal polynomials are used, there is usually less accumulation of rounding errors in the calculations of the estimates of the parameters. A disadvantage of orthogonal polynomials is that they are not convenient for sequential estimation; one cannot easily add another observation. See Section 6.7. A further disadvantage is that polynomials, orthogonal or not, are limited in their ability to fit functions. There is a tendency to think that orthogonal polynomials can be used to fit any single independent variable data, provided the degree is high enough. The model might be impractical, however, if the degree is higher than the fourth or fifth, say. The intervals between the data points could have predicted $\hat{Y}$ values that are quite oscillatory. For curves that seem to require even higher degrees, splines are recommended [13].

6.4 FACTORIAL EXPERIMENTS

6.4.1 Introduction

In Chapter 5 we found that the best design for estimating β_1 of Model 3 is one in which one half of the observations are taken at the smallest permissible value of X_i and the other half at the largest. This is a recommended procedure when the model is known to be linear in X_i and when the standard assumptions indicated by 1111--11 are valid. In other cases there may be several independent variables $x_{1i}, x_{2i}, \ldots, x_{ri}$ which are commonly called *factors*. The selection of prescribed values or *levels* of the factors constitutes a design. There are many possible designs discussed in the statistical literature. The one to be discussed below is the complete factorial design with two levels for each factor. For one factor as in Model 3, the two levels are the lowest and highest permissible values of X_i.

A factor is termed *quantitative* when its possible levels can be ranked in magnitude or *qualitative* when they cannot. Examples of quantitative factors are temperature, pH, Reynolds number, and time. Examples of qualitative factors are the type of catalyst used and the presence or absence of a particular additive.

6.4.2 Two-Level Factorial Design

Consider the model

$$Y_i = \beta_0 + \beta_1 x_{i1} + \beta_2 x_{i2} + \cdots + \beta_r x_{ir} + \varepsilon_i \tag{6.4.1}$$

where each x_{ij} can assume only two values, termed "high" and "low." There are then 2^r possible combinations of factor levels which constitutes a complete *2^r factorial design*.

Example 6.4.1

In a certain chemical process the effects of three operating variables on the overall process yield are to be studied. These are temperature (x_{i1}), type of catalyst (x_{i2}), and pressure (x_{i3}). Over the ranges of temperature and pressure studied it is known that the yield is linear in both temperature and pressure and thus two levels of each can be chosen. Also two different catalysts are chosen. The following levels are selected:

	Low Level	High Level
Temperature	400°K	420°K
Type of catalyst	*A*	*B*
Pressure	2.0×10^6 N/m^2	2.2×10^6 N/m^2

Observe that two factors, temperature and pressure, are quantitative and the remaining factor, type of catalyst, is qualitative. A 2^3 experimental design for this example is given in Table 6.5. When each factor is set at a certain level, and a test is performed, the result is termed an *experimental run*. The runs are not necessarily performed in the order given; rather, the order of the actual tests should be random.

Table 6.5 A 2^3 Experimental Design for Example 6.4.1

Run Number	Temperature (K)	Catalyst Type	Pressure (N/m^2)
1	400	*A*	2.0×10^6
2	420	*A*	2.0×10^6
3	400	*B*	2.0×10^6
4	420	*B*	2.0×10^6
5	400	*A*	2.2×10^6
6	420	*A*	2.2×10^6
7	400	*B*	2.2×10^6
8	420	*B*	2.2×10^6

6.4.3 Coding the Factors

If $\mathbf{X}$ is composed of column vectors that are orthogonal, $\mathbf{X}^T\mathbf{X}$ is a diagonal matrix. Hence the analysis can proceed in a similar manner as for orthogonal polynomials. A set of orthogonal column vectors can be obtained by coding. For quantitative factors the coding is typically given by

$$\text{Coded temp} = 2\frac{\text{actual temp } - 0.5\text{ (high temp } + \text{ low temp)}}{\text{high temp } - \text{low temp}}$$

which produces either $+1$ or -1. The qualitative factors are also assigned values of $+1$ and -1; for example, catalyst A could be signified by -1 and catalyst B by $+1$. For the case of three factors such as in Example 6.4.1 the $\mathbf{X}$ matrix for the 2^3 factorial design can be given as

$$\begin{matrix} & x_{i0} & x_{i1} & x_{i2} & x_{i3} \end{matrix}$$

$$\mathbf{X} = \begin{bmatrix} 1 & -1 & -1 & -1 \\ 1 & 1 & -1 & -1 \\ 1 & -1 & 1 & -1 \\ 1 & 1 & 1 & -1 \\ 1 & -1 & -1 & 1 \\ 1 & 1 & -1 & 1 \\ 1 & -1 & 1 & 1 \\ 1 & 1 & 1 & 1 \end{bmatrix} \tag{6.4.2}$$

In terms of (6.4.1) and Example 6.4.1, the first column is a set of coefficients of β_0, the second for coded temperature, the third for the coded catalyst type, and the last for coded pressure.

Note that **X** given by (6.4.2) contains orthogonal columns. This matrix can be conceived of as representing a particular design for a two-level complete factorial design for three factors. For one factor the upper left 2×2 matrix in **X** is the **X** matrix for a complete factorial design. For a two-factor, two-level complete factorial design the design matrix, if interactions are assumed zero, is the upper left 4×3 matrix. For more than three factors, the pattern of the elements of **X** in (6.4.2) can be repeated. For example, if there are four factors a fifth column would be added which would be a vector of eight -1's followed by eight $+1$'s; the second column would continue the $-1, 1, -1, 1$ pattern for a total of $2^4=16$ terms, etc.

6.4.4 Inclusion of Interaction Terms in the Model

The complete 2^r factorial design can be used in cases for which interaction or cross product terms are included in the regression model. An example is

$$Y_i=\beta_0+\beta_1 x_{i1}+\beta_2 x_{i2}+\beta_3 x_{i1}x_{i2}+\varepsilon_i \tag{6.4.3}$$

where $\beta_3 x_{i1}x_{i2}$ is an interaction term. The **X** matrix remains orthogonal. For the model given by (6.4.3), the design matrix **X** can be

$$\mathbf{X}=\begin{bmatrix} 1 & -1 & -1 & 1 \\ 1 & 1 & -1 & -1 \\ 1 & -1 & 1 & -1 \\ 1 & 1 & 1 & 1 \end{bmatrix} \tag{6.4.4}$$

where the fourth column contains the products $x_{i1}x_{i2}$; x_{i1} is given in the second column and x_{i2} in the third.

In such a design as indicated by (6.4.4) there are four observations and four parameters and thus the predicted values $\hat{Y}_i$ exactly equal Y_i. Hence no estimate of σ^2 can be obtained. If each run is replicated the same number of times, then σ^2 can be estimated. One should be careful to perform the experiments in a randomized order.

6.4.5 Estimation

We have designed the orthogonal nature of **X** for use with ordinary least squares. The OLS estimator is given by

$$\mathbf{b}_{\text{OLS}}=(\mathbf{X}^T\mathbf{X})^{-1}\mathbf{X}^T\mathbf{Y} \tag{6.4.5}$$

Since **X** is orthogonal and contains only $+1$ or -1 terms, the inverse of

$\mathbf{X}^T\mathbf{X}$ is easily found. Suppose that there are four parameters (three factors) so that $\mathbf{X}$ is given by (6.4.2) and then

$$\mathbf{X}^T\mathbf{X} = \text{diag}\begin{bmatrix}8 & 8 & 8 & 8\end{bmatrix} = 8\mathbf{I}$$

$$(\mathbf{X}^T\mathbf{X})^{-1} = \frac{1}{8}\mathbf{I}$$

Example 6.4.2

Estimate the parameters using OLS for the experimental design given in Table 6.5 and the Y_i values given by

$$\mathbf{Y}^T = [49 \;\; 62 \;\; 44 \;\; 58 \;\; 42 \;\; 73 \;\; 35 \;\; 69]$$

where Y_i is in percent.

(*a*) Estimate the parameters in a linear model permitting interaction using OLS.

(*b*) Suppose that the standard assumptions designated 11111011 apply. (These include normal, independent, constant variance errors.) Also assume that an estimate of the variance of Y_i is $s^2 = 2.5$ which we shall consider to have come from a separate experiment and to have 20 degrees of freedom. Using the F statistic find a "parsimonious" (in terms of the parameters) model.

Solution

(*a*) The complete model including all the interaction terms can be written in terms of the coded factors z_{ij} as

$$\eta_i = \beta_0 + \beta_1 z_{i1} + \beta_2 z_{i2} + \beta_3 z_{i3} + \beta_4 z_{i1} z_{i2} + \beta_5 z_{i1} z_{i3} + \beta_6 z_{i2} z_{i3} + \beta_7 z_{i1} z_{i2} z_{i3}$$

where z_{i1}, z_{i2}, and z_{i3} correspond respectively to coded temperature, catalyst, and pressure. The $\mathbf{X}$ matrix then becomes

$$\mathbf{X} = \begin{bmatrix} 1 & \overbrace{-1 \quad -1 \quad -1}^{\text{Main effects}} & \overbrace{1 \quad 1 \quad 1 \quad -1}^{\text{Interaction terms}} \\ 1 & 1 \quad -1 \quad -1 & -1 \quad -1 \quad 1 \quad 1 \\ 1 & -1 \quad 1 \quad -1 & -1 \quad 1 \quad -1 \quad 1 \\ 1 & 1 \quad 1 \quad -1 & 1 \quad -1 \quad -1 \quad -1 \\ 1 & -1 \quad -1 \quad 1 & 1 \quad -1 \quad -1 \quad 1 \\ 1 & 1 \quad -1 \quad 1 & -1 \quad 1 \quad -1 \quad -1 \\ 1 & -1 \quad 1 \quad 1 & -1 \quad -1 \quad 1 \quad -1 \\ 1 & 1 \quad 1 \quad 1 & 1 \quad 1 \quad 1 \quad 1 \end{bmatrix}$$

which contains eight orthogonal columns. The inverse of $\mathbf{X}^T\mathbf{X}$ is $(1/8)\mathbf{I}$ and the $\mathbf{X}^T\mathbf{Y}$ terms are obtained simply by taking sums of the Y_i terms each multiplied by a plus or minus one as indicated in $\mathbf{X}^T$. The resulting predicted equation is

$$\hat{Y}_i = 54 + 11.5z_{i1} - 2.5z_{i2} + 0.75z_{i3} + 0.5z_{i1}z_{i2} + 4.75z_{i1}z_{i3}$$
$$- 0.25z_{i2}z_{i3} + 0.25z_{i1}z_{i2}z_{i3}$$

Since there are eight observations and eight parameters, the calculated $\hat{Y}_i$ values and the measured values Y_i are equal; hence the sum of residuals is zero.

(*b*) The effect of each term in the model can be examined independently because the design is orthogonal. This is an important benefit of such designs. From (6.2.14a, b) the reduction in the sum of squares is given by

$$\mathbf{b}^T\mathbf{X}^T\mathbf{Y} = \mathbf{b}^T\mathbf{X}^T\mathbf{X}\mathbf{b} = 8\mathbf{b}^T\mathbf{I}\mathbf{b} = 8\sum_{k=0}^{7} b_k^2$$

where $\mathbf{X}^T\mathbf{X} = 8\mathbf{I}$ is used. The sum $\mathbf{Y}^T\mathbf{Y}$ is 24624. The residual sum of squares (R), reduction in R, and the F statistic are given in Table 6.6. The F statistic is $(\Delta R/1)$ divided by $s^2(=2.5)$. At the 5% level of significance, $F_{.95}(1,20) = 4.35$. (s^2 was stated to have 20 degrees of freedom.) We have an indication that only the first, second, third, and sixth parameters are needed since only for these is $F > 4.35$. The first parameter is β_0, the second is for coded temperature, the third is for the catalyst,

Table 6.6 Sum of Squares and F ratio for Example 6.4.2

No. of Parameters	R, Residual Sum of Squares	ΔR	F
1	1296	23328	9331.2
2	238	1058	423.2
3	188	50	20
4	183.5	4.5	1.8
5	181.5	2	0.8
6	1	180.5	72.2
7	0.5	0.5	0.2
8	0	0.5	0.2

and the sixth is for the interaction between temperature and pressure. Since the pressure enters through the interaction term, we also include the linear pressure term (fourth parameter). We assume that these provide a parsimonious set of parameters. The model then is

$$\hat{Y}_i = 54 + 11.5z_{i1} - 2.5z_{i2} + 0.75z_{i3} + 4.75z_{i1}z_{i3}$$

or, in terms of the original factors,

$$\hat{Y} = 54 + 11.5\frac{T-410}{10} - 2.5C + 0.75\frac{P-2.1\times10^6}{10^5} + 4.75\frac{T-410}{10}\,\frac{P-2.1\times10^6}{10^5}$$

where T is temperature in K, $C=-1$ for catalyst A, $C=1$ for catalyst B, and P is pressure in N/m^2. This expression can be simplified to

$$\hat{Y}=3656.5-8.825T-2.5C-0.00194P+4.75\times10^{-6}TP$$

The predicted values of $\hat{Y}_i$ are given in Table 6.7 along with the residuals.

Table 6.7 Predicted Values and Residuals for Example 6.4.2

Run No.	Observed Y_i (%)	Predicted $\hat{Y}_i$ (%)	Residual $Y_i-\hat{Y}_i$ (%)
1	49	49	0
2	62	62.5	−0.5
3	44	44	0
4	58	57.5	0.5
5	42	41	1.0
6	73	73.5	−0.5
7	35	36	−1
8	69	68.5	0.5

6.4.6 Importance of Replicates

In the above example the significance of each term in the fitted model was tested using an "external" estimate of the pure error variance. A superior estimate of the pure variance may be obtained by replicating some or all of the runs. In order to preserve the orthogonal character of a 2^r design it is necessary to repeat *all* the 2^r runs the same number of times. It is important, however, that the replicated run is a genuine repeat. Repeatedly measuring the same response does not constitute a genuine replication. It is better to interpose at least one change in a factor level before a run is repeated to obtain an independent response. A random choice of the order in which the runs are performed is important in this connection.

6.4.7 Other Experiment Designs

Many other experimental designs are possible in addition to a complete two-level factorial. For example, if there are two factors, and one is desired at two levels and the other at three, then we would use a 2×3 factorial design. The $\mathbf{X}^T\mathbf{X}$ matrix can also be made diagonal for this case.

Other possible designs are made up of treatments which are selected from among the treatments of a complete factorial design, the selection being made in such a way that the possibilities of eliminating certain parameters are tested in a systematic manner which preserves the possibility of making the remaining tests [14].

6.5 MAXIMUM LIKELIHOOD ESTIMATOR

The treatment of the maximum likelihood (ML) method by some authors leaves the impression that the ML method applies only to cases of noncorrelated errors. In the ML method discussed below, however, correlated errors may be present. Also the errors may have nonconstant variance. With only these two exceptions, the standard assumptions are used; these are denoted 11--1111. Note that the errors are assumed to have a normal probability density. It is noteworthy that ML estimation assumes a great deal of information regarding $\boldsymbol{\varepsilon}$. This is in contrast to OLS estimation.

6.5.1 ML Estimation

For the above assumptions the ML parameter estimator is derived by minimizing the ML loss function given by (6.1.67b). For the linear model $\boldsymbol{\eta}=\mathbf{X}\boldsymbol{\beta}$ we then seek to minimize

$$S_{\mathrm{ML}}=(\mathbf{Y}-\mathbf{X}\boldsymbol{\beta})^T\boldsymbol{\psi}^{-1}(\mathbf{Y}-\mathbf{X}\boldsymbol{\beta}) \tag{6.5.1}$$

with respect to the parameter vector $\boldsymbol{\beta}$. Using (6.1.26b) and (6.1.30) as in Section 6.2.1 results in

$$\boxed{\mathbf{b}_{\mathrm{ML}}=(\mathbf{X}^T\boldsymbol{\psi}^{-1}\mathbf{X})^{-1}\mathbf{X}^T\boldsymbol{\psi}^{-1}\mathbf{Y}} \tag{6.5.2}$$

By using $\mathbf{Y}=\mathbf{X}\boldsymbol{\beta}+\boldsymbol{\varepsilon}$ in (6.5.2) and taking the expected value of $\mathbf{b}_{\mathrm{ML}}$ we get

$$E(\mathbf{b}_{\mathrm{ML}})=\boldsymbol{\beta} \tag{6.5.3}$$

or $\mathbf{b}_{\mathrm{ML}}$ is an unbiased estimator of $\boldsymbol{\beta}$. The covariance matrix of $\mathbf{b}_{\mathrm{ML}}$ can be found as in Section 6.2.3. Let

$$\mathbf{b}_{\mathrm{ML}}=\mathbf{A}(\mathbf{X}\boldsymbol{\beta}+\boldsymbol{\varepsilon}) \qquad \text{where } \mathbf{A}\equiv(\mathbf{X}^T\boldsymbol{\psi}^{-1}\mathbf{X})^{-1}\mathbf{X}^T\boldsymbol{\psi}^{-1} \tag{6.5.4}$$

Then from (6.2.10) the covariance of $\mathbf{b}_{\mathrm{ML}}$ is

$$\operatorname{cov}(\mathbf{b}_{\mathrm{ML}})=\mathbf{A}\boldsymbol{\psi}\mathbf{A}^T=\left[(\mathbf{X}^T\boldsymbol{\psi}^{-1}\mathbf{X})^{-1}\mathbf{X}^T\boldsymbol{\psi}^{-1}\right]\boldsymbol{\psi}\left[\boldsymbol{\psi}^{-1}\mathbf{X}(\mathbf{X}^T\boldsymbol{\psi}^{-1}\mathbf{X})^{-1}\right]$$

$$\boxed{\operatorname{cov}(\mathbf{b}_{\mathrm{ML}})=(\mathbf{X}^T\boldsymbol{\psi}^{-1}\mathbf{X})^{-1}=\mathbf{P}_{\mathrm{ML}}} \tag{6.5.5}$$

Observe that the maximum likelihood estimator is exactly the same as given by the Gauss–Markov theorem (Section 6.1.7). Thus the maximum likelihood method produces a minimum variance unbiased linear estimator for the linear model and the standard assumptions given. The only additional assumption required by ML estimation and not by the Gauss–Markov theorem is the knowledge that the errors have a normal probability density.

The covariance of a collection of predicted points on the regression line or surface represented by $\hat{\mathbf{Y}}=\mathbf{X}_1\mathbf{b}_{\mathrm{ML}}$ is obtained in the same manner, as is (6.2.12a),

$$\operatorname{cov}(\hat{\mathbf{Y}})=\mathbf{X}_1(\mathbf{X}^T\boldsymbol{\psi}^{-1}\mathbf{X})^{-1}\mathbf{X}_1^T=\mathbf{X}_1\mathbf{P}_{\mathrm{ML}}\mathbf{X}_1^T \tag{6.5.6}$$

This expression simplifies to that given by OLS, (6.2.12b), if $\boldsymbol{\psi}=\sigma^2\mathbf{I}$.

The difference between LS and ML estimators given by (6.2.5) and (6.5.2) is clearly the presence of the $\boldsymbol{\psi}^{-1}$ matrix in (6.5.2) but not in (6.2.5). (Recall that $\boldsymbol{\psi}^{-1}$ is the inverse of the covariance matrix of the measurements errors.) If $\boldsymbol{\psi}=\sigma^2\mathbf{I}$, the estimators are exactly the same. When $\boldsymbol{\psi}$ deviates considerably from the $\sigma^2\mathbf{I}$ condition, the LS and ML estimates can be quite different. The LS estimators can have variances that are arbitrarily larger than those given by maximum likelihood. One case in which this can occur is when

$$\boldsymbol{\psi}^{-1}=\operatorname{diag}\left[\sigma_1^{-2}\sigma_2^{-2}\cdots\sigma_n^{-2}\right] \tag{6.5.7}$$

and the σ_i^2 terms are quite disparate in magnitudes such as the set of values of 0.1, 1, 10, 100, 1000, and so on.

Example 6.5.1

For $\boldsymbol{\psi}^{-1}$ given by (6.5.7) derive an expression for the ratio of the variance of b_{LS} to the variance of b_{ML} for the model $\eta_i=\beta X_i$ and simplify the result for $X_i=1$.

Solution

From (6.2.11) the variance of b_{LS} is found to be

$$V(b_{\mathrm{LS}})=\left(\sum_{i=1}^{n}X_i^2\sigma_i^2\right)\left(\sum_{j=1}^{n}X_j^2\right)^{-2}$$

while (6.5.5) yields

$$V(b_{\mathrm{ML}})=\left[\sum_{k=1}^{n}X_k^2\sigma_k^{-2}\right]^{-1}$$

Then the ratio is

$$\frac{V(b_{\text{LS}})}{V(b_{\text{ML}})} = \left(\sum_{i=1}^{n} X_i^2\sigma_i^2\right)\left(\sum_{k=1}^{n} X_k^2\sigma_k^{-2}\right)\left(\sum_{j=1}^{n} X_j^2\right)^{-2}$$

which for $X_i = 1$ reduces to

$$\frac{V(b_{\text{LS}})}{V(b_{\text{ML}})} = \frac{1}{n^2}\left(\sum_{i=1}^{n} \sigma_i^2\right)\left(\sum_{k=1}^{n} \sigma_k^{-2}\right)$$

Example 6.5.2

Investigate the solution in Example 6.5.1 for $\eta=\beta$ and $\sigma_1=1$, $\sigma_2=10$, $\sigma_3=100$, $\sigma_4=1000$, etc.

Solution

For $n=2$, the ratio is $(1/2)^2$ (101) (1.01)=25.5. For $n=3$, it is $(1/3)^2$ (10101) (1.0101)=1133.7. In general the approximate result is $10^{2(n-1)}/(.99n)^2$ which increases rapidly with n. Hence in such cases the ML estimate is far superior to the OLS estimate.

Example 6.5.3

Investigate the solution of Example 6.5.1 for $X_i = 1$, 2, 3, and 4 and for $\sigma_i = 1$, 10, 100, and 1000 respectively.

Solution

The ratio for $n = 1$, 2, 3, and 4 are respectively 1, 16.68, 480.1, and 18610. The ratio does not increase as rapidly for the model $\eta=\beta$ but still indicates the superiority of maximum likelihood estimation compared with ordinary least squares for unequal error variances.

If the measurement errors are correlated, the ψ matrix is not diagonal. Some cases that produce nondiagonal ψ matrices are those involving autoregressive and moving average errors. See Sections 5.13 and 6.9.

Typical terms in $\mathbf{X}^T\psi^{-1}\mathbf{X}$ and $\mathbf{X}^T\psi^{-1}\mathbf{Y}$ for the single response case are

$$\mathbf{X}^T\psi^{-1}\mathbf{X} = \left[\sum_{k=1}^{n}\sum_{l=1}^{n} W_{kl}X_{ki}X_{lj}\right], \qquad i,j=1,2,\ldots,p \tag{6.5.8a}$$

$$\mathbf{X}^T\psi^{-1}\mathbf{Y} = \left[\sum_{k=1}^{n}\sum_{l=1}^{n} W_{kl}X_{ki}Y_l\right], \qquad i=1,\ldots,p \tag{6.5.8b}$$

where W_{kl} is the kl component of ψ^{-1}. If ψ^{-1} is given by (6.5.7) the double summations above can be replaced by single summations as in

$$\mathbf{X}^T\psi^{-1}\mathbf{X}=\left[\sum_{k=1}^{n}\sigma_k^{-2}X_{ki}X_{kj}\right], \qquad i,j=1,2,\ldots,p \tag{6.5.9a}$$

$$\mathbf{X}^T\psi^{-1}\mathbf{Y}=\left[\sum_{k=1}^{n}\sigma_k^{-2}X_{ki}Y_k\right], \qquad i=1,2,\ldots,p \tag{6.5.9b}$$

6.5.2 Estimation of σ^2

One advantage of the ML formulation is that it can provide a direct method for evaluating certain "statistical" parameters such as the variance of the errors. For example, assume that the covariance matrix is known except for the multiplicative constant σ^2 (assumptions 11--1011),

$$\psi=\sigma^2\boldsymbol{\Omega} \tag{6.5.10}$$

where $\boldsymbol{\Omega}$ is completely known but σ^2 is unknown. We start the analysis with the logarithm of the likelihood function as given by (6.1.67a), with $n=N$)

$$\ln L(\boldsymbol{\beta}|\mathbf{Y})=-\frac{1}{2}\left[n\ln 2\pi+\ln(\sigma^{2n}|\boldsymbol{\Omega}|)+S_{\mathrm{ML}}\right] \tag{6.5.11a}$$

where

$$S_{\mathrm{ML}}=\sigma^{-2}(\mathbf{Y}-\boldsymbol{\eta})^T\boldsymbol{\Omega}^{-1}(\mathbf{Y}-\boldsymbol{\eta}) \tag{6.5.11b}$$

Take the derivative with respect to σ^2 and with respect to the parameters $\beta_1,\ldots,\beta_p$ for the linear model $\boldsymbol{\eta}=\mathbf{X}\boldsymbol{\beta}$. Set $\boldsymbol{\beta}=\mathbf{b}_{\mathrm{ML}}$ and $\sigma^2=\hat{\sigma}^2_{\mathrm{ML}}$; then setting the derivatives equal to zero gives

$$-\frac{1}{2}\left[n\left(\hat{\sigma}^2_{\mathrm{ML}}\right)^{-1}-\left(\hat{\sigma}^2_{\mathrm{ML}}\right)^{-2}R^{\Omega}_{\mathrm{ML}}(\mathbf{b}_{\mathrm{ML}})\right]=0 \tag{6.5.12}$$

$$-\frac{1}{2}\left(\hat{\sigma}^2_{\mathrm{ML}}\right)^{-1}\left.\frac{\partial S_{\mathrm{ML}}}{\partial\beta_i}\right|_{\boldsymbol{\beta}=\mathbf{b}_{\mathrm{ML}}}=0, \qquad i=1,2,\ldots,p \tag{6.5.13}$$

where

$$R^{\Omega}_{\mathrm{ML}}(\mathbf{b}_{\mathrm{ML}})=(\mathbf{Y}-\hat{\mathbf{Y}})^T\boldsymbol{\Omega}^{-1}(\mathbf{Y}-\hat{\mathbf{Y}}), \qquad \hat{\mathbf{Y}}=\mathbf{X}\mathbf{b}_{\mathrm{ML}} \tag{6.5.14}$$

Note that the physical or structural parameters, $\boldsymbol{\beta}$, can be estimated directly from (6.5.13) without knowledge of σ^2. Using these estimated values permits the estimation of σ^2 from (6.5.12) as

$$\hat{\sigma}^2_{\mathrm{ML}} = \frac{1}{n} R^{\Omega}_{\mathrm{ML}}(\mathbf{b}_{\mathrm{ML}}) = \frac{1}{n}\left[\mathbf{Y}^T\Omega^{-1}\mathbf{Y} - \mathbf{b}^T_{\mathrm{ML}}\mathbf{X}^T\Omega^{-1}\mathbf{Y}\right] \tag{6.5.15}$$

An advantage of the maximum likelihood method is that it can provide a direct method for estimating σ^2. The estimator $\hat{\sigma}^2_{\mathrm{ML}}$ is unfortunately biased, however. An unbiased estimator for σ^2 is

$$s^2 = \frac{1}{n-p} R^{\Omega}_{\mathrm{ML}}(\mathbf{b}_{\mathrm{ML}}) \tag{6.5.16}$$

See Section 6.8.3 for a derivation. Since (6.5.16) is unbiased, it is recommended for estimating σ^2 when $\psi = \sigma^2\Omega$. A summary of the basic ML estimation equations is given in Appendix B.

Theorem 6.2.4 regarding an F distribution can also be stated for the ML assumptions 11--1011 for a linear model. The statistic

$$\frac{R^{\Omega}_{\mathrm{ML}}\left[\mathbf{b}_{1,\mathrm{ML}}(\boldsymbol{\beta}_2^*), \boldsymbol{\beta}_2^*\right] - R^{\Omega}_{\mathrm{ML}}(\mathbf{b}_{1,\mathrm{ML}}, \mathbf{b}_{2,\mathrm{ML}})}{s^2 q} \tag{6.5.17}$$

has the $F_{1-\alpha}(q, n-p)$ distribution if $\boldsymbol{\beta}_2 = \boldsymbol{\beta}_2^*$. This result can be used to build "parsimonious" models.

Example 6.5.4

The thermal conductivity, k, of Armco iron has been measured using a new method discussed in Beck and Al-Araji [15]. Temperatures between 100 and 362°F (311 and 456 K) were covered along with power inputs between 272 and 602 W as given in Table 6.8. The temperature and power can be considered to be measured so much more accurately than is the measured k that the error will be assumed to be in the k measurement only. Also it is reasonable to assume additive errors. Comparison of values of measurements from different laboratories frequently indicate nonzero mean for the errors. Nevertheless a zero mean value for ϵ_i is assumed.

Each Y_i value given in the last column of Table 6.8 is the average of four values of the conductivity. The values are given to varying significant figures because the original values were given to four figures and sometimes the average yielded four digits and other times more. Notice that there are approximately two power levels, about 275 and 550. The estimated standard deviations found using the four observations for each P and T for the smaller powers is about 0.278 whereas for the higher power the estimated standard deviation is 0.16, smaller by a factor of about 2. Hence the variance for the even-numbered runs of Table 6.8 is about four times those for the odd-numbered ones which are for the larger powers.

Table 6.8 Data for Thermal Conductivity Measurements of Armco Iron for Example 6.5.4

Run No.	Temp. (°F)	Power (W)	Measured Thermal Conductivity (Btu/hr-ft-°F)
1	100	545	41.60
2	90	276	42.345
3	161	602	37.7875
4	149	275	39.5375
5	227	538	36.4975
6	206	274	37.3525
7	270	550	35.785
8	247	274	36.36
9	362	522	34.53
10	352	272	33.915

The objective is to find a parsimonious model of thermal conductivity versus temperature T and power P using maximum likelihood. The errors ε_i are assumed to be independent and to have a normal probability density.

Solution

The assumptions mentioned above of additive, zero mean, independent, and normal errors in Y_i but errorless T_i and P_i can be designated 11011011. We can write the covariance matrix ψ as

$$\psi = \sigma^2 \operatorname{diag}[1 \ \ 4 \ \ 1 \ \ 4 \ \ 1 \ \ 4 \ \ 1 \ \ 4 \ \ 1 \ \ 4]$$

where σ^2 is unknown. In terms of the parameter estimates the statistical parameter σ^2 need not be known. An estimate of σ^2 is given by using the standard deviation values quoted in the example.

A plot of the data is shown by Fig. 6.3. It appears that at least a second degree curve in T may be required and that P may not be needed. From physical considerations it is also expected that k is not a function of P. There are many possible models and a number of ways of proceeding. Some possible models are as follows:

1. $k = \beta_1$.
2. $k = \beta_1 + \beta_2 T$.
3. $k = \beta_1 + \beta_2 P$.
4. $k = \beta_1 + \beta_2 T + \beta_3 P$.
5. $k = \beta_1 + \beta_2 T + \beta_3 T^2$.
6. $k = \beta_1 + \beta_2 T + \beta_3 T^2 + \beta_4 T^3$.

One way to build the model is to start with the simplest and add one term at a time. As one progresses, the need of adding terms can be assessed by examining the

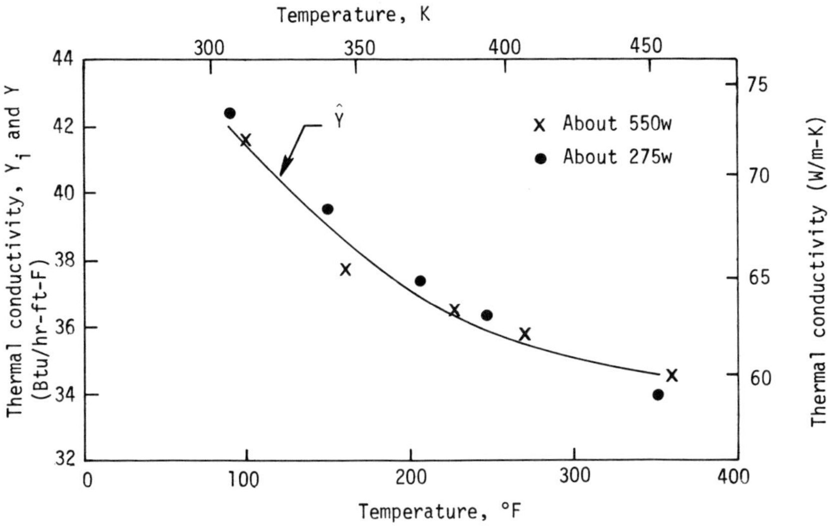

Figure 6.3 Thermal conductivities measured for Armco iron and used in Example 6.5.4.

reduction in the R_{ML}^{Ω} value and by utilizing the F test based on Theorem 6.2.4.

With ψ as given above the parameters were estimated using (6.5.2) with the components given by (6.5.9). The mean square, s^2, is found from (6.5.16). The results of the calculations including the F statistic $(=\Delta R/s^2)$ are tabulated in Table 6.9. These F values can be compared with those given by $F_{.95}(1, n-p)$ which are 5.99, 5.59, 5.32, and 5.12 for $n-p=6$, 7, 8, and 9, respectively. From a comparison of these F values with those in Table 6.9, it appears that the power factor, P_i, need not be used. Also Theorem 6.2.4 leads us to select Model 5 because Model 6 has an unnecessary additional term since $F=5.729<5.99=F_{.95}(1,6)$. However, since these values of 5.729 and 5.99 are so close, it might be that another

Table 6.9 Results of Calculations of R and F for Example 6.5.4

Model No.	No. of Parameters	Degrees of Freedom	R_{ML}^{Ω}	Mean Square, s^2	ΔR^a	F
1	1	9	40.0078	4.445	8769.36	1973
2	2	8	4.47506	0.5594	35.5327	63.52
3	2	8	39.91282	4.9891	0.09495	0.01903
4	3	7	4.18133	0.5973	0.29373	0.4918
5	3	7	1.13092	0.1616	3.34413	20.70
6	4	6	0.57852	0.09642	0.55241	5.729

[a] $\Delta R_1=\mathbf{Y}^T\mathbf{Y}-R_{\text{ML},1}^{\Omega}$, $\Delta R_2=R_{\text{ML},1}^{\Omega}-R_{\text{ML},2}^{\Omega}$, $\Delta R_3=R_{\text{ML},1}^{\Omega}-R_{\text{ML},3}^{\Omega}$, $\Delta R_4=R_{\text{ML},2}^{\Omega}-R_{\text{ML},4}^{\Omega}$, etc.

experiment or a larger sample size might show significance, that is, indicate Model 6 in preference to Model 5.

For Model 5 the regression equation for k is (in English units)

$$\hat{Y}=47.5191-0.07084T+9.669\times 10^{-5}T^2$$

and the estimated covariance matrix of the estimated parameter vector is

$$\text{est. cov}(\mathbf{b}_{\text{ML}})=s^2\begin{bmatrix} 6.741 & -6.133\times 10^{-2} & 1.242\times 10^{-4} \\ & 6.080\times 10^{-4} & -1.282\times 10^{-6} \\ \text{symmetric} & & 2.796\times 10^{-9} \end{bmatrix}$$

where $s^2=0.1616$. The variances of $b_{1,\text{ML}}$, $b_{2,\text{ML}}$, and $b_{3,\text{ML}}$ are the diagonal terms. The estimated standard errors (std. dev.) for them are

$$\text{est. s.e.}(b_{1,\text{ML}})=[0.1616(6.741)]^{1/2}=1.04$$

$$\text{est. s.e.}(b_{2,\text{ML}})=\left[0.1616(6.080\times 10^{-4})\right]^{1/2}=0.00991$$

$$\text{est. s.e.}(b_{3,\text{ML}})=2.13\times 10^{-5}$$

The sizes of the estimated standard errors of the $b_{i,\text{ML}}$'s do not reveal the relative importance of the terms in describing k. The estimated standard error of $b_{1,\text{ML}}$ is 1.04, whereas one standard error in $b_{3,\text{ML}}$ gives a value of 2.6 for the T^2 term of $\hat{Y}$ for $T=350°\text{F}$. Hence the effects on k due to uncertainty in $b_{1,\text{ML}}$ and $b_{3,\text{ML}}$ are about the same.

Example 6.5.5

Find the estimated standard error in $\hat{Y}_i$ for the data of Example 6.5.4.

Solution

The estimated standard error of $\hat{Y}_i$ is found by taking the square root of the diagonal terms of (6.5.6) with $\mathbf{P}_{\text{ML}}$ replaced by est. cov($\mathbf{b}_{\text{ML}}$) which is displayed just above. For arbitrary T_i an expression for est. $V(\hat{Y}_i)$, the ith diagonal term of est. cov($\hat{Y}_i$), is

$$\text{est. } V(\hat{Y}_i)=\left[P_{11}^*+2T_iP_{12}^*+T_i^2(P_{22}^*+2P_{13}^*)+2T_i^3P_{23}^*+T_i^4P_{33}^*\right]s^2$$

where $P_{ij}^*=P_{ij}/\sigma^2$. From above, note that a typical P_{ij}^* is $P_{11}^*=6.741$. Also $s^2=0.1616$. The est. s.e.($\hat{Y}_i$), which is the square root of est. $V(\hat{Y}_i)$, is displayed as the curve in Fig. 6.4.

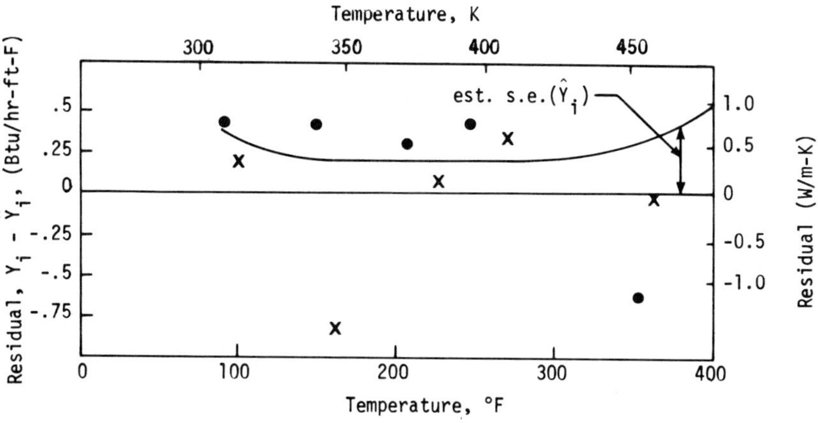

Figure 6.4 Residuals of thermal conductivity values for Example 6.5.4.

6.5.3 Expected Values of S_{ML} and R_{ML}

The expected value of S_{ML} (parameters not estimated) given by (6.5.1) using (6.1.50) can be written with $\boldsymbol{\psi}$ being known as

$$E(S_{\text{ML}}) = E\left[(\mathbf{Y}-\mathbf{X}\boldsymbol{\beta})^T\boldsymbol{\psi}^{-1}(\mathbf{Y}-\mathbf{X}\boldsymbol{\beta})\right] = E\left[\boldsymbol{\varepsilon}^T\boldsymbol{\psi}^{-1}\boldsymbol{\varepsilon}\right]$$

$$= \operatorname{tr}\left[\boldsymbol{\psi}^{-1}\operatorname{cov}\boldsymbol{\varepsilon}\right] = \operatorname{tr}\left[\boldsymbol{\psi}^{-1}\boldsymbol{\psi}\right] = n \tag{6.5.18}$$

where n is the number of observations.

In finding $E(R_{\text{ML}})$, let

$$R_{\text{ML}} = (\mathbf{Y}-\mathbf{X}\mathbf{b}_{\text{ML}})^T\boldsymbol{\psi}^{-1}(\mathbf{Y}-\mathbf{X}\mathbf{b}_{\text{ML}}) \tag{6.5.19}$$

and use $\mathbf{A}$ defined in (6.5.4). Then (6.5.19) can be written as

$$R_{\text{ML}} = \mathbf{Y}^T(\mathbf{I}-\mathbf{X}\mathbf{A})^T\boldsymbol{\psi}^{-1}(\mathbf{I}-\mathbf{X}\mathbf{A})\mathbf{Y} = \mathbf{Y}^T\boldsymbol{\psi}^{-1}(\mathbf{I}-\mathbf{X}\mathbf{A})\mathbf{Y} \tag{6.5.20a}$$

since

$$\mathbf{X}^T\boldsymbol{\psi}^{-1}(\mathbf{I}-\mathbf{X}\mathbf{A}) = \mathbf{X}^T\boldsymbol{\psi}^{-1}\left[\mathbf{I}-\mathbf{X}(\mathbf{X}^T\boldsymbol{\psi}^{-1}\mathbf{X})^{-1}\mathbf{X}^T\boldsymbol{\psi}^{-1}\right] = \mathbf{0} \tag{6.5.20b}$$

Introducing $\mathbf{Y} = \mathbf{X}\boldsymbol{\beta} + \boldsymbol{\varepsilon}$ in (6.5.20a) yields

$$R_{\text{ML}} = \boldsymbol{\varepsilon}^T\boldsymbol{\psi}^{-1}(\mathbf{I}-\mathbf{X}\mathbf{A})\boldsymbol{\varepsilon} \tag{6.5.21}$$

where (6.5.20b) and the following relation are used

$$(\mathbf{I}-\mathbf{XA})\mathbf{X}=\left(\mathbf{I}-\mathbf{X}(\mathbf{X}^T\psi^{-1}\mathbf{X})^{-1}\mathbf{X}^T\psi^{-1}\right)\mathbf{X}=\mathbf{0} \tag{6.5.22}$$

From (6.5.21), R_{ML} can be given by

$$R_{\mathrm{ML}}=\boldsymbol{\varepsilon}^T\psi^{-1/2}\left[\mathbf{I}-\psi^{-1/2}\mathbf{X}(\mathbf{X}^T\psi^{-1}\mathbf{X})^{-1}\mathbf{X}^T\psi^{-1/2}\right]\psi^{-1/2}\boldsymbol{\varepsilon} \tag{6.5.23}$$

Then taking the expected value of (6.5.23) and utilizing (6.1.50) we find

$$E(R_{\mathrm{ML}})=\mathrm{tr}\left\{\left[\mathbf{I}-\psi^{-1/2}\mathbf{X}(\mathbf{X}^T\psi^{-1}\mathbf{X})^{-1}\mathbf{X}^T\psi^{-1/2}\right]\mathrm{cov}(\psi^{-1/2}\boldsymbol{\varepsilon})\right\}$$

$$=\mathrm{tr}(I)-\mathrm{tr}\left[(X^T\psi^{-1}\mathbf{X})^{-1}(\mathbf{X}^T\psi^{-1}\mathbf{X})\right]$$

$$E(R_{\mathrm{ML}})=n-p \tag{6.5.24}$$

The above expression for $E(R_{\mathrm{ML}})$ can be utilized to check the validity of the model and the ψ matrix. It is important to have a check because in many cases there is some uncertainty in $\boldsymbol{\eta}$ or in ψ. For the assumptions used above (11--1111), R_{ML} has a χ^2 distribution with $n-p$ degrees of freedom; this then can provide a statistical test. Table 2.14 gives the χ_1^2 value for which $P(\chi^2<\chi_1^2)$ is a specified value. For example, there is a column of values of χ^2 for which the probability of χ^2 being less than χ_1^2 is equal to .95, or

$$P(\chi^2<\chi_1^2)=.95 \tag{6.5.25}$$

Example 6.5.6

For a second set of data similar to those given in Example 6.5.4, ML estimation has been used. There are 10 separate measurements of the thermal conductivity as a function of temperature and power input. The model used is the one found from the preceding analysis,

$$k=\beta_1+\beta_2T+\beta_3T^2$$

and the errors in k are assumed to be independent and the ψ matrix based on the results of Example 6.5.4 is

$$\psi=0.16\,\mathrm{diag}[1\ \ 4\ \ 1\ \ 4\ \ 1\ \ 4\ \ 1\ \ 4\ \ 1\ \ 4]$$

For the analysis based on these data, this model and ψ yielded $R_{\mathrm{ML}}=13.52$. Assume that the standard assumptions indicated by 11011111 are valid and investigate the "goodness of fit" using the χ^2 distribution.

Solution

Since there are 10 observations ($n = 10$) and three parameters, $n - p = 7$ degrees of freedom for which a table of the χ^2 distribution gives

$$P(\chi^2 < 14.07) = .95, \qquad P(\chi^2 < 2.167) = .05$$

This means that only in 5% of similar cases would $R_{ML} = \chi^2$ exceed 14.07. Also only in 5% of the cases would R_{ML} be less than 2.107. Since $2.167 < 13.52 < 14.07$, we have an indication at the 10% level of significance that the model and ψ are satisfactory.

An important source of error which could cause R_{ML} to be either too small or too large (compared to an interval based in the χ^2 distribution) is an incorrect choice of ψ. For example, ψ might be assumed to be diagonal while the measurement errors are correlated causing ψ to be nondiagonal. For this reason it is always advisable to investigate if the residuals suggest correlation among the errors. Such correlation can be inherent in the measurements themselves or a result of selection of an incorrect model.

Often the model is dictated by physical considerations. In other cases there may be several models that fit equally well.

6.6 LINEAR MAXIMUM A POSTERIORI ESTIMATOR (MAP)

6.6.1 Introduction

Maximum a posteriori estimation utilizes prior information regarding the parameters in addition to information regarding the measurement errors. Inclusion of prior parameter information can have the beneficial effect of reduction of variances of parameter estimators.

In Section 5.4 two MAP cases are considered. The first is for random parameters. One convenient way to visualize this case is to think of some product that is produced in "batches," each of which is different. The prior information is relative to the mean and variance of these batches. The measurements $\mathbf{Y}$, however, deal with a specific (new) batch.

In the second MAP case, the information is visualized as coming from subjective information—belief of an investigator regarding the parameters. This could include mean and variances. In this case the parameters are not random, for example, a parameter of interest might be a fundamental constant such as the speed of light. The *knowledge* about the parameters is probabilistic. This probabilistic information leads us to view the parameters as having probability distributions, similar to the random parameter case.

In both cases or views the derived parameter estimators are formally the same but the meaning of the various terms may be different. In order to be succinct the case of random parameters is treated first and the results for subjective information are given without derivations.

6.6.2 Assumptions

Let us consider the case of MAP estimation involving a random parameter vector. We shall investigate estimation for the standard assumptions denoted 11--1112. Each assumption except for the last is the same as used for ML estimation in Section 6.5. Mathematically the assumptions can be given as

$$\mathbf{Y}=E(\mathbf{Y}|\boldsymbol{\beta})+\boldsymbol{\varepsilon}, \qquad \boldsymbol{\varepsilon}\sim N(0,\boldsymbol{\psi}) \tag{6.6.1a,b}$$

$$V(X_{ij})=0,\ \boldsymbol{\beta}\sim N(\boldsymbol{\mu}_\beta,\mathbf{V}_\beta), \qquad \operatorname{cov}(\boldsymbol{\beta},\boldsymbol{\varepsilon})=\mathbf{0} \tag{6.6.1c,d,e}$$

where $\boldsymbol{\psi}$, $\boldsymbol{\mu}_\beta$, and $\mathbf{V}_\beta$ are known. Note that the distributions for both $\boldsymbol{\beta}$, the random parameter column vector, and $\boldsymbol{\varepsilon}$ are normal.

6.6.3 Estimation Involving Random Parameters

In MAP estimation the estimated parameter vector is the one that maximizes the probability density $f(\mathbf{Y}|\boldsymbol{\beta})$. This density is related to $f(\boldsymbol{\beta}|\mathbf{Y})$ and that for the random parameter, $f(\boldsymbol{\beta})$, by

$$f(\boldsymbol{\beta}|\mathbf{Y})=\frac{f(\mathbf{Y}|\boldsymbol{\beta})f(\boldsymbol{\beta})}{f(\mathbf{Y})} \tag{6.6.2}$$

which is a form of Bayes's theorem. Recall that $f(\mathbf{Y}|\boldsymbol{\beta})$ is the same density used in maximum likelihood estimation; for the given assumptions it is given by (6.1.66). For the conditions indicated by (6.6.1d), $f(\boldsymbol{\beta})$ is

$$f(\boldsymbol{\beta})=(2\pi)^{-p/2}|\mathbf{V}_\beta|^{-1/2}\exp\left[-\frac{1}{2}(\boldsymbol{\beta}-\boldsymbol{\mu}_\beta)^T\mathbf{V}_\beta^{-1}(\boldsymbol{\beta}-\boldsymbol{\mu}_\beta)\right] \tag{6.6.3}$$

where p is the number of parameters. The probability density $f(\mathbf{Y})$ need not be explicitly given since it is not a function of $\boldsymbol{\beta}$.

The maximum of $f(\boldsymbol{\beta}|\mathbf{Y})$ given by (6.6.2) occurs at the same parameter values as does the maximum of its natural logarithm,

$$\ln\left[f(\boldsymbol{\beta}|\mathbf{Y})\right]=-\frac{1}{2}\left[(n+p)\ln 2\pi+\ln|\boldsymbol{\psi}|+\ln|\mathbf{V}_\beta|+S_{\text{MAP}}\right]-\ln f(\mathbf{Y}) \tag{6.6.4a}$$

$$S_{\text{MAP}}\equiv(\mathbf{Y}-\boldsymbol{\eta})^T\boldsymbol{\psi}^{-1}(\mathbf{Y}-\boldsymbol{\eta})+(\boldsymbol{\beta}-\boldsymbol{\mu}_\beta)^T\mathbf{V}_\beta^{-1}(\boldsymbol{\beta}-\boldsymbol{\mu}_\beta) \tag{6.6.4b}$$

When the only parameters of interest occur in $\boldsymbol{\beta}$, not in $\boldsymbol{\psi}$, for example, maximizing $f(\boldsymbol{\beta}|\mathbf{Y})$ can be accomplished by minimizing S_{MAP}.

Following the usual procedure of taking the matrix derivatives with respect to $\boldsymbol{\beta}$, etc., one finds using (6.6.4b), (6.1.33), and (6.1.26b),

$$\nabla_\beta S_{\text{MAP}}|_{\mathbf{b}_{\text{MAP}}} = 2\left[-\mathbf{X}^T\boldsymbol{\psi}^{-1}\mathbf{Y} + \mathbf{X}^T\boldsymbol{\psi}^{-1}\mathbf{X}\mathbf{b}_{\text{MAP}} - \mathbf{V}_\beta^{-1}\boldsymbol{\mu}_\beta + \mathbf{V}_\beta^{-1}\mathbf{b}_{\text{MAP}}\right] = \mathbf{0} \tag{6.6.5}$$

Solving for the estimator $\mathbf{b}_{\text{MAP}}$ yields

$$\mathbf{b}_{\text{MAP}} = \mathbf{P}_{\text{MAP}}\left[\mathbf{X}^T\boldsymbol{\psi}^{-1}\mathbf{Y} + \mathbf{V}_\beta^{-1}\boldsymbol{\mu}_\beta\right], \qquad \mathbf{P}_{\text{MAP}}^{-1} \equiv \mathbf{X}^T\boldsymbol{\psi}^{-1}\mathbf{X} + \mathbf{V}_\beta^{-1} \tag{6.6.6a, b}$$

Notice the definition of $\mathbf{P}_{\text{MAP}}$ given by (6.6.6b). By adding and subtracting $2\mathbf{X}^T\boldsymbol{\psi}^{-1}\mathbf{X}\boldsymbol{\mu}_\beta$ from (6.6.5), the additional expression of

$$\boxed{\mathbf{b}_{\text{MAP}} = \boldsymbol{\mu}_\beta + \mathbf{P}_{\text{MAP}}\mathbf{X}^T\boldsymbol{\psi}^{-1}(\mathbf{Y} - \mathbf{X}\boldsymbol{\mu}_\beta)} \tag{6.6.6c}$$

can be given for $\mathbf{b}_{\text{MAP}}$. In this form the second term on the right side can be considered as a correction to the known mean value of the random parameter vector $\boldsymbol{\mu}_\beta$; it is a result of the new information, $\mathbf{Y}$, for a given "batch." For the case of Model 2, $\eta_i = \beta_1 X_i$, (6.6.6a, b, c) reduces to (5.4.11a, b). Since $\mathbf{Y} = \mathbf{X}\boldsymbol{\beta} + \boldsymbol{\varepsilon}$, $E(\boldsymbol{\beta}) = \boldsymbol{\mu}_\beta$, and $E(\boldsymbol{\varepsilon}) = \mathbf{0}$, use of (6.6.6c) shows that the expected value of $\mathbf{b}_{\text{MAP}}$ is

$$E(\mathbf{b}_{\text{MAP}}) = \boldsymbol{\mu}_\beta \tag{6.6.7}$$

and thus $\mathbf{b}_{\text{MAP}}$ is a biased estimator. Even so it is recommended whenever appropriate owing to the reduced error covariance matrix as shown below. An example of MAP estimation is given in Section 6.7.4 in connection with the sequential MAP method.

The covariance matrix of interest is that of $\mathbf{b}_{\text{MAP}} - \boldsymbol{\beta}$ as explained in connection with (5.4.12) and 5.4.13); $\mathbf{b}_{\text{MAP}} - \boldsymbol{\beta}$ is the difference between the estimator and the parameter vector for the *particular* batch. Utilizing (6.6.6a) we can write

$$\mathbf{b}_{\text{MAP}} - \boldsymbol{\beta} = \mathbf{P}_{\text{MAP}}\mathbf{X}^T\boldsymbol{\psi}^{-1}(\mathbf{X}\boldsymbol{\beta} + \boldsymbol{\varepsilon}) - \boldsymbol{\beta} + \mathbf{P}_{\text{MAP}}\mathbf{V}_\beta^{-1}\boldsymbol{\mu}_\beta \tag{6.6.8a}$$

$$= (\mathbf{P}_{\text{MAP}}\mathbf{X}^T\boldsymbol{\psi}^{-1}\mathbf{X} - \mathbf{I})\boldsymbol{\beta} + \mathbf{P}_{\text{MAP}}\mathbf{X}^T\boldsymbol{\psi}^{-1}\boldsymbol{\varepsilon} + \mathbf{P}_{\text{MAP}}\mathbf{V}_\beta^{-1}\boldsymbol{\mu}_\beta \tag{6.6.8b}$$

Now taking the covariance of $\mathbf{b}_{\text{MAP}} - \boldsymbol{\beta}$, the error covariance matrix of

$\mathbf{b}_{\text{MAP}}$, yields

$$\operatorname{cov}(\mathbf{b}_{\text{MAP}}-\boldsymbol{\beta})=(\mathbf{P}_{\text{MAP}}\mathbf{X}^T\boldsymbol{\psi}^{-1}\mathbf{X}-\mathbf{I})\mathbf{V}_\beta(\mathbf{P}_{\text{MAP}}\mathbf{X}^T\boldsymbol{\psi}^{-1}\mathbf{X}-\mathbf{I})^T$$
$$+\mathbf{P}_{\text{MAP}}\mathbf{X}^T\boldsymbol{\psi}^{-1}\boldsymbol{\psi}\boldsymbol{\psi}^{-1}\mathbf{X}\mathbf{P}_{\text{MAP}} \tag{6.6.9}$$

where (6.1.40) is used. Expanding the right side of (6.6.9) produces

$$\operatorname{cov}(\mathbf{b}_{\text{MAP}}-\boldsymbol{\beta})=\mathbf{P}_{\text{MAP}}\mathbf{X}^T\boldsymbol{\psi}^{-1}\mathbf{X}\mathbf{V}_\beta\mathbf{X}^T\boldsymbol{\psi}^{-1}\mathbf{X}\mathbf{P}_{\text{MAP}}-\mathbf{P}_{\text{MAP}}\mathbf{X}^T\boldsymbol{\psi}^{-1}\mathbf{X}\mathbf{V}_\beta$$
$$-\mathbf{V}_\beta\mathbf{X}^T\boldsymbol{\psi}^{-1}\mathbf{X}\mathbf{P}_{\text{MAP}}+\mathbf{V}_\beta+\mathbf{P}_{\text{MAP}}\mathbf{X}^T\boldsymbol{\psi}^{-1}\mathbf{X}\mathbf{P}_{\text{MAP}} \tag{6.6.10a}$$
$$=-\mathbf{V}_\beta\mathbf{X}^T\boldsymbol{\psi}^{-1}\mathbf{X}\mathbf{P}_{\text{MAP}}+\mathbf{V}_\beta$$
$$=-\mathbf{V}_\beta\mathbf{X}^T\boldsymbol{\psi}^{-1}\mathbf{X}\mathbf{P}_{\text{MAP}}+\mathbf{V}_\beta\mathbf{P}_{\text{MAP}}^{-1}\mathbf{P}_{\text{MAP}} \tag{6.6.10b}$$
$$=\mathbf{V}_\beta\left[-\mathbf{X}^T\boldsymbol{\psi}^{-1}\mathbf{X}+\mathbf{P}_{\text{MAP}}^{-1}\right]\mathbf{P}_{\text{MAP}}$$
$$=\mathbf{V}_\beta\left[-\mathbf{X}^T\boldsymbol{\psi}^{-1}\mathbf{X}+\mathbf{X}^T\boldsymbol{\psi}^{-1}\mathbf{X}+\mathbf{V}_\beta^{-1}\right]\mathbf{P}_{\text{MAP}} \tag{6.6.10c}$$

and thus the covariance of $\mathbf{b}_{\text{MAP}}-\boldsymbol{\beta}$ is

$$\boxed{\operatorname{cov}(\mathbf{b}_{\text{MAP}}-\boldsymbol{\beta})=\mathbf{P}_{\text{MAP}}=\left[\mathbf{X}^T\boldsymbol{\psi}^{-1}\mathbf{X}+\mathbf{V}_\beta^{-1}\right]^{-1}} \tag{6.6.11}$$

This is valid for the stated assumptions which are denoted 11--1112. Note that the effect of the random parameter behavior disappears as $\mathbf{X}^T\boldsymbol{\psi}^{-1}\mathbf{X}$ is made large, which usually results from a large number of measurements.

A summary of the MAP estimation equations is given in Appendix B.

6.6.4 Estimation with Subjective Information

Consider now the case of constant parameters with subjective information about them. This information can be expressed in probabilistic terms if we imagine that the parameters have probability distributions, that is, are random. A close analogue can be drawn with the random parameter case. Now $\boldsymbol{\mu}_\beta$ becomes the prior parameter vector based on belief and $\mathbf{V}_\beta$ is the covariance matrix for a normal distribution. That is, our knowledge or belief regarding $\boldsymbol{\beta}$ is expressed by the probability density, $f(\boldsymbol{\beta})=N(\boldsymbol{\mu}_\beta,\mathbf{V}_\beta)$. We can conceive that this knowledge has been developed from investigation of many tests on similar "batches," from literature values or from other sources. This information is independent of that obtained from the new information contained in $\mathbf{Y}$. With these assumptions and those used regarding $\mathbf{Y}$ above, the assumptions are denoted 11--1113. The estimator and covariance matrix are then the same as derived in Section 6.6.3. Now, however, $\boldsymbol{\mu}_\beta$ and $\mathbf{V}_\beta$ have different meanings.

Let us briefly consider certain implications of the MAP estimators for subjective prior information. Suppose first that the prior information is very poor. This implies that the $\mathbf{V}_\beta$ matrix has large diagonal components. If $\mathbf{V}_\beta$ has large diagonal components, $\mathbf{b}_{\text{MAP}}$ given by (6.6.6a) approaches the $\mathbf{b}_{\text{ML}}$ expression given by (6.5.2) and $\text{cov}(\mathbf{b}_{\text{MAP}}-\boldsymbol{\beta})$ approaches $\text{cov}(\mathbf{b}_{\text{ML}})$ given by (6.5.5). Hence a computer program developed for MAP estimation could be also used for ML estimation (for the assumptions denoted 11--1111). The same program could be also used for OLS estimation by letting $\mathbf{V}_\beta$ have large diagonal components and by replacing $\boldsymbol{\psi}$ by $\sigma^2\mathbf{I}$. If $\boldsymbol{\psi}$ were equal to $\sigma^2\mathbf{I}$ and all the standard assumptions were valid, we would have

$$\mathbf{b}_{\text{MAP}}=\mathbf{b}_{\text{ML}}=\mathbf{b}_{\text{LS}}$$

If the standard assumptions are not valid, the same estimator for OLS given by (6.2.5) is obtained from (6.6.6) by simply replacing $\boldsymbol{\psi}^{-1}$ by $\mathbf{I}$ and by setting $\boldsymbol{\mu}_\beta=\mathbf{0}$ and $\mathbf{V}_\beta^{-1}=\mathbf{0}$; in this case, $\text{cov}(\mathbf{b}_{\text{MAP}}-\boldsymbol{\beta})$ will not yield the correct relation for $\text{cov}(\mathbf{b}_{\text{LS}})$, however.

In (6.6.6) all components of $\mathbf{b}_{\text{MAP}}$ can be uniquely found if

$$|\mathbf{P}_{\text{MAP}}^{-1}|=|\mathbf{X}^T\boldsymbol{\psi}^{-1}\mathbf{X}+\mathbf{V}_\beta^{-1}|\neq 0 \tag{6.6.12}$$

Hence for MAP estimations it is neither necessary that $|\mathbf{X}^T\mathbf{X}|\neq 0$ nor that $n\geqslant p$. In both ML and LS estimation, $\boldsymbol{\beta}$ cannot be estimated if $|\mathbf{X}^T\mathbf{X}|=0$ which is the case if $n<p$ or if there is linear dependence among the sensitivity coefficients. The fact that $\mathbf{b}_{\text{MAP}}$ estimates can be obtained for n as small as 1 is used in the sequential method discussed in the next section.

6.6.5 Uncertainty in $\boldsymbol{\psi}$

One major difficulty in using ML and MAP estimation is that the standard assumptions may not be valid. In particular, the $\boldsymbol{\psi}$ matrix may not be known. There are certain checks and corrections that can be used when there is some uncertainty in $\boldsymbol{\psi}$.

One check involves the expected value of S_{MAP} which is called the *prior* value because the new data are not yet used to obtain parameter estimates. Using (6.1.50), $E(S_{\text{MAP}})$ becomes

$$E(S_{\text{MAP}})=\text{tr}\left[\boldsymbol{\psi}^{-1}\boldsymbol{\psi}\right]+\text{tr}\left[\mathbf{V}_\beta^{-1}\mathbf{V}_\beta\right]=n+p \tag{6.6.13}$$

Now we also know for ML that $E(S_{\text{ML}})=n$ and $E(R_{\text{ML}})=n-p$. Hence it is reasonable to assume that the expected value of R_{MAP}, the minimum value of S_{MAP}, is about equal to n. Again, if $n \gg p$, the difference between $n+p$ and n is relatively small. Hence for many cases R_{MAP} can be considered to have a χ^2 distribution with n degrees of freedom. (This is true provided the assumptions designated 11--1112 or 11--1113 are valid.) A related example is given by Example 6.5.6.

Suppose next that there is uncertainty in ψ; let ψ be equal to $\sigma^2\boldsymbol{\Omega}$, where σ^2 is unknown and $\boldsymbol{\Omega}$ is known. Both σ^2 and $\boldsymbol{\beta}$ can be estimated by maximizing $f(\boldsymbol{\beta}|\mathbf{Y})$ with respect to σ^2 to find its MAP estimator and with respect to $\boldsymbol{\beta}$ to find the mode of the distribution of the $\boldsymbol{\beta}$. After taking the derivative of $\ln f(\boldsymbol{\beta}|\mathbf{Y})$ given by (6.6.4) with respect to $\boldsymbol{\beta}$ and σ^2 and setting the resulting equations equal to zero where $\boldsymbol{\beta}=\mathbf{b}_{\text{MAP}}$ and $\sigma^2=\hat{\sigma}^2_{\text{MAP}}$, we obtain the set of $p+1$ equations given by (6.6.14a, b),

$$\hat{\sigma}^{-2}_{\text{MAP}}\left[-\mathbf{X}^T\boldsymbol{\Omega}^{-1}\mathbf{Y}+\mathbf{X}^T\boldsymbol{\Omega}^{-1}\mathbf{X}\mathbf{b}_{\text{MAP}}\right]-\mathbf{V}_\beta^{-1}\boldsymbol{\mu}_\beta+\mathbf{V}_\beta^{-1}\mathbf{b}_{\text{MAP}}=0 \quad (6.6.14a)$$

$$n\left(\hat{\sigma}^2_{\text{MAP}}\right)^{-1}-\left(\hat{\sigma}^2_{\text{MAP}}\right)^{-2}Q^{\Omega}_{\text{MAP}}=0 \quad (6.6.14b)$$

$$Q^{\Omega}_{\text{MAP}}=(\mathbf{Y}-\mathbf{X}\mathbf{b}_{\text{MAP}})^T\boldsymbol{\Omega}^{-1}(\mathbf{Y}-\mathbf{X}\mathbf{b}_{\text{MAP}}) \quad (6.6.14c)$$

Unfortunately these equations are no longer linear in $\hat{\sigma}^2_{\text{MAP}}$ and $\mathbf{b}_{\text{MAP}}$. Nevertheless we can solve them to obtain

$$\mathbf{b}_{\text{MAP}}=\boldsymbol{\mu}_\beta+\left[\{\mathbf{X}^T\boldsymbol{\Omega}^{-1}\mathbf{X}\}+\hat{\sigma}^2_{\text{MAP}}\mathbf{V}_\beta^{-1}\right]^{-1}\mathbf{X}^T\boldsymbol{\Omega}^{-1}(\mathbf{Y}-\mathbf{X}\boldsymbol{\mu}_\beta) \quad (6.6.15a)$$

$$\hat{\sigma}^2_{\text{MAP}}=\frac{Q^{\Omega}_{\text{MAP}}}{n} \quad (6.6.15b)$$

The nonlinearity has not been removed, but two different approaches are apparent from these equations. First, if it happens that $\hat{\sigma}^2_{\text{MAP}}\mathbf{V}_\beta^{-1}$ is known, the nonlinearity disappears and a direct solution is given for $\mathbf{b}_{\text{MAP}}$ and $\hat{\sigma}^2_{\text{MAP}}$. This case might occur when two (or more) sets of data are analyzed separately but there is a common σ^2 for all the data. Second, an iterative procedure is suggested by (6.6.15). If an initial guess for $\hat{\sigma}^2_{\text{MAP}}$ is available, it is used in (6.6.15a), and then an improved value of $\hat{\sigma}^2_{\text{MAP}}$ is found using (6.6.15b), whereupon this value is used in (6.6.15a), etc., until the changes in $\mathbf{b}_{\text{MAP}}$ and $\hat{\sigma}^2_{\text{MAP}}$ are negligible. If the initial value of $\hat{\sigma}^2_{\text{MAP}}$ were zero, the first estimator for $\boldsymbol{\beta}$ would be $\mathbf{b}_{\text{ML}}$. At the other extreme of $\hat{\sigma}^2 \to \infty$, $\mathbf{b}_{\text{MAP}}=\boldsymbol{\mu}_\beta$. In this iterative procedure, note that some matrix products, namely, those in braces in (6.6.15), need be evaluated only once.

For another case with uncertainty in ψ, see Problem 6.29, where $\psi=\sigma^2\boldsymbol{\Omega}$ and $\mathbf{V}_\beta=\sigma^2\mathbf{V}$ with σ^2 unknown and $\boldsymbol{\Omega}$ and $\mathbf{V}$ known.

6.7 SEQUENTIAL ESTIMATION

6.7.1 Introduction

The sequential estimation procedures developed in this section refer to continually updating parameter estimates as new observations are added. One of the most important advantages of this method is that matrix inverses may not be needed. Another is that the computer memory storage can be greatly reduced. Moreover, the method can be utilized to produce an "on-line" method of parameter estimation for dynamic processes. These and other advantages are discussed further at the end of this section.

The mathematical form derived for MAP estimation, (6.6.6), includes those derived for ML and OLS estimation. For ML estimation with the standard assumptions of 11--1111, (6.6.6) mathematically reduces to (6.5.2) if $\mathbf{V}_\beta^{-1} \to \mathbf{0}$. For the subjective prior information case, this corresponds to no prior information. In this case the value of $\boldsymbol{\mu}_\beta$ is unimportant (provided $\mathbf{V}_\beta^{-1}\boldsymbol{\mu}_\beta = \mathbf{0}$). For (6.6.6) to reduce to the estimator given for OLS estimation, (6.2.5), we may set $\boldsymbol{\psi} = \sigma^2\mathbf{I}$ and $\mathbf{V}_\beta^{-1} = \mathbf{0}$ in (6.6.6). Whether or not these assumptions are valid, the OLS estimator is obtained. If the assumptions denoted 11111-11 (σ^2 need not be known) are valid, then the estimates obtained using $\mathbf{b}_{\text{MAP}}$ will equal those given by $\mathbf{b}_{\text{ML}}$ and $\mathbf{b}_{\text{OLS}}$. In the sequential procedure we use the fact the ML and OLS estimates can be very closely approximated as indicated above if the $\mathbf{V}_\beta$ matrix is diagonal with large diagonal components. The sequential procedure also includes ML and OLS estimation when a set of data has been analyzed to estimate the parameters and then later this information is combined with more data; the information for the first set of data summarized by $\mathbf{b}_{\text{ML}}$ and $(\mathbf{X}^T\boldsymbol{\psi}^{-1}\mathbf{X})^{-1}$ (or $\mathbf{b}_{\text{OLS}}$ and $(\mathbf{X}^T\mathbf{X})^{-1}$) can be mathematically treated in the same manner as $\boldsymbol{\mu}_\beta$ and $\mathbf{V}_\beta$ in MAP estimation. See Section 5.3.4.

Two different sequential procedures are given. The first is the direct method; it involves matrix inverses of dimensions $p \times p$. In the alternate, and recommended, formulation the inverses have dimensions $m \times m$ where m is the number of responses at each "time." In the case of a single response, m is equal to one and thus results in only scalar inverses being required.

6.7.2 Direct Method

Since the MAP estimator can mathematically include ML and OLS estimators and since estimates can be obtained for n as small as one, the sequential estimator given by (6.6.6) is used as a building block for the sequential method.

One important assumption for *sequential* MAP and ML estimation is that the measurements are independent in "time." That is, for the multiresponse case, ψ can be partitioned into the diagonal matrix

$$\psi = \text{diag}[\mathbf{\Phi}_1 \; \mathbf{\Phi}_2 \; \cdots \; \mathbf{\Phi}_n] \tag{6.7.1}$$

where $\mathbf{\Phi}_i$ is $m \times m$ and m is the number of observations taken at each time. The measurements at each time may be correlated since $\mathbf{\Phi}_i$ need not be diagonal. If ordinary least squares estimation is used, the measurement errors in $\mathbf{Y}$ may be correlated in time since in OLS estimation the matrix ψ in (6.6.6) is replaced by $\mathbf{I}$; in this case $\mathbf{P}_{\text{MAP}}$ would *not* yield the covariance of $\mathbf{b}_{\text{OLS}}$.

Sequential MAP and ML estimation can be used when ψ is not given by (6.7.1) but a transformation of the measurements is necessary to produce pseudo-observations that are uncorrelated in time. See Section 6.9.

A sequential estimator can be derived by letting

$$\mathbf{b} \rightarrow \mathbf{b}_{i+1}, \quad \boldsymbol{\mu}_\beta \rightarrow \mathbf{b}_i, \quad \mathbf{Y} \rightarrow \mathbf{Y}_{i+1}, \quad \mathbf{P} \rightarrow \mathbf{P}_{i+1}$$

$$\mathbf{V}_\beta \rightarrow \mathbf{P}_i, \quad \mathbf{X} \rightarrow \mathbf{X}_{i+1}, \quad \psi \rightarrow \mathbf{\Phi}_{i+1} \tag{6.7.2}$$

and introducing into (6.6.6) to find

$$\mathbf{b}_{i+1} = \mathbf{b}_i + \mathbf{P}_{i+1}\mathbf{X}_{i+1}^T\mathbf{\Phi}_{i+1}^{-1}[\mathbf{Y}_{i+1} - \mathbf{X}_{i+1}\mathbf{b}_i] \tag{6.7.3}$$

$$\mathbf{P}_{i+1} = [\mathbf{X}_{i+1}^T\mathbf{\Phi}_{i+1}^{-1}\mathbf{X}_{i+1} + \mathbf{P}_i^{-1}]^{-1} \tag{6.7.4}$$

The i subscript refers to "time" (or whatever the independent variable in terms of which measurements are being added). Thus $\mathbf{b}_{i+1}$ is an estimator for all p parameters based on the data $\mathbf{Y}_1, \mathbf{Y}_2, \ldots, \mathbf{Y}_{i+1}$ as well as on the prior information, if any. In the sequential procedure (6.7.3,4) are used for $i = 1, 2, \ldots, n$. In the above equation $\mathbf{X}_{i+1}$ is an $m \times p$ matrix and $\mathbf{Y}_{i+1}$ is an $m \times 1$ vector. In order to use the above formulation, it is necessary to invert the $p \times p$ matrix $\mathbf{P}_{i+1}$ and the $m \times m$ matrix $\mathbf{\Phi}_{i+1}$ at each time.

6.7.3 Sequential Method Using Matrix Inversion Lemma

The labor in finding the inverses of the matrices in (6.7.4) can be reduced if $m < p$ by using the matrix identities

$$\mathbf{P}_{i+1} = [\mathbf{X}_{i+1}^T\mathbf{\Phi}_{i+1}^{-1}\mathbf{X}_{i+1} + \mathbf{P}_i^{-1}]^{-1}$$

$$= \mathbf{P}_i - \mathbf{P}_i\mathbf{X}_{i+1}^T(\mathbf{X}_{i+1}\mathbf{P}_i\mathbf{X}_{i+1}^T + \mathbf{\Phi}_{i+1})^{-1}\mathbf{X}_{i+1}\mathbf{P}_i \tag{6.7.5a}$$

$$\mathbf{P}_{i+1}\mathbf{X}_{i+1}^T\mathbf{\Phi}_{i+1}^{-1} = \mathbf{P}_i\mathbf{X}_{i+1}^T(\mathbf{X}_{i+1}\mathbf{P}_i\mathbf{X}_{i+1}^T + \mathbf{\Phi}_{i+1})^{-1} \tag{6.7.5b}$$

See Appendix 6B for a derivation of these equations; (6.7.5a) is known as the *matrix inversion lemma*. Note that even though $\mathbf{P}_{i+1}$ is a $p \times p$ matrix, the matrix that must be inverted on the right sides of (6.7.5a, b) is $m \times m$. By introducing (6.7.5) into (6.7.3, 4) we obtain

$$\mathbf{A}_{i+1} = \mathbf{P}_i \mathbf{X}_{i+1}^T \tag{6.7.6a}$$

$$\mathbf{\Delta}_{i+1} = \mathbf{\Phi}_{i+1} + \mathbf{X}_{i+1} \mathbf{A}_{i+1} \tag{6.7.6b}$$

$$\mathbf{K}_{i+1} = \mathbf{A}_{i+1} \mathbf{\Delta}_{i+1}^{-1} \tag{6.7.6c}$$

$$\mathbf{e}_{i+1} = (\mathbf{Y}_{i+1} - \mathbf{X}_{i+1} \mathbf{b}_i) \tag{6.7.6d}$$

$$\mathbf{b}_{i+1} = \mathbf{b}_i + \mathbf{K}_{i+1} \mathbf{e}_{i+1} \tag{6.7.6e}$$

$$\mathbf{P}_{i+1} = \mathbf{P}_i - \mathbf{K}_{i+1} \mathbf{A}_{i+1}^T \tag{6.7.6f}$$

where $\mathbf{K}_{i+1}$ is sometimes called the *gain matrix*. This gives a general sequential procedure that can be used for OLS, WLS, Gauss–Markov, ML, and MAP estimation. The same computer program can be used for each.

Parenthetically we note that the same computer program can also provide a *filter*. That is, the estimator $\mathbf{b}_{i+1}$ can be used to find the best estimate of $\mathbf{Y}_{i+1}$, designated $\hat{\mathbf{Y}}_{i+1}$, based on all the data until and including time $i+1$. Notice that $\hat{\mathbf{Y}}_{i+1} \equiv \mathbf{X}_{i+1} \mathbf{b}_{i+1}$ is not the same vector as would be obtained from using *all* the data $(i = 1, 2, \ldots, n)$ to evaluate $\mathbf{b}$; when we use all the data, as we usually do, the $\hat{\mathbf{Y}}_i$ values are termed *smoothed* values rather than filtered values.

In starting the sequential procedure given by (6.7.6) the $\mathbf{b}_0$ and $\mathbf{P}_0$ matrices are required. For MAP estimation $\mathbf{b}_0$ is $\boldsymbol{\mu}_\beta$ and $\mathbf{P}_0 = \mathbf{V}_\beta$. For ML and OLS estimation $\mathbf{b}_0$ may be set equal to a zero column vector and $\mathbf{P}_0$ is made to be a diagonal matrix; the jth diagonal term of $\mathbf{P}_0$ should be large compared with $b_{j,n}$. The $\mathbf{P}_0$ matrix for the ML and OLS cases is discussed further below. WLS and Gauss–Markov estimates are obtained in a similar way.

Another expression for $\mathbf{P}_{i+1}$ given by Mendel [16, p. 128] is

$$\mathbf{P}_{i+1} = \left[\mathbf{I} - \mathbf{K}_{i+1} \mathbf{X}_{i+1}\right] \mathbf{P}_i \left[\mathbf{I} - \mathbf{K}_{i+1} \mathbf{X}_{i+1}\right]^T + \mathbf{K}_{i+1} \mathbf{\Phi}_{i+1} \mathbf{K}_{i+1}^T \tag{6.7.7}$$

This expression can be shown to be equal to that given by (6.7.6f) by introducing the definitions of $\mathbf{K}_{i+1}$ and $\mathbf{\Delta}_{i+1}$. It is true that (6.7.7) is a more time-consuming expression to evaluate than (6.7.6f), but Mendel shows that it is less sensitive to propagation of errors in $\mathbf{K}$ than is (6.7.6f).

6.7.3.1 Estimation with Only One Observation at Each Time ($m=1$)

An important simplification occurs in the sequential form given (6.7.6) when there is a single observation at each time. This is because $\boldsymbol{\Delta}_{i+1}$ is a scalar and thus its inverse is a scalar. Also note that

$$\mathbf{Y}_{i+1}\rightarrow Y_{i+1},\qquad \boldsymbol{\Phi}_{i+1}\rightarrow\sigma_{i+1}^2,\qquad \mathbf{X}_{i+1}=\left[X_{i+1,1}\cdots X_{i+1,p}\right]$$

where σ_{i+1}^2 is the variance of Y_{i+1} for ML and MAP estimation, but is replaced by unity for OLS estimation.

The sequential procedure for $m=1$ implied by (6.7.6) is

$$A_{u,i+1}=\sum_{k=1}^{p}X_{i+1,k}P_{uk,i}\tag{6.7.8a}$$

$$\Delta_{i+1}=\sigma_{i+1}^2+\sum_{k=1}^{p}X_{i+1,k}A_{k,i+1}\tag{6.7.8b}$$

$$K_{u,i+1}=\frac{A_{u,i+1}}{\Delta_{i+1}}\tag{6.7.8c}$$

$$e_{i+1}=Y_{i+1}-\sum_{k=1}^{p}X_{i+1,k}b_{k,i}\tag{6.7.8d}$$

$$b_{u,i+1}=b_{u,i}+K_{u,i+1}e_{i+1}\tag{6.7.8e}$$

$$P_{uv,i+1}=P_{uv,i}-K_{u,i+1}A_{v,i+1},\qquad v=1,2,\ldots,p\tag{6.7.8f}$$

where $u=1,2,\ldots,p$. It is important to observe that there are no simultaneous equations to solve or nonscalar matrices to invert with this method. This is a somewhat surprising result and it is true for any value of $p\geqslant 1$. This procedure does require starting values for $\mathbf{b}$ and $\mathbf{P}$, however.

Example 6.7.1

Give a set of equations based on (6.7.8) for two parameters that is appropriate for a small programmable calculator. Also indicate the memory locations.

Solution

Before the calculations, values can be stored for b_1, b_2, P_{11}, P_{12}, P_{22}, σ^2, X_1, and X_2. The first five are for "time" index zero whereas X_1 and X_2 are for index 1, that is,

$X_{1,1}$ and $X_{1,2}$. The memory registers can be assigned as follows:

0	1	2	3	4	5	6	7	8	9
b_1	b_2	P_{11}	P_{12}	P_{22}	σ^2	X_1	X_2	A_1	A_2

Later in the calculations, register 5 can be used for Δ and then e/Δ.

A set of equations and storage locations are as follows:

$$A_1 = X_1 P_{11} + X_2 P_{12} \qquad \text{STO 8}$$

$$A_2 = X_1 P_{12} + X_2 P_{22} \qquad \text{STO 9}$$

$$\Delta = A_2 X_2 + A_1 X_1 + \sigma^2 \qquad \text{STO 5}$$

$$P_{11} = -\frac{A_1^2}{\Delta} + P_{11} \qquad \text{STO 2}$$

$$P_{12} = -\frac{A_1 A_2}{\Delta} + P_{12} \qquad \text{STO 3}$$

$$P_{22} = -\frac{A_2^2}{\Delta} + P_{22} \qquad \text{STO 4}$$

$$\frac{e}{\Delta} = \frac{Y - X_1 b_1 - X_2 b_2}{\Delta} \qquad \text{STO 5}$$

$$b_1 = A_1 \frac{e}{\Delta} + b_1 \qquad \text{STO 0}$$

$$b_2 = A_2 \frac{e}{\Delta} + b_2 \qquad \text{STO 1}$$

The i subscript has been dropped but it is implied; for example, in the P_{11} equation, P_{11} on the right is at time i whereas P_{11} on the left is at time $i+1$. The above set of equations are used for each value of i. A special storage location for Y is not necessary because it is read in and used as needed.

Example 6.7.2

Using sequential ordinary least squares, estimate the two parameters for observations **Y** and sensitivity matrix **X** given by

$$\mathbf{Y} = \begin{bmatrix} 11 \\ 4 \\ 25 \end{bmatrix} \qquad \mathbf{X} = \begin{bmatrix} 4 & 3 \\ 1 & 2 \\ 10 & 1 \end{bmatrix}$$

Let $\mathbf{b}_0 = \mathbf{0}$ and $\mathbf{P}_0 = 10^5\mathbf{I}$, $10^{10}\mathbf{I}$, and $10^{13}\mathbf{I}$. (These large $\mathbf{P}_0$ values simulate no prior information.)

Solution

Since OLS is to be used, $\sigma_i^2 = 1$ for $i = 1,2,3$. The equations to be used are given in Example 6.7.1.

$$A_{1,1} = 4(10^5) + 3(0) = 4 \times 10^5, \qquad A_{2,1} = 4(0) + 3(10^5) = 3 \times 10^5$$

$$\Delta_1 = 3 \times 10^5(3) + 4 \times 10^5(4) + 1 = 2{,}500{,}001$$

The rest of the calculations for the first time are given in the third column of Table 6.10 along with results for times 2 and 3. The calculations were performed using a Texas Instruments SR-56 programmable calculator which has a 12 digit accuracy. For this problem the calculator accuracy can be important since subtractions of nearly identical large values occur while calculating the $P_{uv,2}$ values. The parameters are only slightly affected for a large range of $\mathbf{P}_0$ matrices such as from 10^3 to 10^{10} for this example. If the $\mathbf{P}_0$ matrix is $K\mathbf{I}$ where $K > 10^{13}$, however, the $\mathbf{P}_i$ matrices are $\mathbf{0}$ for $i \geqslant 2$ and the parameters do not change after $\mathbf{b}_2$.

Table 6.10 Results for Example 6.7.2 Using TI SR-56 Programmable Calculator

Quantity	Exact Values	$\mathbf{P}_0 = 10^5\mathbf{I}$	$\mathbf{P}_0 = 10^{10}\mathbf{I}$	$\mathbf{P}_0 = 10^{13}\mathbf{I}$
$P_{11,1}$	—	36000.0256	3.6×10^9	3.6×10^{12}
$P_{12,1}$	—	−47999.9808	-4.8×10^9	-4.8×10^{12}
$P_{22,1}$	—	64000.0144	6.4×10^9	6.4×10^{12}
$b_{1,1}$	—	1.759999296	1.76	1.76
$b_{2,1}$	—	1.319999472	1.32	1.32
$P_{11,2}$	.52	0.51999427	0.542	0
$P_{12,2}$	−.56	−.55999339	−0.572	0
$P_{22,2}$	.68	0.67999228	0.679	0
$b_{1,2}$	2	1.9999952	2.0	2.0
$b_{2,2}$	1	1.0000044	1.0	1.0
$P_{11,3}$	0.0131826742	0.0131826666	0.01311537163	0
$P_{12,3}$	−0.0225988701	−0.0225988336	−0.02206035	0
$P_{22,3}$	0.1101694915	0.1101692781	0.107167128	0
$b_{1,3}$	2.436911488	2.43691129	2.436373456	2.0
$b_{2,3}$	0.5367231638	0.5367231198	0.5462544	1.0

Physically $\mathbf{P} = \mathbf{0}$ implies that the variance of the parameters is zero and thus nothing more can be learned from additional data if $\mathbf{P} = \mathbf{0}$; hence the parameters do not change with time for $\mathbf{P}_0 = 10^{13}\mathbf{I}$ after the second data point. However, $\mathbf{P}$ is effectively zero in this example only because of our method and the limited accuracy of the calculator. Hence though $\mathbf{P}_0$ can be selected from a large range of values to simulate no prior information, it can be made too large.

Table 6.11 is given to illustrate the relative errors in the parameters at the third data point for different values of $\mathbf{P}_0$. The large values of $\mathbf{P}_0=10^3\mathbf{I}$ to $10^9\mathbf{I}$ lead to accurate estimates. Small and large values of K in $\mathbf{P}_0=K\mathbf{I}$ can lead, however, to relatively inaccurate parameter values. Small values imply prior parameter estimates are accurately known, which is not compatible with OLS estimation. Small or large values of K should be compared with the values of the square of the parameters. In the present case the parameters are about unity so that $K \leqslant 1$ is termed "small" and $K \geqslant 10^3$ may be termed large. Another indication that K is chosen sufficiently large is that K is large compared with the largest diagonal term of $\mathbf{P}_i$ for $i \geqslant 2$ (for two parameters).

Table 6.11 Relative Errors in $b_{1,3}$ and $b_{2,3}$ for Example 6.7.2

	Relative Errors in	
K in $\mathbf{P}_0=K\mathbf{I}$	$b_{1,3}$	$b_{2,3}$
1	-8.14×10^{-3}	-7.56×10^{-3}
10^3	-8.24×10^{-6}	-7.56×10^{-6}
10^5	-8.13×10^{-8}	-8.20×10^{-8}
10^7	4.58×10^{-7}	-2.16×10^{-5}
10^9	1.35×10^{-5}	5.40×10^{-4}
10^{10}	-2.21×10^{-4}	0.0178
10^{11}	-6.02×10^{-3}	0.348
10^{12}	1.94×10^{-3}	-0.148
10^{13}	-0.1792	0.863

It can be shown that K in $\mathbf{P}_0=K\mathbf{I}$ is too large for the two-parameter case when $\sigma_1^2/\Delta_1=10^{-n_c}$ and n_c is greater than the number of significant figures used by the computer or calculator. It is not difficult to show that

$$\frac{\sigma_1^2}{\Delta_1}=\frac{\sigma_1^2}{K\left[X_{1,1}^2+X_{1,2}^2\right]+\sigma_1^2} \tag{6.7.9}$$

Let σ_1^2/Δ_1 be equal to or greater than 10^{-n_c}, n_c being the number of significant calculational digits. Also let $K=10^{n_k}$ where K is large. Then for K not too large, we should have

$$n_k \leqslant n_c-\log\left(\frac{X_{1,1}^2+X_{1,2}^2}{\sigma_1^2}\right) \tag{6.7.10}$$

Using the values for the above example and $n_c=12$, we find $n_k \leqslant 10.6$. In other words, K should be less that $10^{10.6}$ in order not to be too large. This is consistent with the results of Tables 6.10 and 6.11. To be not near the critical number of significant figures, four less are recommended, that is, $\mathbf{P}_0=10^6\mathbf{I}$ in this case.

6.7.3.2 *Sequential Analysis of Example 5.2.4*

Computer programs can be readily written based on (6.7.8). One advantage is that no separate method is needed for the solution of a set of simultaneous algebraic equations. Moreover, the procedure is readily modified for any number of parameters p. For two parameters a small programmable calculator can also be used.

A computer program was written to estimate, using sequential OLS, the parameters in the model $\eta_i = \beta_1 + \beta_2 X_i$ for the data of Example 5.2.4. Ordinary least squares analysis implies that the variance σ_i^2 is constant, prior parameter values are unknown, and the diagonal components of $\mathbf{P}_0$ must be large. Since OLS analysis is unaffected by the choice of σ_i^2, replace σ_i^2 by 1 for $i = 1,\ldots,9$. For simplicity, let the initial values of β_1 and β_2 be zero, as no prior information is given.

If rough estimates were available for the parameters, then the diagonal components of $\mathbf{P}_0$ could be chosen about 10^3 to 10^8 times larger. If we do not have this information, then for this two-parameter case, (6.7.10) can be used; since $X_{1,1} = 1$ and $X_{2,1} = 0$, we find that $n_k \leqslant n_c$. Thus if a computer with 15 significant digit accuracy is available, $\mathbf{P}_0$ should have diagonal terms less than 10^{15}. It would be safer to reduce $\mathbf{P}_0$ by four orders of magnitude, however, say to 10^{11}. Shown in Table 6.12 are those obtained using $\mathbf{P}_0 = 10^8\mathbf{I}$ but the parameters are identical to the seven decimal places given to those for $\mathbf{P}_0 = K\mathbf{I}, 10^8 < K < 10^{11}$, for a 15 significant digit computer. Actually the values in Table 6.12 after $i = 1$ are exactly the same as those given by the usual least squares procedure if the data were first analyzed for the first two data points, then the first three, etc.

One way to check if $\mathbf{P}_0$ is made large enough is to repeat the calculation with $\mathbf{P}_0$ made larger. This is not efficient, however. Another way is to compare the diagonal components of $\mathbf{P}_0$ and $\mathbf{P}_n$, the matrix for all the data.

Table 6.12 Sequential Analysis of Example 5.2.4

i	b_1	b_2	P_{11}	P_{12}	P_{22}
0	0	0	10^8	0	10^8
1	0.2580000	0.0	1.0000000	0.0	10^8
2	0.2580000	0.17080000	1.0000000	−0.1000000	0.0200000
3	0.1281667	0.2097500	0.8333333	−0.0500000	0.0050000
4	0.4197000	0.1660200	0.7000000	−0.0300000	0.0020000
5	0.8238000	0.1256100	0.6000000	−0.0200000	0.0010000
6	0.9545238	0.1158057	0.5238095	−0.0142857	0.0005714
7	0.7957500	0.1253321	0.4642857	−0.0107143	0.0003571
8	1.1195833	0.1091405	0.4166667	−0.0083333	0.0002381
9	1.2864667	0.1019883	0.3777778	−0.0066667	0.0001667

From (6.6.6b) we can write

$$\mathbf{P}_n = \left[\mathbf{X}^T\boldsymbol{\psi}^{-1}\mathbf{X} + \mathbf{P}_0^{-1}\right]^{-1} \tag{6.7.11}$$

which for the OLS analysis above becomes

$$\mathbf{P}_n = \left[\mathbf{X}^T\mathbf{X} + K^{-1}\mathbf{I}\right]^{-1} = \left[\mathbf{I} + \frac{(\mathbf{X}^T\mathbf{X})^{-1}}{K}\right]^{-1}(\mathbf{X}^T\mathbf{X})^{-1} \tag{6.7.12}$$

Now as $K \to \infty$, $\mathbf{P}_n \to (\mathbf{X}^T\mathbf{X})^{-1}$. Then as this condition is approached, the diagonal components of $\mathbf{X}^T\mathbf{X}$, and hence those of $\mathbf{P}_n$, must be small compared to K. Consequently we can check if $\mathbf{P}_0 = K\mathbf{I}$ is large enough by comparing the diagonal components of $\mathbf{P}_n$ with K. Note that in Table 6.12 the P_{11} and P_{22} values for $i \geqslant 2$ are much less than $K = 10^8$.

Some further advantages are given below of the sequential method compared to the usual OLS analysis illustrated by Example 5.2.4. Each advantage relates to the ability of the sequential method to provide more information than is apparent from the usual OLS analysis. First, the effect of adding a single observation is apparent. For example, the effect of the fourth observation is to make b_1 much larger than if only the first three observations are used. Second, decreasing changes in the parameters with i show that each new observation tends to contribute less new information than the previous one. See b_2 versus i in Table 6.12.

Third, time variations of the parameters can yield insight into the accuracy of the measurements and/or adequacy of the regression function. For example, b_1 seems to be increasing with the i index whereas b_2 is more constant. This increase in b_1 could be due to inaccurate data or to actual time dependence of the parameter. (In this example we know that the former is the case because the regression function used is the correct one.) Owing to the larger variation of b_1 we suspect that the relative errors in the estimate b_1 are greater than in the estimate b_2. The possible time dependence of b_1 could be further investigated by adding measurements or by repeating the analysis with a new set of data. If the increase in b_1 persists, then a change in the regression model would be indicated.

Fourth, some conclusions can be drawn from the time variation of the parameters without any prior statistical knowledge of the measurement errors. If, however, there is statistical knowledge more can be learned.

The sequential method also yields time variation of components of $\mathbf{P}$. Note that P_{11} is decreasing much more slowly than P_{22}. If the measurement errors are independent and have constant variance (or more precisely 1111-01-), P_{11} is proportional to $V(b_1)$ and P_{22} to $V(b_2)$. Hence the measurements for $i>2$ in this example are more effective in reducing errors in b_2 than in b_1.

The decrease in P_{11} and P_{22} with i shown in Table 6.12 is necessary as indicated by the equations in Example 6.7.1 (because $A_1^2/\Delta \geqslant 0$ and $A_2^2/\Delta \geqslant 0$). Physically this is reasonable because added measurements increase the available information, which results in the diagonal components of **P** decreasing or at least not increasing.

6.7.4 Sequential MAP Estimation

Another advantage of the sequential procedure is that the same procedure (and thus computer program) can be used for MAP estimation as well as for WLS, Gauss–Markov, ML, and OLS estimation provided certain standard assumptions are valid. Sets of assumptions permitting sequential analysis are 11-11112 and 11-11113. The condition that the measurement errors be uncorrelated in time is particularly important.

Example 6.7.3

An engineer has been given the task of measuring the thermal conductivity k of a new electrical resistance heating wire. A linear curve of k versus temperture T is needed. Based on his experience with similar alloys he feels that the model $\eta=\beta_1+\beta_2 T$ is reasonable with prior estimates of β_1 and β_2 of $\mu_1=12$ W/m-°C and $\mu_2=0.01$ W/m-°C^2 where T is in °C. He estimates that the covariance matrix for these values is

$$\mathbf{V}_\beta=\begin{bmatrix} 2 & -0.002 \\ -0.002 & 10^{-5} \end{bmatrix}$$

and the prior distribution is normal.

For the new alloy he obtained the following measurements. The error in the temperature level can be neglected but the standard deviation in each measurement of k is about 0.2. Also the errors are independent and normal.

i	T (°C)	Measured Value of k (W/m-°C)
1	20	11.47
2	20	10.94
3	21	11.15
4	100	11.85
5	150	12.55
6	200	13.18
7	250	13.48
8	297	13.90
9	300	14.54
10	302	14.36

Notice that the measurements tend to be concentrated at the extreme temperatures of 20 and 300°C. If there were no uncertainty in the adequacy of the linear in T model and if there were no prior information, the optimum design would consist of one-half of the measurements being at each extreme T. The experimenter compromised by putting most of the measurements at the extremes but some intermediate values were included.

Estimate sequentially the parameters in the model $k=\beta_1+\beta_2 T$ with and without the prior information.

Solution

This problem can be viewed as being one involving subjective prior information. The prior means of b_1 and b_2 are 12 and 0.01, respectively. The $\mathbf{P}_0$ elements are $P_{11}=2$, $P_{12}=-0.002$, and $P_{22}=10^{-5}$. The σ_i^2 values are 0.04. The algorithm given in Example 6.7.1 can be used to estimate the parameters with $X_1=1$ and X_2 being the T values. The results are given in Table 6.13. Notice that the first two observations (both of which are at $T=20$°C) yield estimates of b_1 and b_2 which are near the final values. The variance of b_1 which is given by P_{11} reduces considerably as a result of the first two observations. This is not true, however, for b_2 since P_{22} decreases only slightly. This result is reasonable because b_2 represents the slope of k which, in the absence of prior information, requires measurements at two or more different T_i values. With all the observations used, both P_{11} and P_{22} have decreased considerably, indicating that the new measurements substantially reduced the experimenter's uncertainty.

Table 6.14 gives typical results for sequential estimation with no prior information. Using (6.7.10) it is found that $\mathbf{P}_0=10^4\mathbf{I}$ is large but not too large for 12 significant figure accuracy. In contrast with the prior information case, the first two observations do not yield estimates that are reasonable. The reduction in the $\mathbf{P}$ matrix is negligible from the first to second observation. This is because both are at

Table 6.13 Sequential Estimates Using Prior Information for Example 6.7.3

i	b_1	$b_2\times10^2$	$P_{11}\times10^2$	$P_{12}\times10^5$	$P_{22}\times10^7$
0	12	1	200	−200	100
1	11.271	1.0669	4.399	−20.37	83.50
2	10.997	1.0921	2.386	−18.52	83.33
3	10.971	1.0934	1.719	−18.18	83.32
4	10.973	.9591	1.718	−17.56	42.58
5	10.961	1.0228	1.634	−13.33	21.20
6	10.940	1.0803	1.513	−9.916	11.58
7	10.960	1.0409	1.393	−7.557	6.930
8	10.979	1.0127	1.290	−5.977	4.512
9	10.933	1.0813	1.246	−5.318	3.521
10	10.922	1.0977	1.222	−4.953	2.982

Table 6.14 Sequential Estimates for No Prior Information and $\mathbf{P}_0 = 10^4\mathbf{I}$ for Example 6.7.3; $\sigma_i^2 = 0.04, \mathbf{b}_0 = \mathbf{0}$

i	b_1	$b_2 \times 10^2$	$P_{11} \times 10^2$	$P_{12} \times 10^5$	$P_{22} \times 10^7$
0	0	0	10^6	0	10^{11}
1	.0286	57.207	9.98×10^5	-4.99×10^7	2.49×10^8
2	.0279	55.885	9.98×10^5	-4.99×10^7	2.49×10^8
3	12.274	-5.349	2475.9	-1.22×10^5	5.99×10^5
4	11.018	.8318	2.361	-33.82	84.02
5	10.967	1.0054	1.875	-17.28	27.78
6	10.935	1.0824	1.627	-11.27	13.24
7	10.958	1.0399	1.455	-8.127	7.475
8	10.977	1.0113	1.328	-6.258	4.732
9	10.928	1.0833	1.276	-5.510	3.652
10	10.916	1.1001	1.247	-5.104	3.075

$T = 20°$C. Reasonable values of b_1 and b_2 appear only at $i \geqslant 4$. It is only at $i = 4$ that T changes to 100°C after being near $T = 20°$C for the first three observations. Both b_1 and b_2 can be estimated for the linear model $k = \beta_1 + \beta_2 T$ only if measurements are made at 2 or more T values. Notice that P_{jk} components in Table 6.14 decrease in magnitude more rapidly than for Table 6.13. Hence as the number of observations increases, the importance of the prior information diminishes. Prior information always reduces parameter uncertainty, however.

6.7.5 Multiresponse Sequential Parameter Estimation

When several ($m > 1$) dependent variables are measured at the same time, it is sometimes possible to renumber them so that in effect $m = 1$. This can be done if OLS estimation is being used or if ML and MAP estimation is used and the measurements are independent *at each* time as well as with time. As an example consider the temperature data given in Table 7.14 where there are eight measurements made at each time. Assume that a sequential OLS analysis is to be performed. The temperature measurements of this table can be described either by

$$Y_j(i); \qquad j = 1, 2, \ldots, 8, \quad i = 1, 2, \ldots, n$$

or

$$Y_k; \qquad k = 1, 2, \ldots, 8, 9, 10, \ldots, 8n-2, 8n-1, 8n$$

By using the latter numbering, the problem is changed from one with $m = 8$ to $m = 1$. See Problem 6.26.

6.7.6 Ridge Regression Estimation

Starting about 1960 A. E. Hoerl and R. W. Kennard [17–21] developed a procedure called *ridge analysis*, which is a graphical method for depicting the characteristics of second-order regression functions having many independent variables. To a related procedure Hoerl gave the name "ridge regression," which he and Kennard have pointed out can have a general Bayesian interpretation. Hence the Bayesian estimation procedure which we call maximum a posteriori is related to ridge regression.

For the standard assumptions implied by 1111--11, OLS estimation provides the estimator given by (6.2.5). This estimator among all linear unbiased estimators provides the minimum variance (for these assumptions). The covariance matrix of $\mathbf{b}_{LS}$ is given by (6.2.11). For convenience, Hoerl and Kennard scale the independent variables so that $\mathbf{X}^T\mathbf{X}$ has diagonal elements all equal to one. If the eigenvalues of $\mathbf{X}^T\mathbf{X}$ are denoted $\lambda_j, j=1,2,\ldots,p$, then a seriously "ill-conditioned" (relatively small $|\mathbf{X}^T\mathbf{X}|$) problem is characterized by the smallest eigenvalue $\lambda_{\min}$ being very much smaller than unity. Hoerl and Kennard have noted that OLS estimation provides inadequate estimators for an ill-conditioned problem since $\sigma^2/\lambda_{\min}$ is a lower bound for the average squared distance between $\mathbf{b}_{LS}$ and $\boldsymbol{\beta}$. Thus for such cases $\mathbf{b}_{LS}$ is expected to be far from the true vector $\boldsymbol{\beta}$ with the absolute values of the elements of $\mathbf{b}_{LS}$ being too large.

The ridge regression estimator is given by

$$\mathbf{b}^* = (\mathbf{X}^T\mathbf{X} + K\mathbf{I})^{-1}\mathbf{X}^T\mathbf{Y} \tag{6.7.13}$$

for $K \geqslant 0$. With $\mathbf{X}^T\mathbf{X}$ scaled to have unity diagonal terms, values of K in the range of 10^{-4} to 1 are typical. There is an "optimum" value of K for any problem; Hoerl and Kennard discuss methods for selecting K. The MAP estimator given by (6.6.6c) yields the same estimates as (6.7.13) if $\boldsymbol{\mu}_\beta$ is replaced by $\mathbf{0}$, $\boldsymbol{\psi}^{-1}$ by $\mathbf{I}$, and $\mathbf{V}_\beta^{-1}$ by $K\mathbf{I}$. This has the effect of introducing the subjective prior information that the mean parameter values are zero; then the estimates given by (6.7.13) have smaller absolute values as K becomes larger. Hence as indicated by Theorem 2 of Marquardt's paper [21], (6.7.13) has the potential of reducing the inflated OLS parameter estimates found in ill-conditioned cases. Though the estimator given by (6.7.13) provides biased estimates, Hoerl and Kennard [17] have demonstrated that there exists a $K>0$ such that $E[(\mathbf{b}^*-\boldsymbol{\beta})^T(\mathbf{b}^*-\boldsymbol{\beta})] < E[(\mathbf{b}_{LS}-\boldsymbol{\beta})^T(\mathbf{b}_{LS}-\boldsymbol{\beta})]$ provided $\boldsymbol{\beta}^T\boldsymbol{\beta}$ is bounded. Evidently the MAP formulation has many different interpretations and uses.

6.7.7 Comments and Conclusions on the Sequential Estimation Method

In this subsection some observations regarding the sequential method are given.

1. The method is general as it includes MAP, ML, Gauss–Markov, WLS, and OLS estimators. When MAP and ML estimators are used it is necessary that the measurement errors $\boldsymbol{\varepsilon}$ be additive, normal, and independent in "time" designated by i. If the observations are not independent in time, sometimes transformations can be made to obtain new dependent variables that are independent. In Appendix 6A it is shown how certain autoregressive (AR) models for the observation errors can be treated by constructing independent combinations of the observations. It is assumed that the statistical parameters of σ^2 and ρ are known. See also Section 6.9.2.2.

   If (1) σ^2 is unknown for $\boldsymbol{\psi}=\sigma^2\boldsymbol{\Omega}$ where $\boldsymbol{\Omega}$ is known and (2) there is no prior information, the sequential procedure with σ^2 replaced by 1 can be used to estimate the parameters. After $\mathbf{b}_n$ is found, σ^2 can be estimated using s^2; the estimated covariance matrix would then be $\mathbf{P}_n$ times s^2.
2. If the problem can be formulated so that there is only one independent observation at each i, only a scalar needs to be inverted regardless of how many parameters are present. There are no simultaneous equations to solve.
3. The method readily extends to more than one unknown parameter. The summations in (6.7.8) can be easily programmed for an arbitrary value of p, the number of parameters in the $\mathbf{b}$ vector.
4. An examination of the parameters as a function of the index i can yield information that is not readily available otherwise. First, the models are usually chosen to contain parameters that are constant with time. If inspection of the parameters indicates that there is a time dependence (as in Table 6.12), then the adequacy of the model is questioned. Second, one can obtain an immediate "feel" of the effect of an additional observation which does not depend upon any statistical knowledge of the probability densities. The change in the parameters becomes less as more observations are used.
5. Good practice usually entails an inspection of the residuals. For the linear parameter case the true residuals are not obtained directly in a sequential manner. The residuals $\mathbf{e}$ based on the final parameter values ($\mathbf{b}_n$) are $\mathbf{e}=\mathbf{Y}-\hat{\mathbf{Y}}=\mathbf{Y}-\mathbf{X}\mathbf{b}_n$. These values are not the same as those calculated based on $\mathbf{b}_i$.
6. The sequential method provides at each i a filtered estimate of $\tilde{Y}_i$ as

$\tilde{Y}_i = \mathbf{X}_i \mathbf{b}_i$. These are based on the data until time i and can be used in an on-line analysis. The sequential method given in this section can be related to the discrete Kalman filter [16, p. 159].

7. The sequential MAP estimator can also be interpreted as providing a ridge regression estimator which can be helpful when the data are ill-conditioned, that is, $|\mathbf{X}^T\mathbf{X}|$ is nearly zero.

6.8 MATRIX FORMULATION FOR CONFIDENCE INTERVALS AND REGIONS

Much more information can be conveyed regarding parameters by specifying confidence intervals or regions in addition to parameter estimates. Whenever possible and appropriate it is recommended that confidence regions be presented in addition to estimates.

In order to present meaningful confidence regions it is necessary that the underlying assumptions be valid. Two assumptions frequently violated in scientific work are that the errors have zero mean and that the errors are uncorrelated. Erroneously taking these assumptions to be true has led many to present overly small confidence intervals. Physical parameters have been presented by different experimenters with each successive estimate being outside the preceding confidence interval. This has happened so often that one should be very careful to check his underlying assumptions before presenting his results. Further discussion of assumptions is given in Sections 5.10–5.14 and Section 6.9. Presentation of confidence regions is recommended but must be carefully and honestly given.

For additive, zero mean, normal measurement errors the joint probability density for the parameter vector $\mathbf{b}$ can be written as

$$f(\mathbf{b}) = (2\pi)^{-p/2}|\mathbf{V}_b|^{-1/2}\exp\left[-\frac{1}{2}(\mathbf{b}-\boldsymbol{\beta})^T\mathbf{V}_b^{-1}(\mathbf{b}-\boldsymbol{\beta})\right] \tag{6.8.1}$$

where $\mathbf{V}_b$ is the covariance matrix of $\mathbf{b}$. It is given by (6.2.11) for ordinary least squares, by (6.5.5) for maximum likelihood, and by (6.6.11) for MAP estimation. (The assumptions are different for each case.) For convenience in representing cov($\mathbf{b}$) [or cov($\mathbf{b}-\boldsymbol{\beta}$) for MAP cases], let us use $\mathbf{P}$ for each case,

$$\mathbf{V}_b = \mathbf{P} = [P_{ij}] \tag{6.8.2}$$

To obtain confidence regions, the covariance matrix of $\boldsymbol{\varepsilon}$, that is, $\boldsymbol{\psi}$, should be known at least within a multiplicative constant. Section 6.8.1

gives confidence *intervals*. Section 6.8.2 provides a derivation of a confidence region provided ψ is completely known. For the more general case of $\psi=\sigma^2\Omega$ where σ^2 is unknown and Ω known, a confidence region analysis is given in Section 6.8.3.

6.8.1 Confidence Intervals

A confidence interval can be found for each parameter β_k through the use of the kth diagonal term in $\mathbf{P}$ and the t distribution (if it applies). See Theorem 6.2.3. Suppose that the measurement errors are additive, zero mean, and normal. Also let there be no errors in the independent variables and no prior information regarding the parameters. Also let $\psi=\sigma^2\Omega$ where σ^2 is unknown and Ω known. These assumptions are designated 11--1011. Suppose that OLS or ML has been used and σ^2 was replaced by any constant c^2 and the matrix $\tilde{\mathbf{P}}$ was calculated; for example, for ML it is

$$\tilde{\mathbf{P}}=c^2(\mathbf{X}^T\Omega^{-1}\mathbf{X})^{-1},\ c\neq 0$$

The tilde ($\tilde{}$) is used because σ^2 is unknown. With the above assumptions, for OLS or ML we can give the estimated standard error of b_k as

$$\text{est. s.e.}(b_k)=\left(\tilde{P}_{kk}\right)^{1/2}\frac{s}{c} \tag{6.8.3}$$

where s^2 is the estimated value of σ^2. Then the $100(1-\alpha)\%$ confidence interval is given by

$$b_k-\text{est. s.e.}(b_k)t_{1-\alpha/2}(n-p)<\beta_k<b_k+\text{est. s.e.}(b_k)t_{1-\alpha/2}(n-p) \tag{6.8.4}$$

where $t_{1-\alpha/2}(n-p)$ is the t statistic for $n-p$ degrees of freedom. For 95% confidence and $n-p=10$, we find in Table 2.15, $t_{.975}(10)=2.23$. For a nonlinear example, see Example 7.7.2.

It should be noted that there is considerable danger in constructing confidence *regions* from confidence intervals found as above, since such a procedure can yield highly inaccurate confidence regions.

6.8.2 Confidence Regions for Known ψ

In this section ψ and thus $\mathbf{P}$ are assumed to be known. Consider the matrix product in the exponent of (6.8.1) and set it equal to r^2 since it is a nonnegative scalar,

$$(\mathbf{b}-\boldsymbol{\beta})^T\mathbf{P}^{-1}(\mathbf{b}-\boldsymbol{\beta})=r^2 \tag{6.8.5}$$

Let us think of the hyperellipsoid as being centered at the origin with coordinates being $b_1-\beta_1,\ldots,b_p-\beta_p$. Let l^2 be some specific value. For $r^2 \leqslant l^2$, (6.8.5) represents the interior of a hyperellipsoid. At $r=l$, (6.8.5) produces hypersurfaces of constant probability density. A method for determining values for l is derived below. Although (6.8.5) with $r=l$ describes the ellipsoid, for many purposes a more convenient description can be given in terms of the directions and lengths of the axes of the ellipsoid. To transform (6.8.5) in such a way as to provide such a description, we first find the eigenvalues of $\mathbf{P}$ or $\mathbf{P}^{-1}$ since the eigenvalues of $\mathbf{P}$ are simply the reciprocals of those of $\mathbf{P}^{-1}$.

For convenience in the following derivation, let $\mathbf{P}^{-1}$ be designated $\mathbf{C}$ whose eigenvalues are found by solving the determinantal equation

$$\begin{bmatrix} C_{11}-\lambda & C_{12} & \cdots & C_{1p} \\ C_{12} & C_{22}-\lambda & \cdots & C_{2p} \\ \vdots & \vdots & & \vdots \\ C_{1p} & C_{2p} & \cdots & C_{pp}-\lambda \end{bmatrix}=0 \qquad (6.8.6)$$

The eigenvalues of $\mathbf{C}$ are designated $\lambda_1,\lambda_2,\ldots,\lambda_p$. For convenience in numbering the λ_i's, let $\lambda_i \leqslant \lambda_j$ for $i<j$.

Let $\mathbf{C}=\mathbf{P}^{-1}$ and let $\mathbf{C}$ be given by

$$\mathbf{C}=\mathbf{e}\boldsymbol{\lambda}\mathbf{e}^T \qquad (6.8.7)$$

where

$$\boldsymbol{\lambda}=\operatorname{diag}\left[\lambda_1\lambda_2\cdots\lambda_p\right] \qquad (6.8.8)$$

and where $\mathbf{e}$ is a $p\times p$ matrix,

$$\mathbf{e}=\begin{bmatrix} e_{11} & e_{12} & \cdots & e_{1p} \\ e_{21} & e_{22} & \cdots & e_{2p} \\ \vdots & \vdots & & \vdots \\ e_{p1} & e_{p2} & \cdots & e_{pp} \end{bmatrix}=\left[\mathbf{e}_1\mathbf{e}_2\cdots\mathbf{e}_p\right] \qquad (6.8.9)$$

The vector components of $\mathbf{e}$ are orthogonal and of unit length (i.e., orthonormal),

$$\mathbf{e}^T\mathbf{e}=\mathbf{I} \quad \text{or} \quad \mathbf{e}_i^T\mathbf{e}_i=1 \qquad (6.8.10\text{a,b})$$

which implies that

$$\mathbf{e}^T=\mathbf{e}^{-1} \tag{6.8.10c}$$

Postmultiplying (6.8.7) by $\mathbf{e}$ gives

$$\mathbf{Ce}=\mathbf{e}\boldsymbol{\lambda}\mathbf{e}^T\mathbf{e}=\mathbf{e}\boldsymbol{\lambda} \tag{6.8.11}$$

Introducing the vector components of $\mathbf{e}$ into (6.8.11) yields

$$\mathbf{Ce}_i=\lambda_i\mathbf{e}_i \tag{6.8.12}$$

which can be considered to comprise a set of p linear homogeneous equations with the unknowns $e_{1i}, e_{2i}, \ldots, e_{pi}$. For example, for $p=3$ we have

$$(C_{11}-\lambda_i)e_{1i}+C_{12}e_{2i}+C_{13}e_{3i}=0 \tag{6.8.13a}$$

$$C_{12}e_{1i}+(C_{22}-\lambda_i)e_{2i}+C_{23}e_{3i}=0 \tag{6.8.13b}$$

$$C_{13}e_{1i}+C_{23}e_{2i}+(C_{33}-\lambda_i)e_{3i}=0 \tag{6.8.13c}$$

which constitutes three equations with three unknowns, but since these equations are homogeneous, there are at most two independent equations. A third equation is found from (6.8.10b),

$$e_{1i}^2+e_{2i}^2+e_{3i}^2=1 \tag{6.8.14}$$

Then e_{1i}, e_{2i}, and e_{3i} would usually be found from a solution of (6.8.13a, b) and (6.8.14).

A new coordinate vector can be defined by

$$\mathbf{h}\equiv\mathbf{e}^T(\mathbf{b}-\boldsymbol{\beta}) \qquad \text{or} \qquad h_i=\mathbf{e}_i^T(\mathbf{b}-\boldsymbol{\beta}) \tag{6.8.15}$$

Then introducing (6.8.7) and (6.8.15) in (6.8.5) produces

$$r^2=(\mathbf{b}-\boldsymbol{\beta})^T\mathbf{e}\boldsymbol{\lambda}\mathbf{e}^T(\mathbf{b}-\boldsymbol{\beta})=\mathbf{h}^T\boldsymbol{\lambda}\mathbf{h}=\sum_{i=1}^{p}\lambda_i h_i^2 \tag{6.8.16}$$

Using thc further transformation

$$z_i^2=\lambda_i h_i^2 \tag{6.8.17}$$

we can write

$$r^2=\mathbf{z}^T\mathbf{z}=z_1^2+z_2^2+\cdots+z_p^2 \tag{6.8.18}$$

The probability of a point $\mathbf{z}$ (or $\mathbf{b}-\boldsymbol{\beta}$) lying inside the hypersphere $l^2 \geqslant r^2$, where l is some fixed value, is found by using the above transformation in (6.8.1) to obtain

$$P(r^2 \leqslant l^2) = \int\int \cdots \int (2\pi)^{-p/2} \exp\left[-\frac{r^2}{2}\right] dz_1 \, dz_2 \cdots dz_p \qquad (6.8.19)$$

The integration is performed over the interior of the hypersphere described by (6.8.18). Note that

$$|\mathbf{V}_b|^{-1/2} = |P|^{-1/2} = |C|^{1/2} = (\lambda_1 \lambda_2 \cdots \lambda_p)^{1/2} \qquad (6.8.20)$$

is used in deriving (6.8.19). A volume element inside the hyperellipse can be described by

$$dV = \frac{p\pi^{p/2} r^{p-1} \, dr}{\Gamma\left(\frac{p}{2}+1\right)} \qquad (6.8.21)$$

where $\Gamma(\cdot)$ is the gamma function. The using (6.8.21) in (6.8.19) results in

$$P(r^2 \leqslant l^2) = \frac{p 2^{-p/2}}{\Gamma\left(\frac{p}{2}+1\right)} \int_0^l \exp\left(-\frac{r^2}{2}\right) r^{p-1} \, dr \qquad (6.8.22)$$

(If the transformation $r^2 = x$ is used, (6.8.22) is transformed to a form which is the integral of the chi-squared probability density function with p degrees of freedom.) For $p=1$, (6.8.22) gives

$$P(r^2 \leqslant l^2) = \left(\frac{2}{\pi}\right)^{1/2} \int_0^l \exp\left(-\frac{r^2}{2}\right) dr = \operatorname{erf}(l 2^{-1/2}) \qquad (6.8.23a)$$

and for $p=2$

$$P(r^2 \leqslant l^2) = \int_0^l \exp\left(-\frac{r^2}{2}\right) r \, dr = 1 - \exp\left(-\frac{l^2}{2}\right) \qquad (6.8.23b)$$

These probabilities for $l = 1$, 2, and 3 are given in Table 6.15 for several values of the number of parameters p. These three l values are sometimes called the one-, two-, or three-sigma probabilities. Also given in Table 6.15 are the l values associated with the 90 and 95% confidence regions. Other

Table 6.15 Values of Confidence Region Probabilities for Various Numbers of Parameters

No. of parameters, p	Probability for $l=1$	$l=2$	$l=3$	l Value For 90% Confidence	For 95% Confidence
1	.683	.955	.997	1.645	1.960
2	.3935	.8647	.9889	2.146	2.447
3	.1987	.739	.971	2.500	2.795
4	.0902	.594	.939	2.789	3.080
6	.0144	.323	.8264	3.263	3.548
8	.00175	.143	.658	3.655	3.938
10	.00017	.0527	.468	3.998	4.279

values of $l_{1-\alpha}(p)$ can be obtained using the F or χ^2 tables since

$$l_{1-\alpha}(p)=\left[pF_{1-\alpha}(p,\infty)\right]^{1/2}=\left[\chi^2(p)\right]^{1/2} \tag{6.8.24}$$

In summary the confidence region for known $\mathbf{P}$ (and thus ψ) is the interior of the hyperellipsoid,

$$\boxed{(\mathbf{b}-\boldsymbol{\beta})^T\mathbf{P}^{-1}(\mathbf{b}-\boldsymbol{\beta})=l_{1-\alpha}^2(p)} \tag{6.8.25}$$

where $l_{1-\alpha}(p)$ is the l value associated with the $100(1-\alpha)$ % confidence region for p parameters in the model. This region is more conveniently described by locating the principal axes of the hyperellipsoid. The extremes of the axes in terms of the new coordinates $h_1,\ldots,h_p$ are given by (6.8.15). The maximum coordinate values along the new axes are given by

$$h_1=\pm l_{1-\alpha}(p)\lambda_1^{-1/2},\quad h_2=0,\quad h_3=0,\ldots,h_p=0 \quad \text{(major axis)}$$

$$h_1=0,\quad h_2=\pm l_{1-\alpha}(p)\lambda_2^{-1/2},\quad h_3=0,\ldots,h_p=0$$

$$\vdots$$

$$h_1=0,\quad h_2=0,\quad h_3=0,\ldots,h_p=\pm l_{1-\alpha}(p)\lambda_p^{-1/2} \tag{6.8.26}$$

where (6.8.16) is used with $r^2=l_{1-\alpha}^2(p)$. The h_i values depend on $\mathbf{e}_i$, which is found as suggested by (6.8.13a, b) and (6.8.14). We wish to relate these values to points in the $\mathbf{b}-\boldsymbol{\beta}$ coordinates. Equation 6.8.26 can be written in the matrix form

$$\mathbf{H} = \pm l_{1-\alpha}(p)\,\text{diag}\left[\lambda_1^{-1/2}\,\lambda_2^{-1/2}\,\cdots\,\lambda_p^{-1/2}\right] = \mathbf{e}^T\mathbf{B} \qquad (6.8.27)$$

where $\mathbf{B} = [\mathbf{B}_1\ \mathbf{B}_2\ \cdots\ \mathbf{B}_p]$ and $\mathbf{B}_i$ contains the coordinates of $\mathbf{b} - \boldsymbol{\beta}$ for the ith axis. Solving for $\mathbf{B}$ using (6.8.10c) gives

$$\mathbf{B} = \pm l_{1-\alpha}(p)\mathbf{e}\,\text{diag}\left[\lambda_1^{-1/2}\cdots\lambda_p^{-1/2}\right] \qquad (6.8.28)$$

Example 6.8.1

Consider the case of two parameters. Give an algebraic formulation of the confidence region for $\boldsymbol{\beta}$ using $\mathbf{C} = \mathbf{P}^{-1}$.

Solution

Since $\mathbf{C}$ is given by

$$\mathbf{C} = \begin{bmatrix} C_{11} & C_{12} \\ C_{12} & C_{22} \end{bmatrix}$$

we can expand (6.8.25) to

$$(b_1 - \beta_1)^2 C_{11} + 2(b_1 - \beta_1)(b_2 - \beta_2)C_{12} + (b_2 - \beta_2)^2 C_{22} = l_{1-\alpha}^2(2) \qquad (a)$$

However, we need new coordinates h_1 and h_2 so that (a) can be written

$$l_{1-\alpha}^2 = \lambda_1 h_1^2 + \lambda_2 h_2^2 \qquad (b)$$

The eigenvalues λ_1 and λ_2 are found using (6.8.6),

$$(C_{11} - \lambda)(C_{22} - \lambda) - C_{12}^2 = \lambda^2 - \lambda(C_{11} + C_{22}) + (C_{11}C_{22} - C_{12}^2) = 0 \qquad (c)$$

which can be solved for λ_1 (being the smaller) and λ_2,

$$\lambda_1 = \left\{ C_{11} + C_{22} - \left[(C_{11} + C_{22})^2 - 4C_{11}C_{22} + 4C_{12}^2\right]^{1/2} \right\}/2 \qquad (d)$$

$$\lambda_2 = \left\{ C_{11} + C_{22} + \left[(C_{11} + C_{22})^2 - 4C_{11}C_{22} + 4C_{12}^2\right]^{1/2} \right\}/2 \qquad (e)$$

The coordinates h_1 and h_2 are given by (6.8.15),

$$h_1 = e_{11}(b_1 - \beta_1) + e_{21}(b_2 - \beta_2), \qquad h_2 = e_{12}(b_1 - \beta_1) + e_{22}(b_2 - \beta_2) \qquad (f)$$

The e_{ij} components are found using (6.8.10) and (6.8.12) or for e_{11} and e_{21},

$$(C_{11} - \lambda_1)e_{11} + C_{12}e_{21} = 0$$

$$e_{11}^2 + e_{21}^2 = 1$$

which yields

$$e_{11} = -\frac{C_{12}}{C_{11}-\lambda_1} e_{21} \tag{g}$$

$$e_{21} = \left[1 + \frac{C_{12}^2}{(C_{11}-\lambda_1)^2}\right]^{-1/2} \tag{h}$$

Similarly for e_{12} and e_{22}, we have

$$C_{12}e_{12} + (C_{22}-\lambda_2)e_{22} = 0$$

$$e_{12}^2 + e_{22}^2 = 1$$

which has a solution of

$$e_{22} = -\frac{C_{12}}{C_{22}-\lambda_2} e_{12}, \qquad e_{12} = \left[1 + \frac{C_{12}^2}{(C_{22}-\lambda_2)^2}\right]^{-1/2} \tag{i,j}$$

For two-parameter cases it can be shown that $e_{12} = e_{21}$ and that $e_{11} = -e_{22}$. Symmetry can not generally be arranged for $p > 2$.

The end points of the axes are given by (6.8.28),

$$\left[\ (\mathbf{b}-\boldsymbol{\beta}\ )_{\mathrm{maj}}\ (\mathbf{b}-\boldsymbol{\beta}\ )_{\mathrm{min}}\right] = \pm l_{1-\alpha}(2)\begin{bmatrix} e_{11} & e_{12} \\ e_{21} & e_{22} \end{bmatrix}\begin{bmatrix} \lambda_1^{-1/2} & 0 \\ 0 & \lambda_2^{-1/2} \end{bmatrix} \tag{k}$$

since $e_{11}e_{22} - e_{12}e_{21} = \pm 1$. This equation means, for example,

$$(b_1-\beta_1)_{\mathrm{maj}} = \pm l_{1-\alpha}(2)e_{11}\lambda_1^{-1/2}, \qquad (b_2-\beta_2)_{\mathrm{maj}} = \pm l_{1-\alpha}(2)e_{21}\lambda_1^{-1/2} \tag{l}$$

$$(b_1-\beta_1)_{\mathrm{min}} = \pm l_{1-\alpha}(2)e_{12}\lambda_2^{-1/2}, \qquad (b_2-\beta_2)_{\mathrm{min}} = \pm l_{1-\alpha}(2)e_{22}\lambda_2^{-1/2} \tag{m}$$

Example 6.8.2

Find the 95% confidence region for $i=2$ and $i=9$ of Example 5.2.4.

Solution

Consider the $i=2$ case first. The $\mathbf{C}=\mathbf{P}^{-1}$ matrix for OLS estimation with the standard assumptions is $\sigma^{-2}\mathbf{X}^T\mathbf{X}$. Since $\sigma^2 = 1$, we have

$$\mathbf{C} = \begin{bmatrix} n & \Sigma X_i \\ \Sigma X_i & \Sigma X_i^2 \end{bmatrix} = \begin{bmatrix} 2 & 10 \\ 10 & 100 \end{bmatrix}$$

From (d) and (e) of Example 6.8.1 we find $\lambda_1=0.990001$ and $\lambda_2=101.009999$. The components of $\mathbf{e}$ are found from (g), (h), (i), and (j) to be

$$\mathbf{e}=\begin{bmatrix} -0.9949382 & +0.1004886 \\ +0.1004886 & +0.9949382 \end{bmatrix}$$

From (k) the end points of the ellipse are given by

$$(b_1-\beta_1)_{\text{maj}}=\pm(2.447)(-0.9949382)(0.990001)^{-1/2}=\mp 2.447$$

$$(b_2-\beta_2)_{\text{maj}}=\pm 2.447(0.1004886)(0.990001)^{-1/2}=\pm 0.2471$$

$$(b_1-\beta_1)_{\text{min}}=\pm 2.447(0.1004886)(101.009999)^{-1/2}=\pm 0.02447$$

$$(b_2-\beta_2)_{\text{min}}=\pm 2.447(0.9949382)(101.009999)^{-1/2}=\pm 0.2422$$

These are the end points of the ellipse

$$\begin{aligned} l_{1-\alpha}^2(2)=2.447^2&=\lambda_1 h_1^2+\lambda_2 h_2^2 \\ &=0.990001 h_1^2+101.009999 h_2^2 \end{aligned}$$

which is shown as the larger ellipse in Fig. 6.5.

For $i=9$, the C matrix is

$$\mathbf{C}=\begin{bmatrix} 9 & 360 \\ 360 & 20400 \end{bmatrix}$$

for which $\lambda_1=2.646235$ and $\lambda_2=20406.35377$. The $\mathbf{e}$ matrix is

$$\mathbf{e}=\begin{bmatrix} -0.99984429 & 0.0176466 \\ 0.0176466 & 0.99984429 \end{bmatrix}$$

and the end points of the confidence region are

$$(b_1-\beta_1)_{\text{maj}}=\mp 1.504, \qquad (b_2-\beta_2)_{\text{maj}}=\pm 0.0265$$

$$(b_1-\beta_1)_{\text{min}}=\pm 3.02\times 10^{-4}, \qquad (b_2-\beta_2)_{\text{min}}=\pm 0.0171$$

This curve is also shown in Fig. 6.5; it is very narrow indicating less uncertainty in b_2 than for b_1.

The estimated values of b_1 and b_2 using the nine observations in Table 6.12 are 1.286 and 0.10199, respectively, and thus $b_1-\beta_1=0.286$ and $b_2-\beta_2=0.0199$. (The true values of β_1 and β_2 are 1 and 0.1.) This value is outside the 95% confidence region for $i=9$ because the r^2 value given by (6.8.5),

$$(\mathbf{b}-\boldsymbol{\beta})^T\mathbf{P}^{-1}(\mathbf{b}-\boldsymbol{\beta})=(b_1-\beta_1)^2C_{11}+2(b_1-\beta_1)(b_2-\beta_2)C_{12}+(b_2-\beta_2)^2C_{22}$$

is equal to 12.9, which is greater than $l_{1-\alpha}^2(2)=5.99$. One can observe directly from the plot in Fig. 6.5 that the estimates $b_{1,2}$ and $b_{2,2}$ are inside the $i=2$ confidence region. (See the point X at $b_1-\beta_1=-0.75$ in the figure.)

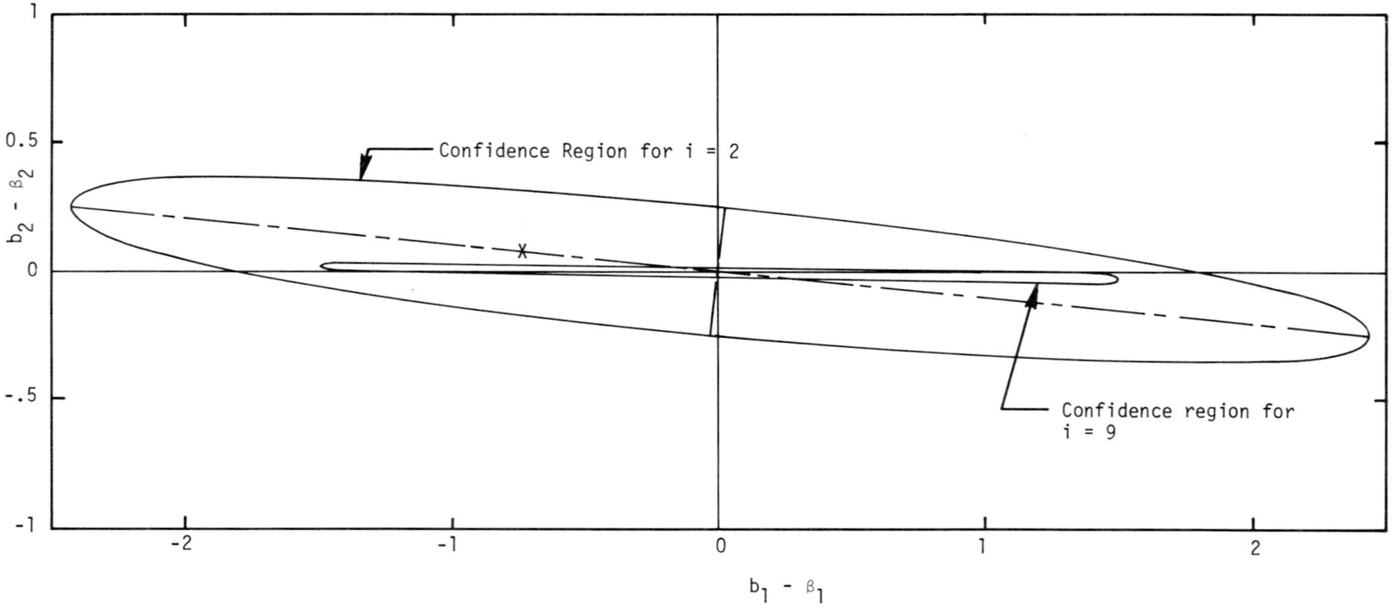

Figure 6.5 Confidence regions for Example 6.8.2.

6.8.3 Confidence Regions for $\psi=\sigma^2\Omega$ with Ω Known and σ^2 Unknown

In the previous subsection we considered the case of known ψ. In this section ψ is equal to σ^2 times Ω where σ^2 is unknown and Ω is known. This case is analyzed by developing an F statistic, which is the ratio of two χ^2 statistics each divided by their respective degrees of freedom. The first χ^2 statistic is $(\mathbf{b}-\boldsymbol{\beta})^T\mathbf{P}^{-1}(\mathbf{b}-\boldsymbol{\beta})$, which has p degrees of freedom. The assumptions are designated 11--1011 which include zero mean, normal measurement errors. In this section we assume that there is no prior information regarding the parameters. Maximum likelihood estimation is assumed.

The second statistic is R_{ML},

$$R_{\mathrm{ML}}=(\mathbf{Y}-\hat{\mathbf{Y}}_{\mathrm{ML}})^T\psi^{-1}(\mathbf{Y}-\hat{\mathbf{Y}}_{\mathrm{ML}})=(\mathbf{Y}-\hat{\mathbf{Y}}_{\mathrm{ML}})^T\Omega^{-1}(\mathbf{Y}-\hat{\mathbf{Y}}_{\mathrm{ML}})\sigma^{-2} \qquad (6.8.29)$$

which we wish to show is a χ^2 statistic with $n-p$ degrees of freedom. As usual, $\psi=\operatorname{cov}(\boldsymbol{\varepsilon})$. Notice that ordinary least squares analysis is not permitted unless $\Omega=\mathbf{I}$ for which case $R_{\mathrm{LS}}=R_{\mathrm{ML}}$. In general, Ω is *not* diagonal. It could result from autoregressive observation errors, for example.

We shall transform R_{ML} to a sum of squares using the identity

$$\Omega=\mathbf{D}\boldsymbol{\Phi}\mathbf{D}^T \qquad (6.8.30)$$

where $\mathbf{D}$ and $\boldsymbol{\Phi}$ are $n\times n$ matrices and $\boldsymbol{\Phi}$ is diagonal. Note that*

$$\Omega^{-1}=(\mathbf{D}^{-1})^T\boldsymbol{\Phi}^{-1}\mathbf{D}^{-1}=[\boldsymbol{\Phi}^{-1/2}\mathbf{D}^{-1}]^T[\boldsymbol{\Phi}^{-1/2}\mathbf{D}^{-1}] \qquad (6.8.31)$$

and thus R_{ML} can be written

$$\begin{aligned} R_{\mathrm{ML}}&=(\mathbf{Y}-\hat{\mathbf{Y}}_{\mathrm{ML}})^T\Omega^{-1}(\mathbf{Y}-\hat{\mathbf{Y}}_{\mathrm{ML}})\sigma^{-2}\\ &=(\mathbf{F}-\hat{\mathbf{H}}_{\mathrm{ML}})^T(\mathbf{F}-\hat{\mathbf{H}}_{\mathrm{ML}})\sigma^{-2} \end{aligned} \qquad (6.8.32)$$

where

$$\mathbf{F}\equiv\boldsymbol{\Phi}^{-1/2}\mathbf{D}^{-1}\mathbf{Y}, \qquad \hat{\mathbf{H}}_{\mathrm{ML}}\equiv\boldsymbol{\Phi}^{-1/2}\mathbf{D}^{-1}\hat{\mathbf{Y}}_{\mathrm{ML}} \qquad (6.8.33)$$

Notice that (6.8.32) is a sum of squares. For the linear-in-the-parameters

*For this case the square root of the diagonal matrix $\boldsymbol{\Phi}$ is a diagonal matrix having elements that are the positive square roots of corresponding elements in $\boldsymbol{\Phi}$.

model $\boldsymbol{\eta}=\mathbf{X}\boldsymbol{\beta}$, (6.8.32) can be written

$$R_{\text{ML}}=(\mathbf{F}-\overset{*}{\mathbf{Z}}\mathbf{b}_{\text{ML}})^T(\mathbf{F}-\overset{*}{\mathbf{Z}}\mathbf{b}_{\text{ML}})\sigma^{-2}$$

$$=\left(\mathbf{F}^T\mathbf{F}-\mathbf{b}_{\text{ML}}^T\overset{*}{\mathbf{Z}}{}^T\mathbf{F}\right)\sigma^{-2},\qquad \overset{*}{\mathbf{Z}}\equiv\boldsymbol{\Phi}^{-1/2}\mathbf{D}^{-1}\mathbf{X} \tag{6.8.34}$$

which has a χ^2 distribution with $n-p$ degrees of freedom. See Theorem 6.2.1. Observe that the covariance matrix of $\mathbf{F}$ is

$$\text{cov}(\mathbf{F})=E\left[\Phi^{-1/2}\mathbf{D}^{-1}\boldsymbol{\varepsilon}\boldsymbol{\varepsilon}^T(\mathbf{D}^{-1})^T\Phi^{-1/2}\right]=\sigma^2\mathbf{I}$$

For convenience, define R^* and $(P^*)^{-1}$ to be

$$R^*\equiv R_{\text{ML}}\sigma^2=(\mathbf{F}-\hat{\mathbf{H}}_{\text{ML}})^T(\mathbf{F}-\hat{\mathbf{H}}_{\text{ML}}) \tag{6.8.35a}$$

$$(\mathbf{P}^*)^{-1}=\mathbf{P}^{-1}\sigma^2 \tag{6.8.35b}$$

Since $\mathbf{F}$ in (6.8.35a) is a random vector satisfying the first five standard assumptions, $\overset{*}{\mathbf{Z}}$ is known, and there is no prior information, (6.8.35a) is analogous to (6.2.19) which can be used to get

$$s^2=\frac{R^*}{n-p} \tag{6.8.36}$$

In order to be consistent, let $(\mathbf{P}^*)^{-1}$ be found for ML estimation or

$$(\mathbf{P}^*)^{-1}=\mathbf{X}^T\boldsymbol{\Omega}^{-1}\mathbf{X} \tag{6.8.37}$$

Now the F statistic is the ratio of two independent random variables, each with a χ^2-divided-by-degrees-of-freedom distribution (see Section 2.8.10). Hence a joint confidence region for the parameter estimates can be found from

$$\frac{(\mathbf{b}_{\text{ML}}-\boldsymbol{\beta})^T(\mathbf{P}^*)^{-1}(\mathbf{b}_{\text{ML}}-\boldsymbol{\beta})/p}{R^*/(n-p)}=F_{1-\alpha}(p,n-p) \tag{6.8.38}$$

or

$$\boxed{(\mathbf{b}_{\text{ML}}-\boldsymbol{\beta})^T\mathbf{X}^T\boldsymbol{\Omega}^{-1}\mathbf{X}(\mathbf{b}_{\text{ML}}-\boldsymbol{\beta})=ps^2F_{1-\alpha}(p,n-p)} \tag{6.8.39}$$

In deriving (6.8.39) we have assumed that $\boldsymbol{\psi}$ is equal to $\sigma^2\boldsymbol{\Omega}$ where $\boldsymbol{\Omega}$ is known. When $\psi = \sigma^2 \mathbf{I}$, (6.8.39) reduces to the more common expression,

$$(\mathbf{b}_{\mathrm{LS}} - \boldsymbol{\beta})^T \mathbf{X}^T \mathbf{X} (\mathbf{b}_{\mathrm{LS}} - \boldsymbol{\beta}) = ps^2 F_{1-\alpha}(p, n-p) \tag{6.8.40}$$

since then $\mathbf{b}_{\mathrm{ML}} = \mathbf{b}_{\mathrm{LS}}$.

When dynamic experiments are performed and automatic digital data acquisition equipment is used, n is usually quite large—possibly several hundred or even thousands. In such cases $n-p$ is large and

$$pF_{1-\alpha}(p, n-p) \approx l_{1-\alpha}^2(p)$$

and then

$$(\mathbf{b}_{\mathrm{ML}} - \boldsymbol{\beta})^T \mathbf{X}^T \boldsymbol{\Omega}^{-1} \mathbf{X} (\mathbf{b}_{\mathrm{ML}} - \boldsymbol{\beta}) \approx \left(\frac{R^*}{n} \right) l_{1-\alpha}^2(p) \tag{6.8.41}$$

For example, for $\alpha = 0.05$, corresponding to the 95% confidence region, and for two parameters ($p = 2$), the values of $2F_{1-\alpha}(p, n-p)$ are 6.18, 6.08, 6.04, and 5.99 for $n = 100$, 200, 400, and ∞, respectively.

6.9 MATRIX ANALYSIS WITH CORRELATED OBSERVATION ERRORS

6.9.1 Introduction

When automatic digital data acquisition equipment is used for dynamic experiments, very large numbers of measurements can be obtained. These observations may not be independent, however. Correlated measurements are frequently obtained when the *same* specimen is tested over some range. An example is the measurement of the electrical resistance of a piece of wire as a function of temperature such as at 20, 21, and 22°C. The measurements at 20 and 21°C may be correlated for a given specimen, but a 20°C value for one specimen probably would not be with a 21°C value for another specimen.

Another example of correlated errors is provided by the case of a cooling billet given by Example 6.2.3. Some of the temperatures recorded are depicted in Fig. 6.1 along with the associated residuals. These data are shown for 96-sec time steps, but observations were actually made at the smaller steps of 12 sec. If all the data between 0 and 1536 sec are used, the regression curve for the 129 observations is very close to that obtained using 17 observations. For the fourth-degree model given in Example 6.2.3, residuals in F° are plotted in Fig. 6.6 for the first 25 data points for $\Delta t = 12$ sec. On the upper side of the horizontal axis is a scale that corresponds to Fig. 6.1. The solid circles show the residuals. There are 11 consecutive negative residuals, followed by 14 positive residuals.

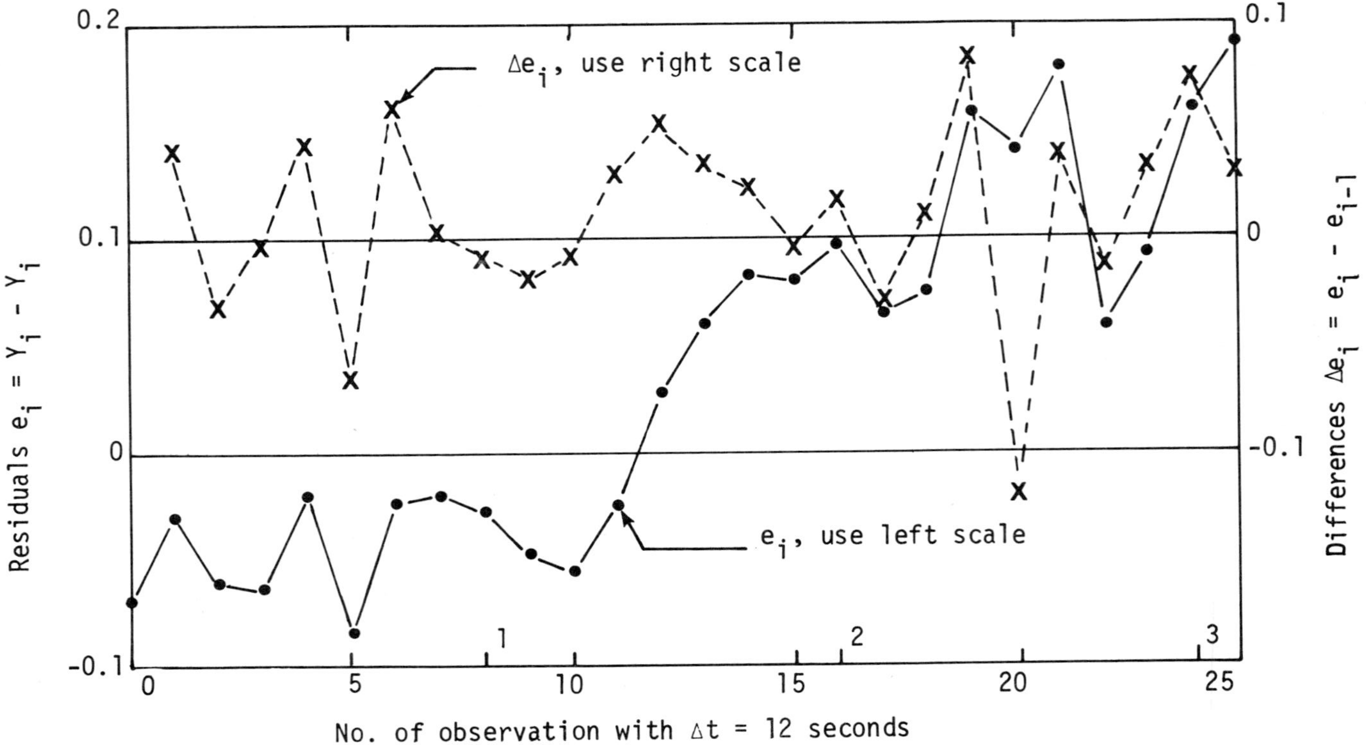

Figure 6.6 Residuals for billet example with $\Delta t = 12$ sec.

Residuals are not precisely independent even if the observation errors are, but for a large number of independent observations, the residuals are nearly independent. This is far from true for the residuals shown in Fig. 6.6; rather, the residuals are highly correlated. A least squares analysis of the measurements containing highly correlated errors may produce satisfactory parameter estimates. There are, however, at least two different dangers in the OLS analysis of highly correlated errors (i.e., ψ having relatively large off-diagonal terms). One of these is that one might present too small confidence regions based on the erroneous assumption of independent errors. Another danger relates to this point and also experimental design when observations are independent and unbiased; doubling the number of them significantly improves the accuracy of the parameter estimates. This may not be true if the additional measurements result in higher correlation between the measurements as in the billet example illustrated by Fig. 6.1 when the time step is halved and the number of measurements is doubled. In such cases one might erroneously design the experiment to have too small a time step.

A simple check to see if the residuals are approximately independent is based on the number of runs. (The number of runs is the number of changes in the signs of the residuals plus one. For example, for $+-+++--$, there are four runs.) For n independent, zero mean random variables the expected number of runs is $(n+1)/2$. For significance tests based on runs, see references 4 and 22. For cumulative errors (see below) the expected number of runs is about $n^{1/2}$. The residuals of Fig. 6.1 exhibit six runs compared with $(n+1)/2=9$. There seems to be no reason to question the independence of residuals. If there are still about six runs when the number of observations is doubled, we should question the independence. Certainly the residuals shown by the solid circles in Fig. 6.6 cannot be considered independent with only two runs while $(n+1)/2=13$. Also shown in Fig. 6.6 as crosses are first differences of the residuals. There are 15 runs out of 24 points, indicating that the first differences are much closer to being independent than the residuals themselves.

6.9.2 Autoregressive Errors (AR)

In this section a first-order, single response, autoregressive model of errors is considered. (A more general analysis is given in Appendix 6A.) Let ε_i, the ith measurement error, be described by

$$\varepsilon_i = \rho\varepsilon_{i-1} + u_i \tag{6.9.1a}$$

$$\varepsilon_0 \equiv 0, \qquad E(u_i) = 0 \tag{6.9.1b}$$

$$E(u_i u_j) = 0 \quad \text{for} \quad i \neq j, \qquad E(u_i^2) = \sigma_i^2 \tag{6.9.1c}$$

for $i=1,2,\ldots,n$. In words, (6.9.1a) states that the present error is a fraction ρ of the previous error plus a zero mean component u_i that is independently distributed. This is called a first order autoregressive process.

It is convenient to relate the $n\times 1$ vectors $\boldsymbol{\varepsilon}$ and $\mathbf{u}$ by

$$\boldsymbol{\varepsilon}=\mathbf{D}u \tag{6.9.2}$$

where $\mathbf{D}$ can be found from (6.9.1a) to be the lower triangular matrix,

$$\mathbf{D}=\begin{bmatrix} 1 & 0 & 0 & \cdots & 0 \\ \rho & 1 & 0 & \cdots & 0 \\ \rho^2 & \rho & 1 & \cdots & 0 \\ \vdots & \vdots & \vdots & & \vdots \\ \rho^{n-1} & \rho^{n-2} & \rho^{n-3} & \cdots & 1 \end{bmatrix} \tag{6.9.3}$$

Also from (6.9.1a) we find [see (6A.5a)] the inverse of $\mathbf{D}$ to be

$$\mathbf{D}^{-1}=\begin{bmatrix} 1 & 0 & 0 & \cdots & 0 & 0 & 0 \\ -\rho & 1 & 0 & \cdots & 0 & 0 & 0 \\ 0 & -\rho & 1 & \cdots & 0 & 0 & 0 \\ \vdots & \vdots & \vdots & & \vdots & \vdots & \vdots \\ 0 & 0 & 0 & \cdots & 0 & -\rho & 1 \end{bmatrix} \tag{6.9.4}$$

which contains a main diagonal of ones and a diagonal just below of $-\rho$.

The covariance matrix of the errors, $\boldsymbol{\psi}$, is given by

$$\boxed{\boldsymbol{\psi}=\operatorname{cov}(\boldsymbol{\varepsilon})=E(\boldsymbol{\varepsilon}\boldsymbol{\varepsilon}^T)=E(\mathbf{D}\mathbf{u}\mathbf{u}^T\mathbf{D}^T)=\mathbf{D}\boldsymbol{\phi}\mathbf{D}^T} \tag{6.9.5a}$$

where (6.1.40) is used and where $\boldsymbol{\phi}$ is defined to be the diagonal matrix

$$\boldsymbol{\phi}\equiv\operatorname{diag}\left[\sigma_1^2\ \sigma_2^2\ \cdots\ \sigma_n^2\right] \tag{6.9.5b}$$

Several classes of $\boldsymbol{\psi}$ matrices can be generated. For $\rho=0$ and $\sigma_i^2=\sigma_u^2$, a constant, the "standard" covariance matrix, $\boldsymbol{\psi}_{ST}$, is found,

$$\boldsymbol{\psi}_{ST}=\sigma_u^2\mathbf{I} \tag{6.9.6}$$

Next, if $\rho=0$, we have $\boldsymbol{\psi}_W$ which is

$$\boldsymbol{\psi}_W=\boldsymbol{\phi}=\operatorname{diag}\left[\sigma_1^2\ \sigma_2^2\ \cdots\ \sigma_n^2\right] \tag{6.9.7}$$

Also, several special autoregressive (AR) cases have error variances as follows

$$\sigma_1^2 = c\sigma_u^2, \quad \sigma_i^2 = \sigma_u^2 \qquad \text{for} \quad i = 2, 3, \ldots, n \tag{6.9.8}$$

for which the ψ matrix, designated ψ_a, becomes

$$\psi_a = \mathbf{D}\phi\mathbf{D}^T$$

$$= \begin{bmatrix} c & c\rho & c\rho^2 & c\rho^3 & \cdots & c\rho^{n-1} \\ & 1+c\rho^2 & \rho(1+c\rho^2) & \rho^2(1+c\rho^2) & \cdots & \rho^{n-2}(1+c\rho^2) \\ & & (1+\rho^2+c\rho^4) & \rho(1+\rho^2+c\rho^4) & \cdots & \rho^{n-3}(1+\rho^2+c\rho^4) \\ & & & & & \vdots \\ \text{symmetric} & & & & & 1+\rho^2+\rho^4 \cdots c\rho^{2(n-1)} \end{bmatrix} \sigma_u^2 \tag{6.9.9}$$

See (6A.18) of Appendix 6A for a derivation. Three special cases associated with (6.9.9) are as follows:

1. ψ_{a1} has $c = (1-\rho^2)^{-1}$; (6A.22a).
2. ψ_{a2} has $c = 1$ and $\rho = 1$; (6A.22b).
3. ψ_{a3} has $c = 1$; (6A.22c).

The ψ_{a1} matrix is the most common of the three special cases. It might be called a "steady-state" case because the diagonal terms are all equal,

$$\sigma_\varepsilon^2 \equiv E(\varepsilon_i^2) = \frac{\sigma_u^2}{1-\rho^2}; \qquad \rho^2 < 1$$

Notice that as $\rho \to 1$, the σ_ε^2 values become much larger than σ_u^2. Physically, ψ_{a1} is appropriate for some process which has been going on a "long" time before the taking of measurements. The other extreme physical situation is for measurements starting when the process starts; this is better described by case 3, ψ_{a3}. In case 3, the variance of ε_i^2 is a minimum at $i = 1$ (provided $\rho > 0$) and gradually increases to the steady-state value of σ_ε^2 just given above.

Case 2 has a simple ψ matrix as given by (6A.22b). Notice, however, that the ith diagonal term is equal to $i\sigma_u^2$. Sometimes this case is considered to be unstable because the variance continually increases. We can use it however, with $\rho = 1$ (although case 1 can not have $\rho = 1$). For case 2 ε_i is found from (6.9.1) to be the sum

$$\varepsilon_i = \sum_{j=1}^{i} u_i$$

For this reason case 2 is called the cumulative error case. The difference of the successive values ε_{i-1} and ε_i is $\varepsilon_i - \varepsilon_{i-1} = u_i$ which is independent from u_j, $j \neq i$. Figure 6.6 shows residuals and differences of residuals for the billet example. Since the residuals are apparently correlated and the differences are not, the cumulative error model for ε_i is better than that of independent ε_i. (Estimation of ρ is discussed in Section 6.9.5.)

6.9.2.1 *OLS Estimation With AR Errors*

Ordinary least squares can always be used irrespective of any of the standard assumptions. In specifying the covariance matrix of the parameters or the confidence regions, some conditions must be known (or assumed). Suppose that the conditions of additive, zero mean, first-order autoregressive errors are valid. Also assume that ψ is known within a multiplicative constant. These conditions are designated 11-2-011.

The covariance matrix of $\mathbf{b}_{\text{LS}}$ is given by (6.2.11)

$$\text{cov}(\mathbf{b}_{\text{LS}}) = (\mathbf{X}^T\mathbf{X})^{-1}\mathbf{X}^T\psi\mathbf{X}(\mathbf{X}^T\mathbf{X})^{-1} \tag{6.2.11}$$

The terms in $\mathbf{X}^T\mathbf{X}$ are given in a detailed form by (6.2.6). The $\mathbf{X}^T\psi\mathbf{X}$ portion of (6.2.11) is a little more difficult to evaluate. Using (6.9.5a) we can write

$$\mathbf{X}^T\psi\mathbf{X} = \mathbf{X}^T\mathbf{D}\phi\mathbf{D}^T\mathbf{X} = (\mathbf{D}^T\mathbf{X})^T\phi\mathbf{D}^T\mathbf{X} \tag{6.9.10}$$

Consider now the $\mathbf{D}^T\mathbf{X}$ matrix product, designated $\tilde{\mathbf{Z}}$,

$$\tilde{\mathbf{Z}} \equiv \mathbf{D}^T\mathbf{X} = \begin{bmatrix} 1 & \rho & \cdots & \rho^{n-3} & \rho^{n-2} & \rho^{n-1} \\ \vdots & \vdots & & \vdots & \vdots & \vdots \\ 0 & 0 & & 1 & \rho & \rho^2 \\ 0 & 0 & \cdots & 0 & 1 & \rho \\ 0 & 0 & \cdots & 0 & 0 & 1 \end{bmatrix} \begin{bmatrix} X_{11} & \cdots & X_{1p} \\ \vdots & & \vdots \\ X_{n-2,1} & \cdots & X_{n-2,p} \\ X_{n-1,1} & \cdots & X_{n-1,p} \\ X_{n1} & \cdots & X_{np} \end{bmatrix} = \left[\tilde{Z}_{ij}\right] \tag{6.9.11a}$$

where the components $\tilde{Z}_{ij}$ can be found using the expressions

$$\tilde{Z}_{nj} = X_{nj}; \qquad \tilde{Z}_{ij} = X_{ij} + \rho\tilde{Z}_{i-1,j} \qquad \text{for} \quad i = n-1, n-2, \ldots, 1; \quad j = 1, 2, \ldots, p \tag{6.9.11b}$$

Notice that the $\tilde{Z}_{ij}$ values are found starting at the "bottom" of each column. Using this notation the $\mathbf{X}^T\psi\mathbf{X}$ product has typical terms, as indicated by the lk term below,

$$\mathbf{X}^T\psi\mathbf{X} = \left[\sum_{i=1}^{n} \tilde{Z}_{il}\tilde{Z}_{ik}\sigma_i^{-2} \right] \tag{6.9.12}$$

Unfortunately simple algebraic expressions for the covariance of $\mathbf{b}_{\mathrm{LS}}$ do not result from (6.2.4) for AR errors; the computer evaluation of the terms is straightforward, however.

Example 6.9.1

Derive an expression for $V(b_{\mathrm{LS}})$ for the simple model $\eta = \beta$ and for $\psi = \psi_{a1}$.

Solution

For this case, $\mathbf{X}^T = [1\,1 \cdots 1]$ and $\mathbf{X}^T\mathbf{X} = n$. The matrix $\tilde{\mathbf{Z}}$ is given by

$$\tilde{\mathbf{Z}}^T = \left[\sum_{i=1}^{n} \rho^{i-1} \quad \sum_{i=1}^{n-1} \rho^{i-1} \cdots \quad 1+\rho+\rho^2 \quad 1+\rho \quad 1 \right]$$

$$= [1-\rho^n \quad 1-\rho^{n-1} \cdots \quad 1-\rho^3 \quad 1-\rho^2 \quad 1-\rho]\frac{1}{1-\rho}$$

and $\tilde{\mathbf{Z}}^T\phi\tilde{\mathbf{Z}}$ is

$$\frac{\sigma_u^{-2}}{(1-\rho)^2}\left[(1-\rho^n)^2 \frac{1}{1-\rho^2} + \sum_{i=1}^{n-1} (1-\rho^i)^2 \right]$$

The result for $V_{a1}(b_{\mathrm{LS}})$, the variance of b_{LS} for $\psi = \psi_{a1}$, can be written in the form

$$V_{a1}(b_{\mathrm{LS}}) = \left[\frac{1}{n} + \frac{2\rho}{n(1-\rho)}\left(1 - \frac{1-\rho^n}{n(1-\rho)} \right) \right] \frac{\sigma_u^2}{1-\rho^2} \tag{6.9.13}$$

For any fixed value of ρ between 0 and 1 and increasing values of n, $V_{a1}(b_{\mathrm{LS}})$ always decreases.

Example 6.9.2

Suppose ρ and the number of observations n are related by

$$\rho = e^{-a/n} \tag{6.9.14}$$

which gives greater correlation between observations for a fixed experimental range as the observations become more "dense." In (6.9.14), a is some constant characteristic of the data. Using the result of Example 6.9.1, investigate $V_{a1}(b_{\mathrm{LS}})$ for $n \to \infty$. Assume that $\sigma_\varepsilon^2 = \sigma_u^2(1-\rho^2)^{-1}$ is held constant.

Solution

In (6.9.13), $\sigma_u^2/(1-\rho^2)$ is replaced by σ_ε^2 and ρ by $e^{-a/n}$. The result is an indeterminant form. After using l'Hospital's rule we obtain

$$\lim_{n\to\infty} V_{a1}(b_{LS}) = \frac{2\sigma_\varepsilon^2}{a}\left[1-\frac{1}{a}(1-e^{-a})\right] \tag{6.9.15}$$

See Fig. 6.7 for a plot of (6.9.15). If the measurements become more "correlated" as n becomes larger in the manner described by (6.9.14), $V_{a1}(b_{LS})$ approaches a constant value for large n rather than going to zero as one obtains from (6.9.13) for ρ = constant and $n\to\infty$.

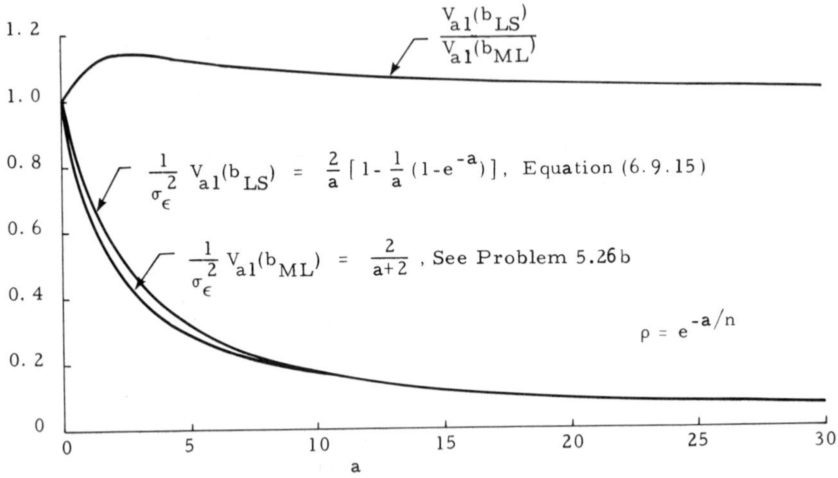

Figure 6.7 Variance of b for $n\to\infty$ found for first order autoregressive errors using least squares and maximum likelihood.

6.9.2.2 ML Estimation With AR Errors

In maximum likelihood or Gauss-Markov estimation, the estimator for the linear model is given by (6.5.2),

$$\mathbf{b}_{ML} = (\mathbf{X}^T\boldsymbol{\psi}^{-1}\mathbf{X})^{-1}\mathbf{X}^T\boldsymbol{\psi}^{-1}\mathbf{Y} \tag{6.5.2}$$

For ML, this equation follows from the assumptions 11--1111.

It simplifies calculations in (6.5.2) if the relation $\boldsymbol{\psi}=\mathbf{D}\boldsymbol{\phi}\mathbf{D}^T$ is used,

$$\mathbf{b}_{ML} = \left(\mathbf{X}^T(\mathbf{D}^{-1})^T\boldsymbol{\phi}^{-1}\mathbf{D}^{-1}\mathbf{X}\right)^{-1}\mathbf{X}^T(\mathbf{D}^{-1})^T\boldsymbol{\phi}^{-1}\mathbf{D}^{-1}\mathbf{Y} \tag{6.9.16a}$$

$$= (\mathbf{Z}^T\boldsymbol{\phi}^{-1}\mathbf{Z})^{-1}\mathbf{Z}^T\boldsymbol{\phi}^{-1}\mathbf{F} \tag{6.9.16b}$$

where

$$\mathbf{Z} \equiv \mathbf{D}^{-1}\mathbf{X}, \qquad \mathbf{F} \equiv \mathbf{D}^{-1}\mathbf{Y} \tag{6.9.16c}$$

Typical terms of **X** and **F** are given by

$$Z_{1j} = X_{1j}, \qquad Z_{ij} = X_{ij} - \rho X_{i-1,j} \qquad \text{for} \quad i = 2, 3, \ldots, n \tag{6.9.17a}$$

$$F_1 = Y_1, \qquad F_i = Y_i - \rho Y_{i-1} \qquad \text{for} \quad i = 2, 3, \ldots, n \tag{6.9.17b}$$

where $\mathbf{D}^{-1}$ displayed by (6.9.4) is used. Note that by replacing Y_i by F_i, the modified observations F_i are uncorrelated with i since from (6.9.1a), $F_i = \eta_i - \rho\eta_{i-1} + u_i$ and the u_i's are uncorrelated.

Another way to define the modified sensitivities and observations is by using

$$\overset{*}{\mathbf{Z}} \equiv \phi^{-1/2}\mathbf{D}^{-1}\mathbf{X}, \qquad \overset{*}{\mathbf{F}} \equiv \phi^{-1/2}\mathbf{D}^{-1}\mathbf{Y} \tag{6.9.18}$$

which permits us to write

$$\mathbf{b}_{\text{ML}} = \left(\overset{*}{\mathbf{Z}}{}^T\overset{*}{\mathbf{F}}\right)^{-1}\overset{*}{\mathbf{Z}}{}^T\overset{*}{\mathbf{F}}, \qquad \text{cov}(\mathbf{b}_{\text{ML}}) = \left(\overset{*}{\mathbf{Z}}{}^T\overset{*}{\mathbf{Z}}\right)^{-1} \tag{6.9.19a, b}$$

where $\mathbf{b}_{\text{ML}}$ has a similar form as given for OLS.

Results for Model $\eta = \beta$

For AR errors for the simple model $\eta = \beta$, the components of **Z** are given by

$$Z_1 = 1, \qquad Z_2 = Z_3 = \cdots = Z_n = 1 - \rho \tag{6.9.20a}$$

For case 1 AR errors, the $\overset{*}{Z}_i$ components for this same model are

$$\overset{*}{Z}_1 = \sigma_\varepsilon^{-1}, \qquad \overset{*}{Z}_2 = \overset{*}{Z}_3 = \cdots = \overset{*}{Z}_n = \left[(1-\rho)/(1+\rho)\right]^{1/2}\sigma_\varepsilon^{-1} \tag{6.9.20b}$$

which are shown in Fig. 6.8 for $\rho = 0$, 0.5, 0.9, and 1.0. The net effect of ρ being between zero and one is to reduce the value of the modified sensitivity compared to X_i. This results in the variance of b being greater than for $\rho = 0$.

For the simple model the F_i values are

$$F_1 = Y_1, \quad F_2 = Y_2 - \rho Y_1, \quad F_3 = Y_3 - \rho Y_2, \cdots \tag{6.9.21a}$$

and the components of ϕ_{a1} are

$$\phi_{a1} = \sigma_u^2 \,\text{diag}\left[(1-\rho^2)^{-1} \; 1 \; 1 \; \cdots \; 1\right] \tag{6.9.21b}$$

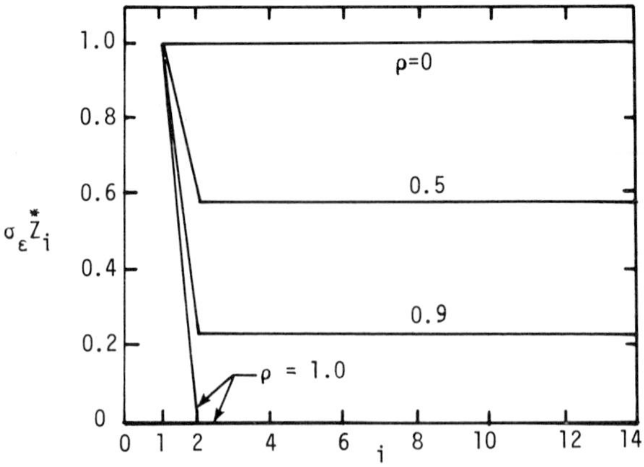

Figure 6.8 Modified sensitivity coefficients for $a1$ errors and the model $\eta=\beta$.

Using (6.9.16b) with (6.9.20a) and (6.9.21) gives for the simple model $\eta=\beta$ and $a1$ errors,

$$b_{\mathrm{ML}}=\frac{(1-\rho^2)Y_1+\sum_2^n(Y_i-\rho Y_{i-1})(1-\rho)}{1-\rho^2+(n-1)(1-\rho)^2} \tag{6.9.22}$$

which has the variance

$$V_{a1}(b_{\mathrm{ML}})=\frac{\sigma_u^2}{1-\rho^2+(n-1)(1-\rho)^2}=\frac{\sigma_\varepsilon^2}{1+(n-1)(1-\rho)/(1+\rho)} \tag{6.9.23}$$

which is depicted in Fig. 6.9 versus n for various ρ values. For any fixed value of ρ between zero and unity, $V_{a1}(b_{\mathrm{ML}})$ always decreases with increasing n values. Physically this represents the case of constant spacing of measurements but an increasing number of them. See Fig. 6.1, for example, which is for the billet problem with time steps of 96 sec. If more measurements are added with the same spacing of 96 sec, n would be increasing while ρ would be fixed, as in Fig. 6.9.

Suppose that the observations become correlated as the time step is reduced and that ρ is related to n, the maximum number of observations, by $\rho=\exp(-a/n)$. Then for variable n and fixed a, we can obtain Fig. 6.10. Here as n becomes large (about 20 for the case of $a=1$), $V_{a1}(b_{\mathrm{ML}})$ approaches a constant value. In this case increasing an already "large" value of n will not significantly improve the accuracy. This case of correlated errors can occur as the time steps are made smaller and smaller

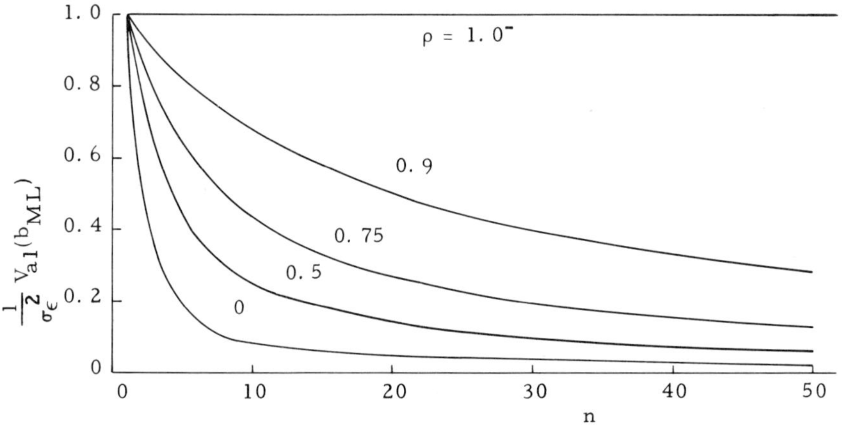

Figure 6.9 Variance of b_{ML} for correlated errors, (6.9.23).

in a dynamic experiment. For example, if in Example 6.2.3 Δt were decreased from 96 sec to 48, then 24, and finally 12, while keeping the same total time of 1536 sec, the observation errors would become more and more correlated. This is demonstrated by the residuals shown in Fig. 6.6.

The asymptotic values of Fig. 6.10 for $n \to \infty$ are given by $2\sigma_\varepsilon^2/(a+2)$. (See Problem 5.26b.) Figure 6.7 depicts this relation as a function of a as well as the ratio of the variances for LS and ML estimators. The maximum

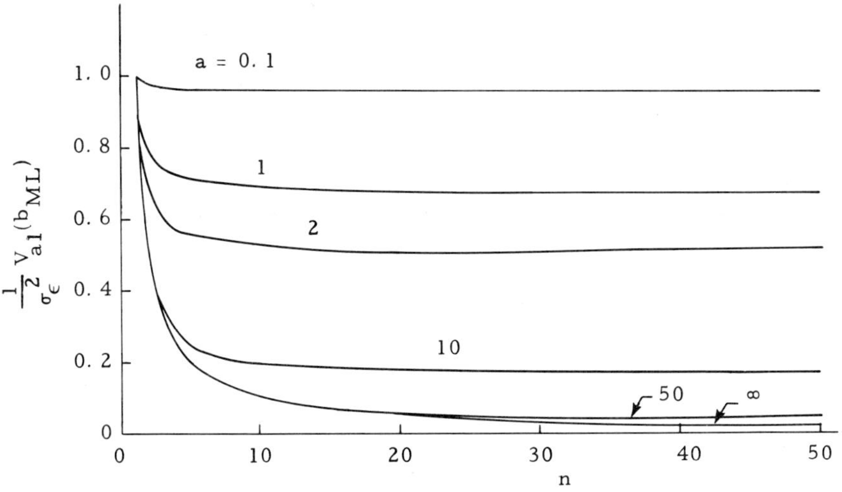

Figure 6.10 Variance of b_{ML} for correlated errors for $\rho = e^{-a/n}$ and "a" fixed, (6.9.23).

ratio is 1.139, occurring at $a=2.7$. For this first-order autoregressive error model and simple physical model example, negligible improvement in accuracy occurs using maximum likelihood (or Gauss–Markov estimation) rather than ordinary least squares. Other cases can be exhibited, however, for which ML estimators are far superior to OLS estimators.

6.9.3 Moving Average Errors (MA)

A brief treatment is given below for first-order moving average errors. A model for first-order moving average errors is

$$\varepsilon_i = u_i - \theta u_{i-1} \tag{6.9.24}$$

$$u_0 = 0, \qquad E(u_i) = 0 \tag{6.9.25}$$

$$E(u_i u_j) = 0 \quad \text{for } i \neq j, \qquad E(u_i^2) = \sigma_i^2 \tag{6.9.26}$$

In matrix form $\boldsymbol{\varepsilon}$ can be written as

$$\boldsymbol{\varepsilon} = \mathbf{D}_m \mathbf{u} \tag{6.9.27}$$

where

$$\mathbf{D}_m = \begin{bmatrix} 1 & 0 & 0 & \cdots & 0 & 0 \\ -\theta & 1 & 0 & \cdots & 0 & 0 \\ 0 & -\theta & 1 & \cdots & 0 & 0 \\ \vdots & \vdots & \vdots & & \vdots & \vdots \\ 0 & 0 & 0 & \cdots & -\theta & 1 \end{bmatrix} \tag{6.9.28}$$

whose inverse is

$$\mathbf{D}_m^{-1} = \begin{bmatrix} 1 & 0 & 0 & 0 & \cdots \\ \theta & 1 & 0 & 0 & \cdots \\ \theta^2 & \theta & 1 & 0 & \cdots \\ \theta^3 & \theta^2 & \theta & 1 & \cdots \\ \vdots & \vdots & \vdots & \vdots & \end{bmatrix} \tag{6.9.29}$$

Analogous to (6.9.5a) $\boldsymbol{\psi}$ is given by

$$\boldsymbol{\psi} = \operatorname{cov}(\boldsymbol{\varepsilon}) = \mathbf{D}_m \boldsymbol{\phi} \mathbf{D}_m^T, \qquad \boldsymbol{\phi} \equiv E(\mathbf{u}\mathbf{u}^T) \tag{6.9.30}$$

In determining the covariance of $\mathbf{b}_{\mathrm{LS}}$ as given by (6.2.11), we have the term

$\mathbf{X}^T\psi\mathbf{X}$. By using (6.9.10) we see that the modified sensitivity given by

$$\overset{*}{\mathbf{Z}}_m \equiv \mathbf{D}_m^T\mathbf{X} \tag{6.9.31}$$

can be convenient to use. The m subscript denotes moving average. Components of $\overset{*}{\mathbf{Z}}_m$ are

$$\overset{*}{Z}_{1j,m} = X_{1j}; \qquad \overset{*}{Z}_{ij,m} = X_{ij} - \theta X_{i-1,j} \quad \text{for } i=2,3,\ldots,n \tag{6.9.32}$$

and for $j=1,2,\ldots,p$. When using ML estimation, we use the modified sensitivity matrix,

$$\mathbf{Z}_m = \mathbf{D}_m^{-1}\mathbf{X} \tag{6.9.33}$$

which has components

$$Z_{1j,m} = X_{1j}, \qquad Z_{ij,m} = X_{ij} + \theta Z_{i-1,j,m} \quad \text{for } i=2,\ldots,n \tag{6.9.34}$$

Notice that the AR $\overset{*}{\mathbf{Z}}$ for LS analysis is similar to the $\mathbf{Z}$ for ML moving average errors, whereas the AR $\mathbf{Z}$ for ML analysis is similar to the LS MA analysis.

6.9.4 Summary of First-Order Correlated Cases for the Model $\eta = \beta$

In Table 6.16 the variances of b are given for six different correlated errors; three of these are for autoregressive errors and the others are for moving average errors. The variances are given for both LS and ML estimators of β. Also the ratio of the variances for LS and ML estimation is given for large values of n, the number of observations.

The following are some conclusions relative to correlated errors.

1. If the measurement errors are actually correlated but are erroneously assumed to be independent, two deleterious effects can result. First, the experimenter might report much more accurate results than he should. Next, the experimental strategy might be based on an incorrect premise. That is, the accuracy of an estimate as measured by the variance always decreases as more independent measurements are included, but this is not necessarily true for correlated errors. Thus the experimenter might take 1000 observations expecting to achieve much more accuracy than if 100 were taken; it might be that the extra 900 measurements are of dubious value for this purpose.
2. For autoregressive errors with $0<\rho\leqslant 1$, the variances of estimates can be much less than for independent errors ($\rho=0$).

Table 6.16 Variance of b for the Model $\eta=\beta$ for Several Types of Correlated Errors

Case	Least Squares Variance, $V(b_{LS})$	Maximum Likelihood Variance, $V(b_{ML})$	$V(b_{LS})/V(b_{ML})$ for large n
$a1$ $\left[c=(1-\rho^2)^{-1}\right]$	$\left[\frac{1}{n}+\frac{2\rho}{n(1-\rho)}\left[1-\frac{1-\rho^n}{n(1-\rho)}\right]\right]\frac{\sigma_u^2}{1-\rho^2}$	$\frac{\sigma_u^2}{(n-1)(1-\rho)^2+1-\rho^2}$	1 for $\rho^2<1$
$a2$ $[c=\rho=1]$	$\frac{(n+1)(2n+1)}{6n}\sigma_u^2$	σ_u^2	$\frac{n}{3}$
$a3$ $[c=1]$	$\left[n-2\rho\frac{1-\rho^n}{1-\rho}+\rho^2\frac{1-\rho^{2n}}{1-\rho^2}\right]\frac{\sigma_u^2}{n^2(1-\rho)^2}$	$\frac{\sigma_u^2}{(n-1)(1-\rho)^2+1}$	1 for $\rho^2<1$
$ma1$ $(-1\leqslant\vartheta\leqslant 1)$	$\frac{(n-1)(1-\theta)^2+1}{n^2}\sigma_u^2$	$\frac{\sigma_u^2(1-\theta)^2}{n-2\theta\frac{1-\theta^n}{1-\theta}+\theta^2\frac{1-\theta^{2n}}{1-\theta^2}}$	1 for $\theta^2<1$
$ma2$ $(\theta=-1)$	$\frac{4n-3}{n^2}\sigma_u^2$	$\frac{2}{n+1}\sigma_u^2$ for n odd	2
$ma3$ $(\theta=1)$	$\frac{\sigma_u^2}{n^2}$	$\frac{6\sigma_u^2}{n(n+1)(2n+1)}$	$\frac{n}{3}$

3. For some correlated error cases, the ML estimators can be greatly superior to those obtained using least squares. See the last column of Table 6.16.
4. When the statistical parameters ρ and θ are known, the estimation procedure for maximum likelihood or generalized least squares is not much more complicated for AR and MA errors than for independent errors.
5. Since the covariance matrix of $\mathbf{b}$ for both ML and LS estimation can be readily given for AR and MA errors, the confidence intervals and regions for known ρ and θ can be developed in exactly the same manner as in Section 6.8.

6.9.5 Simultaneous Estimation of ρ, σ_u^2, and Physical Parameters for the $a1$ Cases

When high speed digital data acquisition equipment is used, the measurements on one specimen are frequently correlated. Neither the best model to describe the correlations nor the statistical parameters such as ρ, θ, and σ_u^2 may be known. In this section the error model is assumed to be the first-order autoregressive designated $a1$. The parameters ρ and σ_u^2 are unknown, however. Maximum likelihood estimation is used because, unlike least squares, it can directly provide estimates of ρ and σ_u^2.

The analysis starts with the logarithm of the likelihood function as given by (6.1.67). For a single response case with the errors as described by (6.9.1), the ψ matrix is

$$\psi_{a1} = \mathbf{D}\phi_{a1}\mathbf{D}^T \tag{6.9.35}$$

where $\mathbf{D}$ is given by (6.9.3) and ϕ_{a1} by

$$\phi_{a1} = \sigma_u^2 \operatorname{diag}\left[(1-\rho^2)^{-1} \quad 1 \quad 1 \quad \cdots \quad 1 \right] \tag{6.9.36}$$

Then the determinant of ψ_{a1} is

$$|\psi_{a1}| = |\mathbf{D}|\,|\phi_{a1}|\,|\mathbf{D}^T| = |\phi_{a1}| = \frac{\sigma_u^{2n}}{1-\rho^2} \tag{6.9.37}$$

Introducing (6.9.37) into (6.1.67a) gives

$$\ln L = -\frac{1}{2}\left[n \ln 2\pi + \ln\left(\frac{\sigma_u^{2n}}{1-\rho^2} \right) + \sigma_u^{-2} S_1 \right] \tag{6.9.38a}$$

$$S_1 = (1-\rho^2)(Y_1 - \eta_1)^2 + \sum_{i=2}^{n} \left[(Y_i - \rho Y_{i-1}) - (\eta_i - \rho\eta_{i-1}) \right]^2 \tag{6.9.38b}$$

In the ML method we seek to minimize $\ln L$ simultaneously with respect to the parameters $\boldsymbol{\beta}$, σ_u^2, and ρ. Necessary conditions for the minimum are that the first derivatives of $\ln L$ with respect to these parameters be equal to zero,

$$\nabla_\beta S_1|_{\beta=\mathbf{b}_{\mathrm{ML}},\rho=\hat{\rho}}=\mathbf{0} \tag{6.9.39a}$$

$$\frac{-n}{\hat{\sigma}_u^2}+\frac{S_1(\mathbf{b}_{\mathrm{ML}},\hat{\rho})}{\hat{\sigma}_u^4}=0 \tag{6.9.39b}$$

$$\frac{2\hat{\rho}}{1-\hat{\rho}^2}+\frac{1}{\hat{\sigma}_u^2}\left.\frac{\partial S_1}{\partial\rho}\right|_{\mathbf{b}_{\mathrm{ML}},\hat{\rho}}=0 \tag{6.9.39c}$$

Solving (6.9.39b) for $\hat{\sigma}_u^2$ yields

$$\hat{\sigma}_u^2=\frac{S_1(\mathbf{b}_{\mathrm{ML}},\hat{\rho})}{n} \tag{6.9.40a}$$

which could be used directly to obtain a (biased) estimate of σ_u^2 if ρ were known. If $\hat{\rho}$ is unknown, (6.9.39c) is employed to find

$$\hat{\rho}=\left[\sum_1^{n-1}e_ie_{i+1}\right]\left[\hat{\sigma}_u^2(1-\hat{\rho}^2)^{-1}+\sum_2^{n-1}e_j^2\right]^{-1},\qquad e_i\equiv Y_i-\hat{Y}_i \tag{6.9.40b}$$

where we have used (6.9.38b) to find

$$\left.\frac{\partial S_1}{\partial\rho}\right|_{\mathbf{b}_{\mathrm{ML}},\hat{\rho}}=2\left(\hat{\rho}\sum_{i=2}^{n-1}e_i^2-\sum_{i=1}^{n-1}e_ie_{i+1}\right) \tag{6.9.41}$$

The solution for $\mathbf{b}_{\mathrm{ML}}$, $\hat{\rho}$, and $\hat{\sigma}_u^2$ is nonlinear even though the model is linear in the parameters other than ρ and σ^2.

Various iterative procedures can be suggested to find $\hat{\rho}$, $\hat{\sigma}_u^2$, and $\mathbf{b}_{\mathrm{ML}}$. One is to first guess a reasonable value of $\hat{\rho}$ (such as $\rho=0$) and then calculate $\mathbf{b}_{\mathrm{ML}}$ using (6.9.39a) or equivalently (6.5.2). Next, (6.9.40a) is used to get a value for $\hat{\sigma}_u^2$ which in turn is used in (6.9.40b) to obtain an improved value of $\hat{\rho}$. The procedure is repeated until $\hat{\rho}$ is essentially unchanged. The converged values of $\hat{\rho}$, $\hat{\sigma}_u^2$, and $\mathbf{b}_{\mathrm{ML}}$ are the desired values.

Example 6.9.3

Give the set of iterative equations for the model $\eta=\beta$ and $a1$ autoregressive errors.

Solution

The estimator for b_{ML} is given by (6.9.22) with ρ replaced by $\hat{\rho}$. The estimated variance of u_i is found using (6.9.40a),

$$\hat{\sigma}_u^2=\left\{(1-\hat{\rho}^2)(Y_1-b_{\mathrm{ML}})^2+\sum_{i=2}^{n}\left[Y_i-\hat{\rho}Y_{i-1}-b_{\mathrm{ML}}(1-\hat{\rho})\right]^2\right\}/n$$

The estimate for $\hat{\rho}$ is given by (6.9.40b) with $\hat{Y}_i$ replaced by b_{ML}. These equations must be solved iteratively even for this simple problem.

Example 6.9.4

For the $a3$ autoregressive case investigate using the Monte Carlo method the estimates of b_1, b_2, $\hat{\rho}$, and $\hat{\sigma}_u^2$ for the model $\eta_i = 100 + 0.1X_i$ where $X_i = 0, 10, 20, \ldots$ and σ_u^2 are chosen from random normal numbers of unit variance and zero mean. For $\rho = 1$, let $n = 10$, 30, 40, 50, and 120. For $n = 60$, let $\rho = -1, -0.5, 0, 0.5$, and 1.

Solution

For this $a3$ case a solution for a given set of measurements is found in a similar manner as for the $a1$ case discussed above. The main difference is that the determinant of ψ_{a3} is σ_u^{2n} rather than the expression given by (6.9.37). The result is the set of equations

$$\mathbf{b}_{\text{ML}} = (\mathbf{Z}^T\mathbf{Z})^{-1}\mathbf{Z}^T\mathbf{F}, \qquad \mathbf{Z} \equiv \mathbf{D}^{-1}\mathbf{X}, \qquad \mathbf{F} \equiv \mathbf{D}^{-1}\mathbf{Y} \tag{a}$$

$$\hat{\sigma}_u^2 = \frac{\sum_1^n (e_i - \hat{\rho} e_{i-1})^2}{n} \qquad \text{for } e_0 \equiv 0 \tag{b}$$

$$\hat{\rho} = \left(\sum_1^{n-1} e_i e_{i+1}\right)\left(\sum_1^{n-1} e_j^2\right)^{-1} \tag{c}$$

The diagonal matrix $\boldsymbol{\phi}_{a3}^{-1}$ is not needed in (a) because it is equal to $\sigma_u^{-2}\mathbf{I}$ and the σ_u^{-2} term cancels. The $\mathbf{Z}$ matrix is equal to

$$\mathbf{D}^{-1}\mathbf{X} = \begin{bmatrix} 1 & & & & \\ -\rho & 1 & & & \\ & -\rho & 1 & & \\ & & \vdots & & \\ & & & -\rho & 1 \end{bmatrix} \begin{bmatrix} 1 & 0 \\ 1 & 10 \\ 1 & 20 \\ \vdots & \vdots \\ 1 & (n-1)10 \end{bmatrix}$$

$$= \begin{bmatrix} 1 & 0 \\ 1-\rho & 10 \\ 1-\rho & 20-10\rho \\ \vdots & \vdots \\ 1-\rho & 10[(n-1)-(n-2)\rho] \end{bmatrix}$$

One array of measurements is generated by using $Y_i = \eta_i + \varepsilon_i$ where $\varepsilon_i = \rho\varepsilon_{i-1} + u_i$ and the u_i are found from a table of normal random numbers with unit variance and zero mean. The Monte Carlo analysis is obtained by generating a large number of sets of b_1, b_2, $\hat{\rho}$, and $\hat{\sigma}_u^2$ where each set corresponds to an array of measurements. Table 6.17 is a summary of results of the Monte Carlo analysis for $\rho=1$ with $n=10$ to 120 and Table 6.18 summarizes results for $n=60$ and $\rho=-1$ to 1. For $n=10$ to 60 there were 34 sets of random data for each n; for $n=120$, 17 sets were used. (In general many more than even 34 sets of data should be used.) The terms with bars over them are average values found from the Monte Carlo simulation.

Table 6.17 Summary of Monte Carlo Investigation for $\rho=1$ for Example 6.9.4

	n				
	10	30	40	50	120
$\overline{\hat{\rho}}$	0.362	0.714	0.783	0.855	0.932
$\overline{\hat{\sigma}_u^2}$	0.803	0.888	0.886	0.928	0.953
$(\bar{b}_1-\beta_1)/\beta_1$	−0.00142	−0.00047	0.00001	0.00039	0.00058
$\overline{\text{est. s.e.}(b_1)/\text{s.e.}(b_1)}$	0.837	0.831	0.837	0.901	0.932
$(\bar{b}_2-\beta_2)/\beta_2$	0.0663	−0.00186	0.0013	0.0045	−0.00003
$\overline{\text{est. s.e.}(b_2)/\text{s.e.}(b_2)}$	0.426	0.327	0.316	0.363	0.360
$\overline{\text{est. cov}(b_1,b_2)}$	−0.0091	−0.0030	−0.0023	−0.0020	−0.0009
$\overline{\text{runs}}/n$ for $\mathbf{D}^{-1}\mathbf{e}$	0.512	0.472	0.486	0.481	0.491

Table 6.18 Summary of Monte Carlo Investigation for $n=60$ for Example 6.9.4

	ρ				
	−1	−0.5	0	0.5	1
$\overline{\hat{\rho}}$	−0.969	−0.485	−0.024	0.430	0.881
$\overline{\hat{\sigma}_u^2}$	0.958	0.966	0.966	0.963	0.933
$(\bar{b}_1-\beta_1)/\beta_1$	0.00013	0.00016	0.00020	0.00014	0.00012
$\overline{\text{est. s.e.}(b_1)/\text{s.e.}(b_1)}$	1.024	1.029	1.005	0.938	0.912
$(\bar{b}_2-\beta_2)/\beta_2$	−0.00048	−0.00060	−0.00080	−0.00093	−0.00048
$\overline{\text{est. s.e.}(b_2)/\text{s.e.}(b_2)}$	0.997	1.004	0.981	0.917	0.353
$\overline{\text{est. cov}(b_1,b_2)}$	−0.00004	−0.00008	−0.00016	−0.00049	−0.00175
$\overline{\text{runs}}/n$ for $\mathbf{D}^{-1}\mathbf{e}$	0.523	0.516	0.500	0.493	0.490

Let us examine Table 6.17. The relative errors in $\hat{\rho}$, $\hat{\sigma}_u^2$, b_1, and b_2 decrease as n is increased. The first two estimates appear to be biased. It is not clear whether b_1 and b_2 are; however, except for $\bar{b}_2$ with $\rho=1$ and $n=10$ the biases indicated by $(\bar{b}_1-\beta_1)/\beta_1$ and $(\bar{b}_2-\beta_2)/\beta_2$ are very small. The average estimated standard error of b_1 divided by the true value is biased but not nearly as much as the same ratio for b_2 which is about 0.36. This means that confidence regions based on these estimated standard errors will be too small, or in other words, the parameter estimates would be presented as being more accurate than they really are. For $\rho=1$ in this example, the true value of the covariance of b_1 and b_2 is zero while Table 6.17 shows small negative values. The average number of runs for the modified residuals, $\mathbf{D}^{-1}(\mathbf{Y}-\hat{\mathbf{Y}})$, is nearly $n/2$, which is close to the$(n+1)/2$ value expected for independent observations.

From Table 6.18 most of the same conclusions as drawn from Table 6.17 are valid. An additional one is that only the cumulative error case ($\rho=1$) for b_2 has a much smaller estimated standard error ratio than for the other ρ values listed. For further discussion, see reference 23.

REFERENCES

1. Hildebrand, F. B., *Methods of Applied Mathematics*, 2nd ed., Prentice-Hall, Inc., Englewood Cliffs, N. J., 1965.
2. Deutsch, R., *Estimation Theory*, Prentice-Hall, Inc., Englewood Cliffs, N. J., 1965.
3. Brownlee, K. A., *Statistical Theory and Methodology in Science and Engineering*, 2nd ed., John Wiley & Sons, Inc., New York, 1965.
4. Draper, N. R. and Smith, H., *Applied Regression Analysis*, John Wiley & Sons, Inc., New York, 1966.
5. Daniel, C. and Wood, F. S., *Fitting Equations to Data*, Wiley-Interscience, New York, 1971.
6. Box, G. E. P. and Draper, N. R., "The Bayesian Estimation of Common Parameters from Several Responses," *Biometrika* **52** (1965), 355–365.
7. Hunter, W. G., "Estimation of Unknown Constants from Multiresponse Data," *Ind. Eng. Chem.* **6** (1967), 461.
8. Welty, J. R., *Engineering Heat Transfer*, John Wiley & Sons, Inc., New York, 1974.
9. Goldfeld, S. M. and Quandt, R. E., *Nonlinear Methods in Econometrics*, North-Holland Publishing Company, Amsterdam, 1972.
10. Himmelblau, D. M., *Process Analysis by Statistical Methods*, John Wiley & Sons, Inc., New York, 1970.
11. Burington, R. S. and May, D. C., *Handbook of Probability and Statistics with Tables*, 2nd ed., McGraw-Hill Book Company, New York, 1970.
12. Beyer, W. H., *Handbook of Tables for Probability and Statistics*, The Chemical Rubber Co., 2nd ed., Cleveland, Ohio, 1968.
13. Rice, J. R., *The Approximations of Functions*, Vol. 1—*Linear Theory*, Addison-Wesley Publishing Co., Reading, Mass., 1964.
14. Myers, R. H., *Response Surface Methodology*, Allyn and Bacon, Inc., Boston, 1971.
15. Beck, J. V. and Al-Araji, S., "Investigation of a New Simple Transient Method of

Thermal Property Measurement," *J. Heat Transfer, Trans. ASME, Ser. C* **96** (1974), 59–64.

16. Mendel, J. M., *Discrete Techniques of Parameter Estimation The Equation Error Formulation*, Marcel Dekker, Inc., New York, 1973.
17. Hoerl, A. E., "Application of Ridge Analysis to Regression Problems," *Chem. Eng. Progr.* **55** (1962), 54–59.
18. Hoerl, A. E., "Ridge Analysis," *Chem. Eng. Progr.*, Symposium Series Vol. **60** (1964), 67–77.
19. Hoerl, A. E. and Kennard, R. W., "Ridge Regression. Biased Estimation for Nonorthogonal Problems," *Technometrics*, **12** (1970), 55–67.
20. Hoerl, A. E. and Kennard, R. W., "Ridge Regression. Applications to Nonorthogonal Problems,': *Technometrics*, **12** (1970), 69–82.
21. Marquardt, D. W., "Generalized Inverses, Ridge Regression, Biased Linear Estimation, and Nonlinear Estimation," *Technometrics*, **12** (1970), 591–612.
22. Swed, F. S. and Eisenhart, C., "Tables for Testing Randomness of Grouping in a Sequence of Alternatives," *Ann. Math. Stat.* **14** (1943), 66–87.
23. Beck, J. V., "Parameter Estimation with Cumulative Errors," *Technometrics* **16** (1974), 85–92.
24. U. S. Bureau of the Census, *Statistical Abstract of the United States* 1974, 95th Annual Edition, Washington, D. C., 1974.

APPENDIX 6A AUTOREGRESSIVE MEASUREMENT ERRORS

Consider the second-order model of autoregressive (AR) measurement errors,

$$\varepsilon_i = \rho_{i1}\varepsilon_{i-1} + \rho_{i2}\varepsilon_{i-2} + u_i \tag{6A.1}$$

for which

$$\varepsilon_0 = \varepsilon_{-1} = 0, \qquad u_i \sim N(0, \sigma_i^2) \tag{6A.2}$$

$$E(u_i u_j) = 0 \qquad \text{for} \quad i \neq j \tag{6A.3}$$

Let us write out the first few terms of (6A.1),

$$\varepsilon_1 = u_1$$

$$\varepsilon_2 = \rho_{21}\varepsilon_1 + u_2 = \rho_{21}u_1 + u_2$$

$$\varepsilon_3 = \rho_{31}\varepsilon_2 + \rho_{32}\varepsilon_1 + u_3 = \rho_{31}\rho_{21}u_1 + \rho_{31}u_2 + \rho_{32}u_1 + u_3$$

$$= (\rho_{31}\rho_{21} + \rho_{32})u_1 + \rho_{31}u_2 + u_3$$

Then in general we can write the $\boldsymbol{\varepsilon}$ vector in terms of the $\mathbf{u}$ vector as

$$\boldsymbol{\varepsilon} = \mathbf{D}_a\mathbf{u} \tag{6A.4a}$$

where

$$\boldsymbol{\varepsilon}^T=[\varepsilon_1\varepsilon_2\cdots\varepsilon_n], \qquad u^T=[u_1u_2\cdots u_n] \tag{6A.4b}$$

$$\mathbf{D}_a=\begin{bmatrix} 1 & & & \\ \rho_{21} & 1 & & \\ \rho_{31}\rho_{21}+\rho_{32} & \rho_{31} & 1 & \\ \vdots & \vdots & \vdots & \cdots \end{bmatrix} \tag{6A.4c}$$

The lower triangular matrix $\mathbf{D}_a$ becomes cumbersome as the dimensions $n\times n$ become large. The inverse of $\mathbf{D}_a$ can be found relatively easily, however. Write out (6A.1) for u_i as

$$u_1=\varepsilon_1$$
$$u_2=-\rho_{21}\varepsilon_1+\varepsilon_2$$
$$u_3=-\rho_{32}\varepsilon_1-\rho_{31}\varepsilon_2+\varepsilon_3$$
$$\vdots$$

or

$$\mathbf{u}=\mathbf{D}_a^{-1}\boldsymbol{\varepsilon} \tag{6A.5a}$$

where $\mathbf{D}_a^{-1}$ is a square matrix with three nonzero diagonals,

$$\mathbf{D}_a^{-1}=\begin{bmatrix} 1 & & & & & \\ -\rho_{21} & 1 & & & & \\ -\rho_{32} & -\rho_{31} & 1 & & & \\ 0 & -\rho_{42} & -\rho_{41} & 1 & & \\ \vdots & \vdots & \ddots & \ddots & \ddots & \\ 0 & 0 & \cdots & -\rho_{n2} & -\rho_{n1} & 1 \end{bmatrix} \tag{6A.5b}$$

The covariance matrix of the errors $\boldsymbol{\psi}$ can be written as

$$\boldsymbol{\psi}=E[\boldsymbol{\varepsilon}\boldsymbol{\varepsilon}^T]=E(\mathbf{D}_a\mathbf{u}\mathbf{u}^T\mathbf{D}_a^T)=\mathbf{D}_a\boldsymbol{\phi}\mathbf{D}_a^T \tag{6A.6a}$$

where $\boldsymbol{\phi}$ is the diagonal matrix

$$\boldsymbol{\phi}=\text{diag}[\sigma_1^2\ \sigma_2^2\ \cdots\ \sigma_n^2] \tag{6A.6b}$$

The inverse of $\boldsymbol{\psi}$ is

$$\boldsymbol{\psi}^{-1}=(\mathbf{D}_a^{-1})^T\boldsymbol{\phi}^{-1}\mathbf{D}_a^{-1} \tag{6A.7}$$

where the inverse of $\mathbf{D}_a$ has the relatively simple form given by (6A.5b).

Because of the form of $\mathbf{D}_a$ which is lower triangular with ones along the diagonal, the determinant of $\mathbf{D}_a$ is unity. Hence the determinant of $\boldsymbol{\psi}$ is

$$|\boldsymbol{\psi}|=|\boldsymbol{\phi}|=\prod_{i=1}^{n}\sigma_i^2 \tag{6A.8}$$

For maximum likelihood estimation we know that

$$\mathbf{b}_{\mathrm{ML}}=(\mathbf{X}^T\boldsymbol{\psi}^{-1}\mathbf{X})^{-1}\mathbf{X}^T\boldsymbol{\psi}^{-1}\mathbf{Y} \tag{6A.9a}$$

$$\operatorname{cov}(\mathbf{b}_{\mathrm{ML}})=(\mathbf{X}^T\boldsymbol{\psi}\mathbf{X})^{-1} \tag{6A.9b}$$

for the linear model $\boldsymbol{\eta}=\mathbf{X}\boldsymbol{\beta}$. By defining

$$\mathbf{Z}_a=\mathbf{D}_a^{-1}\mathbf{X}, \qquad \mathbf{F}_a=\mathbf{D}_a^{-1}\mathbf{Y} \tag{6A.10}$$

The ML relations in (6A.9) can be written in the weighted least squares form of

$$\mathbf{b}_{\mathrm{ML}}=(\mathbf{Z}_a^T\boldsymbol{\phi}^{-1}\mathbf{Z}_a)^{-1}\mathbf{Z}_a^T\boldsymbol{\phi}\mathbf{F}_a \tag{6A.11a}$$

$$\operatorname{cov}(\mathbf{b}_{\mathrm{ML}})=(\mathbf{Z}_a^T\boldsymbol{\phi}_a^{-1}\mathbf{Z}_a)^{-1} \tag{6A.11b}$$

Because of the simple form of the inverse of $\mathbf{D}_a$, the terms in $\mathbf{Z}_a$ and $\mathbf{F}_a$ are

$$Z_{ij,a}=X_{ij}-\rho_{i1}X_{i-1,j}-\rho_{i2}X_{i-2,j} \tag{6A.12a}$$

$$F_{i,a}=Y_i-\rho_{i1}Y_{i-1}-\rho_{i2}Y_{i-2} \tag{6A.12b}$$

Also because $\boldsymbol{\phi}$ is a diagonal matrix, the matrices in (6A.11) can be written in summation form as

$$\mathbf{Z}_a^T\boldsymbol{\phi}^{-1}\mathbf{Z}_a=\left[\sum_{i=1}^{n}Z_{ij,a}Z_{ik}\sigma_i^{-2}\right]; \qquad j=1,2,\ldots,p \quad \text{and} \quad k=1,2,\ldots,p \tag{6A.13a}$$

$$\mathbf{Z}_a^T\boldsymbol{\phi}^{-1}\mathbf{F}_a=\left[\sum_{i=1}^{n}Z_{ij,a}F_{i,a}\sigma_i^{-2}\right]; \qquad j=1,2,\ldots,p \tag{6A.13b}$$

For nonlinear estimation problems the sum of squares function is of interest. It is given by

$$S_{\mathrm{ML}} = (\mathbf{Y}-\boldsymbol{\eta})^T\boldsymbol{\psi}^{-1}(\mathbf{Y}-\boldsymbol{\eta}) = (\mathbf{F}_a-\mathbf{H}_a)^T\boldsymbol{\phi}^{-1}(\mathbf{F}_a-\mathbf{H}_a) \quad (6A.14a)$$

where

$$\mathbf{H}_a = \mathbf{D}_a^{-1}\boldsymbol{\eta} \quad (6A.14b)$$

Again because $\boldsymbol{\phi}$ is diagonal, we can write the relatively simple summation

$$S_{\mathrm{ML}} = \sum_{i=1}^{n} (F_{i,a}-H_{i,a})^2\sigma_i^{-2} \quad (6A.15a)$$

$$= \sum_{i=1}^{n} \left[(Y_i-\rho_{i1}Y_{i-1}-\rho_{i2}Y_{i-2})-(\eta_i-\rho_{i1}\eta_{i-1}-\rho_{i2}\eta_{i-2})\right]^2\sigma_i^{-2} \quad (6A.15b)$$

The above analysis is readily modified for first-order autoregressive cases by setting ρ_{i2} equal to zero. Higher-order cases are also treated simply by adding terms in $Z_{ij,a}$ and $F_{i,a}$.

Covariance Matrices for First-Order Autoregressive Errors with Constant ρ

Some covariance matrices of $\boldsymbol{\psi}$ for the first-order AR case with constant ρ and

$$\sigma_1^2 = c\sigma_u^2, \qquad \sigma_i^2 = \sigma_u^2 \quad \text{for} \quad i = 2,3,\ldots,n \quad (6A.16)$$

are given below. For this case $\mathbf{D}_a$ and $\boldsymbol{\phi}$ are given by

$$\mathbf{D}_a = \begin{bmatrix} 1 & 0 & 0 & 0 & \cdots & 0 \\ \rho & 1 & 0 & 0 & \cdots & 0 \\ \rho^2 & \rho & 1 & 0 & \cdots & 0 \\ \rho^3 & \rho^2 & \rho & 1 & \cdots & 0 \\ \vdots & \vdots & \vdots & \vdots & & \vdots \\ \rho^{n-1} & \cdot & \cdot & \cdot & \rho & 1 \end{bmatrix} \qquad \boldsymbol{\phi} = \mathrm{diag}\left[c\sigma_u^2\ \sigma_u^2\ \cdots\ \sigma_u^2\right]$$

(6A.17)

and then ψ can be multiplied out to get

$$\psi_a = \mathbf{D}_a \phi \mathbf{D}_a^T = \begin{bmatrix} c & c\rho & c\rho^2 & c\rho^3 & \cdots \\ & 1+c\rho^2 & \rho(1+c\rho^2) & \rho^2(1+c\rho^2) & \cdots \\ & & 1+\rho^2+c\rho^4 & \rho(1+\rho^2+c\rho^4) & \cdots \\ & & & 1+\rho^2+\rho^4+c\rho^6 & \cdots \\ & \text{symmetric} & & & \end{bmatrix} \sigma_u^2 \tag{6A.18}$$

Consider the terms in the square matrix in (6A.18). Notice that the diagonal term of the jth row, which we designate G_j, also appears as a product in the jth row for the $(j+1)$th and succeeding terms. For $\rho^2 \neq 1$, G_j is given by

$$G_j = 1+\rho^2+\cdots+\rho^{2(j-1)}+(c-1)\rho^{2(j-1)} = \frac{1-\rho^{2j}}{1-\rho^2}+(c-1)\rho^{2(j-1)} \tag{6A.19}$$

As $j \to \infty$ and $\rho^2 < 1$, G_j approaches the value $(1-\rho^2)^{-1}$. This could be called the "steady-state" value of the variance of ε_i/σ. Evaluating the difference betweeen successive diagonal terms of G_j for any ρ^2 yields

$$\Delta G_j = G_{j+1} - G_j = \rho^{2(j-1)}\left[1 - c(1-\rho^2)\right] \tag{6A.20}$$

At least five cases can be identified using (6A.20):

Case 1.

$$\Delta G_j = 0; \quad c = (1-\rho^2)^{-1}, \quad \rho^2 < 1, \quad (\text{"steady-state" case}). \tag{6A.21a}$$

Case 2.

$$\Delta G_j = 1; \quad c = 1, \quad \rho = 1. \tag{6A.21b}$$

Case 3.

$$\Delta G_j = \rho^{2j}; \quad c = 1. \tag{6A.21c}$$

Case 4.

$$\Delta G_j > 0; \quad c < (1-\rho^2)^{-1}, \quad \rho^2 < 1. \tag{6A.21d}$$

Case 5.

$$\Delta G_j < 0; \quad c > (1-\rho^2)^{-1}, \quad \rho^2 < 1. \tag{6A.21e}$$

In each of these cases except case 2 and if $\rho^2 < 1$, the value of G_∞ is the same, namely, $(1-\rho^2)^{-1}$. In Case 1, G_j is this value of $(1-\rho^2)^{-1}$ for all j. In Case 4, G_j increases monotonically to this value whereas in Case 5, G_1 is the largest value and G_j decreases asymptotically to G_∞. Hence it is possible to have constant, increasing, or decreasing $V(\varepsilon_j)$, depending on whether $c = (1-\rho^2)^{-1}$, $c < (1-\rho^2)^{-1}$, or $c > (1-\rho^2)^{-1}$, respectively. If $\rho = 1$ as in case 2, $G_j = j$, which clearly has no steady-state value; Case 2 is called the cumulative error case. If $\rho^2 > 1$, the G_j values grow very rapidly with j. Some matrices for the above cases are

$$\psi_{a1} = \frac{\sigma_u^2}{1-\rho^2}\begin{bmatrix} 1 & \rho & \rho^2 & \cdots & \rho^{n-1} \\ \rho & 1 & \rho & \cdots & \rho^{n-2} \\ \rho^2 & \rho & 1 & \cdots & \rho^{n-3} \\ \vdots & \vdots & \vdots & & \vdots \\ \rho^{n-1} & \rho^{n-2} & \rho^{n-3} & \cdots & 1 \end{bmatrix} \tag{6A.22a}$$

$$\psi_{a2} = \sigma_u^2\begin{bmatrix} 1 & 1 & 1 & \cdots & 1 \\ 1 & 2 & 2 & \cdots & 2 \\ 1 & 2 & 3 & \cdots & 3 \\ \vdots & \vdots & \vdots & & \vdots \\ 1 & 2 & 3 & \cdots & n \end{bmatrix} \tag{6A.22b}$$

$$\psi_{a3} = \sigma_u^2\begin{bmatrix} 1 & \rho & \rho^2 & \cdots & \rho^{n-1} \\ & 1+\rho^2 & \rho(1+\rho^2) & \cdots & \rho^{n-2}(1+\rho^2) \\ & & 1+\rho^2+\rho^4 & \cdots & \rho^{n-3}(1+\rho^2+\rho^4) \\ & & \vdots & & \vdots \\ \text{symmetric} & & & & \end{bmatrix} \tag{6A.22c}$$

APPENDIX 6B MATRIX INVERSION LEMMA

Important relations for inverses occurring in parameter estimation are derived below. Let $\mathbf{A}$ be $p \times m$ and $\mathbf{B}$ be $m \times p$. Let $\mathbf{I}_m$ be the $m \times m$ identity matrix.

An identity and some rearrangements of it are as follows:

$$-\mathbf{A}(\mathbf{I}_m + \mathbf{BA}) = -(I_p + \mathbf{AB})\mathbf{A} \tag{6B.1}$$

$$-\mathbf{AB} = -(\mathbf{I}_p + \mathbf{AB})\mathbf{A}(\mathbf{I}_m + \mathbf{BA})^{-1}\mathbf{B} \tag{6B.2}$$

$$\mathbf{I}_p = (\mathbf{I}_p + \mathbf{AB}) - (\mathbf{I}_p + \mathbf{AB})\mathbf{A}(\mathbf{I}_m + \mathbf{BA})^{-1}\mathbf{B} \tag{6B.3}$$

Premultiplying (6B.3) by $(\mathbf{I}_p + \mathbf{AB})^{-1}$ yields

$$(\mathbf{I}_p + \mathbf{AB})^{-1} = \mathbf{I}_p - \mathbf{A}(\mathbf{I}_m + \mathbf{BA})^{-1}\mathbf{B} \tag{6B.4}$$

Let $\mathbf{P}$ be defined by

$$\mathbf{P} \equiv \left[\mathbf{X}^T\boldsymbol{\psi}^{-1}\mathbf{X} + \mathbf{V}_\beta^{-1}\right]^{-1} = \left[\mathbf{V}_\beta\mathbf{X}^T\boldsymbol{\psi}^{-1}\mathbf{X} + \mathbf{I}\right]^{-1}\mathbf{V}_\beta \tag{6B.5}$$

Using (6B.4) for (6B.5), with $\mathbf{A} = \mathbf{V}_\beta\mathbf{X}^T\boldsymbol{\psi}^{-1}$ and $\mathbf{B} = \mathbf{X}$, gives

$$\mathbf{P} = \left[\mathbf{I} - \mathbf{V}_\beta\mathbf{X}^T\boldsymbol{\psi}^{-1}\left(\mathbf{I} + \mathbf{X}\mathbf{V}_\beta\mathbf{X}^T\boldsymbol{\psi}^{-1}\right)^{-1}\mathbf{X}\right]\mathbf{V}_\beta$$

$$\boxed{\mathbf{P} = \mathbf{V}_\beta - \mathbf{V}_\beta\mathbf{X}^T\left(\mathbf{X}\mathbf{V}_\beta\mathbf{X}^T + \boldsymbol{\psi}\right)^{-1}\mathbf{X}\mathbf{V}_\beta} \tag{6B.6}$$

This equation is called the *matrix inversion lemma*. Equation 6.7.5a is found from (6B.6) by letting

$$\mathbf{P} \rightarrow \mathbf{P}_{i+1}, \qquad \mathbf{V}_\beta \rightarrow \mathbf{P}_i, \qquad \mathbf{X} \rightarrow \mathbf{X}_{i+1}, \qquad \boldsymbol{\psi} \rightarrow \boldsymbol{\phi}_{i+1}$$

(6B.1) can be written as

$$(\mathbf{I}_p + \mathbf{AB})^{-1}\mathbf{A} = \mathbf{A}(\mathbf{I}_m + \mathbf{BA})^{-1} \tag{6B.7}$$

Using the substitutions suggested above for **A** and **B** results in

$$\boxed{\mathbf{P}\mathbf{X}^T\psi^{-1}=\mathbf{V}_\beta\mathbf{X}^T\left(\mathbf{X}\mathbf{V}_\beta\mathbf{X}^T+\psi\right)^{-1}} \tag{6B.8}$$

PROBLEMS

6.1 Evaluate the following matrix products

(*a*)

$$\begin{bmatrix} 2 & -1 \\ -1 & 2 \end{bmatrix}\begin{bmatrix} 1 & 0 & 1 \\ 1 & -1 & 1 \end{bmatrix}$$

(*b*)

$$\begin{bmatrix} 1 & 1 & 1 & 1 \\ 0 & 1 & 2 & 3 \end{bmatrix}\begin{bmatrix} 1 & 0 \\ 1 & 1 \\ 1 & 2 \\ 1 & 3 \end{bmatrix}$$

(*c*)

$$[a_1\ a_2\ \cdots\ a_n]\begin{bmatrix} b_1 \\ b_2 \\ \vdots \\ b_n \end{bmatrix}$$

(*d*)

$$\begin{bmatrix} b_1 \\ b_2 \\ \vdots \\ b_n \end{bmatrix}[a_1\ a_2\ \cdots\ a_n]$$

(*e*)

$$\begin{bmatrix} c_1 & 0 \\ 0 & c_2 \end{bmatrix}\begin{bmatrix} a_{11} & a_{12} \\ a_{21} & a_{22} \end{bmatrix}$$

(*f*)

$$\begin{bmatrix} a_{11} & a_{12} \\ a_{21} & a_{22} \end{bmatrix}\begin{bmatrix} c_1 & 0 \\ 0 & c_2 \end{bmatrix}$$

6.2 For **X** given by

$$\mathbf{X}=\begin{bmatrix} 1 & 1 & 1 \\ 1 & 2 & 4 \\ 1 & 3 & 9 \\ 1 & 4 & 16 \end{bmatrix}$$

and $\mathbf{D}^{-1}$ with $\rho=0.5$ in (6.9.4), evaluate

$$\mathbf{X}^T\Omega^{-1}\mathbf{X}, \qquad \text{where } \Omega=\mathbf{D}\mathbf{D}^T$$

Choose the proper size for $\mathbf{D}^{-1}$

6.3 Prove for **A** and **B** being nonsingular square matrices

$$(\mathbf{AB})^{-1}=\mathbf{B}^{-1}\mathbf{A}^{-1}$$

6.4 A commonly given method for finding the inverse of a square matrix **A** given by

$$\mathbf{A}=\begin{bmatrix} a_{11} & a_{12} & \cdots & a_{1n} \\ a_{21} & a_{22} & \cdots & a_{2n} \\ \vdots & \vdots & & \vdots \\ a_{n1} & a_{n2} & \cdots & a_{nn} \end{bmatrix}$$

involves the use of *cofactors*. A cofactor A_{ij} is $(-1)^{i+j}D_{ij}$ where D_{ij} is called a minor and is obtained by taking the determinant of the submatrix formed by striking out the row and column corresponding to the element a_{ij}. The inverse of $\mathbf{A}$ contains the elements $A_{ji}/|A|$. Using this expression for the inverse, verify (6.1.10) and (6.1.11).

6.5 For Model 5 of Chapter 5, $\eta_i=\beta_1 X_{i1}+\beta_2 X_{i2}$, show that the matrices $(\mathbf{X}^T\mathbf{X})^{-1}$ and $\mathbf{X}^T\mathbf{Y}$ used in OLS are

$$(\mathbf{X}^T\mathbf{X})^{-1}=\frac{1}{\Delta}\begin{bmatrix} \sum X_{i2}^2 & -\sum X_{i1}X_{i2} \\ -\sum X_{i1}X_{i2} & \sum X_{i1}^2 \end{bmatrix}$$

where $\Delta=(\Sigma X_{i1}^2)(\Sigma X_{i2}^2)-(\Sigma X_{i1}X_{i2})^2$

$$\mathbf{X}^T\mathbf{Y}=\begin{bmatrix} \sum X_{i1}Y_i \\ \sum X_{i2}Y_i \end{bmatrix}$$

6.6 (*a*) For Model 5 of Chapter 5 give the components, $\mathbf{X}^T\boldsymbol{\psi}^{-1}\mathbf{X}$ and $\mathbf{X}^T\boldsymbol{\psi}^{-1}\mathbf{Y}$, of $\mathbf{b}_{\mathrm{ML}}$ given by (6.5.2). Let $\boldsymbol{\psi}^{-1}$ be given by $\mathbf{W}$ where the components are w_{ij} and $w_{ij}\neq 0$ for $i\neq j$.

(*b*) Using the results of (*a*), write out the equation for $b_{1,\mathrm{ML}}$. Check your answer by letting $\boldsymbol{\psi}=\sigma^2\mathbf{I}$ and using the Table 5.3 results for Model 5.

6.7 Find $(\mathbf{X}^T\mathbf{X})^{-1}$ for the cases

(*a*) $Y_i=\beta_1+\beta_2 X_i+\epsilon_i$ with $X_1=1$, $X_2=2$, $X_3=3$, and $X_4=4$.

(*b*) $Y_i=\beta_1 X_i+\beta_2 X_i^2+\epsilon_i$ with $X_i=-2$, -1, 0, 1, and 2.

(*c*) $Y_i=\beta_1 X_{i1}+\beta_2 X_{i2}+\epsilon_i$ for $i=1,2,\ldots,9$.
$X_{11}=X_{41}=X_{71}=-1$; $X_{21}=X_{51}=X_{81}=0$; $X_{31}=X_{61}=X_{91}=1$.
$X_{12}=X_{22}=X_{32}=-1$; $X_{42}=X_{52}=X_{62}=0$; $X_{72}=X_{82}=X_{92}=1$.

6.8 In Welty [8, p. 247] the following are recommended coordinates for natural convection from horizontal cylinders to liquids and gases. Also given are the logarithms to the base 10.

Nu_D	Gr_DPr	$\log Nu_D$	$\log Gr_DPr$
0.490	10^{-4}	−0.3098	−4
0.550	10^{-3}	−0.2596	−3
0.661	10^{-2}	−0.1798	−2
0.841	10^{-1}	−0.0752	−1
1.08	1	0.0334	0
1.51	10	0.1790	1
2.11	10^2	0.3243	2
3.16	10^3	0.4997	3
5.37	10^4	0.7300	4
9.33	10^5	0.9699	5
16.2	10^6	1.2095	6
28.8	10^7	1.4594	7
51.3	10^8	1.7101	8
93.3	10^9	1.9699	9

Let $Y = \log Nu_D$ and $X = \log Gr_DPr$.

(*a*) Using the last seven data pairs in the above data with orthogonal polynomials, find $\alpha_0, \alpha_1, \ldots, \alpha_4$.

Answer. 1.2212, 0.2450, 2.644×10^{-3}, 1.6667×10^{-5}, 5.4545×10^{-5}.

(*b*) Find the sum of squares for each of the models that can be obtained from the results of (*a*).

Answer. 1.6814, 5.95×10^{-4}, 8.21×10^{-6}, 8.15×10^{-6}, 6.80×10^{-6}.

(*c*) Assume that the assumptions 11111011 are valid. At the 5% level of significance give a recommended model in terms of Nu_D and Gr_DPr. How does your model compare with

$$Nu_D = .53(Gr_DPr)^{.25}$$

which is often used for $10^4 < Gr_DPr < 10^9$?

6.9 Repeat Problem 6.8 for the *first* seven data pairs.

6.10 Using all the data of Problem 6.8, repeat Problem 6.8. Use the orthogonal tables in Beyer [12, p. 505] or some other book.

6.11 Using the last seven data pairs of Problem 6.8 with $Y = Nu_D$ and $X = Gr_DPr$, estimate, using OLS,

(*a*) β_1 and β_2 in $Y_i = \beta_1 + \beta_2 X_i + \epsilon_i$.

(*b*) β_1, β_2, and β_3 in $Y_i=\beta_1+\beta_2X_i+\beta_3X_i^2+\epsilon$.

(*c*) Compare the residuals found using the results of (*b*) with those given by the model

$$\hat{\text{Nu}}_{\text{D}}=0.5638(\text{Gr}_{\text{D}}\text{Pr})^{.2450}$$

6.12 Using the data of Problem 6.8 and starting with the last pair of "observations" for $Y=\log \text{Nu}_{\text{D}}$ and $X=\log \text{Gr}_{\text{D}}\text{Pr}$, use OLS sequential estimation for the model

$$Y_i=\beta_1+\beta_2X_i+\epsilon_i.$$

6.13 Repeat Problem 6.12 for the model

$$Y_i=\beta_1+\beta_2X_i+\beta_3X_i^2+\epsilon_i$$

6.14 The following data are a continuation of those in Table 6.2. Using the Y_i data below and orthogonal polynomials, find a satisfactory model utilizing the F test at the 5% level of significance.

Obs. No.	Time (sec)	Y_i (°F)
18	1632	142.93
19	1728	139.34
20	1824	136.04
21	1920	132.94
22	2016	130.07
23	2112	127.39
24	2208	124.88
25	2304	122.46
26	2400	120.18
27	2496	118.04
28	2592	115.97
29	2688	114.13
30	2784	112.35

6.15 Suppose that η is given by

$$\eta=a_0+a_1t+a_2t^2+a_3t^3$$

and that

$$\eta(0)=0, \qquad \eta(1)=1, \qquad \text{and} \quad \frac{d\eta(1)}{dt}=0$$

Show that for these conditions the model becomes

$$\eta=t(2-t)+\beta t(1-t)^2$$

where β is the single parameter.

6.16 The average rainfall, wind velocity, temperature, etc. at any location over a number of years is periodic. Assume that the dependent variable η is the function of t given by

$$\eta = \beta_1 + \beta_2 \cos 2\pi t_i$$

(*a*) From reference 24, Table No. 319, the average wind speed (in mph) at the airport of Great Falls, Montana, is as follows:

Jan.	Feb.	Mar.	Apr.	May	June
15.7	14.8	13.4	13.2	11.4	11.4
July	Aug.	Sept.	Oct.	Nov.	Dec.
10.3	10.5	11.7	13.8	15.0	16.0

Estimate β_1 and β_2 using OLS. Calculate the residuals. (Let the January value be at $t = 0.5/12$, etc).

Answer. 13.1, 2.64

(*b*) Suggest another model that may be able to fit the data better than the one given in (*a*).

6.17 The following data are normal monthly average temperatures (in °F) given by reference 24, Table No. 310, for St. Paul, Minnesota, and San Francisco, California.

	Jan.	Feb.	Mar.	Apr.	May	June
St. Paul	11.8	16.2	28.0	44.6	56.5	66.2
San Fran.	50.9	53.4	54.3	55.3	56.7	58.7
	July	Aug.	Sept.	Oct.	Nov.	Dec.
St. Paul	71.2	69.4	59.1	49.2	31.9	18.2
San Fran.	58.5	59.4	62.2	61.4	57.4	52.0

(*a*) Suggest an appropriate function for η to describe the St. Paul data. The normal minimum monthly temperature is 8–10°F less for all months and the normal maximum temperature is 8–10°F greater for all months.

(*b*) Suggest an appropriate function η to describe the San Francisco data. The normal minimum and maximum monthly temperatures are within ±7°F for all months.

6.18 The model

$$Y_i = \beta_1 + \beta_2 \sin 2\pi t_i + \epsilon_i$$

is proposed for the following data. Very little is known regarding ϵ_i. Estimate

β_1 and β_2 using the sequential method.

t_i	Y_i
0	114
1/12	152
1/4	198
5/12	157
1/2	96
7/12	54
3/4	−10
11/12	51
1	96

Assuming that the standard assumptions hold, estimate the covariance matrix of these estimates.

Answer. for $t_i=1$, $b_1=100.89$, $b_2=103.33$, $P_{11}=4.464$, $P_2=0.0000$, $P_{22}=13.391$

6.19 Use orthogonal polynomials for the Example 6.2.3 data. The components of $\mathbf{X}^T\mathbf{Y}$ are 3390.68, −3302.44, 2608.68, −1448.64, 966.00, and 2347.2. Use F test at 5% level of significance to find the regression line of suitable degree.

6.20 Repeat Problem 6.18 using the standard assumptions and the subjective prior information that

$$\mu_{\beta_1}=100 \qquad \mu_{\beta_2}=50$$

$$\mathbf{V}_\beta=\begin{bmatrix} 200 & 100 \\ 100 & 3000 \end{bmatrix}$$

Estimate β_1 and β_2 using this information and find the covariance matrix of the estimates. The sequential method need not be used.

6.21 Show that

$$\mathbf{b}_{\text{MAP}}-\mathbf{b}_{\text{ML}}=-\mathbf{P}_{\text{ML}}(\mathbf{V}_\beta+\mathbf{P}_{\text{ML}})^{-1}(\mathbf{b}_{\text{ML}}-\boldsymbol{\mu}_\beta).$$

Note that if $\mathbf{V}_\beta$ is sufficiently large compared with $\mathbf{P}_{\text{ML}}$, then $\mathbf{b}_{\text{MAP}}\approx\mathbf{b}_{\text{ML}}$ for any finite $\boldsymbol{\mu}_\beta$.

6.22 Write a FORTRAN or programmable calculator program to calculate the orthogonal coefficients using (6.3.6). Give coefficients for $r=4$, $n=9$, and $i=1,2,\ldots,6$.

6.23 Using the eight measurements at time 0.3 sec of Table 7.14 estimate β_1 in the model $T=\beta_1$. Find the 95% confidence interval. What assumptions are needed?

6.24 Find the 95% confidence region for the data of Example 6.7.3 for the two parameters in the model $k=\beta_1+\beta_2T$. Assume the errors are additive, have zero mean and constant variance, and are uncorrelated and normal. Also let T have negligible errors. There is no prior information regarding the parameters.

6.25 Using the temperature measurements at times 0.3, 0.6, 0.9, and 1.2 sec for all eight thermocouples of Table 7.14, estimate β_1 in the model $T=\eta=\beta_1$. Assume

$$\eta_j(i)=\beta_1+\varepsilon_j(i)$$

$$\varepsilon_j(1)=u_j(1)$$

$$\varepsilon_j(i)=\varepsilon_j(i-1)+u_j(i) \qquad \text{for } j=1,\ldots,8; \quad i=2,3,4$$

$$u_j(i)\sim N(0,\sigma^2),\ \mathrm{E}\left[u_j(i)u_k(1)\right]=0 \text{ except when } i=1 \text{ and } j=k.$$

Use maximum likelihood estimation. Also find the 95% confidence interval.

6.26 The temperatures for thermocouples 5–8 of Table 7.14 can be described by

$$T=\beta_1+\beta_2(t-3.3)^{1/2}$$

from $t=3.3$ to 7.5 sec. β_2 is equal to $2q(\pi k\rho c)^{-1/2}$ with q being heat flux, k thermal conductivity, ρ density, and c specific heat. Using temperatures from thermocouples 5 and 6 from 3.3 to 7.5 sec, estimate β_1 and β_2. Use the sequential estimation method for $m=1$. Calculate and examine the residuals. Discuss your assumptions. How can the model be improved?

6.27 Derive (6.7.9) and (6.7.10).

6.28 The following are temperature measurements taken from Table 7.14:

t, time (sec)	T_5	T_6	T_7	T_8
15	94.56	93.91	94.75	94.17
18	96.52	95.70	96.30	95.96

(*a*) Estimate β in the model $T=92.5+\beta(t-12)$ using OLS estimation.

(*b*) Find the 95% confidence interval for the estimate of β. What assumptions are needed?

6.29 Let the assumptions on the measurement errors be

$$\mathbf{Y}=\mathbf{X}\boldsymbol{\beta}+\boldsymbol{\varepsilon}, \qquad \boldsymbol{\varepsilon}\sim N(\mathbf{0},\sigma^2\boldsymbol{\Omega}) \text{ with } \sigma^2 \text{ unknown and } \boldsymbol{\Omega} \text{ known}$$

$\mathbf{X}$ is errorless. The prior information regarding the random parameter vector is $\boldsymbol{\beta}\sim N(\boldsymbol{\mu}_\beta, \sigma^2\mathbf{V})$ where $\mathbf{V}$ is known. Derive the MAP estimators for $\boldsymbol{\beta}$ and σ^2.

$$\mathbf{b}_{\text{MAP}}=\boldsymbol{\mu}_\beta+\mathbf{P}\mathbf{X}^T\boldsymbol{\Omega}^{-1}(\mathbf{Y}-\mathbf{X}\boldsymbol{\mu}_\beta)$$

$$\hat{\sigma}^2=\frac{1}{n+p}\left\{(\mathbf{Y}-\hat{\mathbf{Y}})^T\boldsymbol{\Omega}^{-1}(\mathbf{Y}-\hat{\mathbf{Y}})+(\mathbf{b}_{\text{MAP}}-\boldsymbol{\mu}_\beta)^T\mathbf{V}^{-1}(\mathbf{b}_{\text{MAP}}-\boldsymbol{\mu}_\beta)\right\}$$

where

$$\mathbf{P}^{-1}=\mathbf{X}^T\boldsymbol{\Omega}^{-1}\mathbf{X}+\mathbf{V}^{-1}, \qquad \hat{\mathbf{Y}}=\mathbf{X}\mathbf{b}_{\text{MAP}}$$

CHAPTER 7

MINIMIZATION OF SUM OF SQUARES FUNCTIONS FOR MODELS NONLINEAR IN PARAMETERS

7.1 INTRODUCTION

This chapter is concerned with methods for minimizing a general sum of squares function when the dependent variable is nonlinear in terms of the parameters. The function to be extremized is assumed to be known although selection of such a function is not a trivial matter and comprises the first optimization problem in parameter estimation. (Extremize means either maximize or minimize.) The extremization of the chosen function, the topic of this chapter, is the second optimization problem of parameter estimation. Other optimization problems relate to optimal design of experiments and optimal designs for discrimination between competing models.

In engineering and science most phenomena are modeled using differential equations. The solution of these equations may be available in closed form or may be obtained through the use of finite difference or finite element computer solutions. Regardless of the method of solution of a differential equation, the model is more often than not a *nonlinear* function of the parameters. The differential equations and boundary and initial conditions may be linear in the usual mathematical sense and still have a solution that is nonlinear in terms of the parameters. See, for examples, the models in Section 7.5.

A problem can be either linear or nonlinear. The nonlinearity in one case can pose more difficulties in obtaining a solution than in another nonlinear case, however.

The extremum found for the sum of squares function when the model is linear in the parameters can be proved to be the correct one for OLS estimation as well as for ML estimation provided the conditions 11--1011 (see Section 6.1.5.2) are satisfied and a unique minimum point exists. Complete assurance* cannot readily be given for nonlinear cases since there may be more than one extremum. If one has reason to doubt that the extremum found is the desired one, it is recommended that contours of constant S be plotted in the region in which the solution is expected. This involves an extensive search. Another possibility is to start the iteration procedure with different sets of initial parameter values.

It should be noted that the same problems associated with ill-conditioning in linear problems also arise in nonlinear estimation problems. In linear problems the theoretical existence of a minimum may be a mirage for ill-conditional cases, that is, those associated with relatively small $|\mathbf{X}^T\mathbf{X}|$ values; see Section 6.7.6. In such cases slight changes in the measurements can cause large movement in the location of the minimum, resulting in large perturbations in the estimated parameter values. Because of this sensitivity for ill-conditioned cases, convergence proofs for nonlinear cases may also be more academic than practical.

Several simple optimum seeking methods are given in this section. In Section 7.4 the Gauss method is given, and in Section 7.6 several modifications of that method are described. Also in Section 7.6 a comparison of several methods is given. Later sections discuss sequential methods and correlated errors.

7.1.1 Trial and Error Search

One of the simplest procedures for extremizing a function is *trial and error*. It is quite inefficient and is not recommended. It is easily described and understood, however.

As for most nonlinear search procedures, the trial and error procedure starts with a set of estimated values of all the parameters. Let the initial parameter estimates be designated $\mathbf{b}^{(0)}$ and for simplicity let there be an associated OLS sum of squares

$$S^{(0)} = \Sigma\left(Y_i - \eta_i^{(0)}\right)^2 \tag{7.1.1}$$

*For some cases convergence proofs are available but they are usually not practical to implement for a number of different reasons. See Bard [5, p. 87].

(A more general function is given in Section 7.3.) Next, another set of parameters $\mathbf{b}^{(1)}$ is chosen more or less arbitrarily and $S^{(1)}$ is calculated. If

$$S^{(1)} < S^{(0)} \tag{7.1.2}$$

then the combination of parameters given by $\mathbf{b}^{(1)}$ must be "better" than $\mathbf{b}^{(0)}$; one might wish to select a $\mathbf{b}^{(2)}$ to be a modification of $\mathbf{b}^{(1)}$ in the same manner that $\mathbf{b}^{(1)}$ was of $\mathbf{b}^{(0)}$. If the inequality in (7.1.2) is reversed, then we would not proceed in the same manner. There are many possible strategies that could be employed to choose different **b**'s. One is to fix all the b_i's except one which is varied until S reaches a relative minimum, at which time this parameter is fixed and another one is varied. This procedure usually requires considering further changes in each parameter after the other parameters have been reestimated.

In the procedure outlined above, there is no rule that *must* be followed in selection of new sets of parameters. Instead one can try any set of parameters that seems reasonable. This can be done interactively using a teletype computer terminal. In so doing one usually finds that this method of solution is inefficient, time-consuming, and not practical. One can, however, obtain a "feel" for the difficulty of extremizing a function, particularly when there is more than one parameter.

7.1.2 Exhaustive Search

Another simple (and also inefficient) procedure is termed an exhaustive search. To illustrate this procedure consider the case of only one unknown parameter, β. The "best" value of this parameter is to be associated with the minimum value of S. Instead of selecting only an initial estimate, a region of β is chosen in which region the minimum value of S is expected to be. Suppose that the parameter β is known to be between 0.5 and 2.0. In an exhaustive search S is calculated at equally spaced values of b in this region (β is the true value and b is an estimated value.) Fig. 7.1*a* shows S for Δb intervals of 0.25. The best $b^{(i)}$ value as indicated by Fig. 7.1*a* is $b^{(3)} = 1$. A more accurate value of b could be found by conducting an exhaustive search with a smaller Δb in the reduced region between $b^{(2)}$ and $b^{(4)}$.

The exhaustive search procedure is undoubtedly expensive but it does have the potential of revealing local minima in addition to the *global* minimum. This is illustrated by Fig. 7.1*b*. The exhaustive search procedure is more likely to produce the global minimum than some other schemes. Irrelevant local minima are encountered in parameter estimation but are not common.

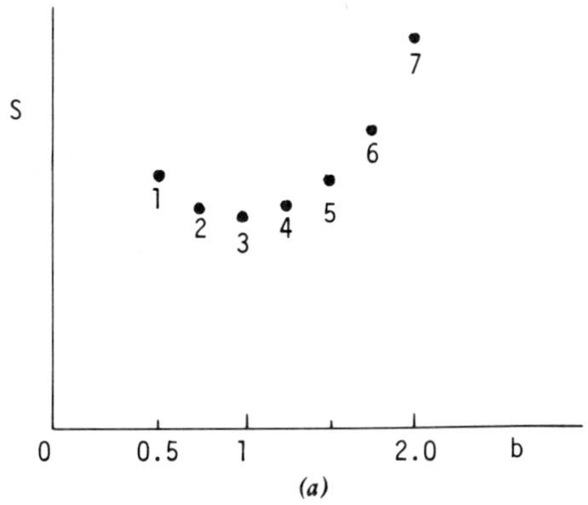

Figure 7.1a Sum of squares in an exhaustive search procedure

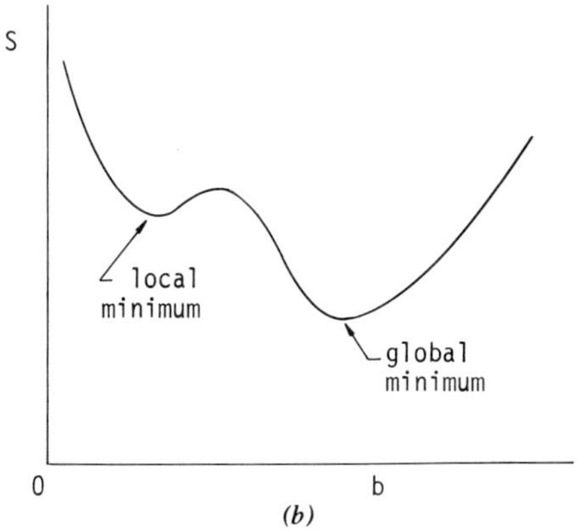

Figure 7.1b S with a local and global minimum.

It should also be noted that the global minimum may not necessarily be the desired one. For example, a certain mechanical device may be known to have a natural frequency about 1 Hz. This frequency is to be more precisely estimated utilizing several measurements of deflection versus time of the device. A local minimum of S would be expected near 1 Hz but the global minimum might occur at a considerably higher frequency. Another example occurs when the functional form of S incorporated in a computer program may have a global minimum at *negative* values of parameters which are not physically possible.

7.1.3 Other Methods

There are many other methods for locating extremes of arbitrary functions. These include direct search, Fibonacci search, gradient methods, random search, Hooke–Jeeves search, simplex exploration, and dynamic programming methods. See a text on optimization such as Beveridge and Schechter [2]. Most of these methods are much more efficient than the two methods just mentioned.

Rather than giving many methods we shall emphasize one basic method and then suggest some modifications of it. This method is called the Gauss method. It has proved to be very effective for a large class of different parameter estimation problems.

7.2 MATRIX FORM OF TAYLOR SERIES EXPANSION

In the Gauss linearization method of minimizing a sum of squares function a Taylor series expansion of a vector is needed. Let $\boldsymbol{\eta}$ be an n vector and a function of the p parameters in the $\boldsymbol{\beta}$ vector. Let $\boldsymbol{\eta}$ have continuous derivatives in the neighborhood of $\boldsymbol{\beta}=\mathbf{b}$. Then the Taylor series for a point $\boldsymbol{\beta}$ near $\mathbf{b}$ begins with the terms

$$\boldsymbol{\eta}(\boldsymbol{\beta})=\boldsymbol{\eta}(\mathbf{b})+\left[\nabla_{\beta}\boldsymbol{\eta}^{T}(\mathbf{b})\right]^{T}(\boldsymbol{\beta}-\mathbf{b})+\cdots \tag{7.2.1}$$

where ∇_{β} is the matrix derivative operator defined by (6.1.21).

7.3 SUM OF SQUARES FUNCTION

In ordinary least squares, weighted least squares, maximum likelihood, and maximum a posteriori estimation, the sum of squares functions to be minimized are generally different. In some cases, however, the MAP sum of squares function reduces to that for ML which in turn reduces to that for OLS estimation. For that reason and for economy in presentation a sum of squares function is given that is appropriate for OLS, WLS, ML, and MAP estimation when appropriately specialized. The function that we consider in this chapter is

$$S=\left[\mathbf{Y}-\boldsymbol{\eta}(\boldsymbol{\beta})\right]^{T}\mathbf{W}\left[\mathbf{Y}-\boldsymbol{\eta}(\boldsymbol{\beta})\right]+(\boldsymbol{\mu}-\boldsymbol{\beta})^{T}\mathbf{U}(\boldsymbol{\mu}-\boldsymbol{\beta})$$

$$=\left[\mathbf{Y}-\boldsymbol{\eta}(\boldsymbol{\beta})\right]^{T}\mathbf{W}\left[\mathbf{Y}-\boldsymbol{\eta}(\boldsymbol{\beta})\right]+(\boldsymbol{\beta}-\boldsymbol{\mu})^{T}\mathbf{U}(\boldsymbol{\beta}-\boldsymbol{\mu}) \tag{7.3.1}$$

Both $\mathbf{W}$ and $\mathbf{U}$ are weighting matrices which are symmetric; $\mathbf{W}$ is positive definite and $\mathbf{U}$ is positive semidefinite. In many cases $\mathbf{W}$ and $\mathbf{U}$ will be assumed to be completely known.

The cases of single and multiresponse for discrete measurements are included in (7.3.1). When a single sensor is used to obtain many observations, we have a single response case. In this situation the observation vector $\mathbf{Y}$ and corresponding vector found from the model, $\boldsymbol{\eta}$, are n vectors. The square matrix $\mathbf{W}$ is $n\times n$. The parameter vector $\boldsymbol{\beta}$ contains p components as does $\boldsymbol{\mu}$; $\mathbf{U}$ is a square matrix of dimensions $p\times p$. Much of this chapter explicitly considers the discrete single response case.

Extensions to multiresponse cases of the algorithms given in Sections 7.4 and 7.6 are not difficult. Section 7.8 provides a sequential method for this case. The multiresponse case occurs when m (>1) measurements are taken at n different times. The observation vector can be written as

$$\mathbf{Y}=\begin{bmatrix}\mathbf{Y}(1)\\ \mathbf{Y}(2)\\ \vdots\\ \mathbf{Y}(n)\end{bmatrix} \qquad \text{where } \mathbf{Y}(i)=\begin{bmatrix}Y_1(i)\\ Y_2(i)\\ \vdots\\ Y_m(i)\end{bmatrix} \tag{7.3.2}$$

Hence the $\mathbf{Y}$ vector contains mn components. The $\boldsymbol{\eta}$ vector can be similarly defined and $\mathbf{W}$ is $mn\times mn$. The $\boldsymbol{\beta}$, $\boldsymbol{\mu}$, and $\mathbf{U}$ matrices remain as given above.

In some situations it is natural to consider continuous rather than discrete measurements in time. This may be either because the measurements are actually continuous or because it is more convenient to analyze them as if they were. Then we would replace (7.3.1) by

$$S=\int_0^{t_f}\left[\mathbf{Y}(t)-\boldsymbol{\eta}(t,\boldsymbol{\beta})\right]^T\mathbf{W}(t)\left[\mathbf{Y}(t)-\boldsymbol{\eta}(t,\boldsymbol{\beta})\right]dt+(\boldsymbol{\mu}-\boldsymbol{\beta})^T\mathbf{U}(\boldsymbol{\mu}-\boldsymbol{\beta}) \tag{7.3.3}$$

where the time limits are 0 and t_f. If the case being considered involves a single response, $\mathbf{Y}(t)$ becomes the scalar $Y(t)$. For the multiresponse case the $\mathbf{Y}(t)$ vector is

$$\mathbf{Y}(t)=\begin{bmatrix}Y_1(t)\\ \vdots\\ Y_m(t)\end{bmatrix} \tag{7.3.4}$$

and $\boldsymbol{\eta}(t,\boldsymbol{\beta})$ is similarly defined. In many algorithms to be given, the summations on a time index can be replaced by integrations over time.

One further modification (or interpretation, depending on one's viewpoint) of (7.3.1) is for situations in which the dependent variables in the system model are not measured directly. This is a case discussed frequently in the systems literature. To illustrate this case assume that the system model is the set of nonlinear, first-order ordinary differential equations,

$$\dot{\mathbf{x}}=\mathbf{f}(t,\mathbf{x},\boldsymbol{\beta},\mathbf{u}) \tag{7.3.5}$$

The dimension of the dependent variable $\mathbf{x}$ is r; $\mathbf{f}$ is some known function of $\mathbf{x}$, t, $\boldsymbol{\beta}$ and $\mathbf{u}$ which could be related to some forcing function or control. It may be that all the components of $\mathbf{x}$ are not measured directly. Instead some linear or nonlinear function of the $\mathbf{x}$ may be measured,

$$\mathbf{Y}=\mathbf{h}\mathbf{x}+\boldsymbol{\varepsilon} \qquad \text{or} \qquad \mathbf{Y}=\mathbf{g}(\mathbf{t},\mathbf{x})+\boldsymbol{\varepsilon} \tag{7.3.6}$$

Here the dimension m of the $\mathbf{Y}_i$ component of the $\mathbf{Y}$ vector would be equal to or less than r. In such cases the dependent variable $\boldsymbol{\eta}$ in (7.3.1) could be replaced by $\mathbf{h}\mathbf{x}$ or $\mathbf{g}(t,\mathbf{x})$.

7.4 GAUSS METHOD OF MINIMIZATION

7.4.1 Derivation

One of the simplest and most effective methods of minimizing the function S is variously called the Gauss, Gauss–Newton, Newton–Gauss, or linearization method; we call it the Gauss method. It is attractive because it is relatively simple and because it specifies direction *and* size of the corrections to the parameter vector. The method is effective in seeking minima that are reasonably well-defined provided the initial estimates are in the general region of the minimum. For difficult cases (i.e., those with indistinct minima) modifications to the Gauss method discussed in Section 7.6 are recommended.

A necessary condition at the minimum of S is that the matrix derivative of S with respect to $\boldsymbol{\beta}$ be equal to zero. For this reason operate upon S using (6.1.30) to get

$$\nabla_{\beta}S=2\left[-\nabla_{\beta}\boldsymbol{\eta}^{T}(\boldsymbol{\beta})\right]\mathbf{W}\left[\mathbf{Y}-\boldsymbol{\eta}(\boldsymbol{\beta})\right]+2\left[-\mathbf{I}\right]\mathbf{U}(\boldsymbol{\mu}-\boldsymbol{\beta}) \tag{7.4.1}$$

Let us use the notation $\mathbf{X}(\beta)$ for the *sensitivity matrix*,

$$\mathbf{X}(\boldsymbol{\beta})\equiv\left[\nabla_{\beta}\boldsymbol{\eta}^{T}(\boldsymbol{\beta})\right]^{T} \tag{7.4.2}$$

so that (7.4.1) set equal to zero at $\boldsymbol{\beta}=\hat{\boldsymbol{\beta}}$ becomes

$$\mathbf{X}^T(\hat{\boldsymbol{\beta}})\mathbf{W}\left[\mathbf{Y}-\boldsymbol{\eta}(\hat{\boldsymbol{\beta}})\right]+\mathbf{U}(\boldsymbol{\mu}-\hat{\boldsymbol{\beta}})=\mathbf{0} \tag{7.4.3}$$

Unfortunately, we cannot easily solve for the estimator $\hat{\boldsymbol{\beta}}$ since $\hat{\boldsymbol{\beta}}$ appears implicitly in $\boldsymbol{\eta}$ and $\mathbf{X}$ as well as appearing explicitly. Suppose that we have an estimate of $\hat{\boldsymbol{\beta}}$ denoted $\mathbf{b}$ and that $\boldsymbol{\eta}$ has continuous first derivatives in $\boldsymbol{\beta}$ and bounded higher derivatives near $\mathbf{b}$. Two approximations are now used in (7.4.3). First, replace $\mathbf{X}(\hat{\boldsymbol{\beta}})$ by $\mathbf{X}(\mathbf{b})$ and second, use the first two terms of a Taylor series for $\boldsymbol{\eta}(\hat{\boldsymbol{\beta}})$ about $\mathbf{b}$. Then (7.4.3) becomes

$$\mathbf{X}^T(\mathbf{b})\mathbf{W}\left[\mathbf{Y}-\boldsymbol{\eta}(\mathbf{b})-\mathbf{X}(\mathbf{b})(\hat{\boldsymbol{\beta}}-\mathbf{b})\right]+\mathbf{U}(\boldsymbol{\mu}-\mathbf{b})-\mathbf{U}(\hat{\boldsymbol{\beta}}-\mathbf{b})\approx\mathbf{0} \tag{7.4.4}$$

Note that this equation is linear in $\hat{\boldsymbol{\beta}}$. If (1) $\boldsymbol{\eta}$ is not too far from being linear in $\boldsymbol{\beta}$ in a region about the solution to (7.4.3) and if (2) this region includes $\mathbf{b}$, the value of $\hat{\boldsymbol{\beta}}$ satisfying (7.4.4) will be a better approximation to the solution (7.4.3) than that provided by $\mathbf{b}$. Assuming these two conditions to be true, (7.4.4) is set equal to zero. In the interest of compactness of notation and to indicate an iterative procedure let

$$\mathbf{b}^{(k)}=\mathbf{b},\qquad \mathbf{b}^{(k+1)}=\hat{\boldsymbol{\beta}},\qquad \boldsymbol{\eta}^{(k)}=\boldsymbol{\eta}(\mathbf{b}),\qquad \mathbf{X}^{(k)}=\mathbf{X}(\mathbf{b}) \tag{7.4.5}$$

Using this notation in (7.4.4) set equal to $\mathbf{0}$ yields p equations in matrix form for $\mathbf{b}^{(k+1)}$,

$$\mathbf{b}^{(k+1)}=\mathbf{b}^{(k)}+\mathbf{P}^{(k)}[\mathbf{X}^{T(k)}\mathbf{W}(\mathbf{Y}-\boldsymbol{\eta}^{(k)})+\mathbf{U}(\boldsymbol{\mu}-\mathbf{b}^{(k)})] \tag{7.4.6a}$$

$$\mathbf{P}^{-1(k)}\equiv\mathbf{X}^{T(k)}\mathbf{W}\mathbf{X}^{(k)}+\mathbf{U} \tag{7.4.6b}$$

which is the Gauss linearization equation. Iteration on k is required for nonlinear models. For linear-in-the-parameters model no iterations are required. Note that for $\eta=\mathbf{X}\boldsymbol{\beta}$, (7.4.6) reduces to the MAP equation (6.6.6a) by setting $\mathbf{b}^{(k)}$ equal to zero. No constraints are included in (7.4.6).

In using (7.4.6) in nonlinear cases an initial estimate of $\boldsymbol{\beta}$, designated $\mathbf{b}^{(0)}$, is needed. With this vector $\boldsymbol{\eta}^{(0)}$ and $\mathbf{X}^{(0)}$ can be calculated, which, in turn, are used in (7.4.6a) to obtain the improved estimate vector $\mathbf{b}^{(1)}$. This completes the first iteration. Then $\boldsymbol{\eta}^{(1)}$ and $\mathbf{X}^{(1)}$ are evaluated so that $\mathbf{b}^{(2)}$ can be found. The iterative procedure continues until there is negligible change in any component of $\mathbf{b}$; one criterion to indicate this is

$$\frac{|b_i^{(k+1)}-b_i^{(k)}|}{|b_i^{(k)}|+\delta_1}<\delta \qquad \text{for } i=1,2,\ldots,p \tag{7.4.7}$$

where δ is a small number such as 10^{-4}. In order to avoid embarrassment if $b_i^{(k)}$ goes to zero, the quantity δ_1 is set equal to another small number such as 10^{-10}. When good initial estimates of the parameters are available and the experiment is well-designed, (7.4.7) is frequently satisfied by the seventh iteration. See Table 7.3 for an example of this. (The fact that (7.4.7) is satisfied does not guarantee that the last $\mathbf{b}^{(k+1)}$ minimizes S, particularly when the minimum is ill-defined.)

As a minimum is being sought, the function S should logically *decrease* from iteration to iteration. One might then include a check in a computer program to see if $S^{(k+1)}$ is less than $S^{(k)}$. If it is not, the procedure could either terminate or the correction of the parameters, $b_i^{(k+1)} - b_i^{(k)}$, could be decreased as discussed in Section 7.6. In some cases, however, a temporary increase in S could permit larger parameter changes to the region of the minimum and actually lead to more rapid convergence.

7.4.2 Components of Gauss Linearization Equation

Consider the sensitivity matrix as defined by (7.4.2); without showing the dependence on $\mathbf{b}^{(k)}$, it can be written for a single response case as

$$\mathbf{X} = \begin{bmatrix} X_{11} & \cdots & X_{1p} \\ \vdots & & \vdots \\ X_{n1} & \cdots & X_{np} \end{bmatrix} = \left[\begin{bmatrix} \dfrac{\partial}{\partial \beta_1} \\ \vdots \\ \dfrac{\partial}{\partial \beta_p} \end{bmatrix} [\eta_1 \cdots \eta_n] \right]^T = \begin{bmatrix} \dfrac{\partial \eta_1}{\partial \beta_1} & \cdots & \dfrac{\partial \eta_1}{\partial \beta_p} \\ \vdots & & \vdots \\ \dfrac{\partial \eta_n}{\partial \beta_1} & \cdots & \dfrac{\partial \eta_n}{\partial \beta_p} \end{bmatrix} \tag{7.4.8}$$

Hence the ij element of $\mathbf{X}^{(k)}$ is

$$X_{ij}^{(k)} = \left(\frac{\partial \eta_i}{\partial \beta_j} \right) \Bigg|_{\mathbf{b}^{(k)}} \tag{7.4.9}$$

This definition of $\mathbf{X}$ is consistent with the linear model. A simple example is $\eta_i = \beta_1 X_{i1} + \beta_2 X_{i2}$ where X_{ij} has the same meaning as in (7.4.9). A model which is nonlinear in a parameter is

$$\eta_i = \beta_1 \exp(\beta_2 t_i) + \beta_3 \tag{7.4.10}$$

Its sensitivity coefficients are

$$X_{i1}=\frac{\partial\eta_i}{\partial\beta_1}=\exp(\beta_2 t_i),\ X_{i2}=\frac{\partial\eta_i}{\partial\beta_2}=\beta_1 t_i\exp(\beta_2 t_i),\ X_{i3}=1 \quad (7.4.11)$$

The matrix $\mathbf{X}^T\mathbf{W}\mathbf{X}$ is a symmetric matrix of dimensions $p\times p$. Let

$$\mathbf{C}\equiv\mathbf{X}^T\mathbf{W}\mathbf{X} \quad (7.4.12)$$

where the C_{ij} element of $\mathbf{C}$ is

$$C_{ij}=\sum_{l=1}^{n}\sum_{r=1}^{n}w_{lr}X_{li}X_{rj}=\Sigma\Sigma w_{lr}\frac{\partial\eta_l}{\partial\beta_i}\frac{\partial\eta_r}{\partial\beta_j} \quad (7.4.13)$$

If the weighting matrix is diagonal, (7.4.13) simplifies to

$$C_{ij}=\sum_{l=1}^{n}w_{ll}X_{li}X_{lj}=\Sigma w_{ll}\frac{\partial\eta_l}{\partial\beta_i}\frac{\partial\eta_l}{\partial\beta_j} \quad (7.4.14)$$

Let the matrix product $\mathbf{X}^T\mathbf{W}(\mathbf{Y}-\boldsymbol{\eta})$ in (7.4.6a) be designated $\mathbf{H}$,

$$\mathbf{H}\equiv\mathbf{X}^T\mathbf{W}(\mathbf{Y}-\boldsymbol{\eta}) \quad (7.4.15)$$

which is a $p\times 1$ vector and has a typical component of

$$H_i=\sum_{l=1}^{n}\sum_{r=1}^{n}w_{lr}X_{li}(Y_r-\eta_r) \quad (7.4.16)$$

For $C_{ij}^{(k)}$ and $H_i^{(k)}$, the quantities X_{li} and η_r are evaluated with $\boldsymbol{\beta}=\mathbf{b}^{(k)}$.

For the simple case of one parameter ($p=1$) the iterative equation is

$$b_1^{(k+1)}=b_1^{(k)}+\frac{H_1^{(k)}+U_1(\mu_1-b_1^{(k)})}{C_{11}^{(k)}+U_1} \quad (7.4.17)$$

An initial estimate of β, designated $b^{(0)}$, must be provided.

Example 7.4.1

Estimate β_1 and β_2 in the model $\eta=\beta_1+(1+\beta_2 t)^2$ using $\mu_1=b_1^{(0)}=2$, $\mu_2=b_2^{(0)}=1$,

$$\boldsymbol{\psi}=\begin{bmatrix}1&1&1\\1&2&2\\1&2&3\end{bmatrix},\quad \boldsymbol{\psi}^{-1}=\begin{bmatrix}2&-1&0\\-1&2&-1\\0&-1&1\end{bmatrix},\quad \mathbf{U}=\begin{bmatrix}0&0\\0&4\end{bmatrix}$$

(This ψ function is associated with cumulative errors in Y_i.) The data are

t_i	-1	0	1
Y_i	1	3	8.5

Use the sum of squares function given by (7.3.1) in which ψ^{-1} is to be used for **W**.

Solution

This problem is solved using (7.4.6) in an iterative manner. Consider the first iteration. The sensitivity coefficients and values are

$$X_{i1}=\frac{\partial\eta_i}{\partial\beta_1}=1, \qquad X_{i2}=\frac{\partial\eta_i}{\partial\beta_2}=2(1+\beta_2 t_i)t_i$$

$$\mathbf{X}^{T(0)}=\begin{bmatrix}1 & 1 & 1\\ 0 & 0 & 4\end{bmatrix}$$

Using this vector we find

$$\mathbf{X}^{T(0)}\psi^{-1}\mathbf{X}^{(0)}=\begin{bmatrix}1 & 1 & 1\\ 0 & 0 & 4\end{bmatrix}\begin{bmatrix}2 & -1 & 0\\ -1 & 2 & -1\\ 0 & -1 & 1\end{bmatrix}\begin{bmatrix}1 & 0\\ 1 & 0\\ 1 & 4\end{bmatrix}=\begin{bmatrix}1 & 0\\ 0 & 16\end{bmatrix}$$

and thus $\mathbf{P}^{-1}$ is

$$\mathbf{P}^{-1(0)}=\mathbf{X}^{T(0)}\psi^{-1}\mathbf{X}^{(0)}+\mathbf{U}=\begin{bmatrix}1 & 0\\ 0 & 16\end{bmatrix}+\begin{bmatrix}0 & 0\\ 0 & 4\end{bmatrix}=\begin{bmatrix}1 & 0\\ 0 & 20\end{bmatrix}$$

which has the inverse

$$\mathbf{P}^{(0)}=\begin{bmatrix}1 & 0\\ 0 & 0.05\end{bmatrix}$$

Another expression needed in (7.4.6) is

$$\mathbf{X}^{T(0)}\psi^{-1}(\mathbf{Y}-\boldsymbol{\eta}^{(0)})+\mathbf{U}(\boldsymbol{\mu}-\mathbf{b}^{(0)})=$$

$$\begin{bmatrix}1 & 1 & 1\\ 0 & 0 & 4\end{bmatrix}\begin{bmatrix}2 & -1 & 0\\ -1 & 2 & -1\\ 0 & -1 & 1\end{bmatrix}\begin{bmatrix}1-2\\ 3-3\\ 8.5-6\end{bmatrix}+\begin{bmatrix}0\\ 0\end{bmatrix}=\begin{bmatrix}-1\\ 10\end{bmatrix}$$

Then we can use (7.4.6a) to complete the first iteration,

$$\mathbf{b}^{(1)}=\begin{bmatrix}2\\ 1\end{bmatrix}+\begin{bmatrix}1 & 0\\ 0 & 0.05\end{bmatrix}\begin{bmatrix}-1\\ 10\end{bmatrix}=\begin{bmatrix}1.0\\ 1.5\end{bmatrix}$$

The initial sum of squares as defined by (7.3.1) has a value of $S^{(0)}=8.25$. After the first iteration we calculate $S^{(1)}$ to be

$$S^{(1)} = [\mathbf{Y} - \boldsymbol{\eta}^{(1)}]^T \boldsymbol{\psi}^{-1} [\mathbf{Y} - \boldsymbol{\eta}^{(1)}] + (\boldsymbol{\mu} - \mathbf{b}^{(1)})^T \mathbf{U} (\boldsymbol{\mu} - \mathbf{b}^{(1)})$$

$$= \begin{bmatrix} 1-1.25 \\ 3-2 \\ 8.5-7.25 \end{bmatrix}^T \begin{bmatrix} 2 & -1 & 0 \\ -1 & 2 & -1 \\ 0 & -1 & 1 \end{bmatrix} \begin{bmatrix} -0.25 \\ 1 \\ 1.25 \end{bmatrix}$$

$$+ \begin{bmatrix} 2-1 \\ 1-1.5 \end{bmatrix}^T \begin{bmatrix} 0 & 0 \\ 0 & 4 \end{bmatrix} \begin{bmatrix} 1 \\ -0.5 \end{bmatrix}$$

$$= 2.6875$$

which is lower than $S^{(0)}$. (If it were not, then a method given in Section 7.6 might be used.)

For the second iteration the calculations proceed in a similar manner. We have

$$\mathbf{X}^{(1)} = \begin{bmatrix} 1 & 1 \\ 1 & 0 \\ 1 & 5 \end{bmatrix}, \quad \mathbf{P}^{-1(1)} = \begin{bmatrix} 1 & 1 \\ 1 & 31 \end{bmatrix}, \quad \mathbf{P}^{(1)} = \frac{1}{30} \begin{bmatrix} 31 & -1 \\ -1 & 1 \end{bmatrix}$$

$$\mathbf{X}^{T(1)} \boldsymbol{\psi}^{-1} (\mathbf{Y} - \boldsymbol{\eta}^{(1)}) + \mathbf{U}(\boldsymbol{\mu} - \mathbf{b}^{(1)}) = \begin{bmatrix} -.25 \\ -.25 \end{bmatrix} + \begin{bmatrix} 0 & 0 \\ 0 & 4 \end{bmatrix} \begin{bmatrix} 1 \\ -.5 \end{bmatrix} = \begin{bmatrix} -.25 \\ -2.25 \end{bmatrix}$$

which results in

$$\mathbf{b}^{(2)} = \begin{bmatrix} 1 \\ 1.5 \end{bmatrix} + \frac{1}{30} \begin{bmatrix} 31 & -1 \\ -1 & 1 \end{bmatrix} \begin{bmatrix} -0.25 \\ -2.25 \end{bmatrix} = \begin{bmatrix} 0.81667 \\ 1.4333 \end{bmatrix}$$

The associated S value is $S^{(2)} = 2.497059$ which is, as it should be, less than $S^{(1)}$.

The above results along with those of the third and fourth iterations are summarized in Table 7.1. The $\Delta b_i^{(k)}$ expression means $\Delta b_i^{(k)} = b_i^{(k+1)} - b_i^{(k)}$. Notice that the results converge quite rapidly since the relative changes in both parameters by the fourth iteration are less in absolute value than 10^{-4} (thus satisfying (7.4.7) if δ is 10^{-4}). There are some changes in sign in the corrections, Δb_i, but no instability is noted.

Table 7.1 Summary of Calculations for Example 7.4.1

$k+1$	$b_1^{(k+1)}$	$b_2^{(k+1)}$	$\Delta b_1^{(k)}/b_1^{(k)}$	$\Delta b_2^{(k)}/b_2^{(k)}$	$S^{(k+1)}$
0	2	1			8.250000
1	1	1.50000	-0.5000	0.500	2.687500
2	0.816667	1.433333	-0.1833	-0.0444	2.497059
3	0.810561	1.435251	-0.748×10^{-2}	0.134×10^{-2}	2.496940
4	0.810630	1.435166	8.56×10^{-5}	-5.85×10^{-5}	2.496939

7.4.3 Comments on Gauss Linearization Equation

Several comments and observations regarding the Gauss linearization equation are given in this section.

(*a*) By letting $\mathbf{W}=\mathbf{I}$ and $\mathbf{U}=\mathbf{0}$, (7.4.6) provides ordinary least squares estimates.

(*b*) If the observation errors $\boldsymbol{\varepsilon}$ satisfy the standard conditions of being additive, zero mean, $\boldsymbol{\psi}$ is known within a multiplicative constant and the independent variable(s) and $\boldsymbol{\beta}$ are nonstochastic (i.e., 11---011), then nonlinear Gauss–Markov estimation can be used by setting $\mathbf{W}=\boldsymbol{\Omega}^{-1}$ and $\mathbf{U}=\mathbf{0}$. (We are using $\boldsymbol{\psi}=\sigma^2\boldsymbol{\Omega}$ where $\boldsymbol{\Omega}$ is completely known.)

(*c*) If in addition to the conditions given in (*b*), the errors are normal, the assumptions are designated 11--1011. Then (7.4.6) provides a nonlinear ML estimator if $\mathbf{W}=\boldsymbol{\Omega}^{-1}$ and $\mathbf{U}=\mathbf{0}$.

(*d*) If the conditions in (*c*) are valid except there is prior information and σ^2 is known, then an MAP estimator is provided by (7.4.6). Suppose that $\boldsymbol{\beta}$ is a random parameter vector with a mean $\boldsymbol{\mu}_\beta$, covariance $\mathbf{V}$, and normal probability density. The assumptions are then designated 11--1112. By letting $\mathbf{W}=\boldsymbol{\psi}^{-1}$, $\boldsymbol{\mu}=\boldsymbol{\mu}_\beta$, and $\mathbf{U}=\mathbf{V}^{-1}$, the corresponding MAP estimator is given by (7.4.6).

(*e*) In order to utilize (7.4.6) to estimate the parameters it is necessary that $\mathbf{P}^{-1}$ have an inverse, that is, P^{-1} be nonsingular. Then its determinant must not be zero or

$$|\mathbf{P}^{-1}|=|\mathbf{X}^T\mathbf{W}\mathbf{X}+\mathbf{U}|\neq 0 \tag{7.4.18}$$

in the region of the minimum. We call this the *identifiability* condition. If this determinant is identically equal to zero, there is, in general, *no* unique point at which the minimum occurs. A method does not give the *complete* location of the minimum when it specifies a *single* parameter point while there is more than one point at which the minimum occurs. Since the Gauss method will not yield any point in this case, one is alerted to the nonexistence of such a point.

For least squares estimation $\mathbf{W}=\mathbf{I}$ and $\mathbf{U}=\mathbf{0}$. For this case it is necessary that

$$\Delta=|\mathbf{X}^T\mathbf{X}|\neq 0 \tag{7.4.19}$$

in the neighborhood of the minimum of S. This determinant is shown to be equal to zero in Appendix A if any column of $\mathbf{X}$ can be expressed as a linear combination of other columns. This condition, linear dependence, can be written as

$$\sum_{j=1}^{p} C_j X_{ij} = 0 \qquad \text{for } i = 1, 2, \ldots, n \text{ for at least one } C_j \neq 0 \qquad (7.4.20)$$

If (7.4.20) is true, then Δ given by (7.4.19) is equal to zero.

This condition of linear dependence is almost satisfied in many more cases than would be expected. In such cases Δ is almost zero; this is what is meant by ill-conditioning. See Section 6.7.6. If (7.4.20) is almost satisfied, the sum of squares function S will have a unique minimum point and thus a unique set of parameters. The minimum point will not be very pronounced, however. As an example consider the sum of squares function

$$S = \left(2 - \beta_1 - e^{-\beta_2}\right)^2 + \left(3 - 2\beta_1 - e^{-2\beta_2}\right)^2$$

which is plotted in Fig. 7.2 for $S = 0, .1, 1$ and 3. The sensitivity matrix for

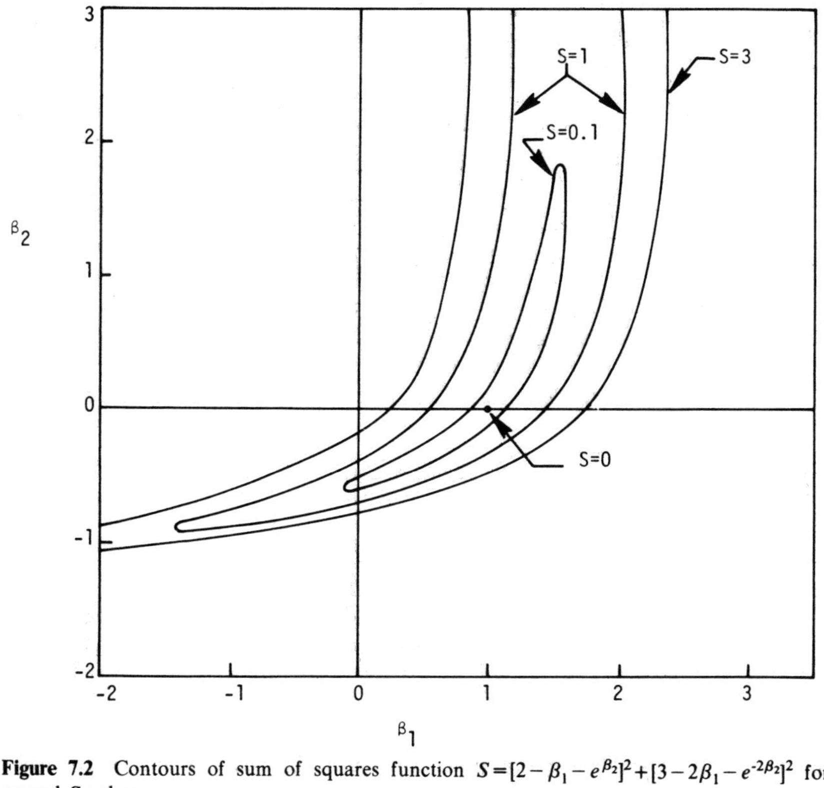

Figure 7.2 Contours of sum of squares function $S = [2 - \beta_1 - e^{\beta_2}]^2 + [3 - 2\beta_1 - e^{-2\beta_2}]^2$ for several S values.

the two points indicated in S is

$$\mathbf{X}=\begin{bmatrix} 1 & -e^{-\beta_2} \\ 2 & -2e^{-2\beta_2} \end{bmatrix}$$

which exhibits linear dependence for either $\beta_2=0$ or $\beta_2\to\infty$. Since the minimum S value is at $\beta_1=1$ and $\beta_2=0$, the condition of linear dependence is almost satisfied in the region of the minimum point. Along the long axis of the $S=0.1$ contour, the change in S is much more gradual than in other directions. Contours that are long, narrow, and curving such as $S=0.1$ are frequently associated with near linear dependence or, equivalently, with Δ being relatively small. Moreover, such contours are typically associated with slow convergence of the Gauss method. For this reason it is important to examine the sensitivity coefficients over the region of interest.

For ML estimation, it is necessary that

$$|\mathbf{X}^T\boldsymbol{\psi}^{-1}\mathbf{X}|\neq 0 \tag{7.4.21}$$

but ML will not lead to a choice of $\mathbf{W}=\boldsymbol{\psi}^{-1}$ which would cause this determinant to be equal to zero if $\mathbf{X}^T\mathbf{X}$ is not equal to zero. Also if $\mathbf{X}^T\mathbf{X}$ is equal to zero, there is no $\boldsymbol{\psi}^{-1}$ which will make (7.4.21) be true. (See Appendix A.) Hence the condition given by (7.4.19) is again the important one.

For MAP estimation with $\mathbf{W}=\boldsymbol{\psi}^{-1}$ and $\mathbf{U}=\mathbf{V}_\beta^{-1}$, it is possible that (7.4.18) may be true even if $|\mathbf{X}^T\mathbf{X}|=0$. Thus if there is prior information, the sensitivity coefficients may not have to be independent to permit estimation using (7.4.6).

(*f*) When convergence to the estimates is attained, the matrix derivative of S goes to zero as indicated by (7.4.4). Since the same terms appear in (7.4.6), the $p\times 1$ vector given next must also go to zero at convergence,

$$\mathbf{X}^{T(k)}\mathbf{W}(\mathbf{Y}-\boldsymbol{\eta}^{(k)})+\mathbf{U}(\boldsymbol{\mu}-\mathbf{b}^{(k)})=\mathbf{0} \tag{7.4.22}$$

This means that every component of this vector must be zero. For cases when $\mathbf{U}=\mathbf{0}$, this results in each H_i given by (7.4.16) being equal to zero. Knowledge of this fact can sometimes aid in checking computer codes that are not yielding converging solutions.

(*g*) Though (7.4.6) has been obtained by using the linear approximation given by (7.2.1), the Gauss equation is not a rigorous first order approximation because a first order series was not used for $\mathbf{X}^{(k+1)}$.

7.4.4 Linear Dependence of Sensitivity Coefficients

As stated in the preceding subsection under point (*e*), the function S for LS or ML estimation has no unique minimum point if the sensitivity coefficients are linearly dependent. It has been found from experience that difficulty encountered in convergence is frequently due to approximate linear dependence. In most of these cases the sensitivity coefficients were not plotted and examined beforehand. Indeed, in many cases the user of a nonlinear least squares program may not realize their importance and thus not even examine them after lack of rapid convergence is apparent. For effective nonlinear estimation, the careful examination of these sensitivity coefficients is imperative. In order to demonstrate what should be inspected, the following discussion is given.

For single response cases with approximately constant standard deviations of the measurements, it is convenient to examine

$$X_{ij}' = \beta_j X_{ij} = \frac{\beta_j \partial \eta_i}{\partial \beta_j} \tag{7.4.23}$$

Note that X_{ij}' has the units of η. Then the magnitude of each sensitivity can be compared with the others as well as with η itself.

For multiresponse cases it is often more meaningful to plot

$$X_{kj}^{+}(i) \equiv \frac{\beta_j}{\sigma_k(i)} \frac{\partial \eta_k(i)}{\partial \beta_j} \tag{7.4.24}$$

which is dimensionless. (Note i refers to "time," j to the parameter, and k to the response.)

In Figs. 7.3 and 7.4 some sensitivities are plotted versus the variable t_i. Those in Fig. 7.3 are linearly dependent but those in Fig. 7.4 are not. The first nine graphs in Fig. 7.3 are for two parameters; it is not difficult to see the linear dependence in the sensitivities in each case. Note that the location of the zero value on the X axes is not arbitrary in most cases. The last three cases are for the three parameters being estimated simultaneously; the linear dependence is less obvious than for two-parameter cases.

The importance of the zero value of X is also shown by Figs. 7.4a,b,c,d, and f which are not linearly dependent cases as drawn, but each becomes dependent if the zero is moved. What zero location would do this in each case?

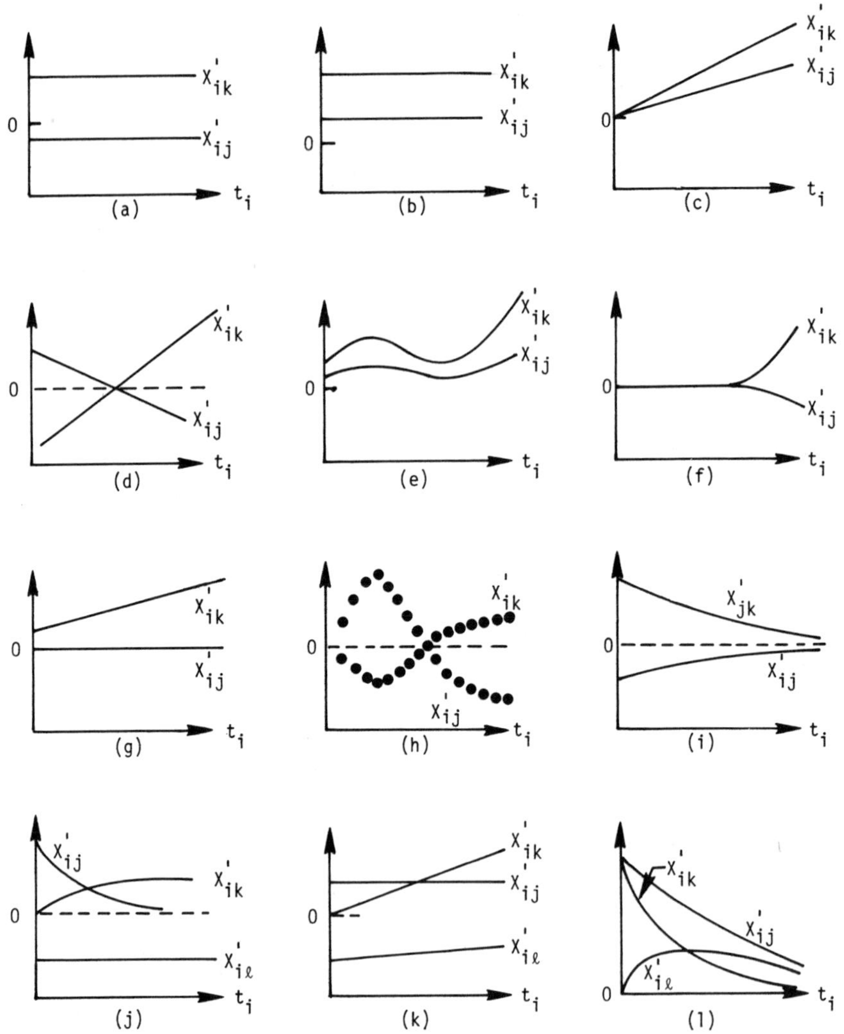

Figure 7.3 Examples of some linearly dependent sensitivity coefficients.

7.5 EXAMPLES TO ILLUSTRATE GAUSS MINIMIZATION METHOD INVOLVING ORDINARY DIFFERENTIAL EQUATIONS

7.5.1 Estimation of a Parameter for a Long Fin

Consider a long fin which has a temperature at its base, $z=0$, of 200°C and which is exposed to a fluid at 100°C, see Fig. 7.5. The differential

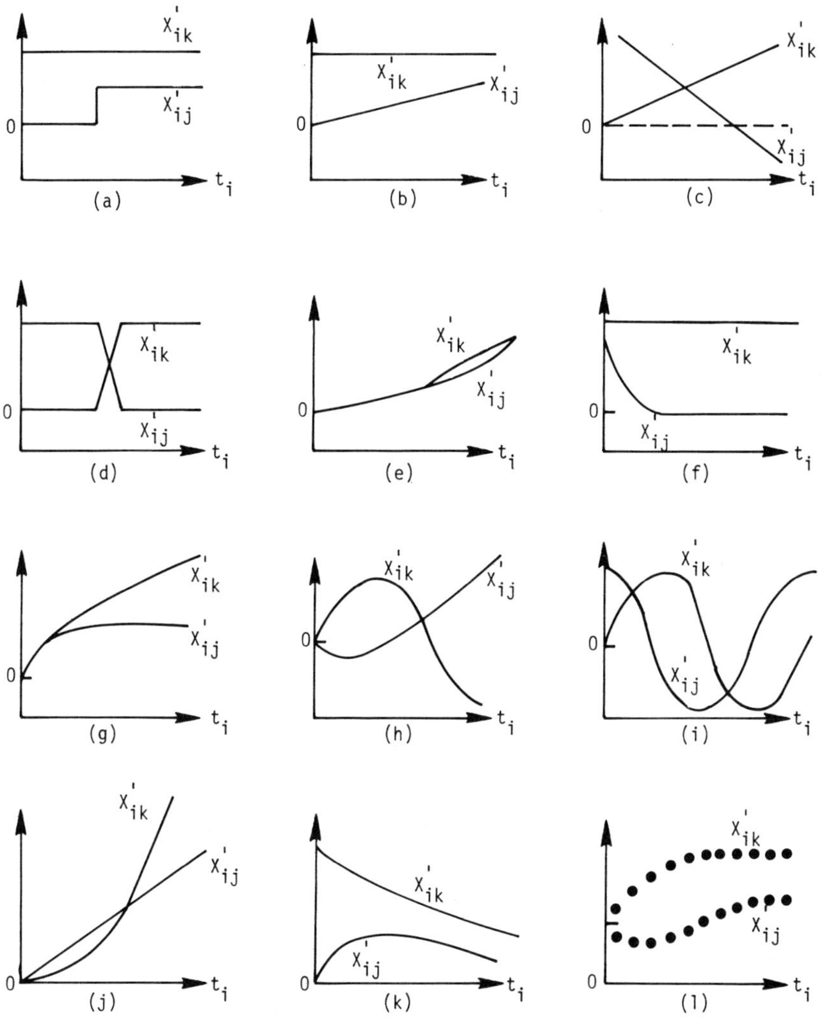

Figure 7.4 Examples of some linearly independent sensitivity coefficients.

equation for this case can be written as

$$\frac{d^2T}{dz^2} = M^2(T - T_\infty) \tag{7.5.1}$$

where M^2 is given by

$$M^2 = \frac{hP}{kA} \tag{7.5.2}$$

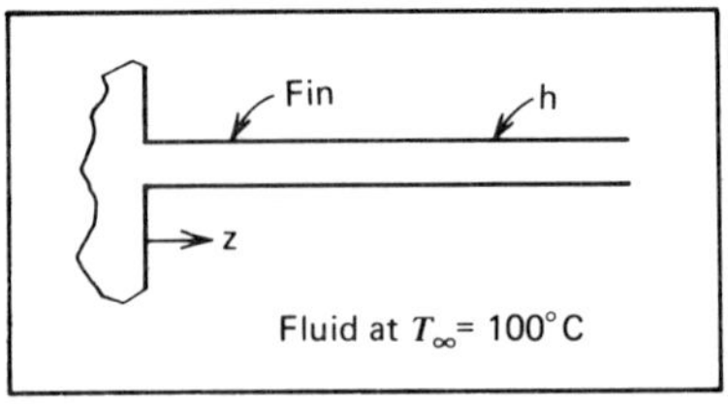

Figure 7.5 Geometry for fin in a fluid.

h is the heat transfer coefficient, k is the thermal conductivity of the fin, A is the fin cross-sectional area normal to z, and P is the perimeter of A.

The boundary conditions are

$$T(0)=T_0, \qquad T(\infty)=T_\infty \tag{7.5.3}$$

Equations 7.5.1 and 7.5.3 give a complete mathematical statement of the classical boundary value problem. The solution for T assuming constant M is

$$T=T_\infty+(T_0-T_\infty)e^{-Mz} \tag{7.5.4}$$

which contains the three parameters T_0, T_∞, and M. In a sense only M is a parameter since T_0 and T_∞ can be considered "states"; that is, they are the temperatures at $z=0$ and infinity, respectively. From examining (7.5.4), we see that T is linear in T_0 and T_∞ but nonlinear in terms of M.

Each of the three parameters (T_0, T_∞, and M) enters the problem in such a manner that all three could be found simultaneously, rather than only in certain combinations. On the other hand, h, P, k, and A can not be found independently but only in combination M. If other boundary conditions were known, a different set of parameters might be found. For example, if the heat flux q at $z=0$ has a known value, the boundary condition is

$$\frac{-k\,dT(0)}{dz}=q \tag{7.5.5}$$

and the parameters M, k, and T_∞ can be simultaneously estimated.

Simulated data for this problem are given in Table 7.2. The assumptions are

$$Y_i=T_i+\varepsilon_i$$

$$V(z_i)=0, \qquad V(Y_1)=0, \qquad V(Y_6)=0$$

$$\varepsilon_i \sim N(0,\sigma^2), \qquad i=2,3,4,5$$

$$E(\varepsilon_i\varepsilon_j)=0, \qquad i\neq j \text{ and } i,j=2,3,4,5$$

The parameter M is nonrandom and there is no prior information. Using our notation, these assumptions are designated 11111-11; the value of σ^2 need not be known to estimate parameters.

Table 7.2 Simulated Data for Fin Example

i	Position z (m)	Temperature Y (°C)
1	0	200
2	0.125	166
3	0.250	144
4	0.375	128
5	0.5	120
6	∞	100

The assumptions $V(Y_1)=V(Y_6)=0$ mean that T_0 and T_∞ are the known values of 200 and 100°C, respectively. The parameter is M. For the standard assumptions given above, OLS and ML estimation provide the same parameter estimates. The weighting matrix $\mathbf{W}$ in (7.3.1) is $\mathbf{W}=\boldsymbol{\psi}^{-1}=\sigma^{-2}\mathbf{I}$ since

$$\operatorname{cov}(\boldsymbol{\varepsilon})=E(\boldsymbol{\varepsilon}\boldsymbol{\varepsilon}^T)=\operatorname{diag}[\sigma^2\ \sigma^2\ \sigma^2\ \sigma^2] \tag{7.5.6}$$

The iterative relation for finding M can be found using (7.4.17) which can be written as

$$\hat{M}^{(k+1)}=\hat{M}^{(k)}+\left[\sum_{i=2}^{5} X_i^{(k)}\left(Y_i-T_i^{(k)}\right)\right]\left[\sum_{j=2}^{5}\left(X_j^{(k)}\right)^2\right]^{-1} \tag{7.5.7}$$

where the sensitivity coefficient is

$$X_i^{(k)}=\frac{\partial T_i^{(k)}}{\partial M}=-(T_0-T_\infty)z_i\exp\left(-\hat{M}^{(k)}z_i\right) \tag{7.5.8}$$

From a knowledge of heat transfer for this particular problem, an initial estimate of M could be given. Many methods are available, however, that use only the given data rather than relying on experience. One of these is used in this example. Since only one parameter is unknown, let us pick a single Y_i. Let us choose the value of $Y_3=144$°C which is nearest the average of T_0 and T_∞. For $T=144$ and $z=0.25$, (7.5.4) yields

$$144=100+(200-100)\exp\left[-\hat{M}^{(0)}(0.25)\right]$$

which can be approximately solved for $\hat{M}^{(0)} = 3.28\text{m}^{-1}$. Using this value, (7.5.4) gives the residual vector of

$$\mathbf{e}^{(0)} = \mathbf{Y} - \mathbf{T}^{(0)} = \begin{bmatrix} 166\text{–}166.36502501 \\ 144\text{–}144.04316545 \\ 128\text{–}129.22925777 \\ 120\text{–}119.39800423 \end{bmatrix} = \begin{bmatrix} -0.36502501 \\ -0.04316545 \\ -1.22925777 \\ 0.60199577 \end{bmatrix}$$

The sum of squares associated with these residuals is $S^{(0)} = 2.008580086$ using $\sigma^2 = 1$.The sensitivity matrix $\mathbf{X}^{(0)}$ and $\mathbf{X}^{T(0)}\mathbf{X}^{(0)}$ are

$$\mathbf{X}^{(0)} = \frac{\partial \mathbf{T}^{(0)}}{\partial M} = \begin{bmatrix} -8.29562813 \\ -11.01079136 \\ -10.96097166 \\ -9.69900211 \end{bmatrix}, \qquad \mathbf{X}^{T(0)}\mathbf{X}^{(0)} = \Sigma X_i^{2(0)} = 404.268514$$

Using these values and $\Sigma X_i^{(0)} e_i^{(0)} = 11.13849884$ in (7.5.7) yields

$$\hat{M}^{(1)} = 3.28 + \frac{11.13849884}{404.268514} = 3.28 + .0275522 = 3.3075522m^{-1}$$

The associated sum of squares is $S^{(1)} = 1.700958954$. Note that

$$\frac{M^{(1)} - M^{(0)}}{M^{(0)}} = \frac{0.0275522}{3.28} = 0.0084$$

which is small compared with unity and thus the initial estimate was very good. This can also be verified from an examination of the small residuals in $\mathbf{e}^{(0)}$. One might desire to minimize S more precisely. Then using the above $M^{(1)}$ value we can find

$$e^{(1)} = \begin{bmatrix} -0.13685534 \\ 0.25916365 \\ -0.92881366 \\ 0.86739235 \end{bmatrix} \qquad \mathbf{X}^{(1)} = \begin{bmatrix} -\ \ 8.26710692 \\ -10.93520909 \\ -10.84830512 \\ -\ \ 9.56630382 \end{bmatrix}$$

$$\hat{M}^{(2)} = 3.3075522 + \frac{0.07570426}{397.123747}$$

$$= 3.3075522 + 0.0001906314 = 3.3077428m^{-1}$$

$$\frac{\hat{M}^{(2)} - \hat{M}^{(1)}}{\hat{M}^{(1)}} = 5.763 \times 10^{-5}, \qquad S^{(2)} = 1.70094504$$

A third iteration yields

$$\mathbf{e}^{(2)} = \begin{bmatrix} -0.13527965 \\ 0.26124785 \\ -0.92674605 \\ 0.86921561 \end{bmatrix}, \qquad \mathbf{X}^{(2)} = \begin{bmatrix} -\ \ 8.26690996 \\ -10.93468804 \\ -10.84752977 \\ -\ \ 9.56539220 \end{bmatrix}$$

$$\hat{M}^{(3)} = 3.3077428 + \frac{0.000198096}{397.0748} = 3.3077433 m^{-1}$$

$$\frac{\hat{M}^{(3)} - \hat{M}^{(2)}}{\hat{M}^{(2)}} = 1.5 \times 10^{-7}, \qquad S^{(3)} = 1.70094500$$

Several observations can be made from the above results. First, M seems to be converging rapidly. The relative corrections are decreasing by a factor smaller than 0.01. The sum of the components in $\mathbf{e}^{(2)}$ is not zero, unlike a model containing a parameter which has a constant sensitivity vector. Note that the residual values changed much more between iterations than the sensitivity vector ($\mathbf{X}$) values. Another observation is that S is *decreasing* as the iterations proceed.

In the example given above the initial $\hat{M}^{(0)}$ value is relatively close to the converged value, resulting in only three iterations being required. Not many iterations are required, however, for a range of initial values as large as 0 to 10 (or even -3 to 10, as indicated by the $\hat{M}^{(0)} = 10$ case) as shown in Table 7.3. Eight or fewer iterations were required to converge to within

Table 7.3 Parameter Values as a Function of Iteration for Various Initial Estimates for Fin Example

Iteration number	$\hat{M}^{(0)}=0$	$\hat{M}^{(0)}=6$	$\hat{M}^{(0)}=8$	$\hat{M}^{(0)}=10$
0	0.00	6.0	8.0	10.0
1	1.8186667	2.1484134	−0.0457249	−3.1824944
2	2.9666083	3.0970338	1.7825492	−1.1017732
3	3.2883466	3.3001212	2.9506108	0.8839079
4	3.3076357	3.3077155	3.2865352	2.4554228
5	3.3077430	3.3077432	3.3076195	3.1918187
6	3.3077433	3.3077433	3.3077430	3.3053045
7			3.3077433	3.3077364
8				3.3077433

seven significant figures. The sum of squares for this example has the same shape as given in Fig. 1.7.

In order to design an experiment to obtain the greatest accuracy (minimum variance of the parameters if there is no bias), the sensitivity coefficients should be plotted and examined *before* the experiment is performed. As a result, one can more intelligently design the experiment in terms of placement of sensors and duration of the experiment. It is suggested in Chapter 8 that a reasonable optimal experiment criterion is to maximize $\Delta=|X^TX|$ for independent errors and subject to constraints of a maximum duration of the experiment and maximum range of the dependent variable.

Figure 7.6 depicts the dimensionless temperature and dimensionless sensitivity coefficient for the example of this section. Note that the sensitivity coefficient starts at zero at $z=0$, increases in magnitude until $Mz=1$, and gradually decreases in magnitude. The Δ criterion for the single parameter M is maximized by selecting the maximum magnitude values of the sensitivity coefficient. If a single measurement is to be utilized in estimation, it should be chosen corresponding to about $Mz=1$ which corresponds to the dimensionless temperature ratio of 0.368. Owing to the flatness of the sensitivity curve shown in Fig. 7.6, little decrease in accuracy would result if the dimensionless ratio were chosen to be as large as 0.5 or as small as 0.25. A Y_i value corresponding to the dimensionless T ratio of 0.5 was chosen in the above example for obtaining the initial estimate of $M=3.28$. For the more common case of many observations, see Chapter 8.

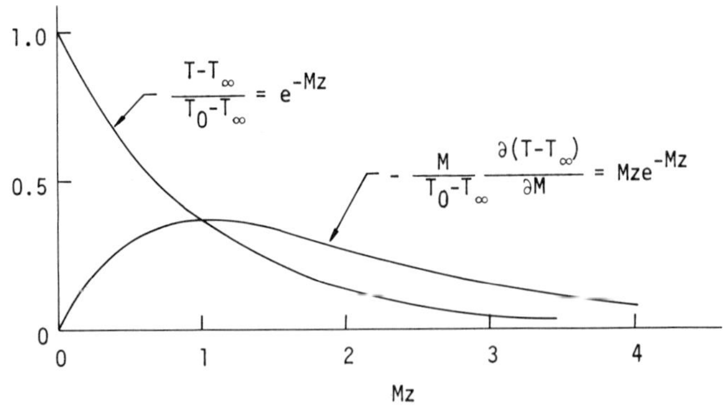

Figure 7.6 Dimensionless temperature and sensitivity coefficient for example of Section 7.5.1.

7.5.2 Example of Estimation of Parameters in Cooling Billet Problem

A similar problem to the preceding one in terms of the differential equation is that of cooling a billet (or any object) that has a negligible temperature variation through it. The temperature of the billet changes after it is placed in a fluid at a different temperature. An analysis of a cooling billet was also given in Section 6.2.

Let T be the temperature of the billet at any time t and let T_∞ be the fluid temperature. A differential equation describing the temperature in this case is

$$\rho c V \frac{dT}{dt} = hA\left(T_\infty - T\right) \tag{7.5.9}$$

where ρ is density of the billet, c is specific heat, V is volume, h is heat transfer coefficient, and A is billet heated area. Various terms could be parameters, but the most common one would be the heat transfer coefficient. For convenience, however, the factor $hA/\rho cV$ is considered as the parameter; several parameters may be formed from it also.

Three cases are investigated below. For each one the initial temperature is T_0 or $T(0) = T_0$. Also, T_∞ is considered to be a constant. In the first case, let $hA/\rho cV$ be the constant parameter β. The solution for T is then

$$\frac{T(t) - T_\infty}{T_0 - T_\infty} = e^{-\beta t} \tag{7.5.10}$$

Another case is for $hA/\rho cV$ being a function of time. A possible function is

$$hA/\rho cV = \beta_1 + \beta_2 t + \beta_3 t^2 \tag{7.5.11}$$

and the solution of the differential equation is

$$\frac{T(t) - T_\infty}{T_0 - T_\infty} = \exp - \left(\beta_1 t + \beta_2 t^2/2 + \beta_3 t^3/3\right) \tag{7.5.12}$$

A third possible model for $hA/\rho cV$ is

$$\frac{hA}{\rho c V} = \beta_1 + \beta_2 \left(T - T_\infty\right)^n \qquad \text{for } n \neq 0 \tag{7.5.13}$$

where n could also be a parameter. The solution in this case is

$$\frac{T(t)-T_\infty}{T_0-T_\infty}=e^{-\beta_1 t}\left[1+\frac{\beta_2}{\beta_1}(T_0-T_\infty)^n(1-e^{-\beta_1 nt})\right]^{-1/n}$$

$$\text{for } n\neq 0 \tag{7.5.14}$$

In each of the models given above, (7.5.10, 12, 14), T is nonlinear in terms of the parameters. Only for the last model of h, (7.5.13), was the differential equation nonlinear.

If the factor $hA/\rho cV$ (or more specifically h) varies during an experiment, (7.5.12) and (7.5.14) provide a number of competing models. For example, β_2 and/or β_3 might be set equal to zero in (7.5.12). In (7.5.14), β_2 might be zero or n might be unity, etc.

The sensitivity coefficients for the second model, (7.5.12), are found to be

$$X_i(t)\equiv\frac{\partial T(t)}{\partial\beta_i}=-(T_0-T_\infty)\left(\frac{t^i}{i}\right)\exp\left[-(\beta_1 t+\beta_2 t^2/2+\beta_3 t^3/3)\right] \tag{7.5.15}$$

where $i=1$ for β_1, etc. For the third model, (7.5.14), the sensitivities are

$$X_1(t)=-(T_0-T_\infty)e^{-\beta_1 t}C^{-1/n}\left\{t+\frac{\beta_2}{nC\beta_1^2}(T_0-T_\infty)^n\left[n\beta_1 te^{-\beta_1 nt}\right.\right.$$
$$\left.\left.-(1-e^{-\beta_1 nt})\right]\right\} \tag{7.5.16}$$

$$X_2(t)=-\frac{(T_0-T_\infty)^{n+1}e^{-\beta_1 t}(1-e^{-\beta_1 nt})}{n\beta_1 C^{1/n+1}} \tag{7.5.17}$$

$$X_3(t)=\frac{\partial T}{\partial n}=-\frac{(T_0-T_\infty)^{n+1}e^{-\beta_1 t}D}{nC^{(n+1)/n}} \tag{7.5.18a}$$

where

$$D=\frac{\beta_2}{\beta_1}\left[(1-e^{-\beta_1 nt})\,ln(T_0-T_\infty)+\beta_1 te^{-\beta_1 nt}\right]-\frac{ln\, C^2}{n(T_0-T_\infty)^n} \tag{7.5.18b}$$

and where C is the expression in the brackets of (7.5.14).

For estimating h the above models are superior to the power series model for T given in Section 6.2 because the above models utilize basic physical laws while the power series for T does not. Whenever possible, models based on the physical mechanisms (called mechanistic models by G. E. P. Box) should be employed.

A further choice based on physical arguments can be made between the power series in t given by (7.5.11) and the temperature-dependent model for $hA/\rho cV$ given by (7.5.13). In many situations the heat transfer coefficient does change with time because some *related* quantity is changing with time rather than the passage of time per se. In the present model, h might change because a billet's temperature changes with time. Physically, the heat transfer coefficient h might account for heat transfer by both natural convection and radiation which could both cause h to be a function of $T-T_\infty$. This suggests that the model (7.5.13) would be superior to (7.5.11).

To illustrate parameter estimation involving the above models, consider again the measurements given in Table 6.2. Results of calculations are summarized in Table 7.4. Ordinary least squares was used with the η values being the T values of (7.5.12) or (7.5.14) and with $T_0=279.59°$F, the T at $t=0$, and $T_\infty=81.5°$F. Models 1, 2, and 3 are for $hA/\rho cV$ given by (7.5.11) with Model 1 being β_1, Model 2 being $\beta_1+\beta_2 t$, and so on. Model 4 is for $hA/\rho cV$ given by (7.5.13) with $n=1$.

Table 7.4 Estimation of Parameter in Models for Cooling Billet Data[a]

Model No.	No. of Parameters	Parameters b_1	b_2	b_3	R	$s=[R/(n-p)]^{1/2}$
1	1	2.70882			38.73731	1.6070
2	2	2.90679	−1.39433		0.7896890	0.23750
3	3	2.90824	−1.41968	0.071344	0.7892203	0.24639
4	2	2.13656	0.0041327		1.162777	0.28819

[a]Units consistent with time in hours.

The h values in units of Btu/hr-ft^2-°F can be found using the appropriate model [(7.5.11) or (7.5.13)] and multiplying by the $\rho cV/A$ value of 0.83432. (This means that b_1' of Fig. 7.7 is 0.83432 times b_1 of Table 7.4). Resulting curves for Models 1, 2, and 3 are shown in 7.7. Also depicted are the Fig. 6.2 results for the temperature power series analysis.

Notice that Models 2 and 3 results are almost identical so that Model 3 is not needed. The results of Model 2 and the power series model are very

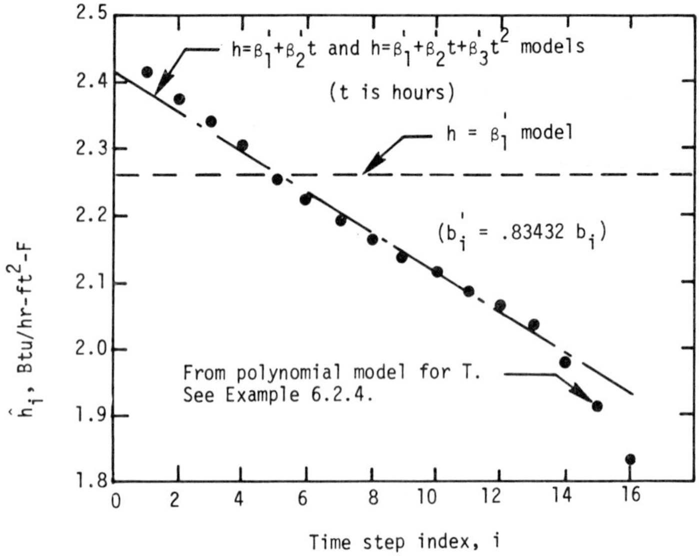

Figure 7.7 Parameter estimates for Models 1, 2, and 3 of Table 7.4 ($i = t_i/\Delta t$).

similar. The constant h given by Model 1 does not apppear to be adequate from inspection of Fig. 7.7 because the other results are quite different and appear to be consistent with each other. Another argument that suggests that Model 1 is not adequate is the large reduction in the sum of squares functions (38.7 to 0.789). This sum R also suggests that Model 3 is not needed because the decrease is very slight between Models 2 and 3; further note that the estimated standard error* of the temperatures actually increases for Model 3. For further related discussion, see Example 7.7.4.

Model 4 results given in Table 7.4 show a slightly increased value of R compared with Model 2 which also has two parameters. Even though R for Model 4 is larger one might prefer Model 4 because it represents a more reasonable physical model as mentioned above.

An attempt to obtain simultaneous estimates of β_1, β_2, and n in (7.5.14) with $n=1$ initially was unsuccessful. A further calculation was performed for Model 4 to estimate just n with the converged values of β_1 and β_2 given in Table 7.4. The value obtained was 1.000006. Since this value is nearly unity, the value previously used, linear dependence in the sensitivity coefficients is suggested. To investigate this previously unsuspected dependence, the sensitivity coefficients were plotted as shown in Fig. 7.8. Notice

*In Table 7.4 the value of $n=16$ was used rather than 17, the number of the observations, because the first value was used to determine T_0.

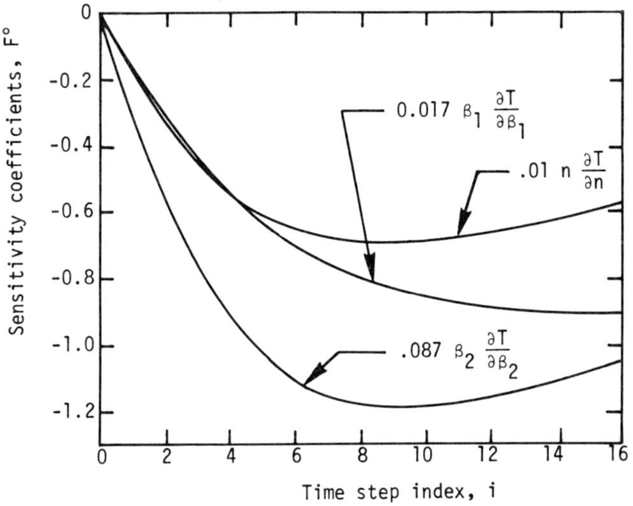

Figure 7.8 Sensitivity coefficients for Model 4 of Table 7.4 for $n=1$.

that the sensitivity coefficients for n and β_2 are very nearly proportional, which tends to make $\mathbf{X}^T\mathbf{X}$ singular and thus the parameters very difficult to estimate simultaneously.

Let us now compare from another point of view these results based on the solution of the describing differential equation with these from the power series method of Section 6.2. The results of this section required the evaluation of two nonlinear parameters whereas the Section 6.2 method requires the estimation of five linear parameters. This is an illustration of the principle of *parsimony* which states that we employ the *smallest possible* number of parameters for adequate representation [22]. Also, note that the estimated standard deviation of the measured temperatures was 0.243°F for the T series, 0.238°F for Model 2, and 0.288°F for Model 4; the small differences between these values indicate that the mechanistic models are good in this case.

After obtaining parameters using OLS or ML, say, it is advisable to examine the residuals to see what assumptions seem to be valid regarding the measurement errors. The residuals for Model 2 are very similar to those given in the upper part of Fig. 6.1. Visual inspection of these residuals does not lead to a contradiction of the assumptions of additive, zero mean, constant variance, independent, and normal errors. Hence accurate parameter estimates would be expected with the least squares method used for this problem.

7.6 MODIFICATIONS OF GAUSS METHOD

The Gauss method has the feature of giving both the direction and the magnitude of the change in the estimate of the parameters of each step in the iteration procedure. Small changes in the parameters in the direction indicated by the Gauss method decrease the sum of squares. Occasionally, however, the size of the change indicated by the method is so large that the successive estimates oscillate and, even worse, the procedure may be unstable. This can result from near-linear dependence of the sensitivity coefficients and/or very poor initial parameter estimates. When the sensitivity coefficients are nearly dependent (which might be termed "overparameterization"), one should consider alternatives in addition to other minimization procedures. One obvious procedure is to decrease the number of parameters being estimated. Another is to redesign the experiment so that the correlation between parameters is reduced. See Chapter 8 for optimal experiment design.

A great many algorithms have been proposed to improve the convergence of the Gauss method. Some of these may be termed modifications to the Gauss method whereas other methods some would call distinctly different methods; in the latter category are the Levenberg [3] and Marquardt [4] methods. We choose to treat these methods as modifications of the Gauss method, however.

This section considers just a few of the possible methods. In this as in other iterative problems there appears to be no end to the possibilities; the ingenuity of various researchers evidenced by the numerous algorithms for this problem is impressive. For a survey, see Bard [1,5].

7.6.1 Box–Kanemasu Interpolation Method

Since the Gauss method depends on a linear approximation to η, in some nonlinear estimation cases the corrections can oscillate with increasing amplitudes and thus lead to nonconvergence. In this section we give the Box–Kanemasu modification of the Gauss method which may converge when the Gauss method does not. The Box–Kanemasu method does not, however, include a check that the sum of squares function S decreases from iteration to iteration. Bard [5] has made the point that all acceptable methods should ensure that S does monotonically decrease to a minimum. This is a reasonable requirement but in some cases it may lead to more calculations than without it; this is illustrated by Table 7.6 which is discussed later. On the other hand, this requirement might improve convergence in other cases. In order to ensure S continually decreases, a modification to the Box–Kanemasu method that has been used by Bard [5] and others is included.

Since the linear approximation is valid over some region, a sufficiently small correction in the direction given by the Gauss method should improve the estimate (i.e., reduce S). Many methods have been proposed which use the direction provided by the Gauss method but modify the step size. We generalize (7.4.6) to

$$\mathbf{b}^{(k+1)} = \mathbf{b}^{(k)} + h^{(k+1)} \Delta_g \mathbf{b}^{(k)} \tag{7.6.1}$$

$$\Delta_g \mathbf{b}^{(k)} = \mathbf{P}^{(k)} \left[\mathbf{X}^{T(k)} \mathbf{W} (\mathbf{Y} - \boldsymbol{\eta}^{(k)}) + \mathbf{U} (\boldsymbol{\mu} - \mathbf{b}^{(k)}) \right] \tag{7.6.2}$$

where $h^{(k+1)}$ is a scalar interpolation factor. Note that this factor may be iteration-dependent. If $h^{(k+1)}$ is set equal to 1, we have the Gauss method.

In one class of methods, a search on h is performed to precisely determine the minimum S along the Gauss direction [5].

Interpolation methods attempt to find good, acceptable values of $h^{(k+1)}$ without bothering to locate precisely the value associated with the minimum S value. Of the many methods possible, one of these is the halving and doubling method [6–8]. The modification that we describe utilizes an equation given by Box and Kanemasu [9]. The modification is more general, however, since (a) the sum of squares function given by (7.3.1) is used rather the OLS S function and (b) a check for decreasing S is included.

In the Box–Kanemasu method, S is approximated at each iteration by

$$S = a_0 + a_1 h + a_2 h^2 \tag{7.6.3}$$

where a_0, a_1, and a_2 are constants characteristic of each iteration. The $h^{(k+1)}$ value is taken where S given by (7.6.3) is minimized.

A second approximation in this method is that $\boldsymbol{\beta}$ is given by

$$\boldsymbol{\beta} = \mathbf{b}^{(k)} + h \Delta_g \mathbf{b}^{(k)} \tag{7.6.4}$$

A minimum of three conditions are needed to find the parameters a_0, a_1, and a_2. One condition is to use the S value at $h=0$, that is, at $\boldsymbol{\beta} = \mathbf{b}^{(k)}$; this S value is designated $S_0^{(k)}$. A second S value, denoted $S_\alpha^{(k)}$, is found at $h = \alpha$. Initially α is set equal to 1.

The third condition for finding the a_i's uses (7.6.4) to find the derivative of S at $h=0$ and in the $\Delta_g \mathbf{b}^{(k)}$ direction. This derivative is

$$\left. \frac{dS}{dh} \right|_{h=0} = \sum_{i=1}^{P} \left. \frac{\partial S}{\partial \beta_i} \right|_{h=0} \left. \frac{\partial \beta_i}{\partial h} \right|_{h=0} = \left. \left[(\nabla_\beta S)^T \frac{d\boldsymbol{\beta}}{dh} \right] \right|_{h=0} \tag{7.6.5}$$

The matrix derivative $\nabla_{\beta} S$ is found from (7.4.1) to be

$$\nabla_{\beta} S\big|_{h=0} = -2\left[\mathbf{X}^{T(k)}\mathbf{W}(\mathbf{Y}-\boldsymbol{\eta}^{(k)})+\mathbf{U}(\boldsymbol{\mu}-\mathbf{b}^{(k)})\right] \tag{7.6.6a}$$

and the derivative of $\boldsymbol{\beta}$ with respect to h is found from (7.6.4) to be

$$d\boldsymbol{\beta}/dh = \Delta_g \mathbf{b}^{(k)} \tag{7.6.6b}$$

Then using (7.6.6a, b) in (7.6.5) yields

$$(dS/dh)\big|_{h=0} = -2G^{(k)} \tag{7.6.7}$$

$$G^{(k)} \equiv \left[\Delta_g \mathbf{b}^{(k)}\right]^T \left[\mathbf{X}^{T(k)}\mathbf{W}\mathbf{e}^{(k)}+\mathbf{U}(\boldsymbol{\mu}-\mathbf{b}^{(k)})\right] \tag{7.6.8a}$$

$$= \left[\Delta_g \mathbf{b}^{(k)}\right]^T \mathbf{P}^{-1(k)} \Delta_g \mathbf{b}^{(k)} \tag{7.6.8b}$$

Note that $G^{(k)}$ is a scalar so that it is also equal to its transpose. From the definition of G it can be proved that $G \geqslant 0$.

Using the three conditions for S yields

$$a_0 = S_0^{(k)}, \qquad a_1 = -2G^{(k)}, \qquad a_2 = \left[S_\alpha^{(k)} - S_0^{(k)} + 2G^{(k)}\alpha\right]\alpha^{-2} \tag{7.6.9}$$

The minimum S is located where the derivative of S [given by (7.6.3)] with respect to h is equal to zero; it occurs at the h value of $-a_1/2a_2$ or

$$\boxed{h^{(k+1)} = G^{(k)}\alpha^2\left[S_\alpha^{(k)} - S_0^{(k)} + 2G^{(k)}\alpha\right]^{-1}} \tag{7.6.10}$$

This h value is used in (7.6.1) to find the $(k+1)$st iterate for the **b** vector. The equation given by Box and Kanemasu [9] is obtained from (7.6.10) by setting $\alpha = 1$. An equation similar to (7.6.10) is given by Hartley [8] and is attributed to Dr. K. Ruedenburg.

There are some restrictions on the use of (7.6.10). These relate to the possible values of a_2. See Fig. 7.9. Three different cases are for $a_2 = 0$, $a_2 < 0$, and $a_2 > 0$ which are discussed individually below.

In each case a condition suggested by Bard [5] is that

$$S_\alpha^{(k)} < S_0^{(k)} \tag{7.6.11}$$

The parameter α is made sufficiently small for this condition to be satisfied. If this inequality is not true for $\alpha = 1$, α is made $\frac{1}{2}$ and the inequality is checked again. Should the inequality require the investigation

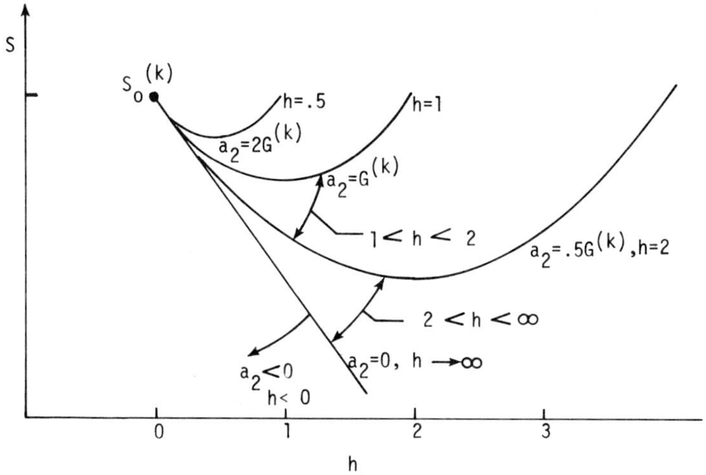

Figure 7.9 Sum of squares versus the h parameter for the Box–Kanemasu method using (7.6.3) for approximating S.

of α values less than 0.01, say, the calculations are terminated. It may be that the problem has been incorrectly programmed; for example, the sensitivity coefficients may be incorrect. It is also possible that the sensitivity coefficients are nearly linearly dependent. In the Box–Kanemasu method, the inequality given by (7.6.11) is not considered and α is always one.

The first case that we consider is for $a_2=0$. If this occurs the S expression as given by (7.6.3) is a straight line which has no minimum at any finite value of h; see Fig. 7.9. Hence for this case we set $h^{(k+1)}=A\alpha$ where A is some constant equal to or slightly larger than unity; one possible value is 1.1.

The second case if for $a_2<0$ which would cause h given by (7.6.10) to be negative. Again we set $h^{(k+1)}=A\alpha$.

The third and most interesting case is for $a_2>0$ as shown in Fig. 7.9. For this case h is calculated using (7.6.10) provided $h^{(k+1)}\leqslant A\alpha$; if inequality is not satisfied, again we set $h^{(k+1)}=A\alpha$.

All three cases can be included by requiring that the inequality,

$$S_\alpha^{(k)} \geqslant S_0^{(k)} - (2-A^{-1})\alpha G^{(k)} \tag{7.6.12}$$

be satisfied in order to use (7.6.10). If it is not satisfied, we suggest that $h^{(k+1)}$ be set equal to $A\alpha$ and the calculation proceeds to the next iteration.

A computer program flow chart incorporating the above constraints is

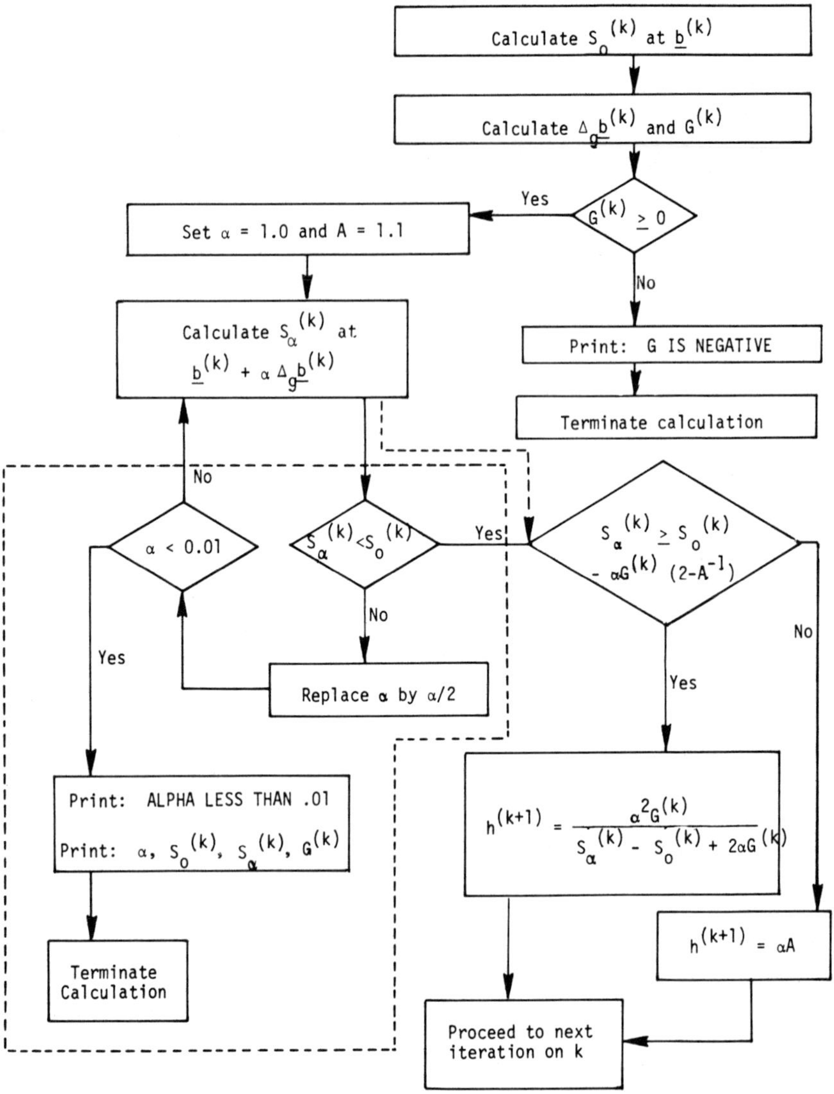

Figure 7.10 Flow chart of a procedure using the Box–Kanemasu equation.

given by Fig. 7.10. In addition to the two inequalities given by (7.6.11) and (7.6.12), there is a check on the sign of G. From the definition of G, it must be positive; thus if it is negative something is incorrect. It is important in such a program to calculate $S_0^{(k)}$ and $S_\alpha^{(k)}$ correctly; the same weighting $\mathbf{W}$ must be used as is used to evaluate $\Delta_g\mathbf{b}^{(k)}$ and the term involving $\mathbf{U}$ in (7.3.1) must be included if it is also implied in the $\Delta_g\mathbf{b}^{(k)}$ calculation.

For the Box–Kanemasu method, the section of the flow chart in Fig. 7.10 that is enclosed in a box with dashed lines would be bypassed and α would be always unity.

Example 7.6.1

For the model $\eta = \beta_1 t + \exp(-\beta_2 t)$ and the observations of 2 at $t=1$ and 3 at $t=2$, use the Box–Kanemasu method for one step starting at $\beta_1 = 1$ and $\beta_2 = 2$. Use OLS.

Solution

The S function for this case is shown in Fig. 7.2. For OLS the matrix correction $\Delta_g \mathbf{b}^{(k)}$ is

$$\Delta_g \mathbf{b}^{(k)} = (\mathbf{X}^{T(k)}\mathbf{X}^{(k)})^{-1}\mathbf{X}^{T(k)}(\mathbf{Y} - \boldsymbol{\eta}^{(k)})$$

For the first iteration, $k=0$ and $\mathbf{X}^T$ is

$$\mathbf{X}^{T(0)} = \begin{bmatrix} 1 & 2 \\ -\exp(-b_2^{(0)}) & -2\exp(-2b_2^{(0)}) \end{bmatrix} = \begin{bmatrix} 1 & 2 \\ -0.1353 & -0.03663 \end{bmatrix}$$

Also $(\mathbf{X}^{T(0)}\mathbf{X}^{(0)})^{-1}$ and $\mathbf{X}^{T(0)}(\mathbf{Y} - \boldsymbol{\eta}^{(0)})$ are

$$(\mathbf{X}^{T(0)}\mathbf{X}^{(0)})^{-1} = \frac{1}{0.05477}\begin{bmatrix} 0.01966 & 0.2086 \\ 0.2086 & 5 \end{bmatrix}$$

$$\mathbf{X}^{T(0)}(\mathbf{Y} - \boldsymbol{\eta}^{(0)}) = \begin{bmatrix} 1 & 2 \\ -0.1353 & -0.03663 \end{bmatrix}\begin{bmatrix} 0.8647 \\ 0.9817 \end{bmatrix} = \begin{bmatrix} 2.828 \\ -0.1529 \end{bmatrix}$$

which give $[\Delta_g \mathbf{b}^{(0)}]^T = [0.4323 \quad -3.195]$. Hence the Gauss parameter estimates are 1.4323 and -1.195 which results in an S value of 123.42. Such a large relative change in b_2 (-3.195 compared with 2.0) suggests that "overshooting" and non-convergence might occur using the Gauss method. The method does converge for this example, however, as shown below.

For the Box–Kanemasu method we check the inequality (7.6.12) for $\alpha = 1$. If the number of observations n equals the number of parameters p, it can be shown that $G^{(k)} = S_0^{(k)}$. In this example $n = p = 2$. Then using $A = 1.1$, (7.6.12) gives

$$S_1^{(0)} = 123.42 \geqslant S_0^{(0)} - (2 - A^{-1})G^{(0)}$$

$$= 1.7113 - (2 - 1.1^{-1})1.7113 = -0.1556$$

so that (7.6.10) can be used to get the small value of

$$h^{(1)} = G^{(0)}\left[S_1^{(0)} - S_0^{(0)} + 2G^{(0)}\right]^{-1} = 0.0137$$

With this value we use (7.6.1) to get

$$\mathbf{b}^{(1)} = \mathbf{b}^{(0)} + h^{(1)}\Delta_g \mathbf{b}^{(0)} = \begin{bmatrix} 1 \\ 2 \end{bmatrix} + 0.0137\begin{bmatrix} 0.4323 \\ -3.195 \end{bmatrix} = \begin{bmatrix} 1.0059 \\ 1.9563 \end{bmatrix}$$

which has an associated S value of 1.6645, a value lower than $S^{(0)}$.

For the modified Box–Kanemasu method we use (7.6.11) to check if $S_1^{(0)}=123.42$ is less than $S_0^{(0)}=1.7113$. Since it is not, α is made equal to $\frac{1}{2}$ and $S_{1/2}^{(0)}$ is calculated at $\mathbf{b}^{(0)}+\frac{1}{2}\Delta_g\mathbf{b}^{(0)}$ to get $S_{1/2}^{(0)}=0.0279$; now the inequality given by (7.6.11) is satisfied. Hence set $\alpha=\frac{1}{2}$. The right side of (7.6.12), which is

$$S_0^{(0)}-(2-A^{-1})\alpha G^{(0)}=1.7113-(2-1.1^{-1})\left(\frac{1}{2}\right)(1.7113)=0.779$$

is found to be greater than $S_{1/2}^{(0)}$ so that (7.6.12) is *not* satisfied. Consequently $h^{(1)}$ is calculated using

$$h^{(1)}=A\alpha=\frac{1.1}{2}=0.55$$

which gives $S^{(1)}=0.00884$. If we seek the location of the minimum S for this iteration, we get $h=0.5336$ and $S=0.000862$. It is partially fortuitous that the modified Box–Kanemasu method happened to yield an h value so near this latter value. As a result, the modified Box–Kanemasu method at the end of the first iteration is much nearer the minimum than the other two methods.

A summary of the h and S values for the three methods for several iterations is given below. The modified Box–Kanemasu method converging most rapidly of the three methods was a direct result of the excellent choice of $h=0.55$ in the first iteration. As is shown in Section 7.6.4.3, the same method is not most efficient for all cases.

Iteration Number	Gauss		Box–Kanemasu		Modified Box–Kanemasu	
	h	S	h	S	h	S
0		1.7113		1.7113		1.7113
1	1.0	1.234×10^{2}	0.0137	1.664	0.550	8.7×10^{-3}
2	1.0	4.559	0.0208	1.595	0.924	8.8×10^{-4}
3	1.0	5.520×10^{-1}	0.036	1.48	0.955	5.2×10^{-5}
4	1.0	6.19×10^{-2}	0.077	1.25	0.951	3×10^{-6}
6	1.0	5.5×10^{-4}	0.638	1.48×10^{-2}	0.944	1.74×10^{-8}
8	1.0	3.1×10^{-6}	0.372	2.8×10^{-4}	0.942	1×10^{-10}
10	1.0	1.37×10^{-8}	0.444	6.6×10^{-5}	0.941	6×10^{-13}
15	1.0	1.4×10^{-14}	0.933	3.9×10^{-8}	0.941	2×10^{-18}

7.6.2 Levenberg Damped Least Squares Method

Levenberg [3] tried to overcome the instability of "overshooting" in the Gauss method by introducing constraints into the minimization of S. The function that Levenberg considered was the OLS sum of square function plus an addition term. By using a WLS function we can generalize

Levenberg's function by using (7.3.1) with $\boldsymbol{\mu}$ being $\mathbf{b}^{(k)}$, $\boldsymbol{\beta}$ being $\mathbf{b}^{(k+1)}$, and $\mathbf{U}$ being replaced by $\lambda\boldsymbol{\Omega}$ where $\boldsymbol{\Omega}$ is a diagonal matrix. Using these definitions in (7.4.6) gives

$$\mathbf{b}^{(k+1)} = \mathbf{b}^{(k)} + (\mathbf{X}^{T(k)}\mathbf{W}\mathbf{X}^{(k)} + \lambda^{(k)}\boldsymbol{\Omega}^{(k)})^{-1}\mathbf{X}^{T(k)}\mathbf{W}\mathbf{e}^{(k)} \qquad (7.6.13)$$

The effects of the $\boldsymbol{\Omega}$ matrix are to reduce the size and to change the direction of the step. Provided there is a unique minimum which is the only stationary point and the iteration procedure given by (7.6.13) convergences, the estimates found would be those sought. The presence of the $\lambda\boldsymbol{\Omega}$ term tends to reduce oscillations or instabilities particularly as the diagonal components of $\lambda\boldsymbol{\Omega}$ are made relatively large compared to the diagonal terms in $\mathbf{X}^T\mathbf{W}\mathbf{X}$.

Box and Kanemasu [9] in describing the changes in the estimates of the parameters in progressing to the minimum of S state that the term $\lambda[\mathbf{b}^{(k)} - \mathbf{b}^{(k+1)}]^T\boldsymbol{\Omega}^{(k)}[\mathbf{b}^{(k)} - \mathbf{b}^{(k+1)}]$ introduces a spherical constraint that causes a spiral path.

Levenberg proved that S decreases in the initial iterations if λ is first large and then allowed to decrease (provided S does not have a stationary point at $\mathbf{b}^{(k)}$). One recommendation of Levenberg was to make

$$\boldsymbol{\Omega} = \mathbf{I} \qquad (7.6.14)$$

Incidently if $\boldsymbol{\Omega} = \mathbf{I}$ and λ is very large, (7.6.13) can be written as

$$\mathbf{b}^{(k+1)} = \mathbf{b}^{(k)} + K\mathbf{X}^{T(k)}\mathbf{W}\mathbf{e}^{(k)}, \qquad K = \lambda^{-1} \qquad (7.6.15)$$

which is called the *method of steepest descent*. This method gives a direction for the step but not a step size. Since the step size is arbitrary, this method can be very inefficient particularly as the minimum is approached. Hence the method of steepest descent is not recommended. Note, however, that it does not require the inverse of a matrix as do the Gauss and Levenberg methods.

Concomitant with $\boldsymbol{\Omega} = \mathbf{I}$, Levenberg suggested the two possibilities of (1) letting λ be a constant value and (2) varying λ as the minimum is approached. One possibility is that the kth value of λ is

$$\lambda^{(k)} = \frac{\mathbf{e}^{T(k)}\mathbf{W}\mathbf{X}^{(k)}X^{T(k)}\mathbf{W}\mathbf{e}^{(k)}}{S^{(k)}} \qquad (7.6.16)$$

We shall call this the *unscaled* Levenberg procedure. Note that $\lambda^{(k)}$ given by (7.6.16) goes to zero as the minimum S is approached because each component of $\mathbf{X}^T\mathbf{W}\mathbf{e}$ goes to zero at the minimum of S. (See point (f) in

Section 7.4.3.) As the minimum is approached S also decreases but the numerator decreases more rapidly.

Box and Kanemasu [9] have made the point that the use of $\boldsymbol{\Omega}=\mathbf{I}$ in (7.6.15) results in the method *not* being invariant under linear transformations of the parameters unlike the Gauss method which is invariant.

Another recommendation of Levenberg was to set $\boldsymbol{\Omega}$ equal to the diagonal terms of $\mathbf{X}^T\mathbf{W}\mathbf{X}$ or

$$\boldsymbol{\Omega}_m = \operatorname{diag}\left[C_{11}\; C_{22}\; \dots\; C_{pp} \right] \tag{7.6.17}$$

where C_{jj} is given by (7.4.13). This choice for $\boldsymbol{\Omega}$ has the effect of making the iteration problem invariant under scale changes in the parameters [9]. For this choice of $\boldsymbol{\Omega}$, the following expression for λ has been suggested by Davies and Whitting [10],

$$\lambda^{(k)} = \frac{\mathbf{e}^{T(k)}\mathbf{W}\mathbf{X}^{(k)}\boldsymbol{\Omega}_m^{(k)}\mathbf{X}^{T(k)}\mathbf{W}\mathbf{e}^{T(k)}}{S^{(k)}} \tag{7.6.18}$$

which also goes to zero as the minimum is approached. Using this expression and (7.6.17) gives what we call the *scaled* Levenberg method.

When $\boldsymbol{\Omega}$ is set equal to $\mathbf{I}$ and when λ is given [10] by

$$\lambda^{(k)} = \frac{3\mathbf{e}^{T(k)}\mathbf{W}\mathbf{X}^{(k)}\boldsymbol{\Omega}_m\mathbf{X}^{T(k)}\mathbf{W}\mathbf{e}^{(k)}}{\mathbf{e}^{T(k)}\mathbf{W}\mathbf{X}^{(k)}\mathbf{X}^{T(k)}\mathbf{W}\mathbf{e}^{(k)}} \tag{7.6.19}$$

we term the associated procedure the *modified* Levenberg method.

Though the Levenberg method and its modifications can remove instability and reduce oscillations, it also can increase considerably the number of iterations in a given case.

7.6.3 Marquardt's Method

A method similar to Levenberg's is the well-known method due to Marquardt [4]. This method uses (7.6.13) with $\boldsymbol{\Omega}$ given by $\boldsymbol{\Omega}_m$ which is defined by (7.6.17), but Marquardt uses a different choice for λ than does Levenberg. Again if $\lambda\boldsymbol{\Omega}_m$ is large compared to $\mathbf{X}^T\mathbf{W}\mathbf{X}$, the parameter correction is in the same direction as given by steepest descent which does not require that $|\mathbf{X}^T\mathbf{W}\mathbf{X}| \neq 0$. It is for this reason that the Levenberg and Marquardt methods are helpful when $\mathbf{X}^T\mathbf{W}\mathbf{X}$ is poorly conditioned at the starting parameter vector but is better conditioned in the neighborhood of the least squares solution (Gallant [11]). Both methods also provide a compromise between the steepest descent and Gauss methods with the initial iterations close to the steepest descent method and the final iterations close to the Gauss method.

Marquardt proposed what Box and Kanemasu call the (λ, ν) algorithm in which $\lambda^{(k)}$ is calculated from

$$\lambda^{(k)} = \frac{\lambda_0}{\nu^k} \tag{7.6.20}$$

The initial value of $\lambda\boldsymbol{\Omega}$, corresponding to the first iteration, is then $\lambda\boldsymbol{\Omega} = \lambda_0\boldsymbol{\Omega}_m/\nu$ where ν is some constant greater than unity. This method supposedly possesses the virtues of the steepest descent and Gauss methods where each is most effective. Marquardt's recommendations have been followed by many and have been incorporated in numerous computer programs.

Though the Marquardt method has been widely used since the publication of Marquardt's paper in 1963, there are still some unresolved questions regarding the effectiveness of this method compared to others. This is discussed further in the next section. Moreover, Box and Kanemasu [9] have presented an analysis that indicates there is no need to compromise the direction of a step to be between those given by steepest descent and Gauss methods. They demonstrate this by showing that the steepest descent and Gauss vectors have the same direction if the parameters are transformed into a linearly invariant metric. They also showed that the constrained minimization in this metric is merely equivalent to using a modification employing (7.6.1,2). Nevertheless, Gallant's observation given above regarding the value of the Marquardt method when the *initial* $\mathbf{X}^T\mathbf{W}\mathbf{X}$ matrix is poorly conditioned is valid.

7.6.4 Comparison of Methods

In addition to the modifications of the Gauss method given above, doubtless many more will be suggested in the future. Moreover, many quite different approaches are presently available. These include derivative–free methods [5], quasi-linearization [12], stochastic approximation [13], and invariant embedding [13]. For our purposes, however, some interpolation scheme such as that of Section 7.6.1 combined with a sequential procedure (such as given in Section 7.8) is usually adequate, particularly if the experiment is carefully designed as discussed in Chapter 8.

The purpose of this section is to provide comparisons of several methods. Assuming that all the methods converge to the same parameter vector, one of the most important considerations in selection of a method is the relative computer time needed. In many parameter estimation problems, the model is a set of ordinary or partial differential equations which may require time-consuming finite-difference methods for solution. In such cases the time used in repeatedly solving the model is much greater than that used in performing the other requisite calculations in parameter estimation. Hence the number of evaluations of the model gives an approximate relative measure of the computer time required for a particular method.

Another criterion for comparison of methods is the power to solve difficult cases. Unfortunately many of the more powerful procedures can take considerably more computer time than a relatively simple method such as the Gauss method. The relative amount of time depends on the problem. Indeed there are cases for which the Gauss method does not converge. Usually at the expense of more computer time and greater programming complexity, computer methods can be evolved to treat more difficult cases. Though there is an advantage in having more powerful computer programs available, such programs can never supplant careful design of experiments. Thus our tendency is to recommend more efficient though less powerful methods not only to save computer time but also to encourage careful experiment design. In a case that is poorly designed not only is the minimum S difficult to locate but the associated parameter values are probably very sensitive to the measurement errors. Good designs yield more accurate parameter estimates *and* require less computer time.

One way to compare methods is to apply them to a large variety of problems and to also vary the initial parameter estimates. Some authors have done this. Usually the cases are not randomly selected. Rather there is a tendency to use as test cases those that have S functions which are difficult to minimize. One such case is given next.

7.6.4.1 *Box–Kanemasu Example*

The model and data investigated by Box and Kanemasu [9] are

$$\eta = \beta_1 \beta_2 \xi_1 (1 + \beta_1 \xi_1 + 5000\xi_2)^{-1} \tag{7.6.21}$$

ξ_1	1	2	1	2
ξ_2	1	1	2	2
Y	0.1165	0.2114	0.0684	0.1159

The independent variables are ξ_1 and ξ_2. The errors in Y are assumed to be additive, to have zero mean, to have constant variance, and to be independent and normal, or in symbols 11111-11. Ordinary least squares and maximum likelihood give the same parameter estimates in this case.

The sensitivity coefficients can be given by

$$\tilde{X}_1 = \frac{\beta_1}{\beta_2}\frac{\partial \eta}{\partial \beta_1} = \frac{\eta}{\beta_2}\left(1 - \frac{\eta}{\beta_2}\right); \qquad X_2 = \frac{\partial \eta}{\partial \beta_2} = \frac{\eta}{\beta_2} \tag{7.6.22}$$

where $\tilde{X}_1$ is used rather than X_1 for its greater convenience. From (7.6.21) we can observe that η is linear in β_2 but nonlinear in β_1. Also the sensitivity coefficients as expressed by (7.6.22) can be plotted or tabulated versus $\beta_1\xi_1$ for ξ_2 equal to 1 and 2; see Table 7.5. The initial β_1 and β_2 values chosen were 300 and 6, respectively, and the converged values are 716.955 and 0.944469. Then in the vicinity of the minimum, the maximum value of $\beta_1\xi_1$ would be about 1500. Notice that in Table 7.5 the $\tilde{X}_1$ and X_2 sensitivities for $\beta_1\xi_1$ less than 1000 are approximately proportional at both $\xi_2 \doteq 1$ and 2. This means that the minimum of S is probably ill-defined. This is demonstrated by the long narrow valley shown in Fig. 7.11a; such cases might pose difficulty in convergence for the Gauss method.

Table 7.5 Table of Sensitivity Coefficients for η Model Given by (7.6.21)

	$\xi_2=1$		$\xi_2=2$	
$\beta_1\xi_1$	$\tilde{X}_1$	X_2	$\tilde{X}_1$	X_2
0	0	0	0	0
250	.0453	.0476	.0238	.0244
500	.0826	.0909	.0453	.0476
1000	.1389	.1667	.0826	.0909
2000	.2041	.2857	.1389	.1667
3000	.2344	.3750	.1775	.2308
100,000	.0453	.9524	.0827	.9091
∞	0	1	0	1

Another indication that the minimum S is poorly defined can be obtained by examining the $(\mathbf{X}^T\mathbf{X})^{-1}$ matrix for the final b's; it is

$$\mathbf{C}=(\mathbf{X}^T\mathbf{X})^{-1}=\begin{bmatrix} 2.38096\times 10^9 & -2.57548\times 10^6 \\ \text{symmetric} & 2.79758\times 10^3 \end{bmatrix}$$

The correlation coefficient of the estimators is approximated by $C_{12}/(C_{11}C_{22})^{1/2}=-0.997909$. Since the absolute value of this number is very near unity, the two parameters are shown to be highly correlated.

Figure 7.11b shows a comparison of results given by Box and Kanemasu [9] for the Marquardt method described in Section 7.6.3, the modified Gauss method involving halving and doubling mentioned in Section 7.6.1 and the Box–Kanemasu method of Section 7.6.1. The number of times, n_f, that the function η had to be evaluated is plotted versus λ_0, the initial value

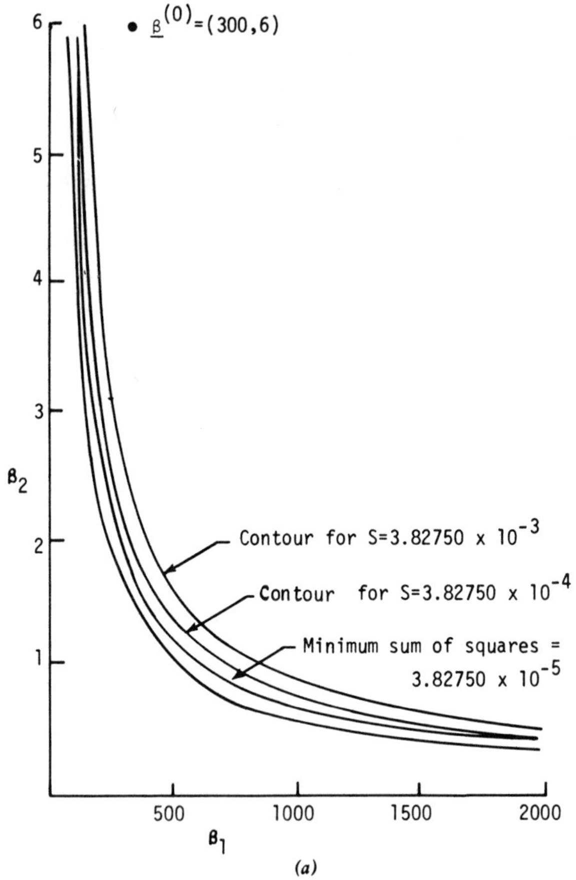

(a)

Figure 7.11a Sum of squares contours for the example of Section 7.6.4.1. (Reprinted by permission of Prof. G. E. P. Box.)

of λ chosen in the Marquardt method. (Note that the number of iterations is not equivalent to the number of function evaluations.)

The following conclusions can be drawn from Fig. 7.11b:

1. The Box–Kanemasu method is superior to any of those investigated.
2. The next best method is usually the halving–doubling method.
3. The Marquardt method is usually the slowest of those considered.

These conclusions should not be overgeneralized because only one particular example was considered. From theoretical considerations and from this example, however, Box and Kanemasu found that Marquardt's method

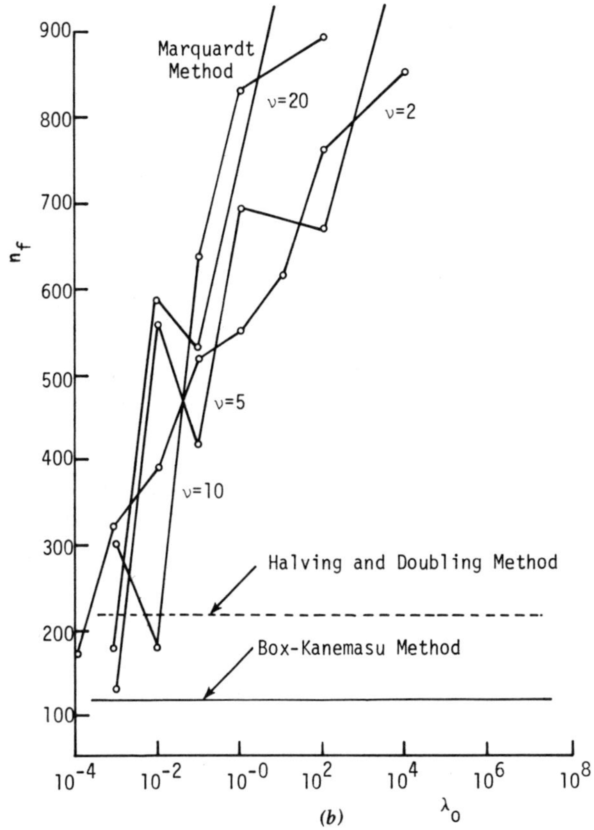

(b)

Figure 7.11b Comparison of the Marquardt, halving and doubling, and Box-Kanesmasu modifications of the Gauss method. (Reprinted by permission of Prof. G. E. P. Box.)

was not superior to the Gauss method with the modifications using (7.6.1,2), with h replaced by $1/(1+\lambda)$, λ being the value that Marquardt would recommend.

7.6.4.2. Bard Comparisons

Y. Bard [1] gave an excellent survey of 13 best known gradient methods including several modifications of the Gauss method. Bard found that several modifications of the Gauss method were better than the Marquardt method. In his book, Bard [5, p. 111] appears to favor a modification of the Box–Kanemasu method which he calls the interpolation–extrapolation method.

7.6.4.3. *Davies and Whitting Comparison*

In the paper by Davies and Whitting [10] a comparison is given of five methods for seven problems considered by Jones [14]. The methods include the Levenberg procedures using λ given by (7.6.16) and (7.6.18). There was negligible difference in the rate of convergence between the two methods. They also used the modified Levenberg method which used (7.6.19) for $\mathbf{\Omega}=\mathbf{I}$ and another equation for λ when $\mathbf{\Omega}=\mathbf{\Omega}_m$; again negligible difference were observed in the rate of convergence. Another method (SPIRAL) is due to Jones [14], who estimates the Levenberg parameter λ in another manner. The first six columns of Table 7.6 contain the results given by Davies and Whitting [10]. We have added the last two columns which were obtained using the Box–Kanemasu method ($\alpha=1$, always) and the modified version which reduces α if the condition given by (7.6.11) is not satisfied. For both methods sequential estimation was used with the initial $\mathbf{P}$ matrix being the large value of $10^7\mathbf{I}$ (or equivalently the small $\mathbf{U}$ of $10^{-7}\mathbf{I}$); this reduces difficulties when $\mathbf{X}^T\mathbf{X}$ is singular.

For all problems except Problem 5, the results of Table 7.6 indicate that the Gauss method is greatly superior to the Marquardt, SPIRAL, Levenberg, and Modified Levenberg methods. A weakness of the Gauss method is indicated by its inability to treat Problem 5; this was caused by the $\mathbf{X}^T\mathbf{X}$ matrix being initially singular. The Gauss method works for Problem 4 which has the same S function but different initial parameter estimates.

The unmodified and modified Box–Kanemasu methods (with $\mathbf{U}=10^{-7}\mathbf{I}$) compare very well with the Gauss method (based on the number of iterations in Table 7.6) for Problems 3, 4, 6, and 7 and are superior for Problem 5. Only one iteration is needed for Problems 6 and 7 which has the linear model of $\eta_i=\beta_i X_{i1}+\beta_2 X_{i2}+\beta_3 X_{i3}$. Hence these methods have the very desirable characteristic of not requiring extensive iterations for linear problems; notice, in contrast, that the other techniques necessitated from 9 to 78 iterations.

For Problem 4 the methods using the Box–Kanemasu equation are about equal to the Gauss method but much better for Problem 5; all the other methods are much less effective for Problems 4 and 5.

It is surprising that for Problems 1 and 2 the Gauss method converges in two iterations whereas all the others take many more. For these two problems the S function is

$$S=(\beta_1^2-\beta_2)^2+(1-\beta_1)^2$$

The two problems have different initial parameter estimates, problem 1 starting with $\beta_1=-1.2$ and $\beta_2=1$. Regardless of the initial parameter

Table 7.6 Number of Iterations Required to Reduce Sum of Squares to Approximately 10^{-14} in Various Test Problems as Given by Davies and Whitting [10]

Problem Number	Marquardt	SPIRAL	Levenberg	Modified Levenberg	Gauss	Box–Kanemasu	Modified Box–Kanemasu
1	92	17	71	100*	2	99	26
2	72	27	69	100*	2	59	103
3	49	39	58	94	9	11	10
4	98	66	45	100*	13	13	14
5	103	76	46	100*	—	15	15
6	61	9	42	60	1	1	1
7	21	13	54	78	1	1	1

*Convergence was not attained at 100 iterations and computation was stopped.

values, the Gauss method results in the correct β_1 and β_2 values in exactly two iterations. The first iteration results in the correct value of β_1 but usually with $S^{(1)}$ being much larger than $S^{(0)}$. In the second iteration in the Gauss method S is reduced to zero. From this example (as in others) we see that $S^{(1)}$ being larger than $S^{(0)}$ does not necessarily mean that the Gauss method will encounter difficulty in convergence. In fact, in this problem the other methods were much less efficient.

In conclusion, although the Gauss method is very competitive in many problems, it may not work because $\mathbf{X}^T\mathbf{X}$ is temporarily singular. For that reason we suggest that the $\mathbf{U}$ be made equal to $C\mathbf{I}$ where C is a very small value compared to the diagonal terms of $\mathbf{X}^T\mathbf{X}$ (or that $\mathbf{P}^{(0)} = K\mathbf{I}$ be used with large K in the sequential method as discussed in Section 6.7). In addition the Box–Kanemasu method is recommended in order to be more certain of finding the parameter values minimizing S, even though in some cases the Gauss method would be more efficient. In cases when the Box–Kanemasu method does not converge, then the modification of it that we have suggested is recommended provided one is convinced that a *unique* minimum *point* exists.

7.7 MODEL BUILDING AND CONFIDENCE REGIONS

Most of the developments discussed in this section utilize an assumption of a multivariate normal distribution of errors. In particular this assumption is used in determining confidence intervals and regions for parameters, locating confidence intervals for Y_i and applying the F test for model building. The assumptions used in this section include additive, zero mean, normal errors in $\mathbf{Y}$. The independent variables are assumed to be errorless. The assumptions are designated 11--1-11.

Unlike linear estimation, the below expressions are all approximate, with the approximation being better for cases which are less nonlinear than others. For a measure of the nonlinearity, see Beale [15] and Guttman and Meeter [16]. The expressions may be approximations due to (1) the sensitivity coefficients being functions of *estimated* parameters, (2) linear approximations to derive equations such as (7.7.1), and (3) the usual estimates for σ^2.

7.7.1 Approximate Covariance Matrix of Parameters

The approximate covariance matrix of the parameters has different forms depending upon the method of estimation. In general, the expressions are similar to comparable linear estimation cases.

For ordinary least squares estimation with the assumptions denoted 11----11, the approximate covariance matrix of $\mathbf{b}_{\mathrm{LS}}$ is analogous to (6.2.11),

$$\operatorname{cov}(\mathbf{b}_{\mathrm{LS}}) \approx (\mathbf{X}^T\mathbf{X})^{-1}\mathbf{X}^T\boldsymbol{\psi}\mathbf{X}(\mathbf{X}^T\mathbf{X})^{-1} = \mathbf{P}_{\mathrm{LS}} \tag{7.7.1}$$

where $\mathbf{X}$ is the sensitivity matrix which is a function of $\mathbf{b}_{\mathrm{LS}}$ and where $\boldsymbol{\psi} = E(\boldsymbol{\varepsilon}\boldsymbol{\varepsilon}^T)$. [Recall that in OLS estimation using (7.4.6) we set $\mathbf{W} = \mathbf{I}$ and $\mathbf{U} = \mathbf{O}$.] With the additional standard assumptions of uncorrelated and constant variance measurement errors and σ^2 being unknown (1111-011), the estimated covariance of $\mathbf{b}_{\mathrm{LS}}$ simplifies to

$$\operatorname{cov}(\mathbf{b}_{LS}) \approx (\mathbf{X}^T\mathbf{X})^{-1}s^2; \qquad s^2 \approx \frac{(\mathbf{Y}-\hat{\mathbf{Y}})^T(\mathbf{Y}-\hat{\mathbf{Y}})}{n-p} \tag{7.7.2}$$

When maximum likelihood estimation is used and the measurement errors are normal, the approximate covariance matrix of $\mathbf{b}_{\mathrm{ML}}$ is

$$\operatorname{cov}(\mathbf{b}_{\mathrm{ML}}) \approx (\mathbf{X}^T\boldsymbol{\psi}^{-1}\mathbf{X})^{-1} = \mathbf{P}_{\mathrm{ML}} \tag{7.7.3}$$

The assumptions are 11--1-11 (see Section 6.1.5). If $\boldsymbol{\psi}$ is known to within the multiplicative constant, σ^2, that is, $\boldsymbol{\psi} = \sigma^2\boldsymbol{\Omega}$ (11--1011), then

$$\operatorname{cov}(\mathbf{b}_{\mathrm{ML}}) \approx (\mathbf{X}^T\boldsymbol{\Omega}^{-1}\mathbf{X})^{-1}s^2 \tag{7.7.4}$$

where s^2 can be estimated using

$$s^2 \approx \frac{(\mathbf{Y}-\hat{\mathbf{Y}})^T\boldsymbol{\Omega}^{-1}(\mathbf{Y}-\hat{\mathbf{Y}})}{n-p} \tag{7.7.5}$$

7.7.2 Approximate Correlation Matrix

The approximate OLS and ML correlation matrices can be obtained using the above equations. The ij element of the correlation matrix is given by

$$r_{ij} = P_{ij}(P_{ii}P_{jj})^{-1/2} \tag{7.7.6}$$

where P_{ij} is a term of (7.7.1), (7.7.2), or (7.7.3) depending on the case. The diagonal elements of $\mathbf{r}$ are all unity and the off-diagonal element must be in the interval $[-1, 1]$. If $\boldsymbol{\psi}$ is known only to within a multiplicative constant σ^2, the σ^2 value need not be known for $\mathbf{r}$ since it cancels in (7.7.6).

Whenever all the off-diagonal elements exceed 0.9 in magnitude, the estimates are highly correlated and tend to be inaccurate. Bacon [17] has suggested that when this is true, a simpler model form than the one originally proposed may be appropriate. A poor experimental design may also be responsible for the high correlations. In that case it is recommended that the sensitivity coefficients be examined and that the experimental design strategies discussed in Chapter 8 be employed. Box [18, 19] has shown, however, that high correlations among the parameters can be due to a large extent to the *nature* of the model itself and thus no experimental design could be expected to yield uncorrelated parameter estimates. See Problem 7.18.

A rule of thumb for anticipating the inaccuracy in calculation is given by Gallant [11]; he suggested that difficulty in computation may be encountered when the common logarithm of the ratio of the largest to smallest eigenvalues of **r** exceeds one-half the number of significant decimal digits used by the computer. See also Appendix A.4.

Example 7.7.1

Use the Gallant criterion for the Box–Kanemasu example of Section 7.6.4.1.

Solution

For that example **r** can be written

$$\mathbf{r} = \begin{bmatrix} 1 & r_{12} \\ r_{12} & 1 \end{bmatrix} \text{where } r_{12} = -0.997909$$

Using (d) and (e) of Example 6.8.1 gives $\lambda_1 = 1 - r_{12}$ and $\lambda_2 = 1 + r_{12}$. The Gallant criterion then is $\log[(1-r_{12})/(1+r_{12})] = 2.98$. Hence in order to solve the Box–Kanemasu data, the Gallant criterion indicates that at least six significant figures be used in the calculation.

7.7.3 Approximate Variance of $\hat{\mathbf{Y}}$

Approximate variances for the predicted values of the model $\hat{\mathbf{Y}}$ for OLS and ML are given by expressions similar to (6.2.12) and (6.5.6).

7.7.4 Approximate Confidence Intervals and Regions

As frequently happens for nonlinear problems, there are several approaches for providing approximate confidence intervals and regions. Some of these are discussed below.

Consider first calculation of approximate confidence intervals. Use the linear approximation given by the Taylor series of (7.2.1) (which is used in the Gauss equation); then, analogous to (6.8.4), we have the approximate

100(1 − α) % confidence interval

$$b_k \pm \text{est. s.e.}(b_k)t_{1-\alpha/2}(v); \qquad k = 1, 2,, \ldots, p \tag{7.7.7}$$

for any one of the parameters where

$$\text{est. s.e.}(b_k) = P_{kk}^{1/2} \tag{7.7.8}$$

and v is the number of degrees of freedom, frequently $n-p$, associated with the estimate of σ^2. The term P_{kk} is the kth diagonal term of $\mathbf{P}$ and represents the variance of b_k. For the assumptions indicated, (7.7.2) is used to find P_{kk} for OLS and (7.7.4) for ML estimation. The expression given by (7.7.8) is approximate not only because s^2 is used for σ^2 in (7.7.2) and (7.7.4), but also because the sensitivity coefficients are approximate since they are functions of $\mathbf{b}$.

Example 7.7.2

The 90% confidence interval is to be found for each of the parameters of the model and for the data

$$\eta = \beta_1 \exp(\beta_2 t) \qquad \begin{array}{cccc} t & 0 & 0 & 1 \\ Y & 1 & 2 & 3 \end{array}$$

Assume that the standard assumptions designated by 11111011 are valid. The minimum sum of squares is 0.5 and occurs at $b_1 = 1.5$ and $b_2 = \ln 2 = 0.693$.

Solution

The sensitivities, sensitivity matrix, and $\mathbf{X}^T\mathbf{X}$ are

$$X_{i1} = \exp(\beta_2 t_i), \qquad X_{i2} = \beta_1 t_i \exp(\beta_2 t_i)$$

$$\mathbf{X}^T(\mathbf{b}) = \begin{bmatrix} 1 & 1 & 2 \\ 0 & 0 & 3 \end{bmatrix}, \qquad \mathbf{X}^T\mathbf{X} = \begin{bmatrix} 6 & 6 \\ 6 & 9 \end{bmatrix}$$

Owing to the standard assumptions, OLS and ML give the same estimates and approximate covariance matrix of the parameters. Using (7.7.2) gives

$$\mathbf{P} \approx \frac{1}{54\text{-}36} \begin{bmatrix} 9 & -6 \\ -6 & 6 \end{bmatrix} \frac{0.5}{1} = \begin{bmatrix} \frac{1}{4} & \frac{-1}{6} \\ \frac{-1}{6} & \frac{1}{6} \end{bmatrix}$$

From t tables we get $t_{.95}(1) = 6.31$ for 90% confidence (see Table 2.15). Then, for

90% confidence intervals, (7.7.7) and (7.7.8) yield

$$b_1 \pm (0.25)^{1/2}(6.31) = 1.5 \pm 3.16 \qquad \text{or} \qquad -1.66 < \text{b}_1 < 4.66$$

$$b_2 \pm (1/6)^{1/2}(6.31) = 0.693 \pm 2.58 \qquad \text{or} -1.89 < \beta_2 < 3.27$$

Generally confidence *intervals* provide poor approximations to the confidence *region* even for linear cases. This is even more true for this nonlinear case, as is shown below. In addition to the approximations that are present in this example, the reader should be aware that this example was constructed for pedagogical simplicity and not because it is a realistic example. We can expect accurate results from the equations given above for real cases only when the assumptions are nearly valid and when the number of observations n is large.

Gallant has given a Monte Carlo study [11] for a certain nonlinear model with four parameters that showed small differences using the above method for determining the confidence intervals compared to a more exact method (which involves the likelihood ratio). He indicated, however, that when **P** is an ill-conditioned matrix the likelihood ratio method can yield considerably shorter confidence intervals than those using the above method, which is based on the asymptotic normality of the least squares estimator. We will describe this more powerful (but in application more time-consuming) procedure in relation to the confidence regions.

Let us now consider two methods of finding confidence regions. Again a number of standard assumptions are needed including knowledge of the probability density which we shall assume to be normal. Assume that the conditions denoted 11--1011 are valid. Then using a Taylor series approximation for η the $100(1-\alpha)\%$ confidence region for both OLS and ML estimation is given by

$$(\mathbf{b}-\boldsymbol{\beta})^T(\mathbf{P}^*)^{-1}(\mathbf{b}-\boldsymbol{\beta}) = ps^2F_{1-\alpha}(p, n-p) \tag{7.7.9}$$

where for OLS and ML estimation $(\mathbf{P}^*)^{-1}$ is

$$(\mathbf{P}^*_{\text{LS}})^{-1} \approx \mathbf{X}^T\mathbf{X}(\mathbf{X}^T\boldsymbol{\Omega}\mathbf{X})^{-1}\mathbf{X}^T\mathbf{X} \tag{7.7.10a}$$

$$(\mathbf{P}^*_{\text{ML}})^{-1} \approx \mathbf{X}^T\boldsymbol{\Omega}^{-1}\mathbf{X} \tag{7.7.10b}$$

In (7.7.10a) the sensitivity matrices are functions of $\mathbf{b}_{\text{LS}}$ whereas those in (7.7.10b) depend upon $\mathbf{b}_{\text{ML}}$. In giving (7.7.9, 10), results of Section 6.8.3 are used along with (6.2.11) and (6.5.5). The approximate confidence contours given by (7.7.9) are ellipsoids.

A more natural confidence region is found using a likelihood ratio [20]. It is the nonlinear analogue to the F statistic given by (6.2.24)

$$\frac{[S(\beta)-R]/p}{R/(n-p)} \approx F_{1-\alpha}(p,n-p) \qquad (7.7.11)$$

where p is present instead of q because a confidence region is needed for all the parameters. (R is the minimum S.) Usually (7.7.11) does *not* produce ellipsoids in parameter space. The contours are along the lines of constant likelihood ratio but the *confidence level* $1-\alpha$ is approximate [15, 16]. The approximation enters in (7.7.11) because the numerator and denominator no longer contain *independent* χ^2 distributions. If we are dealing with a model which is nonlinear in the parameters, computation of confidence contours using (7.7.11) is more difficult and time-consuming than that using (7.7.9) in which a linear approximation is incorporated. For example, if there are two parameters, a separate numerical search problem is involved to obtain β_1 for each choice of β_2.

Example 7.7.3

The 50, 75, and 90% confidence regions are to be found for β_1 and β_2 for the model and data of Example 7.7.2. Use both methods given above.

Solution

Method Based on (7.7.9)
For this method $(\mathbf{P}^*_{\mathrm{LS}})^{-1}$ is $\mathbf{X}^T\mathbf{X}$ since $\boldsymbol{\Omega}=\mathbf{I}$ for the assumption given. The value of s^2 is $R/(n-p)=0.5/(3-2)=0.5$. Then using the $\mathbf{X}^T\mathbf{X}$ matrix of Example 7.7.2 in (7.7.9) gives

$$[b_1-\beta_1 \;\; b_2-\beta_2]\begin{bmatrix} 6 & 6 \\ 6 & 9 \end{bmatrix}\begin{bmatrix} b_1-\beta_1 \\ b_2-\beta_2 \end{bmatrix} = 2(0.5)F_{1-\alpha}(2,1) \qquad (a)$$

Finding the confidence region using this expression is readily accomplished using the results of Example 6.8.1. Note that $\mathbf{C}$ of that example is the square matrix in (a) given above. From (d) and (e) of Example 6.8.1 the eigenvalues are found to be $\lambda_1 = [15-(153)^{1/2}]/2 = 1.31534$, $\lambda_2 = [15+(153)^{1/2}]/2 = 13.68466$. Then using ($g$)–($j$) of Example (6.8.1) gives

$$\mathbf{e} = \begin{bmatrix} -0.788205 & 0.615412 \\ 0.615412 & 0.788205 \end{bmatrix}$$

Since $\mathbf{e}$ is orthonormal and since $ps^2=1$, we can write analogous to (*l*) and (*m*),

$$(b_1-\beta_1)_{maj} = \pm F_{1-\alpha}^{1/2}(2,1)e_{22}\lambda_1^{-1/2} = \pm 0.68726 F_{1-\alpha}^{1/2}(2,1)$$

$$(b_2-\beta_2)_{maj} = \mp F_{1-\alpha}^{1/2}(2,1)e_{12}\lambda_1^{-1/2} = \mp 0.53660 F_{1-\alpha}^{1/2}(2,1)$$

$$(b_1-\beta_1)_{min} = \mp F_{1-\alpha}^{1/2}(2,1)e_{21}\lambda_2^{-1/2} = \mp 0.16636 F_{1-\alpha}^{1/2}(2,1)$$

$$(b_2-\beta_2)_{min} = \pm F_{1-\alpha}^{1/2}(2,1)e_{11}\lambda_2^{-1/2} = \mp 0.21307 F_{1-\alpha}^{1/2}(2,1)$$

The three confidence regions can be found by using these equations with the F values which are $F_{.50}(2,1)=1.5$, $F_{.75}(2,1)=7.5$, and $F_{.9}(2,1)=49.5$. They are shown in Fig. 7.12 as dashed ellipses that are centered at $\beta_1=1.5$ and $\beta_2=0.693$. Notice that the 90% confidence region extends as far as $(b_1-\beta_1)_{maj}=\pm.68726(49.5)^{1/2}=\pm 4.84$ and $(b_2-\beta_2)_{maj}=\pm 3.78$. Since $4.84/1.5=3.22$ and $3.78/.693=5.45$, the confidence region is relatively large and the parameters must be considered

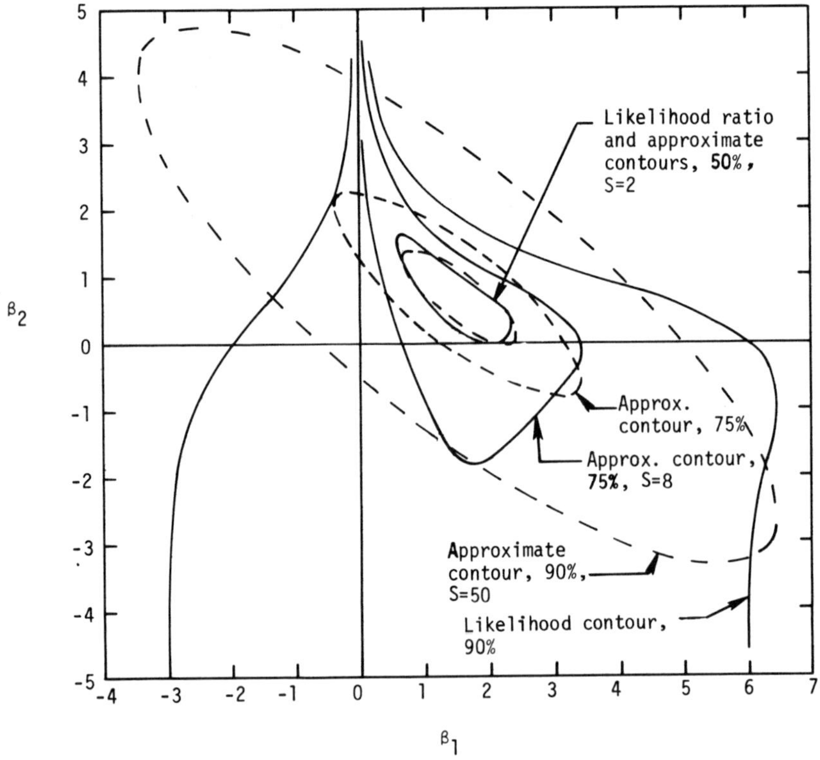

Figure 7.12 Likelihood ratio and approximate confidence contours for 50, 75, and 90%. S is given in Example 7.7.3 as equation (*b*).

inaccurate or *ill-determined*. If the confidence region were sufficiently small so that

$$p^{1/2}sF_{.95}^{1/2}e_{22}\lambda_1^{-1/2}/b_1 \ll 1 \qquad \text{and} \qquad p^{1/2}sF_{.95}^{1/2}e_{12}\lambda_1^{-1/2}/b_2 \ll 1$$

we would say that the parameters are *well-determined*.

From the above equations the ratio of the major to minor axes of the ellipses is found to be $(\lambda_2/\lambda_1)^{1/2}=3.23$ which indicates that the ellipses are neither extremely narrow nor approach a circle. The major axis forms an angle of $\tan^{-1}(e_{12}/e_{22})=-37.98°$ with the $b_1-\beta_1$ axis. Along this angle the S function varies most slowly. Another way of thinking about the major axis is to note that it can be described by

$$\delta\beta_2=\left(\frac{e_{12}}{e_{22}}\right)\delta\beta_1=-0.7808\,\delta\beta_1$$

or

$$\delta\beta_2+0.7808\,\delta\beta_1=0$$

this means that we can estimate the sum $\beta_2+0.78\beta_1$ more accurately than β_2 or β_1.

One should distinquish between the ill-determined condition provided by the confidence region and difficulty in convergence indicated by high parameter correlations or equivalently by a large value of the Gallant criterion. Difficult convergence is related to the character of $\mathbf{X}^T\mathbf{X}$ (and thus near linearity of sensitivity coefficients), but not to the accuracy of the measurements (the s^2 value). The confidence region depends on *both*. For related discussion, see reference 21.

Method Based on (7.7.11)

For this equation the S function is needed; it is

$$S=(1-\beta_1)^2+(2-\beta_1)^2+(3-\beta_1e^{\beta_2})^2 \qquad (b)$$

Because of the relative simplicity of this function, we can solve for $\exp(\beta_2)$ in terms of β_1 and S; then taking the natural logarithm of $\exp(\beta_2)$ gives for β_2

$$\beta_2=\ln\left(\frac{3\pm(S-5+6\beta_1-2\beta_1^2)^{1/2}}{\beta_1}\right) \qquad (c)$$

In more realistic examples a search for β_2 would be needed for fixed β_1 and S. Since $R=0.5$, $n=3$, and $p=2$, (7.7.11) becomes

$$S=F_{1-\alpha}(2,1)+0.5 \qquad (d)$$

By introducing this value of S into (c) and solving for β_2 for a range of β_1 values, we obtain the confidence regions shown as solid lines in Fig. 7.12.

Unlike the dashed confidence regions which are all ellipses with the same ratio of major to minor diameters, these are quite different in shape particularly as S becomes large, or equivalently, as the percent confidence approaches 100. It is significant, however, that at the smallest S value (corresponding to 50%) the true shape of the confidence region does approach the same ellipses given by (7.7.9). From Fig. 7.12 note that a likelihood confidence region tends to be larger than the corresponding approximate region. If a likelihood ratio contour had been inside the associated approximate contour, then the latter would be suggested as a conservative estimate of the likelihood ratio confidence region. Though this condition is not satisfied in this case, it may be in some cases. See Bard [5, p. 209].

For better-designed experiments and for more observations (usually $n \gg 3$), the two methods show much better agreement. For an example, see the Box and Hunter paper [22].

It is instructive to make a comparison of a confidence region constructed using the confidence intervals given in Example 7.7.2 with the confidence regions shown in Fig. 7.12. Consider the 90% case which covers the region $-1.66 < \beta_1 < 4.66$ and $-1.89 < \beta_2 < 3.27$. This region contains some of the 90% confidence regions and additional areas, but also there are areas in the more correct confidence regions not covered. This indicates that a confidence region constructed from confidence intervals may not be a good approximation of the approximate regions; a poor approximation to the likelihood ratio confidence region is provided particularly as the percent confidence approaches 100.

7.7.5 Model Building Using the F test

The F statistic given by (6.2.24),

$$F = \frac{\left[R(\mathbf{b}_1(\boldsymbol{\beta}^*_2), \boldsymbol{\beta}^*_2) - R(\mathbf{b}_1, \mathbf{b}_2) \right] / q}{R(\mathbf{b}_1, \mathbf{b}_2)/(n-p)} \tag{7.7.12}$$

and used for confidence regions can also be used for nonlinear model building. As pointed out below (7.7.11), this expression is approximate for nonlinear models. The assumptions denoted 11--1011 are used in connection with (7.7.12). Recall that (7.7.12) implies two groups of parameters; $\boldsymbol{\beta}_1$ is assumed needed and $\boldsymbol{\beta}_2$ may or may not be needed. There are p parameters in all with $\boldsymbol{\beta}_1$ having $p-q$ and $\boldsymbol{\beta}_2$ having q. The objective is to see if the $\boldsymbol{\beta}_2$ parameters should be included in the model. $\boldsymbol{\beta}^*_2$ may contain values suggested by a competing model. (See discussion following Theorem 6.2.4.) When all p parameters are simultaneously estimated, the minimum sum of squares is designated $R(\mathbf{b}_1, \mathbf{b}_2)$; if the standard assumptions of 11111011 are valid OLS can be used, but if the errors neither have a constant variance nor are independent (11001011), then maximum likelihood estimation should be used.

The statistic F, (7.7.12), is utilized by comparing its value with $F_{1-\alpha}(q, n-p)$. If $F > F_{1-\alpha}(q, n-p)$, then we have an indication that the parameters $\boldsymbol{\beta}_2$ are needed in the model and should not be the values implied by $\boldsymbol{\beta}^*_2$.

Example 7.7.4

Using results of the estimates of parameters for the cooling billet problem, determine the 95% confidence model selecting from those given by (7.5.11, 12). Assume that the standard assumptions indicated by 11111011 are valid.

Solution

The F test can be used in this example in a manner that is very similar to that used for linear problems. Using the sums of squares given in Table 7.4 we can construct Table 7.7. The F values are compared with $F_{.95}(1, n-p) = 4.54$, 4.60, and 4.67 for $n-p$ equal to 15, 14, and 13, respectively. Since 0.0077 is less than 4.67 but 672.8 is not less than 4.6, we have an indication that the two parameters β_1 and β_2 in (7.5.11) are needed but the β_3 parameter is not.

Table 7.7 Sum of Squares and F Statistic for Example 7.7.4

Model No.	No. of Parameters	Degrees of Freedom	R	Mean Square, $s^2 = R/(n-p)$	ΔR	F
1	1	15	38.73731	2.582		
2	2	14	0.7896890	0.05641	37.9476	672.8
3	3	13	0.7892203	0.06071	0.000469	0.0077

7.8 SEQUENTIAL ESTIMATION FOR MULTIRESPONSE DATA

There are a number of advantages of sequential estimation including the development of simpler, yet more general computer programs and the possibility of providing more insight into model building. For other advantages, see Section 6.7.7.

The process is considered to be sequential with time because dynamic experiments are so common. However, there may be cases when one might wish to have some other independent variable, such as position, to be the sequential variable. At each time the experiment is considered to involve $m(\geqslant 1)$ different responses. The m different responses could include different dependent variables such as temperature, concentration, velocity, or voltage; they could also represent measurements from m similar sensors.

Two sequential methods are given in this section. Both utilize the Gauss method and can incorporate the Box–Kanemasu modification.

The first sequential method is simply a direct adaptation of (7.4.6). It would be recommended when a sequential method is being used to reduce the memory requirements in a computer program and the sequential values of the parameters are *not* of interest.

The second method is similar to the linear sequential method of Section 6.7. The measurements must either be actually independent or can be treated as if they are. The latter situation is the case, for example, when OLS is used. Moreover, autoregressive and moving average correlated errors can be analyzed by using certain differences which are uncorrelated. Provided the errors are or can be treated as being uncorrelated, this method is recommended when $m \leqslant p$ because the order of the matrix inversion is reduced. Only a scalar needs to be inverted if $m=1$. Advantage can be taken of this fact by renumbering the observations as if they were all at different times even though physically $m>1$.

Prior information can be included through the use of $\boldsymbol{\mu}$ and $\mathbf{U}$.

7.8.1 Assumptions

The standard assumptions of additive, zero mean, and noncorrelated errors are used. The errors have a known covariance matrix $\boldsymbol{\psi}$. The independent variables are considered to be nonstochastic. The parameters are assumed to be constant, that is, not time variable or random. These assumptions are designated 11-1-11-. The assumptions regarding the measurement errors can be written as

$$\mathbf{Y}=\mathbf{E}(\mathbf{Y}|\boldsymbol{\beta})+\boldsymbol{\varepsilon}=\boldsymbol{\eta}+\boldsymbol{\varepsilon} \tag{7.8.1}$$

$$E(\boldsymbol{\varepsilon})=\mathbf{0} \tag{7.8.2}$$

$$\begin{aligned} E\left[\varepsilon_j(u)\varepsilon_k(v)\right] &= \sigma_j^2(u) \qquad \text{for } u=v, j=k \\ &= 0 \qquad \text{otherwise} \end{aligned} \tag{7.8.3}$$

The measurement vector $\mathbf{Y}$ is composed of the m-dimensional vectors $\mathbf{Y}(i)$,

$$\mathbf{Y}=\begin{bmatrix} \mathbf{Y}(1) \\ \mathbf{Y}(2) \\ \vdots \\ \mathbf{Y}(n) \end{bmatrix} \text{where} \mathbf{Y}(i)=\begin{bmatrix} Y_1(i) \\ Y_2(i) \\ \vdots \\ Y_m(i) \end{bmatrix} \tag{7.8.4}$$

The associated error vector $\boldsymbol{\varepsilon}$ and dependent variable vector $\boldsymbol{\eta}$ are similarly composed of the m-dimensional vectors $\boldsymbol{\varepsilon}(i)$ and $\boldsymbol{\eta}(i)$ where $i=1,,\ldots,n$.

The covariance matrix of the errors for the given assumptions is the diagonal matrix ψ,

$$\psi = \text{diag}\left[\boldsymbol{\Phi}(1)\ldots\boldsymbol{\Phi}(n)\right] \qquad \text{where } \boldsymbol{\Phi}(i) = \text{diag}\left[\sigma_1^2(i)\ldots\sigma_m^2(i)\right] \quad (7.8.5)$$

Consistent with the above assumptions is estimation using the assumptions 11---011 which are the assumptions used in Gauss–Markov estimation for linear models. With the additional assumption of normal errors, the results are equivalent to those obtained using maximum likelihood. The sequential procedure given below could also be used with OLS estimation by simply replacing $\boldsymbol{\Phi}$ by $\mathbf{I}$.

7.8.2 Direct Method

Equations (7.4.6) apply to the multiresponse case. In order to reduce the plethora of subscripts and superscripts we write (7.4.6) as

$$\mathbf{b}^* = \mathbf{b} + \mathbf{P}\left[\mathbf{X}^T\boldsymbol{\Phi}^{-1}\mathbf{e} + \mathbf{U}(\boldsymbol{\mu} - \mathbf{b})\right] \quad (7.8.6)$$

$$\mathbf{e} \equiv \mathbf{Y} - \boldsymbol{\eta}, \qquad \mathbf{P} \equiv \left[\mathbf{X}^T\boldsymbol{\Phi}^{-1}\mathbf{X} + \mathbf{U}\right]^{-1} \quad (7.8.7)$$

where $\mathbf{X}$ and $\boldsymbol{\eta}$ are evaluated at $\mathbf{b}$ which stands for $\mathbf{b}^{(k)}$. The vector $\mathbf{b}^*$ is the same as $\mathbf{b}^{(k+1)}$ in (7.4.6a). Recall that k is an iteration index which changes only after all the data ($m \times n$ observations) have been considered.

The sensitivity matrix in (7.8.6, 7) can be partitioned so that

$$\mathbf{X} = \begin{bmatrix} \mathbf{X}(1) \\ \mathbf{X}(2) \\ \vdots \\ \mathbf{X}(n) \end{bmatrix} \qquad \text{where } \mathbf{X}(i) = \begin{bmatrix} X_{11}(i) & X_{12}(i) & \cdots & X_{1p}(i) \\ X_{21}(i) & X_{22}(i) & \cdots & X_{2p}(i) \\ \vdots & \vdots & & \\ X_{m1}(i) & X_{m2}(i) & \cdots & X_{mp}(i) \end{bmatrix}$$

(7.8.8)

Notice that $X_{jk}(i)$ is the sensitivity coefficient for the jth dependent variable and kth parameter and at the ith time,

$$X_{jk}(i) = \frac{\partial \eta_j(i)}{\partial \beta_k} \quad (7.8.9)$$

A typical term of the triple product $\mathbf{X}^T\boldsymbol{\Phi}^{-1}\mathbf{X}$ of (7.8.7) can be indicated by

$$\mathbf{X}^T\mathbf{\Phi}^{-1}\mathbf{X} = \left[\sum_{t=1}^{n} \sum_{j=1}^{m} X_{jk}(t) X_{jl}(t) \sigma_j^{-2}(t) \right]; \qquad k,l = 1,2,,\ldots,p \quad (7.8.10)$$

Rather than waiting to evaluate the sums in $\mathbf{X}^T\mathbf{\Phi}^{-1}\mathbf{X}$ after a complete iteration, a running summation can be performed to reduce the computer memory requirements. Let us define $C_{sl}(i)$ as being

$$C_{sl}(i) = U_{sl} + \sum_{t=1}^{i} \sum_{j=1}^{m} X_{js}(t) X_{jl}(t) \sigma_j^{-2}(t) \qquad (7.8.11)$$

Then the sl term of the $\mathbf{P}^{-1}$ matrix can be constructed from

$$C_{sl}(i+1) = C_{sl}(i) + \sum_{j=1}^{m} X_{js}(i+1) X_{jl}(i+1) \sigma_j^{-2}(i+1) \qquad (7.8.12)$$

which forms a recursion relation with the starting value of $C_{sl}(0) = U_{sl}$. Both s and l go from 1 to p.

A similar recursion relation can be given for the components of the vector, $\mathbf{X}^T\mathbf{\Phi}^{-1}\mathbf{e}$; one component is

$$d_s(i+1) = d_s(i) + \sum_{j=1}^{m} X_{js}(i+1) e_j(i+1) \sigma_j^{-2}(i+1) \qquad (7.8.13)$$

which has the starting condition of

$$d_s(0) = \sum_{j=1}^{p} U_{sj}(\mu_j - b_j) \qquad (7.8.14)$$

The range of s is 1 to p.

The procedure to find the new parameter vector is first to use (7.8.12) and (7.8.13). Then the $(i+1)$th matrix of $\mathbf{P}$ is found by finding the inverse,

$$\mathbf{P}(i+1) = \begin{bmatrix} C_{11}(i+1) & \cdots & C_{1p}(i+1) \\ \vdots & & \vdots \\ C_{1p}(i+1) & \cdots & C_{pp}(i+1) \end{bmatrix}^{-1} \qquad (7.8.15)$$

Finally the parameter vector is found using

$$\mathbf{b}^*(i+1) = \mathbf{b} + \mathbf{P}(i+1)\mathbf{d}(i+1) \qquad (7.8.16)$$

Note that $\mathbf{b}^*(i+1)$ is the parameter vector at the $(i+1)$th time in the kth iteration where k has the same meaning as (7.4.6). The vector $\mathbf{b}$ in (7.8.16) is $\boldsymbol{\mu}$ for the first iteration and $\mathbf{b}^*(n)$ of the preceding iteration for the subsequent iterations on k.

If only the final converged parameter values are of interest, it is necessary to invert the $\mathbf{P}$ matrix just once for each complete iteration; in other words, it would be necessary to just find $\mathbf{P}(n)$ and then use

$$\mathbf{b}^*(n)=\mathbf{b}+\mathbf{P}(n)\mathbf{d}(n) \tag{7.8.17}$$

At the end of an iteration the **b** vector is replaced by **b***(n) and another iteration on k is begun with **X** being evaluated at the new **b**. The process is continued until convergence is attained.

7.8.3 Sequential Method Using the Matrix Inversion Lemma

The usually preferred sequential method involves the matrix inversion lemma. Using expressions in the preceding section we can write

$$\mathbf{b}^*(i+1)=\mathbf{b}+\mathbf{P}(i+1)\left[\mathbf{d}(i)+\mathbf{X}^T(i+1)\mathbf{\Phi}^{-1}(i+1)\mathbf{e}(i+1)\right] \tag{7.8.18}$$

$$\mathbf{P}(i+1)=\left[\mathbf{P}^{-1}(i)+\mathbf{X}^T(i+1)\mathbf{\Phi}^{-1}(i+1)\mathbf{X}(i+1)\right]^{-1} \tag{7.8.19}$$

where typical terms are

$$\mathbf{X}^T(i+1)\mathbf{\Phi}^{-1}(i+1)\mathbf{e}(i+1)=\left[\sum_{j=1}^{m} X_{js}(i+1)e_j(i+1)\sigma_j^{-2}(i+1)\right] \tag{7.8.20}$$

$$\mathbf{X}^T(i+1)\mathbf{\Phi}^{-1}(i+1)\mathbf{X}(i+1)=\left[\sum_{j=1}^{m} X_{js}(i+1)X_{jl}(i+1)\sigma_j^{-2}(i+1)\right] \tag{7.8.21}$$

The range of s and l in these last two equations is 1 to p.

Using (6.7.5a) and (6.7.5b) we can write analogous to (6.7.6a, b, c, f).

A.	$\mathbf{A}(i+1)=\mathbf{P}(i)\mathbf{X}^T(i+1)$	(7.8.22a)
B.	$\mathbf{\Delta}(i+1)=\mathbf{\Phi}(i+1)+\mathbf{X}(i+1)\mathbf{A}(i+1)$	(7.8.22b)
C.	$\mathbf{K}(i+1)=\mathbf{P}(i)\mathbf{X}^T(i+1)\mathbf{\Delta}^{-1}(i+1)$	(7.8.22c)
E.	$\mathbf{P}(i+1)=\mathbf{P}(i)-\mathbf{K}(i+1)\mathbf{X}(i+1)\mathbf{P}(i)$	(7.8.22d)

Utilizing (6.7.5b) and (7.8.22c) in (7.8.18) yields after some rearrangement

D. $$\mathbf{b}^*(i+1)=\mathbf{b}^*(i)+\mathbf{K}(i+1)\left\{\mathbf{e}(i+1)-\mathbf{X}(i+1)\left[\mathbf{b}^*(i)-\mathbf{b}\right]\right\}$$

(7.8.22e)

At each time i, the above five equations are to be used in the order A, B, C, D, and E.

It is important to understand that three different parameter vectors appear in (7.8.22e). First, $\mathbf{b}^*(i+1)$ is the estimated vector for the $(i+1)$th time and at the $(k+1)$th iteration. Second, $\mathbf{b}^*(i)$ is for the ith time and $(k+1)$th iteration. Finally, $\mathbf{b}$ is the vector found at the nth time of the kth iteration; the vector $\mathbf{b}$ does *not* change during an iteration over the time index i. Note that $\mathbf{P}(i+1)$, $\mathbf{X}(i+1)$, $\boldsymbol{\Delta}(i+1)$ and $\mathbf{e}(i+1)=\mathbf{Y}(i+1)-\boldsymbol{\eta}(i+1)$ are functions of $\mathbf{b}$ and *not* $\mathbf{b}^*(i)$.

When there is prior information, each iteration starts with $\mathbf{b}=\boldsymbol{\mu}$ and $\mathbf{P}(0)=\mathbf{V}_\beta=\mathbf{U}^{-1}$. A detailed example is given in Section 7.9.1.

When there is no prior information, a special starting procedure is needed at $i=1$ for each of the k iterations. As suggested in Section 6.7, we can select $\mathbf{P}(0)$ to be a diagonal matrix with relatively large terms. Since $\mathbf{P}$ represents the approximate variance–covariance matrix of the parameter vector, $\boldsymbol{\beta}$, a $\mathbf{P}(0)$ with large diagonal components indicates poor prior information. In general, $\mathbf{P}(0)$ is fixed from iteration to iteration (i.e., over the k index). Provided no problems are encountered with round-off, $\mathbf{P}(0)$ could have any diagonal values that are large (by a factor of 10^6, say) compared to the corresponding values of $\mathbf{P}(n)$.

7.8.3.1 Sequential Method for $m=1$, p Arbitrary

An important application of the previous analysis is for $m=1$ because the only inverse that must be found, (that of $\boldsymbol{\Delta}$), becomes the inverse of a scalar. As shown in Section 6.7.5, uncorrelated multiresponse measurements can be renumbered so that the $m=1$ analysis can be employed.

From (7.8.22) evaluated for $m=1$ and p arbitrary we find

$$A_u(i+1)=\sum_{j=1}^{p} X_j(i+1)P_{uj}(i), \qquad u=1,2,\ldots,p \tag{7.8.23a}$$

$$\Delta(i+1)=\sigma^2(i+1)+\sum_{j=1}^{p} X_j(i+1)A_j(i+1) \tag{7.8.23b}$$

$$K_u(i+1)=A_u(i+1)/\Delta(i+1), \qquad u=1,2,\ldots,p \tag{7.8.23c}$$

$$H(i+1)=Y(i+1)-\eta(i+1)-\sum_{j=1}^{p} X_j(i+1)[b^*_j(i)-b_j] \tag{7.8.23d}$$

$$b^*_u(i+1)=b^*_u(i)+K_u(i+1)H(i+1), \qquad u=1,\ldots,p \tag{7.8.23e}$$

$$P_{uv}(i+1)=P_{uv}(i)-K_u(i+1)A_v(i+1), \qquad u,v=1,\ldots,p \tag{7.8.23f}$$

The above scalar equations should be used in the order given. These equations can be programmed in a straightforward manner, but one must remember that three **b** vectors are involved and that the **P**(0) matrix does not vary from iteration to iteration. Only two sets of **b** storage locations are needed, however, one is for $\mathbf{b}=\mathbf{b}^*(n)$ and the other is for $\mathbf{b}^*(i)$ *and* $\mathbf{b}^*(i+1)$ since the latter replaces the former as calculations proceed. Since $m=1$, only one subscript is needed for $\mathbf{X}_j(i+1)$ and none is needed for $H(i+1)$. When programming, the $i+1$ index need not be carried.

The sequential procedure given by (7.8.23) reduces to that given by the linear analysis given by (6.7.8) by setting $\mathbf{b}=\mathbf{0}$.

For MAP estimation the starting value of **b** is $\boldsymbol{\mu}$ and the starting **P**(0) matrix is $\mathbf{U}^{-1}=\mathbf{V}_\beta$ for *each* iteration. All the sensitivity coefficients for the first iteration (which includes $i=1,2,,\ldots,n$) and each $\eta(i+1)$ is evaluated using $\mathbf{b}=\boldsymbol{\mu}$. For subsequent iterations **X** and η are evaluated at $\mathbf{b}(n)$, the final vector of the previous iteration. See the Section 7.9.1 example.

When there is no prior information, follow the procedure described below (7.8.22).

7.8.4 Correlated Errors With Known Correlation Parameters

If the measurement errors in **Y** are additive and correlated in time, the above analyses can be used with slight modifications for the autoregressive and moving average cases discussed in Section 6.9 and Appendix 6A, provided the correlation parameters are known. The analyses can be performed by replacing **X** by $\mathbf{Z}=\mathbf{D}^{-1}\mathbf{X}$ and **e** by $\mathbf{f}=\mathbf{D}^{-1}\mathbf{e}$. For example, for first-order autoregressive errors, the terms in the matrices would be

$$Z_{jk}(i)=X_{jk}(i)-\rho_j(i)X_{jk}(i-1) \tag{7.8.24a}$$

$$f_j(i)=e_j(i)-\rho_j(i)e_j(i-1) \tag{7.8.24b}$$

for $i=1,2,,\ldots,n$ and where $X_{jk}(0)$ and $e_j(0)$ are defined to be equal to zero.

7.9 EXAMPLES UTILIZING SEQUENTIAL ESTIMATION

The purpose of this section is to present four different examples in which a sequential method is used. In the first example, detailed results are given to illustrate the sequential MAP method based on the matrix inversion lemma. The physical problem involves heat conduction in a finite plate. The second problem, involving an ordinary differential equation, is the cooling billet problem previously considered. The third problem uses simulated data for heat conduction in a semi-infinite body; the errors are correlated. The final example is a realistic example involving about 1000 actual measurements; it is again for heat conduction in a finite plate. Except for the second example each of the problems in this section involves multiresponse data.

7.9.1 Simple MAP Example Involving Multiresponse Data

This example provides a detailed analysis of a maximum a posteriori sequential estimation procedure for some multiresponse data. The data are abstracted from the extensive values contained in Table 7.14. The physical problem is that of a flat plate heated on one side and insulated on the other. More extensive use of these data is discussed in Section 7.9.4 where a more complete description of the problem is given. In that section the results of estimation for the thermal conductivity, k, and the specific heat–density product, c, are given; a finite difference procedure is used to calculate the dependent variable (temperature). In this section a limited amount of data is used; $K=k^{-1}$ and $\alpha(=k/c)$ are estimated; an analytical solution is used for the dependent variable; and subjective prior information is used.

The measurements for this case are given in Table 7.8 and other known information is given in Table 7.9. The temperature measurements in Table 7.8 can be considered to be additive, zero mean, constant variance, uncorrelated, and normal. Since at each time there are two measurements at each location, an estimate of pure error can be obtained. Using the fact that $V(\varepsilon_1-\varepsilon_2)=2\sigma^2$ (for assumptions 1111----), an estimate of σ^2 from the data of Table 7.8 can be found to be about 0.0625. The errors in the time measurements are negligible compared to those in the temperature rise —that is, those above the initial temperature which is about 81.69°F. The assumptions given above and that of subjective prior information is denoted 11111113.

The temperatures corresponding to η are calculated using the solution given by (8.5.25), which can also be differentiated with respect to K and α to get the sensitivity coefficients. For the times given in Table 7.8, the first three terms at most are needed in the summation. Values for the sensitivity coefficients of K and α, designated X_1 and X_2, respectively, are given in

Table 7.8 Data for Section 7.9.1 Problem

Time (sec)	Time Measured from Start of Heating (hr)	T_1 (°F) ($x=L$)	T_4 (°F) ($x=L$	T_5 (°F) ($x=0$)	T_6 (°F) ($x=0$
15.3	12/3600	85.83	85.37	94.73	94.48
18.3	15/3600	87.46	87.16	96.44	96.03

Table 7.9 Known Conditions for Section 7.9.1 Problem

$\mu_1(=K)=45^{-1}=0.022222$ hr-ft-°F/Btu	$q=9612$ Btu/ft²-hr
$\mu_2(=\alpha)=0.8$ ft²/hr	$L=1/12$ ft
$V_{\beta,11}=4\times10^{-6}$(hr-ft-°F/Btu)²	$T_0=81.69$°F
$V_{\beta,12}=V_{\beta,21}=0$	
$V_{\beta,22}=0.01$ ft⁴/hr²	

Table 7.10. The corresponding calculated values of temperature, denoted η, are also given. All the values of X_1, X_2, and η in Table 7.10 are based on the parameter values of $K=\mu_1$ and $\alpha=\mu_2$. The information in Table 7.10 is given in an order appropriate for sequential estimation with a single observation being added at one "time." Other orders are possible, however.

Table 7.10 Sensitivity Coefficients and η for the First Iteration of the Section 7.9.1 Problem

i	Time (hr)	x	T	$X_1(i)$	$X_2(i)$	$\eta(i)$
1	12/3600	L	T_1	177.8	8.158	85.640
2	12/3600	L	T_4	177.8	8.158	85.640
3	12/3600	0	T_5	570.9	8.930	94.377
4	12/3600	0	T_6	570.9	8.930	94.377
5	15/3600	L	T_1	252.4	10.49	87.299
6	15/3600	L	T_4	252.4	10.49	87.299
7	15/3600	0	T_5	650.0	10.87	96.136
8	15/3600	0	T_6	650.0	10.87	96.136

In Table 7.11 are the sequential values for the first iteration. These values use the sensitivity coefficients given in Table 7.10 (except to more significant figures). Notice that all the $X_j(i)$'s and $\eta(i)$ are evaluated for the *same* values of $K=\mu_1$ and $\alpha=\mu_2$; the sensitivity coefficients and η in Table 7.11 are *not* evaluated using the updated values of $b_1(i)$ and $b_2(i)$. The sequential procedure used is given by (7.8.23), which is identical to that given by (6.7.8) for the linear case. For the nonlinear case, however, iteration is required.

The starting values of b_1, b_2, P_{11}, P_{12}, and P_{22} for this MAP analysis are μ_1, μ_2, $V_{\beta,11}$, $V_{\beta,12}$, and $V_{\beta,22}$, respectively. These starting values are used for *every* iteration as can be observed from the starting conditions in Table 7.12, which is for the second iteration. In this iteration, however, all the

Table 7.11 Sequential Estimation of K and α in Problem of Section 7.9.1, First Iteration. Based on Parameter Values $K=\mu_1=0.022222$ and $\alpha=\mu_2=0.8$

i	Time (sec)	x	T	$b_1\times 10^2$	b_2	$P_{11}\times 10^6$	$P_{12}\times 10^6$	$P_{22}\times 10^6$
0				2.2222	0.8	4	0	10000
1	12	L	T_1	2.2380	0.81814	3.408	−67.89	2211
2	12	L	T_4	2.2188	0.79603	3.386	−70.47	1915
3	12	0	T_5	2.3074	0.78031	0.5543	−20.23	1024
4	12	0	T_6	2.2836	0.78452	0.3988	−17.48	975.0
5	15	L	T_1	2.2701	0.79407	0.3321	−12.79	645.0
6	15	L	T_4	2.2815	0.78603	0.3030	−10.74	500.9
7	15	0	T_5	2.2873	0.78492	0.2372	−9.48	476.8
8	15	0	T_6	2.2654	0.78912	0.2064	−8.89	465.5

Table 7.12 Sequential Estimation of K and α in Problem of Section 7.9.1, Second Iteration. Based on Parameter Values $K=b_1^{(1)}(n)$ $=0.022654$ and $\alpha=b_2^{(1)}(n)=0.78912$

i	Time (sec)	x	T	$b_1\times 10^2$	b_2	$P_{11}\times 10^6$	$P_{12}\times 10^6$	$P_{22}\times 10^6$
0				2.2222	0.8	4	0	10000
1	12	L	T_1	2.2372	0.81793	3.446	−66.17	2103
2	12	L	T_4	2.2188	0.79589	3.425	−68.63	1809
3	12	0	T_5	2.3077	0.78075	0.5606	−19.87	979.6
4	12	0	T_6	2.2838	0.78481	0.4039	−17.20	934.2
5	15	L	T_1	2.2701	0.79421	0.3354	−12.53	615.6
6	15	L	T_4	2.2816	0.78632	0.3056	−10.50	476.6
7	15	0	T_5	2.2874	0.78525	0.2393	−9.28	454.2
8	15	0	T_6	2.2654	0.78930	0.2083	−8.71	443.8

values of $\cdot X_1(i)$, $X_2(i)$, and $\eta(i)$ are found based on the values of the *previous* iteration, $b_1^{(1)}(n)=0.022654$ and $b_2^{(1)}(n)=0.78912$.

If this case had not been one involving prior information, then the initial values of $b_1^{(2)}(0)$ and $b_2^{(2)}(0)$ for the second iteration would be simply the final values of the first iteration at which values the sensitivity coefficients and η are also evaluated. The $P_{kl}^{(2)}(0)$ values would not be different, however (unless one wishes to use $P_{kl}^{(2)}(0)$ to introduce the Levenberg or Marquardt modifications).

The iterations in general continue until negligible changes in $b_j^{(k)}(n)$ occur. In this case only two iterations ($k=2$) apparently are needed.

7.9.2 Cooling Billet Problem

The data for the cooling billet problem are given in Table 6.2; results of analyses of these data are given in Sections 6.2, 7.5.2, and 7.7.5. Though converged parameter values for a differential equation model are given in the last two sections, sequential results are not given. The purpose of this section is to demonstrate the power of a sequential procedure to provide insight for model building for the cooling billet case.

Consider first results of the power series models for h given by (7.5.11). The values of b_1 for Model 1 are depicted in Fig. 7.13. (Recall that Model 1 is $hA/\rho cV = \beta_1$.) The value of $b_1(i)$ is the approximation to β_1 using the data for times through t_i; the estimate is approximate in that the sensitivity coefficients were evaluated at the converged parameter values using *all the data*. Because the b_1 values of Model 1 continually decrease in Fig. 7.13 and because $b_1(i)$ is the average until t_i, the actual time variation of h is probably larger than that shown.

Parameters for Model 2 ($hA/\rho cV = \beta_1 + \beta_2 t$) are shown in Figs. 7.13 and 7.14. The b_1 values shown in Fig. 7.13 are relatively constant with time; this is particularly true for the last five time steps. For the last half of the time steps shown in Fig. 7.14, b_2 is relatively constant. This constancy of parameters suggests adequacy of Model 2.

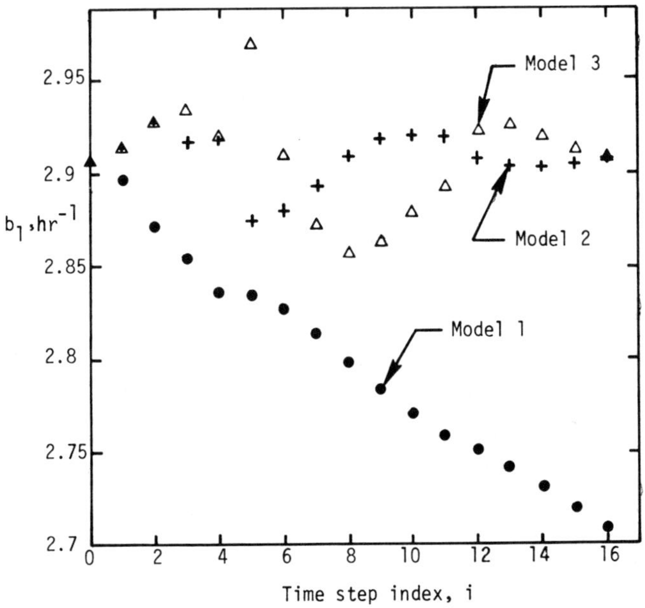

Figure 7.13 Estimates of β_1 for Models 1, 2, and 3 for the cooling billet data.

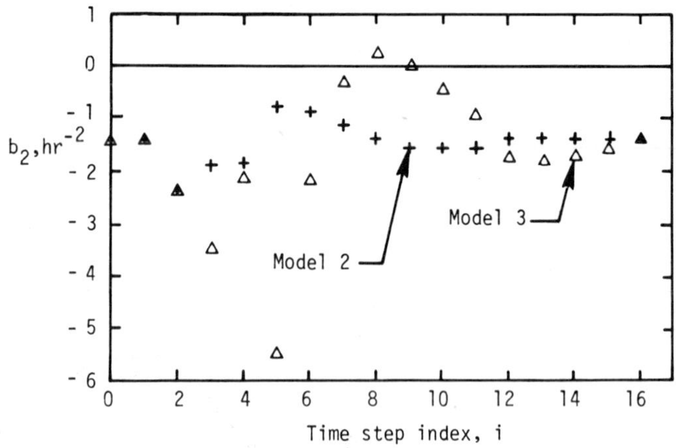

Figure 7.14 Estimates of β_2 for Models 2 and 3 for the cooling billet data.

For Model 3 ($hA/\rho cV = \beta_1 + \beta_2 t + \beta_3 t^2$), the b_1 values are relatively stable in Fig. 7.13, those of b_2 in Fig. 7.14 less so, and those in Fig. 7.15 showing b_3 are quite oscillatory. The oscillatory nature of b_3, passing through zero several times, suggests that Model 3 is not appropriate.

Sometimes one wonders about the effects on the parameters of additional data. Without obtaining the data one cannot be certain. However, the sequential method applied to existing data should give insight into the effects. Additional data would be expected to continue trends already observed in the parameters. For example, Fig. 7.15 suggests that b_3 might continue to oscillate about zero with decreasing amplitude.

The sequential results for β_1 and β_2 in Model 4, $hA/\rho cV = \beta_1 + \beta_2(T - T_\infty)$, are shown in Fig. 7.16. The b_1 values are relatively constant but the b_2 values are not. The 50% variation of b_2 in the last five steps suggests that the model should be further examined by using additional data or related models should be tried. Physically the model is quite attractive but some improvement is needed such as adding an exponent n on $T - T_\infty$; this was tried but was unsuccessful as mentioned in Section 7.5.2 although additional data might be needed also to estimate n.

7.9.2.1 Other Possible Models

In addition to the models mentioned above, many more could be suggested. For example, one extension of Models 1, 2, and 3 is to investigate other terms such as $\beta_4 t^3$. Another approach is to always use the linear-in-t model (Model 2) but to use different β pairs (such as β_{1i}, β_{2i} for the ith time interval) for successive time intervals [23–25].

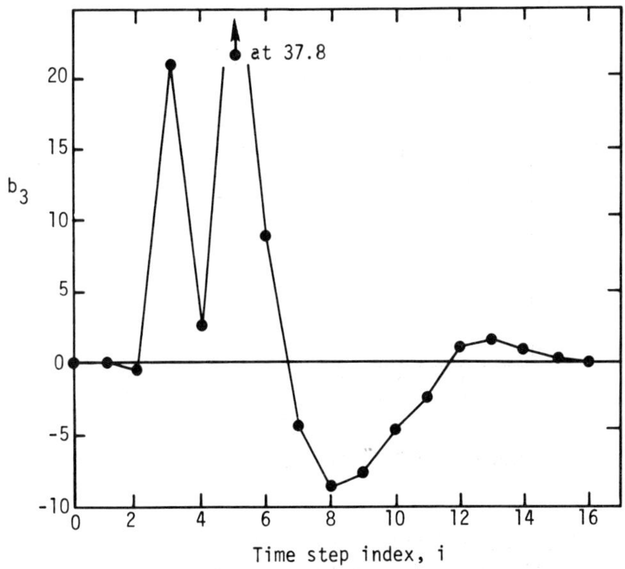

Figure 7.15 Estimates of β_3 for Model 3 for cooling billet data.

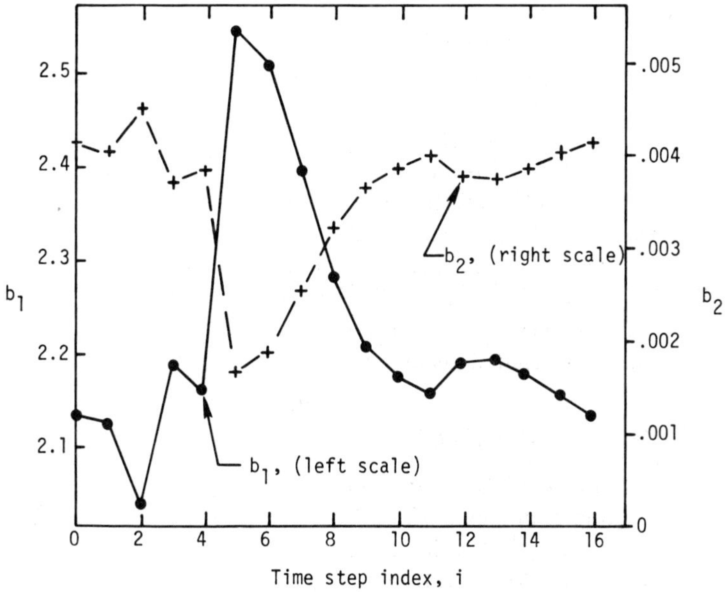

Figure 7.16 Sequential parameter estimates for Model 4 for cooling billet data.

7.9.3 Semi-Infinite Body Heat Conduction Example

Consider a semi-infinite, heat-conducting body initially at the temperature T_0 of 100°F (311 K) and whose surface temperature takes a step increase to $T_\infty = 200$°F (366 K). The temperature is measured every 9 sec until 252 sec at the two locations $x_1 = 0.125$ in. (0.00318 m) and $x_2 = 0.25$ in. (0.00635 m) where x is measured from the heated surface. This example is of the Monte Carlo type because the data are simulated using the known value of thermal diffusivity α, equal to 0.01 ft^2/hr (2.58×10^{-7}m^2/sec) and because 15 different sets of random errors are considered. The first five cases use independent, normal errors with unit variance. The measurement errors for the first case are taken from the first two columns of Table XXIII of [26], the second case uses those from the next two columns, etc.

Correlated errors are obtained for the first-order autoregressive process a1 described in Section 6.9.2; the error ε_i at time t_i is given by

$$\varepsilon_i = \rho\varepsilon_{i-1} + u_i \tag{7.9.1}$$

where the $\rho = 0.5$ for cases 6–10 and $\rho = 0.9$ for the remaining cases. The u_i random variable values are the same values taken from [26] as mentioned above.

A mathematical statement of the physics of the problem is

$$\alpha \frac{\partial^2 T}{\partial x^2} = \frac{\partial T}{\partial t} \tag{7.9.2}$$

$$T(0,t) = T_\infty, \; T(x,0) = T_0, T(\infty,t) = T_0 \tag{7.9.3}$$

which has the exact solution,

$$\frac{T(x,t) - T_0}{T_\infty - T_0} = \operatorname{erfc} \frac{x}{2(\alpha t)^{1/2}} \qquad \text{where } \operatorname{erfc}(y) = \frac{2}{\pi^{1/2}} \int_y^\infty e^{-u^2} du \tag{7.9.4}$$

Note that $T(x,t)$ is a nonlinear function of α. Because there is no characteristic length or time in this problem, the dimensionless temperature given by (7.9.4) and the α sensitivity [given by (8.5.4)] can be plotted versus the single dependent variable $\tau = \alpha t / x^2$, see Fig. 8.12.

The parameter α was estimated using nonlinear ML sequential estimation for the multiresponse data mentioned above (two locations at each of which 28 measurements were simulated). In most cases only three iterations were needed to converge to $\hat{\alpha}$ after starting with the initial estimate of 0.006 ft^2/hr which is low compared to the true value of 0.01 ft^2/hr.

Estimated α values for the 15 cases are given in the third column of

Table 7.13. The first five cases tabulated which are for $\rho=0$ are more accurate on the average than the next five which are for $\rho=0.5$ which are in turn more accurate (on the average) than the last 5 cases ($\rho=0.9$). Hence we can conclude that correlated errors can substantially reduce the accuracy of the parameter estimates.

Table 7.13 Results for Estimating Thermal Diffusivity for Cases with Noncorrelated and Correlated Errors

Case	ρ	$\hat{\alpha}\times 10^4$ (ft^2/hr)	R	s^2	$\mathbf{Z}^T\mathbf{Z}$ $\times 10^{-6}$	est. s.e.($\hat{\alpha}$) $\times 10^4$
1	0	99.71	59.39	1.080	197	0.740
2	0	100.64	54.15	.985	192	0.716
3	0	100.17	58.71	1.067	195	0.740
4	0	99.29	52.55	.955	199	0.693
5	0	98.56	56.49	1.191	202	0.768
6	0.5	99.49	60.85	1.106	52.4	1.45
7	0.5	101.88	53.76	.977	49.7	1.40
8	0.5	100.39	59.2	1.076	51.3	1.45
9	0.5	98.69	52.7	.958	53.4	1.34
10	0.5	97.22	56.3	1.024	55.3	1.36
11	0.9	98.6	64.67	1.176	4.275	5.25
12	0.9	115.3	53.40	.971	3.09	5.61
13	0.9	101.7	59.37	1.079	4.00	5.19
14	0.9	97.2	55.15	1.003	4.40	4.97
15	0.9	91.0	61.90	1.125	5.06	4.72

The minimum sum of squares R also has a tendency to increase in variability with increasing ρ. The values of R were calculated using the converged values of $\hat{\alpha}$ in (7.3.1) with $\mathbf{U}=0$ and $\mathbf{W}=\boldsymbol{\Omega}^{-1}$; rather than using $\boldsymbol{\Omega}^{-1}$ directly, however, it was introduced by replacing $\mathbf{X}$ by $\mathbf{Z}$ and $\mathbf{e}$ by $\mathbf{f}$ as mentioned in Section 7.8.4.

The mean sum of squares was formed using $s^2=R/(n-p)$ where $n=2\times 28=56$ and $p=1$, in this example. The estimated standard error $\hat{\alpha}$ was found using (6.8.3) which can be written for this *one*-parameter case as

$$\text{est. s.e.}(\alpha)=\left[\mathbf{Z}^T\mathbf{Z}\right]^{-1/2}s \tag{7.9.5}$$

These values increase by a factor of about 7 from $\rho=0$ to 0.9. In each set of five cases for a given ρ value, the true α value of 100×10^{-4} ft^2/hr is in the interval given by $\hat{\alpha}\pm$est. s.e.($\hat{\alpha}$) in three of the five cases. This is consistent with the 67% confidence interval that could be calculated using the t distributtion.

7.9.4 Analysis of Finite Heat-Conducting Body with Multiresponse Experimental Data

7.9.4.1 Description of Experiment

In this experiment two adjacent, identical Armco iron cylindrical specimens (1 in. (0.0254 m) thick and 3 in. (0.0762 m) in diameter) were heated by a single flat heater placed between them. The heat flow was in the axial direction. All surfaces except those where the heater was located were insulated. Four thermocouples (numbered 5, 6, 7, and 8) were carefully attached to the heated surface of each specimen and four (numbered 1–4) were attached in the opposite flat insulated surfaces. The sensors were located at angles of 0, 90, 180, and 270°. The sensors at 0 and 180° were electrically averaged as were also those at 90 and 270°. This then provided eight temperature histories with four being at a heated surface and four being at an insulated surface.

Before the start of an experiment the specimens were allowed to come to a uniform temperature. The heating period lasted 15.3 sec after which the specimens attained a higher equilibrium temperature. For further discussion of the experiments, see reference 27.

The data used to find parameters in this case are given in Table 7.14. An IBM 1800 computer was used to digitize the analogue signals produced by the thermocouples. Typical results for the heated and insulated surfaces are shown in Fig. 7.17; the heated surface temperature increases only during heating whereas the insulated surface temperature increases after heating and finally approaches the same equilibrium temperature as the heated surface.

Table 7.14 Measured Temperatures for Finite Heat-Conducting Specimen

Time	Temperature (°F) for Numbered Thermocouples							
(sec)	TC 1	TC 2	TC 3	TC 4	TC 5	TC 6	TC 7	TC 8
0.3	81.67	81.37	81.36	81.37	81.49	81.58	81.69	81.68
0.6	81.67	81.29	81.52	81.61	81.41	81.74	81.60	81.60
0.9	81.75	81.45	81.52	81.45	81.49	81.66	81.44	81.77
1.2	81.75	81.12	81.44	81.53	81.41	81.66	81.52	81.93
1.5	81.59	81.45	81.60	81.61	81.33	81.66	81.60	81.60
1.8	81.50	81.45	81.52	81.53	81.41	81.50	81.69	81.77
2.1	81.50	81.37	81.52	81.61	81.33	81.66	81.60	81.60
2.4	81.42	81.37	81.52	81.37	81.58	81.58	81.69	81.77
2.7	81.67	81.37	81.44	81.45	81.58	81.74	81.60	81.60
3.0	81.59	81.29	81.44	81.61	81.58	81.58	81.44	81.68
3.3	81.67	81.29	81.52	81.69	81.66	81.74	81.60	81.77

Table 7.14 (Cont.)

Time	Temperature (°F) for Numbered Thermocouples							
(sec)	TC 1	TC 2	TC 3	TC 4	TC 5	TC 6	TC 7	TC 8
3.6	81.59	80.96	81.36	81.45	82.96	83.21	83.24	83.64
3.9	81.67	80.96	81.44	81.69	84.03	84.03	84.22	84.78
4.2	81.59	80.80	81.28	81.53	85.01	84.68	85.11	85.19
4.5	81.59	80.80	81.28	81.45	85.50	85.58	85.77	85.85
4.8	81.42	80.96	81.36	81.37	85.99	85.83	86.34	86.42
5.1	81.59	80.88	81.28	81.45	86.31	86.48	86.75	86.91
5.4	81.67	80.96	81.36	81.61	86.80	86.81	87.16	87.07
5.7	81.67	81.04	81.36	81.53	87.29	87.30	87.56	87.56
6.0	81.75	81.04	81.36	81.45	87.62	87.62	87.97	87.89
6.3	81.83	81.21	81.44	81.61	88.03	87.87	88.14	88.37
6.6	81.91	81.21	81.60	81.53	88.36	88.27	88.54	88.62
6.9	81.83	81.53	81.60	81.61	88.60	88.68	88.87	88.70
7.2	81.99	81.61	81.85	81.77	89.01	88.93	89.36	89.19
7.5	81.91	81.37	81.77	82.02	89.25	89.01	89.60	89.35
7.8	82.16	81.78	81.93	82.02	89.50	89.25	89.77	89.60
8.1	82.24	82.02	82.18	82.10	89.91	89.74	89.77	89.84
8.4	82.40	82.02	82.09	82.26	90.07	89.91	90.18	90.09
8.7	82.48	82.43	82.34	82.35	90.32	90.07	90.50	90.25
9.0	82.65	82.68	82.42	82.35	90.56	90.23	90.83	90.33
9.3	82.73	82.68	82.75	82.59	90.97	90.56	91.07	90.58
9.6	82.81	82.43	82.58	82.59	91.21	90.64	91.16	90.99
9.9	82.97	83.17	82.91	82.92	91.46	90.89	91.40	90.74
10.2	83.14	83.25	83.07	82.92	91.62	91.29	91.65	90.99
10.5	83.22	83.57	83.40	83.08	91.87	91.29	91.81	91.23
10.8	83.46	83.74	83.48	83.33	92.03	91.46	92.22	91.39
11.1	83.71	83.82	83.56	83.24	92.28	91.70	92.38	91.72
11.4	83.71	84.31	83.73	83.57	92.69	92.03	92.46	91.72
11.7	83.87	84.39	83.97	83.57	92.77	92.19	92.71	91.96
12.0	84.03	84.72	84.22	83.73	92.93	92.27	92.87	92.21
12.3	84.20	84.96	84.38	83.90	93.26	92.44	93.28	92.13
12.6	84.36	84.88	84.38	84.22	93.34	92.60	93.52	92.29
12.9	84.52	85.21	84.54	84.22	93.50	92.93	93.52	92.62
13.2	84.52	84.96	84.79	84.63	93.67	93.25	93.77	92.78
13.5	84.85	85.05	84.71	84.55	93.75	93.17	93.93	92.94
13.8	85.10	85.05	84.87	84.63	94.07	93.58	94.18	92.94
14.1	85.10	84.96	85.11	84.71	93.99	93.58	94.09	93.27
14.4	85.42	85.05	85.28	84.96	94.24	93.91	94.42	93.52
14.7	85.50	85.29	85.20	85.20	94.40	93.91	94.42	93.84
15.0	85.75	85.37	85.28	85.04	94.56	93.91	94.75	94.17
15.3	85.83	85.21	85.52	85.37	94.73	94.48	94.99	94.25
15.6	85.75	85.62	85.85	85.28	94.97	94.48	95.07	94.41
15.9	85.91	85.78	85.93	85.77	95.22	94.64	95.07	94.82
16.2	86.16	86.03	85.85	85.94	95.30	94.56	95.40	94.82
16.5	86.32	86.03	86.26	86.02	95.38	95.05	95.73	95.07

Table 7.14 (Cont.)

Time (sec)	Temperature (°F) for Numbered Thermocouples							
	TC 1	TC 2	TC 3	TC 4	TC 5	TC 6	TC 7	TC 8
16.8	86.57	86.11	86.34	86.10	95.71	95.13	95.89	95.23
17.1	86.73	86.35	86.34	86.35	95.71	95.29	95.89	95.31
17.4	87.06	86.43	86.58	86.51	96.03	95.54	95.97	95.47
17.7	87.14	86.52	86.83	86.67	96.28	95.54	96.14	95.72
18.0	87.22	86.52	86.91	86.83	96.52	95.70	96.30	95.96
18.3	87.46	86.92	86.99	87.16	96.44	96.03	96.54	96.04
18.6	87.63	87.01	87.32	87.24	96.77	96.03	96.71	96.37
18.9	87.71	87.17	87.40	87.32	95.87	95.29	95.81	95.47
19.2	87.87	87.50	87.48	87.57	94.56	94.31	94.50	94.33
19.5	87.87	87.66	87.73	87.49	93.99	93.74	93.85	93.52
19.8	88.12	87.66	88.05	87.90	93.67	93.25	93.36	93.19
20.1	88.52	88.07	88.14	88.06	93.18	92.76	93.12	93.03
20.4	88.61	88.07	88.22	88.06	92.93	92.52	92.79	92.45
20.7	88.77	88.23	88.54	88.55	92.52	92.19	92.38	92.29
21.0	88.77	88.56	88.54	88.30	92.36	91.95	92.14	92.05
21.3	88.93	89.31	89.79	89.55	92.11	91.70	92.05	91.88
21.6	89.18	88.39	88.79	88.63	91.79	91.62	91.89	91.80
21.9	89.10	88.72	88.87	88.55	91.79	91.46	91.65	91.72
22.2	89.26	88.64	89.11	88.96	91.54	91.38	91.40	91.31
22.5	89.42	88.80	89.20	89.04	91.62	91.05	91.40	91.39
22.8	89.59	88.88	89.44	89.12	91.46	90.89	91.24	91.15
23.1	89.75	88.88	89.44	89.37	91.38	90.97	91.16	91.07
23.4	89.91	89.13	89.69	89.37	91.13	90.81	91.07	90.74
23.7	89.83	89.21	89.44	89.37	91.13	90.72	91.16	90.99
24.0	89.91	89.29	89.52	89.37	91.13	90.89	90.91	90.90
24.3	89.91	89.29	89.60	89.53	91.13	90.81	90.67	90.74
24.6	90.08	89.37	89.77	89.53	90.89	90.72	90.75	90.74
24.9	89.91	89.37	89.69	89.45	90.97	90.56	90.75	90.66
25.2	90.16	89.37	89.77	89.61	90.64	90.56	90.75	90.74
25.5	90.16	89.54	89.79	89.61	90.81	90.48	90.67	90.82
25.8	90.08	89.54	89.85	89.77	90.72	90.48	90.67	90.58
26.1	90.16	89.62	90.01	89.77	90.89	90.40	90.58	90.58
26.4	90.24	89.70	89.85	89.85	90.56	90.32	90.58	90.66
26.7	90.16	89.54	89.77	89.85	90.64	90.48	90.58	90.41
27.0	90.16	89.62	89.93	89.85	90.64	90.32	90.50	90.41
27.3	90.32	89.62	90.18	89.94	90.40	90.56	90.42	90.50
27.6	90.24	89.62	89.85	89.85	90.56	90.40	90.42	90.17
27.9	90.40	89.62	90.01	89.85	90.56	90.23	90.50	90.33
28.2	90.24	89.62	90.01	89.69	90.56	90.32	90.58	90.50
28.5	90.40	89.70	90.09	89.94	90.48	90.15	90.26	90.33
28.8	90.40	89.70	90.09	89.77	90.40	90.15	90.26	90.50
29.1	90.24	89.62	90.01	89.85	90.40	90.23	90.34	90.33
29.4	90.48	89.78	90.01	89.85	90.56	90.23	90.26	90.25
29.7	90.40	89.78	90.01	89.94	90.48	90.40	90.34	90.25
30.0	90.48	89.62	89.93	89.77	90.48	90.32	90.34	90.50

Owing to the statistical design of the experiment there are two types of replicated measurements. First, until the heating starts at 3.3 sec, the temperatures are the same (within experimental error) for all times and thermocouples. Second, there are four sensors at both the heated and insulated surfaces. The standard deviations of the temperatures until 3.0 sec for any given sensor is about 0.1F° (0.06 K); the standard deviation of the average temperatures of the eight sensors is also about 0.1°F (0.06 K). During heating and at the heated surface, the standard deviations of temperature at fixed times are almost 0.4F° (0.22 K). These values of the standard deviation provided estimates of the pure error.

Even though the temperature calibration for the data shown in Table 7.14 was carefully performed, there are slight differences in the initial temperatures readings given by the various sensors. In an attempt to reduce bias, corrections of −0.06, +0.18, +0.04, +0.03, +0.09, −0.09, −0.06, and −0.17F° were respectively added to measurements given by sensors 1 through 8.

Another consideration in the design of the experiment was to achieve a relatively large signal-to-noise ratio while still not having such a large variation in temperature that the parameters would change. A maximum temperature rise about 15F° (8.33 K) satisfies both conditions; using the pure error standard deviation value of 0.1F° (0.06 K), the maximum signal-to-noise ratio is 150.

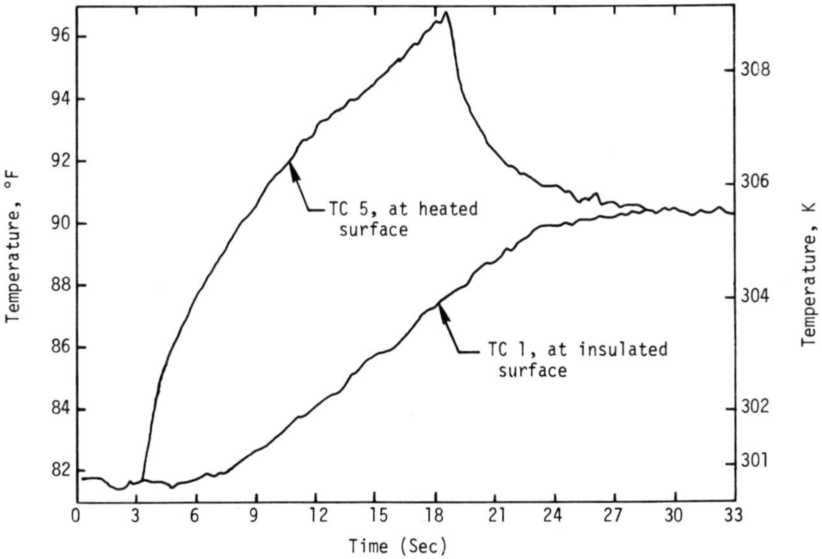

Figure 7.17 Typical temperature histories for the finite heat-conducting specimens.

7.9.4.2 *Physical Model of Heat-Conducting Body*

The temperature in each specimen can be mathematically described by

$$k\frac{\partial T}{\partial x^2}=c\frac{\partial T}{\partial t} \tag{7.9.6}$$

$$-k\frac{\partial T(0,t)}{\partial x}=q(t), \qquad \frac{\partial T(L,t)}{\partial x}=0 \tag{7.9.7}$$

$$T(x,0)=81.54°\text{F} \tag{7.9.8}$$

where k is thermal conductivity and c is the density–specific heat product. In both specimens the coordinate x is measured from the heated surface. The specimen thickness L was $1/12$ f (0.0254 m). The parameters are k and c. The heat flux $q(t)$ was zero until $t=3.3$ sec and was zero after $t=18.6$ sec; the value of $q(t)$ in the interval $3.3 \leqslant t \leqslant 18.6$ sec was 2.67 Btu/ft^2-sec (30,300 W/m^2). One can derive the exact solution of this problem in terms of infinite series. However, PROGRAM PROPERTY (developed at MSU) which utilizes finite differences in the solution of (7.9.6–7.9.8) was used to obtain the parameters.

The sensitivity coefficients shown in Fig. 7.18 were also obtained by using finite differences as indicated by (7.10.1). Note that those for c are

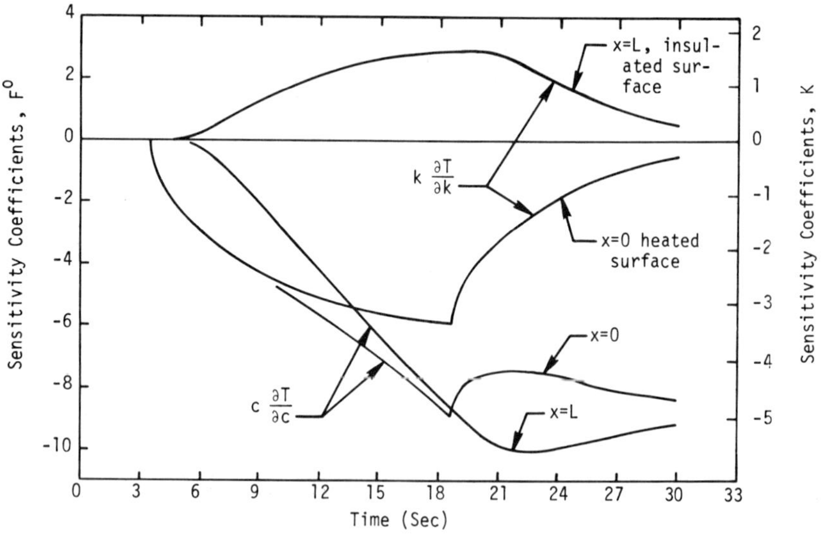

Figure 7.18 Sensitivity coefficients for the finite heat-conducting specimens.

always negative whereas those for k are both positive and negative. Hence the sensitivities are not linearly dependent and no difficulty would be anticipated in estimating the parameters.

7.9.4.3 Parameter Estimates

For the data given, the ordinary least squares parameter estimates are $k=43.343$ Btu/hr-ft-F (75.014 W/m-K) and $c=55.596$ Btu/ft^3-F (3728.6 kJ/m^3-C). If the data for only part of the interval had been used, the k and c values would have been modified by the additive changes Δk and Δc given in Fig. 7.19. Note that the changes in k in the last third of the time interval is less than 1% of the final k value; the c value changes about 2% in the same interval. These small values suggest that the model is correct. For times less than 7 sec the corrections become very large which would be expected since the sensitivity coefficients shown in Fig. 7.18 are approximately linearly dependent for such small times.

The residuals for all eight temperature histories are graphed in Figs. 7.20 and 7.21. Evidently, the measurement errors cannot be considered to be independent. From reference 28, p. 97, the mean number of runs (changes of sign plus 1) of the residuals is about $n/2$ for independent observations.

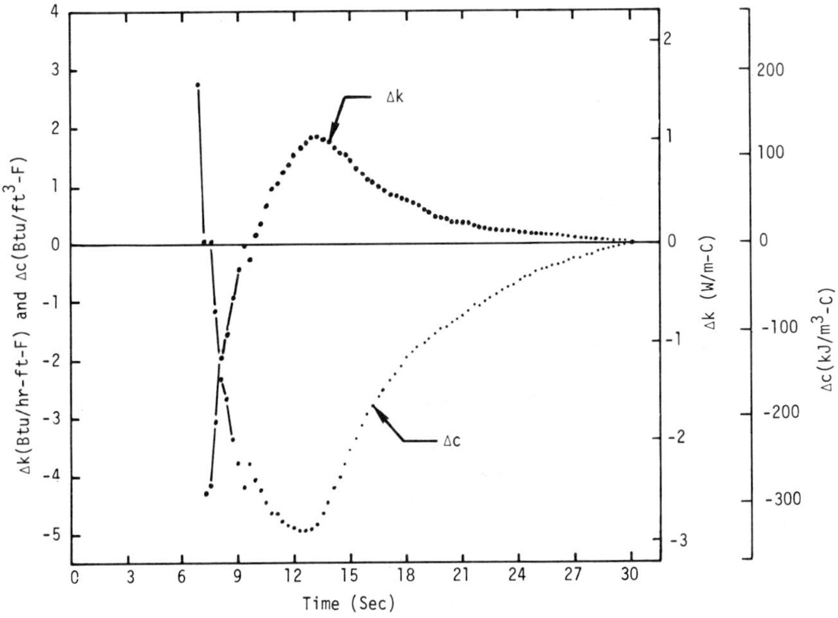

Figure 7.19 Sequential corrections to the parameters for finite heat-conducting example.

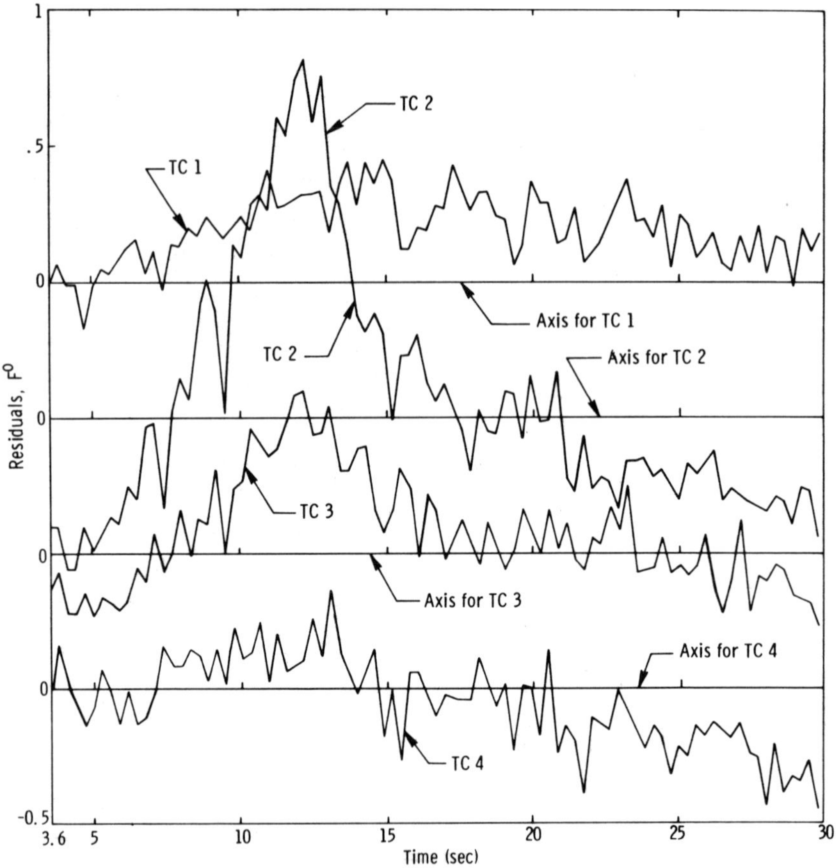

Figure 7.20 Residuals for thermocouples 1–4, which are located at the insulated surface of the specimen.

There are 89 observations for each thermocouple and the number of runs varies from 4 to 23 with the average being 12.75. This value is considerably less than $89/2=44.5$. For first-order cumulative errors, the expected number of runs is about $n^{1/2}$, which is 9.4 in this case. Hence the measurement errors would be more appropriately modeled as being cumulative rather than as being independent. (An analysis for ρ similar to that in Section 6.9.5 modified for nonlinear parameters gives $\rho \approx 0.87$.)
If the usual equation for estimating s^2 is used, we have

$$s^2 = \frac{1}{nm-1} \sum_{i=1}^{n} \sum_{j=1}^{m} \left[Y_j(i) - \hat{T}_j(i) \right]^2$$

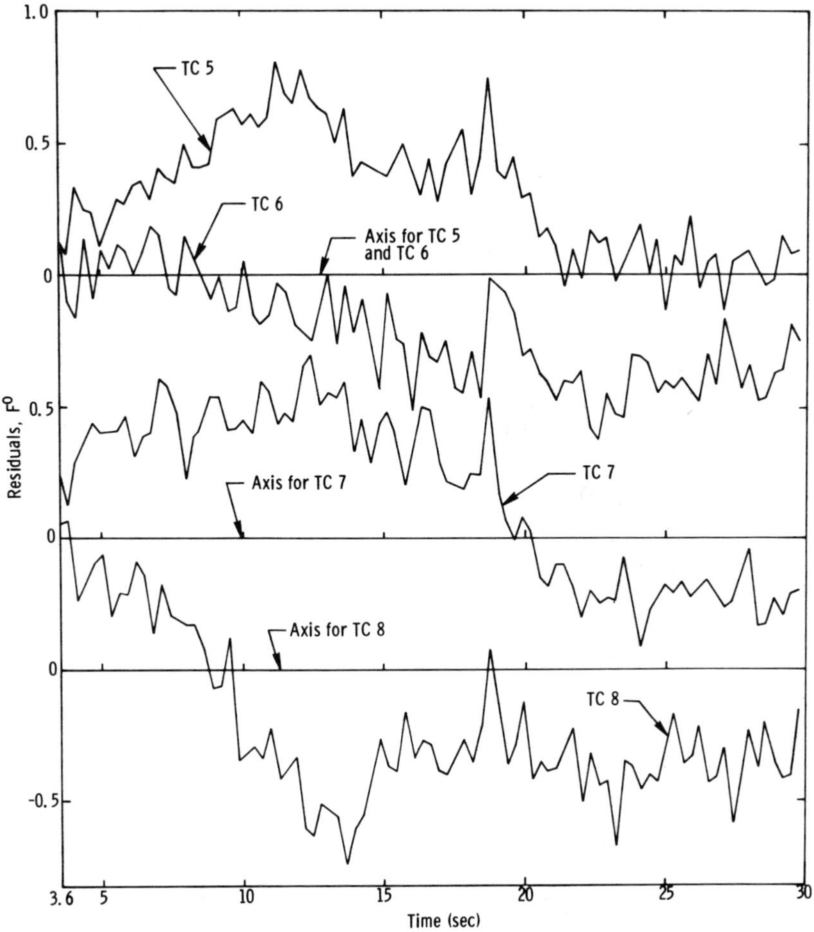

Figure 7.21 Residuals for thermocouples 5–8, which are located at the heated surface of the specimen.

where $n=89$ and $m=8$. The resulting s value is 0.339, which is different than the pure error estimates which were near 0.1 partly because the residuals are highly correlated.

In finishing this analysis we should calculate the confidence region for k and c. But since the confidence region is discussed above for other cases, it is not given. Note, however, that various basic assumptions should be checked first. The standard assumptions of additive, zero mean, constant variance, normal measurement errors seem reasonable. Also, the time and location measurement errors are probably negligible and no prior information is used. The assumption of independent errors is not appropriate.

These assumptions are designated 11101011. If a first-order autoregressive model is used, ρ is found to be about 0.95. Using this ρ, the covariance matrix of the errors ψ is approximated by $\mathbf{D}\phi\mathbf{D}^T$ as indicated in Appendix 6A. For this case in which OLS is used, an approximate confidence interval could be constructed using (7.7.9) and (7.7.10a) and as discussed following these equations.

A final comment is made regarding the estimation procedure. Data from all eight thermocouples were used and eight sets of residuals were found. Because the measurements at $x=0$ and $x=L$ are both repeated four times, the same parameter values would be found using the average values at each time at these two locations. In other words, instead of having $m=8$, we would have $m=2$. Justification for this simplification is given in Section 5.5.

7.10 SENSITIVITY COEFFICIENTS

There are several methods of determining the sensitivity coefficients. One can evaluate them by generating sensitivity equations by differentiation of the model equations. One can also use finite differences which are simpler in terms of computer programming and are usually adequate, although there are cases when direct solution of the sensitivity equations is needed. This section discusses both methods. Even though one may evaluate the sensitivity coefficients utilizing finite differences, the study of the sensitivity equations can yield insights into the sensitivity coefficients themselves.

7.10.1 Finite Difference Method

One finite difference approximation for the sensitivity for the ith observation, jth parameter, and lth dependent variable is the forward difference approximation,

$$\frac{\partial \eta_l(i)}{\partial \beta_j} = X_{lj}(i) \approx \frac{\eta_l(b_1,\ldots,b_j+\delta b_j,\ldots,b_p)-\eta_l(b_1,\ldots,b_j,\ldots,b_p)}{\delta b_j} \tag{7.10.1}$$

where δb_j is some relatively small quantity. One possible value of δb_j is given by

$$\delta b_j = 0.0001\, b_j \tag{7.10.2}$$

This simple procedure is frequently quite satisfactory. Note, however, that if the initial value of b_j is chosen to be zero or if b_j approaches zero (7.10.2) can lead to difficulty.

Bard [5, p. 118] gives a brief discussion leading him to recommend the value of

$$|\delta b_j| = \left[2\sqrt{2}\, \varepsilon \left| \frac{S}{C_{jj}} \right| \right]^{1/2} \tag{7.10.3}$$

where ε is the relative error in the computed values of S and where C_{jj} is defined by (7.4.13). He also recommends that lower and upper limits be placed on the values given by (7.10.3) such as $10^{-5}|b_j| \leqslant \delta b_j \leqslant 10^{-2}|b_j|$. Two sources of error contribute to inaccuracies in approximating $X_{lj}(i)$: (1) the rounding error when two closely spaced values of η are subtracted, and (2) the truncation error due to the inexact nature of (7.10.1). In order to reduce the truncation error the constant 0.0001 in (7.10.2) is made small. If, however, it is too small then the rounding error becomes important. A more accurate finite difference approximation is the central difference scheme,

$$X_{lj}(i) \approx \frac{\eta_l(b_1,\ldots,b_j+\delta b_j,\ldots,b_p) - \eta_l(b_1,\ldots,b_j-\delta b_j,\ldots,b_p)}{2\delta b_j} \tag{7.10.4}$$

Unfortunately this approximation requires almost twice as many values of η_l to be calculated as does (7.10.1). For this reason the use of (7.10.4) is not recommended.

If there is uncertainty due to the use of (7.10.1) one could repeat the final iteration by replacing (7.10.1) by a backward difference approximation (replacing δb_j by $-\delta b_j$) and then averaging the parameter estimates from the forward and backward difference approximations.

7.10.2 Sensitivity Equation Method

When the model is a set of first-order ordinary differential equations or a partial differential equation, the solution frequently cannot be written explicitly. In such cases explicit expressions for the sensitivities cannot be given. One can, however, derive sensitivity equations which can be solved separately from the model.

Consider, for example, the model

$$k\frac{\partial^2 T}{\partial x^2} = c\frac{\partial T}{\partial t} \tag{7.10.5}$$

where k and c are parameters. The boundary and initial conditions are

$$q = -k\left.\frac{\partial T}{\partial x}\right|_{x=0}, \qquad \left.\frac{\partial T}{\partial x}\right|_{x=L} = 0, \qquad T(x,0) = T_i \tag{7.10.6}$$

In this case T represents the dependent variable η. Differentiating (7.10.5) and (7.10.6) with respect to k yields

$$k\frac{\partial^2 X_1}{\partial x^2}+\frac{\partial^2 T}{\partial x^2}=c\frac{\partial X_1}{\partial t} \tag{7.10.7a}$$

$$0=k\frac{\partial X_1}{\partial x}\bigg|_{x=0}+\frac{\partial T}{\partial x}\bigg|_{x=0}, \qquad \frac{\partial X_1}{\partial x}\bigg|_{X=L}=0, \qquad X_1(x,0)=0 \tag{7.10.7b}$$

where $X_1\equiv\partial T/\partial k$. Notice that (7.10.7a) contains a nonhomogeneous term which is contained in (7.10.5). Solution of (7.10.7) provides a solution to the sensitivity coefficient for k. Unlike the use of differences such as in (7.10.1) which utilizes the program calculating η, a special computer program must be written for calculating X_1 using (7.10.7); in many cases this would involve a finite difference computer program. The sensitivity equation for c is found by differentiating (7.10.5,6) with respect to c,

$$k\frac{\partial^2 X_2}{\partial x^2}=c\frac{\partial X_2}{\partial t}+\frac{\partial T}{\partial t} \tag{7.10.8a}$$

$$\frac{\partial X_2}{\partial x}\bigg|_{x=0}=0, \qquad \frac{\partial X_2}{\partial x}\bigg|_{x=L}=0, \qquad X_2(x,0)=0 \tag{7.10.8b}$$

where $X_2\equiv\partial T/\partial c$.

From a knowledge of the solutions of the above equations one can predict several results. First, the basic problem (7.10.5) and (7.10.6) is linear and so are the sensitivity problems for X_1 and X_2. Second, X_1 is a function of k and X_2 is a function of c. Hence T is a nonlinear function of k and c. Third, in general X_1 and X_2 will not be zero at the boundaries $x=0$ and L. Next, in (7.10.8), the only nonhomogeneous term is $\partial T/\partial t$ which, for monotonically increasing time values of T, will be positive; such a term acts like a heat sink which will cause X_2 to be always negative (provided T is monotonically increasing). Fifth, the $\partial^2 T/\partial x^2$ in (7.10.7a) which is equal to $(k/c)\,\partial T/\partial t$ simulates a heat *source* but the $\partial T/\partial x$ at $x=0$ term in (7.10.7b) simulates a heat *sink* at $x=0$. Hence X_1 could be either positive or negative and would in general be smaller in magnitude than X_2.

Another point relates to the relationships between T, T_i, X_1, and X_2. By multiplying (7.10.7) by k and by multiplying (7.10.8) by c and then adding the corresponding equations together, one can show that

$$T(x,t)-T_i=-kX_1(x,t)-cX_2(x,t) \tag{7.10.9}$$

Insight regarding the sensitivity coefficients can be obtained from this equation. For example, the larger $T(x,t)-T_i$, the larger on the average are the magnitudes of X_1 and X_2. (7.10.9) also suggests that both sensitivity coefficients need not be calculated because T must always be calculated.

A final point is that in this case the parameters can be transformed so that T is linear in one of them. This is true for the parameters $\alpha = k/c$ and $K = k^{-1}$; with these parameters the model (7.10.5) and (7.10.6) becomes

$$\alpha \frac{\partial^2 T}{\partial x^2} = \frac{\partial T}{\partial t} \tag{7.10.10a}$$

$$qK = -\left.\frac{\partial T}{\partial x}\right|_{x=0}, \qquad \left.\frac{\partial T}{\partial x}\right|_{x=L} = 0, \qquad T(x,0) = T_i \tag{7.10.10b}$$

The sensitivity coefficient for K, denoted X_3, is found from a solution of

$$\alpha \frac{\partial^2 X_3}{\partial x^2} = \frac{\partial X_3}{\partial t} \tag{7.10.11a}$$

$$q = -\left.\frac{\partial X_3}{\partial x}\right|_{x=0}, \qquad \left.\frac{\partial X_3}{\partial x}\right|_{x=L} = 0, \qquad X_3(x,0) = 0 \tag{7.10.11b}$$

Notice that X_3 is independent of K and hence is linear in K. By multiplying (7.10.11) by K and then comparing with (7.10.10), one can find

$$T(x,t) - T_i = KX_3(x,t) \tag{7.10.12}$$

and thus X_3 (x,t) can be very simply found from $T(x,t)$ using this equation.

In summary, we have given two basic methods for evaluating sensitivity coefficients when the model is not an algebraic form but rather involves solution of differential equations. It is assumed that these equations cannot be readily solved to obtain closed forms since otherwise algebraic forms can be written. We imagine that these equations are solved numerically using finite differences or some other method. The first method of evaluating the sensitivities involves only a computer program to approximate the differential equations for the model; (7.10.1), which requires only dependent variable values, is used. In general this is the simplest and recommended method. The second method involves numerically approximating the sensitivity equations which derived from the model. Examples of sensitivity equations are given by (7.10.7) and (7.10.8). Even though the sensitivity equations are not solved, inspection of these equations can sometimes provide considerable insight and/or relations between the sensitivity coefficients.

REFERENCES

1. Bard, Y., "Comparison of Gradient Methods for the Solution of Nonlinear Parameter Estimation Problems," *SIAM J. Numer. Anal.* **7** (1970), 157–186.
2. Beveridge, G. S. G. and Schechter, R. S., *Optimization: Theory and Practice*, McGraw-Hill Book Company, New York, 1970.
3. Levenberg, K., "A Method for the Solution of Certain Non-linear Problems in Least Squares," *Quart. Appl. Math.* **2** (1944), 164–168.
4. Marquardt, D. W., "An Algorithm for Least Squares Estimation of Nonlinear Parameters," *J. Soc. Ind. Appl. Math.* **11** (1963), 431–441.
5. Bard, Y., *Nonlinear Parameter Estimation*, Academic Press, Inc., New York, 1974.
6. Box, G. E. P., "Use of Statistical Methods in the Elucidation of Physical Mechanisms," *Bull. Inst. Intern. Stat.*, **36** (1958), 215–255.
7. Booth, G. W. and Peterson, T. I., *Non-linear Estimation*, IBM Share Program Pa. No. 687 W1 NL1, 1958.
8. Hartley, H. O., "The Modified Gauss–Newton Method for the Fitting of Nonlinear Regression Functions by Least Squares," *Technometrics* **3** (1961), 269–280.
9. Box, G. E. P. and Kanemasu, H., "Topics in Model Building, Part II, On Non-linear Least Squares," Tech. Rep. No. 321, University of Wisconsin, Dept. of Statistics, Madison, Wis., Nov. 1972.
10. Davies, M. and Whitting, I. J., "A Modified Form of Levenberg's Correction," in *Numerical Methods for Non-Linear Optimization*, edited by F. A. Lootsma, Academic Press, London, 1972, pp. 191–201.
11. Gallant, A. R., "Nonlinear Regression," *Am. Stat.* **29** (1975), 73–81.
12. Seinfeld, J. H. and Lapidus, L., *Mathematical Models in Chemical Engineering, Vol. 3., Process Modeling, Estimation and Identification*, Prentice-Hall, Inc., Englewood Cliffs, N.J., 1974.
13. Graupe, D., *Identification of Systems*, Van Nostrand-Reinhold Company, New York, 1972.
14. Jones, A.; "SPIRAL—a New Algorithm for Nonlinear Parameter Estimation Using Least Squares," *Comput. J.* **13** (1970), 301–308.
15. Beale, E. M. L., "Confidence Regions in Nonlinear Estimation," *J. Roy. Stat. Soc.* **B22** (1960), 41–76.
16. Guttman, I. and Meeter, D. A., "On Beale's Measures of Nonlinearity," *Technometrics* **7** (1965), 623–637.
17. Bacon, D. W. and Henson, T. L., "Statistical Design and Model Building," Course notes, Dept. of Chemical Engineering, Queen's University, Kingston, Ontario.
18. Box, G. E. P., "Fitting Empirical Data," *Ann. N. Y. Acad. Sci.*, **86** (1960), 792.
19. Box, G. E. P., "Some Notes on Nonlinear Estimation," Tech. Rep. No. 25, University of Wisconsin, Dept. of Statistics, Madison, Wis., 1964.
20. Gallant, A. R., "The Power of the Likelihood Ratio Test of Location in Nonlinear Regression Models," *J. Am. Stat. Assoc.* **70** (1975), 198–203.
21. Box, G. E. P. and Hunter, J. S., "A Confidence Region for the Solution of a Set of Simultaneous Equations with an Application to Experimental Design," *Biometrika* **41** (1954), 190–199.

22. Box, G. E. P. and Hunter, W. G., "A Useful Method for Model-Building," *Technometrics* **4** (1962), 301–318.
23. Beck, J. V., "Determination of Optimum, Transient Experiments for Thermal Contact Conductance," *Int. J. Heat Mass Transfer* **12** (1969), 621–633.
24. Beck, J. V., "Transient Sensitivity Coefficients for the Thermal Contact Conductance," *Int. J. Heat Mass Transfer* **10** (1967), 1615–1617.
25. Van Fossen, G. J., Jr., "Design of Experiments for Measuring Heat-Transfer Coefficients with a Lumped-Parameter Calorimeter," *NASA Tech. Note NASA TN D-7857*, Jan. 1975.
26. Burington, R. S. and May, D. C., *Handbook of Probability and Statistics with Tables*, 2nd ed., McGraw-Hill Book Company, New York, 1970.
27. Farnia, K., "Computer-Assisted Experimental and Analytical Study of Time/Temperature-Dependent Thermal Properties of the Aluminum Alloy 2024-T351," Ph.D. Thesis, Dept. of Mechanical Engineering, Michigan State University, 1976.
28. Draper, N. R. and Smith, H., *Applied Regression Analysis*, John Wiley & Sons, Inc., New York, 1966.

PROBLEMS

7.1 Using OLS with the Gauss linearization method, estimate β in the model $\eta = 377/(1+\beta t)$ for the data

t_i	0.25	0.5	0.75	1	2	3
Y_i	150	90	70	55	30	20

(*a*) Start with $\beta = 6$ being the initial estimate.

Answer. 6.0648.

(*b*) Start with $\beta = 3$ being the initial estimate.
(*c*) Start with $\beta = 12$ being the initial estimate.

7.2 Using OLS with the Gauss linearization method, estimate β_1 and β_2 in the model $\eta = \beta_1/(1+\beta_2 t)$ for the data given in Problem 7.1.

(*a*) Start with $\beta_1 = 377$ and $\beta_2 = 5.8$ being the initial estimates.
(*b*) Start with $\beta_1 = 300$ and $\beta_2 = 4$ being the initial estimates.
(*c*) Start with $\beta_1 = 600$ and $\beta_2 = 8$ being the initial estimates.
(*d*) Start with $\beta_1 = 600$ and $\beta_2 = 4$ being the initial estimates.

7.3 Using OLS with the Gauss linearization method, estimate a_0 and a_1 for the model and data in Example 5.2.3.

7.4 For the model $T = 81.5 + 198.3e^{-\beta t}$, estimate using OLS and the Gauss method the parameter β for the data given in Table 6.2. Use as the data the Y_i values for observations 5, 9, 13, and 17.

Answer. 7.49611×10^{-4}.

7.5 Repeat Problem 7.1 using the Box–Kanemasu modification.

7.6 Repeat Problem 7.4 using the Box–Kanemasu modification.

7.7 Repeat Problem 7.1*a* using the sequential method of Section 7.8.3.1.

7.8 For the model $T=\beta_1+\beta_2 e^{-\beta_3 t}$, estimate the parameters β_1, β_2, and β_3 for the Y_i data given in Table 6.2. Use a computer program incorporating the sequential method. Use OLS.

Answer. 100.60, 178.83, 0.000882.

7.9 For the model given by

$$\mathrm{Nu}=\beta_1 \mathrm{Re}^{\beta_2+\beta_3 \log \mathrm{Re}}$$

estimate the parameters using a sequential computer program with OLS for the data given in Example 6.2.2. (Do *not* take logarithms of both sides.)

7.10 Repeat Problem 7.9 using as the sum of squares function,

$$S=\Sigma\left(1-\frac{\eta_i}{Y_i}\right)^2$$

7.11 Using the Box–Kanemasu approximation and the initial estimate of $M^{(0)}=10$, estimate M for the problem of Section 7.5.1. Use a programmable calculator or a computer.

7.12 Sometimes two functions are known and one or more parameters in one function is to be adjusted to cause "agreement." An example is $u \tanh u$, which is to be approximated by $u^2/(1+\beta u^2)$ for "small" values of u.

(*a*) By using Taylor series, show that "agreement" is obtained by letting $\beta=\frac{1}{3}$.

(*b*) Suggest a mathematical function which would be minimized to obtain agreement in this case.

7.13 Another set of functions as in Problem 7.12 is $u^2/[2(1+\beta u^2)]$ and $uI_1(u)/I_0(u)$ where $I_i(u)$ is a modified Bessel function. By using a Taylor series show that "agreement" is obtained by letting $\beta=\frac{1}{8}$.

7.14 The following measurements and associated standard deviations are for the model,

$$\eta_i=\beta_1 \exp(-\beta_2 t_i)$$

t_i	0	0.125	0.25	0.375	0.5
Y_i	101	66	44	28	20
σ_i	1	3	6	2	1

Use maximum likelihood estimation to estimate β_1 and β_2 assuming the standard assumptions indicated by 11011111 are valid. Let the initial values of β_1 and β_2 be 100 and 3, respectively. Use a sequential method of solution. Also find the covariance matrix of the estimates and the minimum sum of squares.

Answer. $b_1 = 100.922$, $b_2 = 3.2877$, $P_{11} = 0.9744$, $P_{12} = 0.0245$, $P_{22} = 0.007933$, $S_{\text{Min}} = 0.8502820$.

7.15 When there are large numbers of equally spaced measurements in the region $0 \leqslant t \leqslant 1$, the OLS sum of squares function is proportional to the integral,

$$S = \int_0^1 (Y - \eta)^2 \, dt$$

For the special case of $Y = t^6 + t^7$ and $\eta = \beta_1 t^6 + \beta_2 t^7$, show that

$$S = \frac{z_1^2}{13} + \frac{z_2^2}{15} + \frac{z_1 z_2}{7}$$

where $z_i = 1 - \beta_i$. Use the method given in Section 6.8 to find the major and minor axes of this curve. Plot for $S = 1$ and 10.

7.16 For the model $\eta = \beta_1 t + \exp(-\beta_2 t)$ and the data

t_i	0	1	2
Y_i	1	2	3

and using the OLS sum of squares function, show that for a fixed value of S the value of β_1 is related to β_2 and S by

$$\beta_1 = \left\{ 8 - e^{-\beta_2} - 2e^{-2\beta_2} \pm \left[5S - (1 - e^{-\beta_2})^4 \right]^{1/2} \right\} / 5$$

(*a*) Plot the contour of $S = 0.1$ in the β_1, β_2 plane.

(*b*) From the discriminant in the expression for β_1 show that the β_2 value can be as small as $-\ln[1 + (5S)^{1/4}]$ and can be as large as $-\ln[1 - (5S)^{1/4}]$ provided $S < 0.67$. If $S > 0.97$, what is the maximum value of β_2? What condition(s) must be given with respect to β_2 to insure that the contour for fixed S is closed?

(*c*) Derive an expression for the β_1 value at the end point(s) of an S contour. (At the end of the contour, the discriminant is equal to zero.) For $S = 1$, show that the end point is $\beta_1 = -1.3898$ and $\beta_2 = -0.9144$.

(*d*) Expand S for small values of β_2 to obtain $S \approx 5[1 - \beta_1 + \beta_2]^2$ and plot for $S = 0.1$. Compare with the result of part (*a*).

(*e*) Show that $|\mathbf{X}^T\mathbf{X}| = [2e^{-\beta_2}(1 - e^{-\beta_2})]^2$. Find the location of the β_2 values corresponding to the minimum and maximum values of $|\mathbf{X}^T\mathbf{X}|$. Plot $|\mathbf{X}^T\mathbf{X}|$ versus β_2 for $-1.5 < \beta_2 < 4$.

7.17 Using the data of the preceding problem and the initial values of $\beta_1 = 1.5$ and $\beta_2 = 1.5$, estimate β_1 and β_2 using the Box–Kanemasu modification of the Gauss method.

7.18 Show that the sensitivity coefficients for the model

$$\eta = \frac{1 + \beta_1 t}{\beta_2 + \beta_3 t}$$

are linearly dependent for a certain combination of the parameter values. [*Hint*: write the sensitivities in terms of $z=\beta_3 t/\beta_2$ and $R=\beta_1\beta_2/\beta_3$.] What can you conclude from this linear dependence? What new parameters could be selected to eliminate the linear dependence mentioned above?

7.19 Repeat Problem 6.26 for the model

$$T=\beta_1+\beta_2(t-\beta_3)^{1/2}$$

7.20 Using the data in Table 7.14 between 3.6 and 18.0 sec for temperature histories from thermocouple 1 (which is at $x=L$) and thermocouple 5 (which is at $x=0$) estimate k and α in the model given by (8.5.25) with $T_0=81.66$°F, $L=1/12$ f, and $q=2.67$ Btu/ft^2-sec. Let t in (8.5.25) be the times in Table 7.14 minus 3.3. In other words, 3.6 sec in Table 7.14 corresponds to time 0.3 sec in (8.5.25). Use as initial estimates $k=40$ Btu/hr-ft-°F and $\alpha=1$ ft^2/hr. (Be careful with units.) Use OLS with the Gauss or other method.

7.21 Repeat Problem 7.20 but use the average of temperatures 1, 2, 3, and 4 instead of 1 and the average of 5, 6, 7, and 8 instead of 5.

7.22 Derive (7.5.16) and (7.5.17).

7.23 Derive (7.8.18).

7.24 Verify the sensitivity coefficient values given in Fig. 7.8 using the approximate equation, (7.10.1). A programmable calculator or computer should be used. Investigate using values of $\delta b_j=\varepsilon b_j$ where ε is equal to (a) 0.01, (b) 0.001, and (c) 0.0001.

CHAPTER 8

DESIGN OF OPTIMAL EXPERIMENTS

8.1 INTRODUCTION

Carefully designed experiments can result in greatly increased accuracy of the estimates. This has been demonstrated by various authors, but special mention should be made of the work of G. E. P. Box and collaborators. See, for example, Box and Lucas [1] and Box and Hunter [2]. An important work on optimal experiments is the book by Fedorov [3].

In many areas of research, great flexibility is permitted in the proposed experiments. This is particularly true with the present ready accessibility of large-scale digital computers for analysis of the data and automatic digital data acquisition equipment for obtaining the data. This means that transient, complex experiments can be performed that involve numerous measurements for many sensors. With this great flexibility comes the opportunity of designing experiments to obtain the greatest precision of the required parameters. A common measure of the precision of an estimator is its variance; the smaller the variance, the greater the precision. Information regarding the variances and covariances is included in determination of confidence regions. We shall utilize the minimum confidence region to provide a basis for the design of experiments having minimum variance estimators.

The design of optimal experiments is complicated by the necessity of adding practical considerations and constraints. The best design for a particular case, for example, might have certain unique restrictions on the dependent variable vector $\boldsymbol{\eta}$ or on the independent variables such as time or position. The optimal design problem involves two parts: (1) the determination of an objective function together with its constraints and (2) the extremization of the objective function.

When we say that we desire to find the optimal experiment, we wish to determine the conditions under which each observation should be taken in order to extremize a certain optimal criterion. For example, the best duration of the experiment may be needed or the optimal placement of sensors may be required. In cases involving partial differential equations the optimal boundary and initial conditions may also be needed. Many of these cases are illustrated in subsequent sections.

In most of this chapter it is assumed that the form of the model is known although it contains unknown parameters. If a form in terms of a finite number of parameters is not known, the search for an optimal strategy may be quite different. This involves discrimination, which is discussed in Section 8.9.

8.2 ONE PARAMETER EXAMPLES

In order to illustrate optimal design, some one-parameter linear and nonlinear examples are given in this section. The standard conditions designated 11111-11 (see Section 6.1.5.2) are considered to be valid.

8.2.1 Linear Examples for One Parameter

8.2.1.1 Model $\eta_i = \beta X_i$ with No Constraints

Consider first the case of the linear model $\eta_i = \beta X_i$. Owing to the standard assumptions, ordinary least squares (OLS) and maximum likelihood (ML) yield the same estimator and variance of

$$b = \frac{\sum_{j=1}^{n} Y_j X_j}{\Delta}, \qquad V(b) = \frac{\sigma^2}{\Delta} \quad \text{where} \quad \Delta = \sum_{i=1}^{n} X_i^2 \tag{8.2.1}$$

Observe that minimization of the variance of b implies the maximization of Δ. Note that Δ is maximized by (a) making the maximum value of $|X|$ as large as possible, (b) concentrating all the n measurements at the maximum permissible value of $|X|$, and (c) making n as large as possible.

This case illustrates the necessity of certain practical constraints. The maximum value of $|X|$ must be finite if $|\eta|$ is to be finite. Next it must be decided what constraints, if any, are to be placed on the measurements in a fixed X range. If there are none, the optimal solution is to concentrate all the measurements at the maximum $|X|$. In other cases where X is a function of time, measurements at equal time intervals might be dictated by the capabilities of the measuring equipment. The latter case is emphasized in this text because of its common occurrence. Furthermore, equal spacing of measurements usually provides more information for checking the validity of the model and the statistical assumptions than does concentrating the measurements at the maximum $|X|$.

8.2.1.2 Model $\eta=\beta X(t)$ for Fixed Large n and Equally Spaced Measurements

The model $\eta=\beta X$ can represent cases where X is any known function of time. (The word "time" is used but the results could also apply for other variables such as position, temperature, etc.) For a large number of observations, Δ can be approximated by

$$\Delta=\sum_{i=1}^{n} X^2(t_i)=\frac{1}{\Delta t}\sum_{i=1}^{n} X^2(t_i)\Delta t\approx\frac{n}{t_n}\int_0^{t_n} X^2(t)\,dt \qquad (8.2.2)$$

where $t_n=n\Delta t$ and the measurements are assumed to be uniformly spaced in t over $0\leqslant t\leqslant t_n$.

Example 8.2.1

Compare the value of Δ associated with n measurements of $\eta=\beta Ct_n^m$ and for $\eta_i=\beta C\,(it_n/n)^m$ with $i=1,2,\ldots,n$, $t_i\geqslant t_j$ for $i>j$, and where m is a nonnegative exponent. Let n be large. C is an arbitrary constant which plays no role in this problem but is included for later use for scaling η. Notice that the first case has all the measurements concentrated at the location of the maximum η of the second case.

Solution

For the first case the sensitivity X is Ct_n^m and then Δ obtained from (8.2.1) is $nC^2t_n^{2m}$. For the second case, (8.2.2) yields

$$\Delta=\sum_{i=1}^{n} C^2\left(\frac{it_n}{n}\right)^{2m}\cong nt_n^{-1}\int_0^{t_n} C^2t^{2m}\,dt=\frac{nC^2t_n^{2m}}{2m+1}$$

In both cases Δ is proportional to $nC^2t_n^{2m}$ and thus is made larger by increasing n,

C^2, or t_n. The ratio of the first Δ to the second is $2m+1$. Hence for all models with $m>0$, β can be estimated more accurately by concentrating all the measurements at the maximum η; this becomes more apparent as m is increased in value. For dynamic experiments, however, measurements uniformly spaced in time are usually more appropriate than concentrated ones.

In this section the constraint of a fixed but large number of observations, n, is investigated. In any practical experiment, n must be finite. No constraint is placed on the magnitude of $X(t)$ or, equivalently, on η. For the case of n equally spaced measurements starting at $t=0$ and ending at t_n, the criterion for optimum measurements for the linear model $\eta=\beta X(t)$ is to maximize

$$\Delta^n = \frac{\Delta}{n} = t_n^{-1} \int_0^{t_n} [X(t)]^2\, dt \tag{8.2.3}$$

with respect to t_n. It is assumed that n is large and the standard assumptions 11111-11 are valid. Notice that Δ^n is a function of t_n but not n; t_n is now simply the maximum t.

A necessary condition for Δ^n to be a maximum with respect to the duration of the experiment t_n is that

$$\frac{\partial \Delta^n}{\partial t_n} = 0 = \left(-t_n^{-2} \int_0^{t_n} [X(t)]^2\, dt + \frac{1}{t_n} [X(t_n)]^2 \right)\bigg|_\tau \tag{8.2.4}$$

where τ is the time t_n that maximizes Δ^n; (8.2.4) can also be written as

$$[X(\tau)]^2 = \tau^{-1} \int_0^{\tau} (X)^2\, dt = \Delta^n(\tau); \qquad \tau \neq 0 \tag{8.2.5}$$

This expression is interesting because it provides insights into conditions that permit an optimum time to exist. In words, (8.2.5) states that at the optimum time, the square of the sensitivity must equal the average value of the squared sensitivity.

Example 8.2.2

The velocity distribution for laminar flow between parallel plates separated by the distance H is

$$u = u_0 \frac{y}{H}\left(1 - \frac{y}{H}\right)$$

where u_0 is the maximum velocity and y is the distance measured from one wall. Suppose that value u_0 is the parameter of interest and that u is to be measured at equal intervals from $y=0$ to where Δ^n is maximized. Find the y value to maximize

Δ^n. This would provide an optimal experiment for this case provided the standard assumptions are valid relative to errors in u and the y measurements are errorless.

Solution

The optimal distance y can be found by using (8.2.5) with t being y/H and $X(t)=t(1-t)$; the optimal value of $\tau=y/H$ is then found from

$$\tau^2(1-\tau)^2=\tau^{-1}\int_0^{\tau}t^2(1-t)^2\,dt=(\tau^2/3)-0.5\tau^3+0.2\tau^4$$

which simplifies to the algebraic equation $24\tau^2-45\tau+20=0$; this is a simple quadratic equation which can be solved for $\tau=0.724$ and 1.15 but only the first is physically possible. Hence the optimal maximum y is $0.724H$.

Now (8.2.5) is a necessary, but not sufficient condition. It is also true at relative maxima and minima as well as the true maximum. For the maxima there are a number of possibilities with respect to (8.2.5). First, it might not be satisfied at any finite time, thus indicating that there is no maximum at a finite time. Next, it might be satisfied at all time τ, indicating that the maximum is attained at all times $t \geqslant 0$. Also (8.2.5) could be satisfied at one as well as many values of τ. Each of these cases is illustrated below.

Some general observations can be drawn from (8.2.5). Visualize a function $|X(t)|$ that is zero at $t=0$, increases monotonically with t until some time $t_{\max}$, and decreases monotonically to zero. Such an $X(t)$ is shown in Fig. 8.1. Here

$$X(t)=t\exp(1-t) \tag{8.2.6}$$

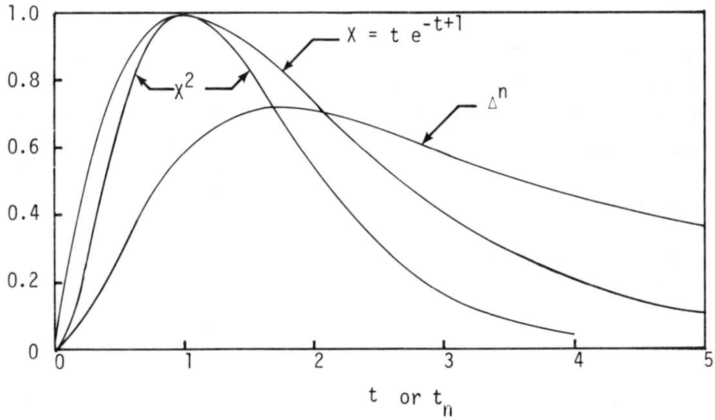

Figure 8.1 Sensitivity coefficient, X^2, and Δ^n for $\eta=\beta t\exp(1-t)$. The only constraint is that of fixed large n.

As long as $|X|$ is increasing, the instantaneous X^2 must be larger than the average X^2 and consequently a maximum in Δ^n cannot occur when $|X|$ is monotonically increasing. After the maximum X^2, the instantaneous value of X^2 decreases, but the average continues to rise for a while; for any $|X|$ function reaching a maximum monotonically and then decreasing, the maximum Δ^n must be at some time greater than the time at which $|X|$ has a maximum.

Consider now the $X(t)$ function given by (8.2.6) which is shown in Fig. 8.1 along with Δ^n and X^2. The maximum of Δ^n is at $\tau=1.691817$ (see Problem 8.1). Notice that the X^2 crosses Δ^n at its maximum as indicated by (8.2.5).

Condition (8.2.5) can be readily used for other cases. Several special cases are now considered; see Figs. 8.2*a* and 8.2*b*. Case 1 is for a constant X and thus the average of $(X)^2$ is equal to $(X)^2$ at all times; hence all times can represent optimum conditions. Case 2 is for X an exponential which increases asymptotically to unity. The maximum Δ^n occurs only at infinite t, but a maximum is closely approximated in a finite time. Case 3 is for a decaying exponential which has a maximum at $t=0$. Case 4 has a monotonically increasing X and as a consequence a monotonically increasing Δ^n. Further cases are shown in Figs. 8.3*a* and 8.3*b*. Case 5 is a cosine which has a maximum Δ^n at $t=0$. Case 6 is the sine function which has a maximum Δ^n at $t=2.2467$. Both these sinusoidal cases have numerous maxima or minima, but only one global maximum. The final case, case 7, has its function X depicted in Fig. 8.3*a*, and its Δ^n shown in Fig. 8.3*b*; it is first positive, becomes negative and asymptotically approaches -0.5. This case has a local maximum of Δ^n near $t=0.4$, but the true maximum occurs at infinity.

Example 8.2.3

A point on a rotating wheel is observed normal to its axis and is seen to move a distance s with respect to the axis. The known model is $s=\beta \sin \omega t$ where ω is known angular velocity. The measurements of t can be assumed to be errorless, but those of s satisfy the standard conditions. A large number of uniformly spaced measurements are to be taken starting at $t=0$. For an optimal test to estimate β, what should be the duration of the test?

Solution

The conditions for Fig. 8.3*b* are satisfied. The Δ^n for this case is Δ_6^n which has a global maximum at $\omega t=2.2467$ radians. Hence the duration of the test should be $t_n=2.2467/\omega$.

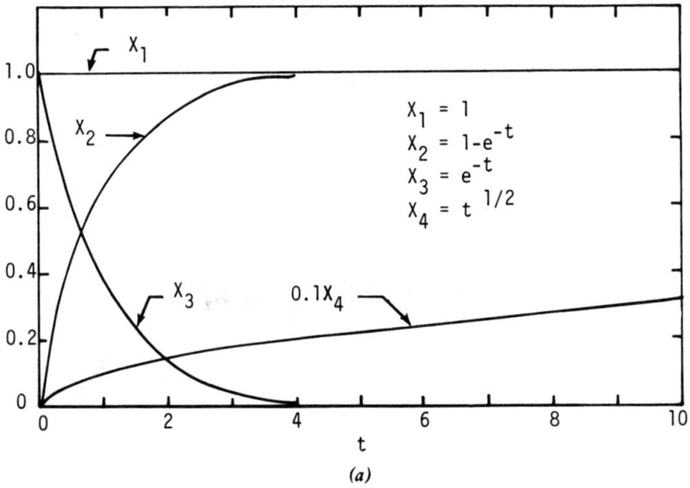

Figure 8.2a Sensitivity coefficients for $X_1 = 1$, $X_2 = 1 - e^{-t}$, $X_3 = e^{-t}$, and $X_4 = t^{1/2}$.

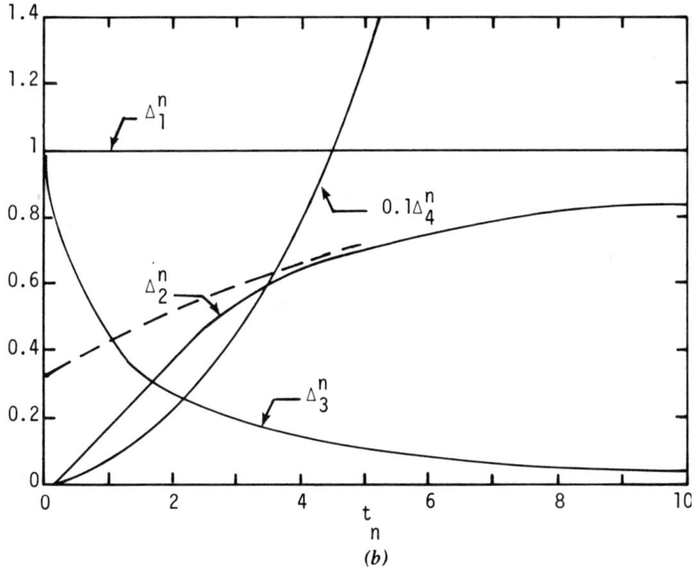

Figure 8.2b Δ^n criterion for the sensitivities given in Fig. 8.2a. The only constraint is that of a fixed large n.

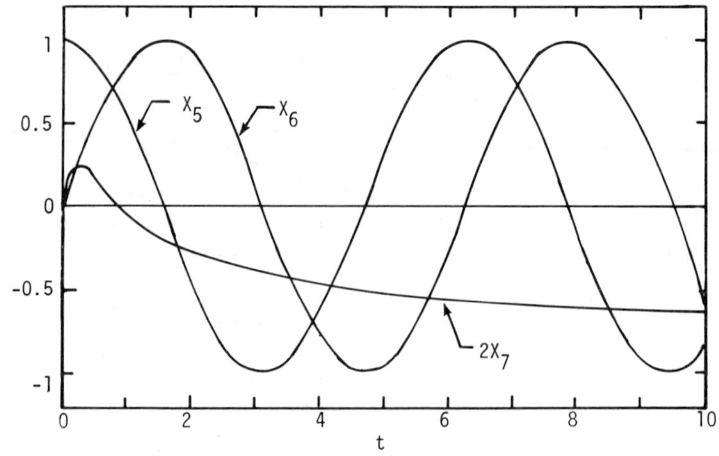

(a)

Figure 8.3a Sensitivity curves for $\cos t$, $\sin t$, and another function defined in Fig. 8.3b.

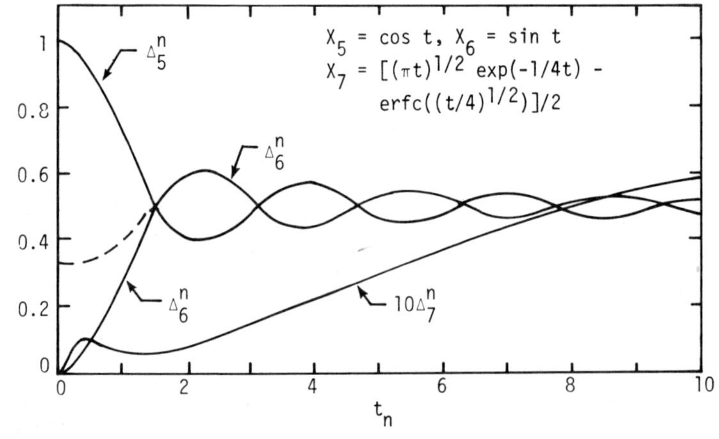

(b)

Figure 8.3b Δ^n criterion for the sensitivities given in Fig. 8.3a. The only constraint is that of a fixed large n.

8.2.1.3 Model $\eta = \beta X(t)$ for Fixed Large n and Fixed Maximum Value of $|\eta|$

For some models of η, there are upper bounds of $|\eta|$ implicit in the model. Examples are the η models of $\beta C \sin t$, $\beta C \cos t$, $\beta C e^{-2t}$, and $\beta C \tanh 3t$. In each of these cases the largest possible $|\eta|$ is $|\beta C|$. In other cases, such as for $\eta = \beta C t$, there are no implicit limits on $|\eta|$ if the t range is not restricted. For both types of η's the maximum $|\eta|$ might be specified to be $\eta_{\max}$ by appropriately adjusting C. For the model $\eta=\beta C\sin t$ and $t>0$, C would be equal to $\eta_{\max}/\beta\sin t_n$ for $0<t_n<\pi/2$ and $\eta_{\max}/\beta$ for $t>\pi/2$. For another example, let η be temperature, t time, and Q the rate of energy input; for some physical conditions $\eta=\beta Qt$ and the maximum temperature (η) is known and is to be attained by adjusting the energy input Q. In the following analysis the maximum $|\eta|$ is to be $\eta_{\max}$ for both types of models; this is equivalent to prescribing the maximum $|X|$ since $\eta_{\max}=|\beta||X|_{\max}$.

The derivation of a criterion starts with Δ^n which includes the constraint of fixed large n for measurements uniformly spaced in t. The problem is to maximize Δ^n subject to the constraint of the maximum X^2 being equal to exactly $X^2_{\max}$. Let $X(t)=Cf(t)$ where C is to be adjusted to make $\max X^2=X^2_{\max}$. Let $X_{\max}$ be the positive square root of $X^2_{\max}$. Then $X^2_{\max}=C^2f^2_{\max}$ where $f^2_{\max}$ designates the maximum f^2 value. We can write Δ^n as

$$\Delta^n=t_n^{-1}\int_0^{t_n}X^2\,dt=t_n^{-1}\int_0^{t_n}C^2f^2\,dt=\frac{1}{t_n}\int_0^{t_n}X^2_{\max}\left(\frac{f}{f_{\max}}\right)^2dt$$

Observe that $(X/X_{\max})^2=(f/f_{\max})^2$ is independent of C. Then for arbitrary values of C (or $X_{\max}$) the criterion to maximize is

$$\Delta^+=\Delta^nX^{-2}_{\max}=t_n^{-1}\int_0^{t_n}\left[X^+(t)\right]^2dt;\quad X^+(t)\equiv\frac{X(t)}{X_{\max}}\tag{8.2.7}$$

Although X^+ is indicated to have a dependence only on t, it may also depend on t_n. For example, for $0<t<t_n<\pi/2$, $X^+=\sin t/\sin t_n$ for $\eta=\beta C\sin t$.

As the first example of the use of the criterion given by (8.2.7), let $X(t)=Ct^m$, $t>0$, $C>0$, and then $X_{\max}=Ct_n^m$. Hence $X^+=(t/t_n)^m$ where m is an exponent equal to or greater than zero. Using (8.2.7), Δ^+ for this case is

$$\Delta^+=t_n^{-1}\int_0^{t_n}\left(\frac{t}{t_n}\right)^{2m}dt=\frac{1}{2m+1}\tag{8.2.8}$$

Note that this result is independent of C and t_n, unlike the similar case treated in Example 8.2.1 where no constraints are used. It is also unlike the result of the single constraint of fixed large n; see Δ_4^n in Fig. 8.2. The result given by (8.2.8) means that there is *no unique* optimum time t_n for $X=Ct^m$ and $X_{\max}$ being the same value in each case.

For the case of $X = C\sin t$ the use of (8.2.7) yields

$$\Delta^+ = \frac{1}{t_n}\int_0^{t_n}\left(\frac{\sin t}{\sin t_n}\right)^2 dt = \frac{2 - t_n^{-1}\sin 2t_n}{4\sin^2 t_n} \qquad \text{for } 0 < t_n < \frac{\pi}{2}$$

$$= \frac{1}{4}(2 - t_n^{-1}\sin 2t_n) \qquad \text{for } \frac{\pi}{2} \leqslant t_n \tag{8.2.9}$$

The latter portion of the t_n curve is the same as given by Δ_6^n in Fig. 8.3*b* and the 0 to $\pi/2$ portion is shown as the dashed curve in the same figure. Note that the maximum is unaffected by the constraint of a fixed range of η. The same is true for the maximum Δ^n for the $X = C\cos t$ curve (see Δ_5^n in Fig. 8.3*b*).

Based on the above examples the shape of the Δ^+ curve may or may not be affected by the η range constraint. Also the location of the maximum might or might not be changed.

8.2.2 One-Parameter Nonlinear Cases, $\eta = \eta(\beta, t)$

For one-parameter nonlinear cases with the standard assumptions of 11111-11 valid, the variance of the estimator b of β in the model $\eta = \eta(\beta, t)$ is approximately

$$V(b) \cong \sigma^2\left[\sum_{i=1}^{n} X_i^2\right]^{-1} \qquad \text{where } X_i = \left.\frac{\partial\eta(\beta, t_i)}{\partial\beta}\right|_{\beta = b} \tag{8.2.10}$$

Again the optimal experiment involves minimizing the sum of the squares of the sensitivity coefficients. As in the linear case the optimal unconstrained experiment would involve locating all the observations at the maximum possible $|X|$. An analysis is given below for cases for which it is more practical to use uniformly spaced measurements in t.

The constraints of (1) a fixed number n of equally spaced measurements between $t = 0$ and t_n and (2) a maximum value of $|\eta|$, designated $\eta_{\max}$, are to be included in the analysis. For large n (8.2.2) permits writing

$$\Delta = \frac{n}{t_n}\int_0^{t_n} X^2\,dt = \frac{n\eta_{\max}^2}{\beta^2}\left[\frac{1}{t_n}\int_0^{t_n}(X^+)^2(\eta_m^+)^{-2}\,dt\right], \tag{8.2.11}$$

$$X^+ \equiv \frac{\beta}{\eta_{\text{nom}}}\frac{\partial\eta}{\partial\beta}, \qquad \eta_m^+ \equiv \frac{\eta_{\max}}{\eta_{\text{nom}}} \tag{8.2.12}$$

where η_{nom} is some nominal value of η which is chosen to make η_m^+ a function of t_n only and not of t or η_{max}; η_{nom} itself is not a function of t_n. Note that maximizing Δ subject to the constraints of fixed n and η_{max} is equivalent to maximizing the term inside the brackets of (8.2.11), which is defined to be

$$\Delta^+ \equiv \frac{1}{t_n}\int_0^{t_n}(X^+)^2(\eta_m^+)^{-2}\,dt \tag{8.2.13}$$

For linear models this criterion is equivalent to that given by (8.2.7).

A necessary condition to maximize Δ^+ with respect to t_n is found from $\partial\Delta^+/\partial t_n = 0$ which yields

$$\Delta^-(\tau) = \left[X^+(\tau)\right]^2\left\{1 + \left[2\tau/\eta_m^+(\tau)\right]\left[d\eta_m^+(\tau)/dt_n\right]\right\} \tag{8.2.14}$$

where τ is the value of t_n maximizing Δ^+ and $\Delta^-(\tau)$ is defined by

$$\Delta^-(\tau) = \tau^{-1}\int_0^{\tau}(X^+)^2\,dt \tag{8.2.15}$$

As an example of a nonlinear model let η be given by

$$\eta = C\exp(-\beta t) \tag{8.2.16}$$

for which it is convenient to make η_{nom} equal to C. Then X^+ becomes

$$X^+ = -\beta t\exp(-\beta t) \tag{8.2.17}$$

which has a maximum amplitude at $t^+ = \beta t = 1$ as shown in Fig. 8.4. For this case the maximum η occurs at $t = 0$ so that η_m^+ is equal to unity. The

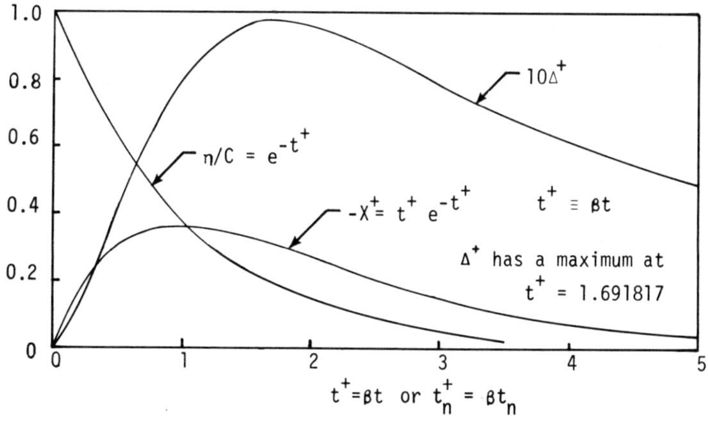

Figure 8.4 Sensitivity coefficient, η/C, and Δ^+ for $\eta = C\exp(-\beta t)$.

location of the maximum Δ^+ defined by (8.2.13) is at $\beta t = 1.691817$. Unlike linear parameter estimation cases, the possible dependence of t_n on β complicates optimal design in nonlinear cases. For the present case this difficulty is not as severe as it first may seem, however. Notice that Δ^+ shown in Fig. 8.4, though having a unique maximum, has values within 80% of its maximum for the large range of $t^+ = \beta t$ between one and three. Hence in this example an accurate initial estimate of β is not necessary to obtain a good experiment design. Another approach is to note that since the optimal Δ^+ occurs when $\eta / C \cong 0.184$, the duration could be selected to make η / C approximately this value.

Another example which has a model related to (8.2.16) is

$$\eta = C\left[1 - \exp(-\beta t)\right], \qquad \eta_{\text{nom}} = C \tag{8.2.18}$$

which has the sensitivity X^+ of $\beta t \exp(-\beta t)$. In this example, η initially increases with time so that $\eta_m^+ = 1 - \exp(-\beta t_n)$. Using (8.2.14) for determining the optimal duration τ gives

$$\Delta^-(\tau) = \frac{\left[\beta\tau \exp(-\beta\tau)\right]^2}{1 + \left[2\beta\tau \exp(-\beta\tau)\right]\left[1 - \exp(-\beta\tau)\right]^{-1}} \tag{8.2.19}$$

where Δ^- is the same as Δ^+ of Fig. 8.4. This condition is satisfied only near $\tau = 0$ where $\Delta^+ = \frac{1}{3}$. For small t values η is approximately $C\beta t$ which has this Δ^+ value [see (8.2.8)], whereas for larger t values Δ^+ decreases. Again it is observed that the constraint on the maximum value of $|\eta|$ changes the optimal conditions.

Example 8.2.4

Consider again the example of the cooling billet investigated in Example 6.2.3 and Sections 7.5.2 and 7.9.2. The billet was heated to the temperature T_0 and then allowed to cool in open air at a temperature $T_\infty \cong 81.5°\text{F}$ (301 K). Though several models were considered for this billet, let us consider here only Model 1, which is [see (7.5.10)] $(T - T_\infty)/(T_0 - T_\infty) = \exp(-\beta t)$. The parameter to be estimated is β. The optimal duration of the experiment is to be found for a large number of measurements uniformly spaced in t starting at $t = 0$. The initial temperature difference $T_0 - T_\infty$ can be set by simply heating the billet before placing it in the air. The air temperature T_∞ is a fixed, known value. The temperature T must be between T_∞ and T_0.

Solution

The model can be considered to be

$$\eta = T - T_\infty = C\exp(-\beta t) \qquad \text{where } C = T_0 - T_\infty$$

This model is identical to (8.2.16) and η_{nom} is also C. Furthermore, $\eta_m^+ = \eta_{\max}/n_{\text{nom}} = 1$. With the condition of a large number of equally spaced measurements, this example is the same as considered for (8.2.16) for which we found the optimal experiment to have the duration $\tau \cong 1.69/\beta$. This time τ can be compared with the value actually used. From Fig. 7.13 the b_1 value is about 2.7 hr^{-1} and thus $\tau \cong 1.69/2.7 \cong 0.63$ hr. From Table 6.2 the maximum time is 1536 sec or 0.427 hr; this time corresponds to $\beta t = 1.15$ at which time Δ^+ in Fig. 8.4 is about 80% of its maximum value. Hence for estimating β in Model 1, the experiment was well designed. For the more complicated models discussed in Section 7.5.2, the optimal duration may be different.

8.2.3 Iterative Search Method

One obvious way to maximize Δ, Δ'', or Δ^+ is to plot it versus t_n and then observe the value of t_n which maximizes the selected Δ function. A more direct procedure is to linearize in a similar fashion as in the Gauss method. Let us illustrate the method by considering Δ''. Assume that a maximum exists and that an estimate of the optimal τ is $\tau^{(k)}$. A necessary condition at the maximum Δ'' is given by (8.2.5). Expanding both sides using a truncated Taylor series gives

$$[X^{(k)}]^2 + 2X^{(k)}\frac{\partial X^{(k)}}{\partial t_n}\Delta\tau^{(k)} = [\tau^{(k)}]^{-1}\int_0^{\tau^{(k)}} [X^{(k)}]^2\,dt$$

$$-[\tau^{(k)}]^{-2}\int_0^{\tau_0}[X^{(k)}]^2\,dt\,\Delta\tau^{(k)} + [\tau^{(k)}]^{-1}[X^{(k)}]^2\Delta\tau^{(k)} \qquad (8.2.20)$$

which can be solved for the correction $\Delta\tau^{(k)}$ to get

$$\Delta\tau^{(k)} = \tau^{(k)}A^{(k)}\left[A^{(k)} - 2\tau^{(k)}X^{(k)}\frac{\partial X^{(k)}}{\partial t_n}\right]^{-1} \qquad (8.2.21a)$$

$$A^{(k)} \equiv [X^{(k)}]^2 - [\tau^{(k)}]^{-1}\int_0^{\tau^{(k)}}[X^{(k)}]^2\,dt \qquad (8.2.21b)$$

A few points can be made in connection with the iterative procedure for finding the optimal τ given by (8.2.21). First, an initial estimate of $\tau^{(0)}$ is needed. A reasonable value to use is twice the t value at which $|X|$ is a maximum. This is the value that is found for $\tau^{(1)}$ if one starts at $\tau^{(0)}$ corresponding to the maximum value of $|X|$. Second, improved values of τ are given by

$$\tau^{(k+1)} = \tau^{(k)} + \Delta\tau^{(k)} \qquad (8.2.22)$$

Third, in order to be sure that each iteration helps to increase Δ^n, the values of Δ^n should also be calculated and compared as one proceeds. If Δ^n should decrease a smaller $\Delta\tau$ should be selected. (It is also possible that the method is leading to a *minimum* rather than a maximum.) Fourth, τ cannot be negative even though this procedure seeks a maximum in the region $-\infty<\tau<\infty$. The procedure based on (8.2.21) is not appropriate if the maximum Δ^n occurs at the boundary point $\tau=0$ and $\partial\Delta^n/\partial t_n$ is not zero there. See Δ_3^n in Fig. 8.2*b*. Finally, the procedure terminates when $|\Delta\tau^{(k)}|$ is much smaller than $\tau^{(k)}$.

8.3 CRITERIA FOR OPTIMAL EXPERIMENTS FOR MULTIPLE PARAMETERS

8.3.1 General Criteria

When there are two or more parameters to estimate, the choice of a criterion to indicate the optimal design of the experiments is less straightforward than for the case of one parameter. Many criteria have been proposed. They are usually given in terms of $\mathbf{X}^T\mathbf{X}$. For both linear and nonlinear estimation, $\mathbf{X}$ represents the sensitivity matrix. Recall that the covariance matrix of the estimator vector $\mathbf{b}$ is $(\mathbf{X}^T\mathbf{X})^{-1}\sigma^2$ for the standard assumptions of additive, zero mean, constant variance, independent, normal, measurement errors in $\mathbf{Y}$; additional assumptions are that there are no errors in the independent variables and that $\boldsymbol{\beta}$ is a constant parameter vector with no prior information. The value of σ^2 need not be known. These assumptions are designated 11111-11. For these assumptions OLS, Gauss–Markov, ML, and MAP all give the same estimator.

Some of the criteria which have been suggested in terms of $\mathbf{X}^T\mathbf{X}$ are as follows: (*a*) maximization of the determinant of $\mathbf{X}^T\mathbf{X}$ (or equivalently, the maximization of the product of the eigenvalues of $\mathbf{X}^T\mathbf{X}$). (*b*) maximization of the minimum eigenvalue of $\mathbf{X}^T\mathbf{X}$; and (*c*) maximization of the trace of $\mathbf{X}^T\mathbf{X}$. These criteria are listed by Badavas and Saridis [4], who used the second criterion. Additional criteria are listed on p. 52 of Fedorov [3]. McCormack and Perlis [5] used a criterion similar in principle to (*c*). We recommend the first one because it is equivalent to minimizing the hypervolume of the confidence region (provided the assumptions 11111-11 are valid). A criterion similar to $\max|\mathbf{X}^T\mathbf{X}|$ was used by Smith [7] as early as 1918. The best-known early work involving $\max|\mathbf{X}^T\mathbf{X}|$ was reported by Box and Lucas [1] in 1959, however.

Another derivation for the closely related criterion of maximization of $|\mathbf{X}^T\psi^{-1}\mathbf{X}|$ is given in Chapter 11 in Nahi's book [6]. [See (8.3.2) below.] The derivation is based on the Cramer–Rao lower bound which is appealing, he states, since the lower bound does not depend on the knowledge of the specific estimator (LS, ML, etc.) to be used.

As mentioned above, we derive our criterion based on the assumption that we wish to minimize the hypervolume of the confidence region. In so doing it is implied that each parameter is considered in the same manner and that the cost of each measurement is the same. In Section 8.8 the case of only selected estimated parameters being of interest is discussed.

A criterion is derived in Appendix 8A that is valid for the assumptions of additive, zero mean normal errors in Y_i, and errorless dependent variables. Specifically the assumptions are denoted 11--1111. For the OLS, Gauss–Markov, and ML estimators, the criterion is the maximization of the determinant of the covariance matrix of the estimator vector **b**. For the standard assumptions denoted 11111-11, the above estimators all have the same covariance matrix of $(\mathbf{X}^T\mathbf{X})^{-1}\sigma^2$; then the related criterion for optimal experiments is to maximize

$$\Delta_{\text{ST}} = |\mathbf{X}^T\mathbf{X}| \tag{8.3.1}$$

For maximum likelihood estimation and assumptions denoted 11--1011, the criterion is to maximize

$$\Delta_{\text{ML}} = |\mathbf{X}^T\psi^{-1}\mathbf{X}| \tag{8.3.2}$$

For ordinary least squares estimation with the same assumptions, the criterion is to maximize

$$\Delta_{\text{OLS}} = \frac{|\mathbf{X}^T\mathbf{X}|^2}{|\mathbf{X}^T\psi\mathbf{X}|} \tag{8.3.3}$$

The last two expressions are valid for correlated, nonconstant variance measurement errors. The max $|\mathbf{X}^T\mathbf{X}|$ criterion reduces to the Δ criterion utilized in Section 8.2 for one parameter.

The constraint of a fixed number of observations n is also of interest for the multiparameter case. We wish to define a max Δ^n criterion that includes this constraint in such a way that Δ^n is consistently defined with the one-parameter case and also so that a replication of discrete measurements will not change its value. Such a criterion is

$$\max \Delta^n \equiv \max \frac{\Delta}{n^p} \tag{8.3.4}$$

where Δ could be replaced by the expressions in (8.3.1,2,3) depending on the assumptions and estimation method.

When the measurements are uniformly spaced in t between 0 and t_n and n is large, Δ_{ST}^n for $p=2$ is

$$\Delta_{ST}^n = C_{11}C_{22} - C_{12}^2; \qquad C_{ij} \equiv \frac{1}{t_n}\int_0^{t_n} X_i(t)X_j(t)\,dt \tag{8.3.5}$$

where $X_i(t)$ is the sensitivity coefficient for parameter i and time t. The extension to $p>2$ is direct. If, in addition, there is a constraint of the maximum η being specified, one can modify (8.3.5) by replacing the integrals by a typical expression of

$$C_{ij}^+ = (\eta_m^+)^{-2}\frac{1}{t_n}\int_0^{t_n} X_i^+(t)X_j^+(t)\,dt \tag{8.3.6a}$$

where

$$X_i^+(t) \equiv \frac{\beta_i}{\eta_{\text{nom}}}\frac{\partial\eta(t)}{\partial\beta_i}, \qquad \eta_m^+ \equiv \frac{\eta_{\max}}{\eta_{\text{nom}}} \tag{8.3.6b}$$

which are similar to the expressions given in Section 8.2.2. Then for two parameters with the constraints of large n with uniform spacing in t and the maximum η being $\eta_{\max}$, we have the criterion of maximizing

$$\Delta^+ = C_{11}^+C_{22}^+ - (C_{12}^+)^2 \tag{8.3.7}$$

If there are multiresponses in the experiment and measurements are taken with uniform time spacing starting at $t=0$, another Δ^+ criterion must be given. As above, the symbol Δ^+ means that the constraint of the same η range is included in addition to the constraint of uniform Δt. Examples of multiresponse cases involving transient temperature measurements at more than one position are studied in Section 8.5. Let m denote the number of independent responses. This case can be treated by extending the definition of C_{ij}^+ given by (8.3.6a) to

$$C_{ij}^+ = (\eta_m^+)^{-2}\frac{1}{mt_n}\sum_{k=1}^{m}\int_0^{t_n} X_i^+(t,x_k)X_j^+(t,x_k)\,dt \tag{8.3.8}$$

where x_k is used to designate the kth response. By defining C_{ij}^+ in this manner, Δ^+ is unchanged in value if m sensors are located at the same position (or measure the same quantity).

8.3.2 Case of Same Number of Measurements as Parameters ($n=p$)

One possible multiparameter case is when the number of measurements and parameters are equal. Without prior information the minimum number of measurements n needed to estimate p parameters is $n=p$. In this case **X** is a square matrix. This results in the following simplications for Δ_{ST}, Δ_{ML}, and Δ_{OLS},

$$\Delta_{ST} = |\mathbf{X}|^2, \qquad \Delta_{ML} = \Delta_{OLS} = \frac{|\mathbf{X}|^2}{|\psi|} \tag{8.3.9}$$

Note that the *same* criterion is given for *both* ML and OLS estimation for the assumptions of 11--1011. Also observe that the optimal choice of **X** elements are affected by the accuracy of and correlation between the measurements.

Consider now the criterion of maximizing Δ_{ST}, which is equivalent to maximizing the absolute value of

$$|\mathbf{X}| = \begin{vmatrix} X_{11} & X_{12} & \cdots & X_{1p} \\ \vdots & & \vdots & \\ X_{p1} & X_{p2} & \cdots & X_{pp} \end{vmatrix} \tag{8.3.10}$$

since when **X** is a square matrix, $\boldsymbol{\Delta} = |\mathbf{X}^T\mathbf{X}| = |\mathbf{X}|^2$. In the remainder of Section 8.3.2 let $|\mathbf{X}|$ denote the absolute value of the determinant of **X**.

As mentioned above there usually must be constraints on the range of operability (a term used by Atkinson and Hunter [8]). Let $R(\mathbf{f})$ define the region of operability; the **f** vector has elements that can be illustrated by writing the linear model as $\eta = \beta_1 f_1 + \cdots + \beta_p f_p$. However, not all the values of X_{ij} may be attainable, as, for example, when $f_1 = 1$. Let those values which are available for experimentation define the attainable region $R(\mathbf{x})$, a subspace of the p-dimensional **X** space. The design problem then becomes that of selecting n points in $R(\mathbf{x})$ which maximizes $|\mathbf{X}|$. Atkinson and Hunter [8] have shown that the value of the determinant given by (8.3.10) is proportional to the volume of the simplex formed by the origin and the p experimental points. Thus an optimal design is one for which the simplex volume is maximized. It follows, then, that for an $n = p$ design to be optimal, the experimental points must lie on the *boundary* of $R(\mathbf{x})$.

8.3.2.1 Linear Examples for $p = 2$

Constraints of $0 \leqslant f_1 \leqslant 1$ *and* $0 \leqslant f_2 \leqslant 1$

Consider the simple linear model

$$\eta = \beta_1 f_1 + \beta_2 f_2 \tag{8.3.10}$$

with the constraints

$$0 \leqslant f_1 \leqslant 1 \qquad \text{and} \qquad 0 \leqslant f_2 \leqslant 1$$

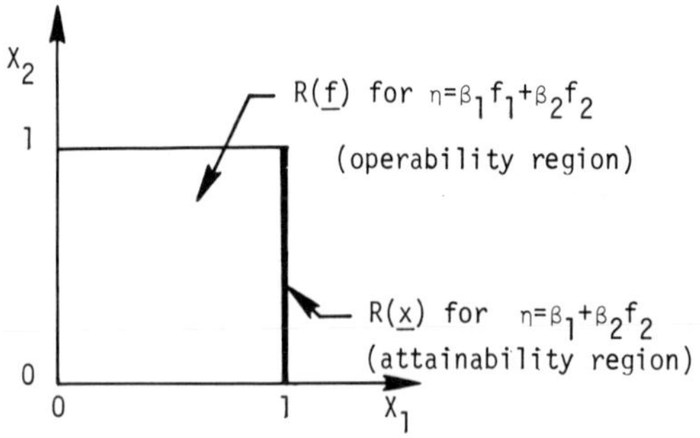

Figure 8.5 Regions of operability and attainability $[R(\mathbf{f})$ and $R(\mathbf{x})]$ for constraints of $0 \leqslant f_1 \leqslant 1$ and $0 \leqslant f_2 \leqslant 1$.

Thus the region of operability $R(\mathbf{f})$ is the unit square shown in Fig. 8.5. For this case the sensitivity coefficients are $X_1 = f_1$ and $X_2 = f_2$ and the absolute value of the determinant of $\mathbf{X}$ is

$$|\mathbf{X}| = |f_{11} f_{22} - f_{12} f_{21}| \tag{8.3.11}$$

where the vertical bars on the right side mean absolute value. For the model $\eta = \beta_1 + \beta_2 f_2$, the attainable region $R(\mathbf{x})$ is the unit vertical line at $X_1 = 1$ shown in Fig. 8.5. From the geometrical interpretation of $\max|\mathbf{X}|$, the optimal design for two experiments consists of 2 points in $R(\mathbf{x})$ which, together with the origin, form a triangle of greatest area. For this case the optimal two points are the extremes of the line, $(X_1, X_2) = (1,0)$ and $(1,1)$. For this design $|\mathbf{X}| = 1$. If the attainable region $R(\mathbf{x})$ happens to be the operability region $R(\mathbf{f})$ which is the unit square, an infinity of designs give the same maximum value of the determinant, namely, one experiment at $(X_1, X_2) = (0,1)$ and the other anywhere between and including $(1,0)$ and $(1,1)$ or one at $(1,0)$ and the other anywhere between and including $(0,1)$ and $(1,1)$. All these designs also give a $|\mathbf{X}|$ value of unity.

Constraint of $0 \leqslant \eta \leqslant \eta_{max}$

Frequently a more realistic constraint than on the f_i values is on the range of η. In this section the case of $0 \leqslant \eta \leqslant \eta_{max}$ is investigated for $n = p = 2$ and the linear model. Three different variations of this case are considered.

Case 1

In this case there are no constraints on the f_i's so that $R(\mathbf{f})$ is equal to $R(\mathbf{x})$. For the model $\eta = \beta_1 f_1 + \beta_2 f_2$ the optimal design points are found from

$$\max|\mathbf{X}| = \max|X_{11} X_{22} - X_{21} X_{12}| = \max|f_{11} f_{22} - f_{21} f_{12}| \tag{8.3.12}$$

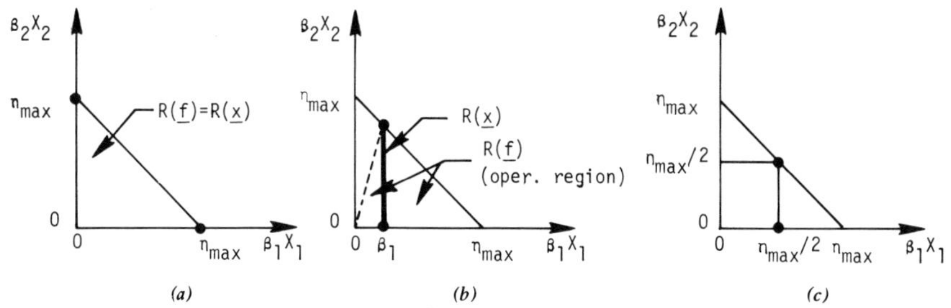

Figure 8.6 Several regions of operability and attainability for constraint of $0 \leqslant \eta \leqslant \eta_{max}$.

while satisfying the condition

$$\max \eta = \eta_{max} = \max|\beta_1 f_1 + \beta_2 f_2| \tag{8.3.13}$$

The region $R(\mathbf{x})$ is the triangular region bounded on one side by the line determined by varying f_1 and f_2 in (8.3.13); see Fig. 8.6a. The largest $|\mathbf{X}|$ value is found by the two points and the origin comprising the largest triangle in $R(\mathbf{x})$. In this example the optimal conditions are $(\beta_1 X_1, \beta_2 X_2) = (\eta_{max}, 0)$ and $(0, \eta_{max})$. This results in $\max|\mathbf{X}|$ being equal to $|\eta_{max}^2/\beta_1\beta_2|$.

Case 2

In this case the operability region $R(\mathbf{f})$ is greater than the attainable region $R(\mathbf{x})$. As an example consider the model $\eta = \beta_1 + \beta_2 f_2$ for which $R(\mathbf{x})$ is the vertical line at $\beta_1 X_1 = \beta_1$ shown in Fig. 8.6b. Hence the two extreme points along $R(\mathbf{x})$ together with the origin form the maximum triangle. The maximum $|\mathbf{X}|$ is $|(\eta_{max} - \beta_1)/\beta_2| = \max|f_2|$, which is made larger by increasing $\max|f_2|$.

Case 3

In the last case η is given by

$$\eta = C(\beta_1 f_1' + \beta_2 f_2') \tag{8.3.14}$$

where C is adjusted to make $\max|\eta| = \eta_{max}$. In symbols, C is

$$C = \frac{\eta_{max}}{\max|\beta_1 f_1' + \beta_2 f_2'|} \tag{8.3.15}$$

Then the maximum $|\mathbf{X}|$ value is

$$\max|\mathbf{X}| = \max\left(C^2|f_{11}' f_{22}' - f_{21}' f_{12}'|\right) = \frac{\eta_{max}^2 \max|f_{11}' f_{22}' - f_{21}' f_{12}'|}{\max(\beta_1 f_1' + \beta_2 f_2')^2} \tag{8.3.16}$$

In order to illustrate this expression, consider the case of $\eta = C(\beta_1 + \beta_2 f_2')$ for which $f_{11}' = f_{21}' = 1$. If we further choose f_{22}' to be equal to or greater than $f_{12}' > 0$, the

$\max|f'_{11}f'_{22}-f'_{21}f'_{12}|$ value is f'_{22} (by setting $f'_{12}=0$) and $\max|\mathbf{X}|$ is

$$\max|\mathbf{X}|=\frac{\eta_{\max}^2}{\beta_2}(\beta_2 f'_{22})(\beta_1+\beta_2 f'_{22})^{-2}=C^2 f'_{22} \tag{8.3.17}$$

which is similar to that given for case 2. Note that now the maximum of $\max|\mathbf{X}|$ is not simply given by the maximum value of f'_{22}. Rather, differentiating (8.3.17) with respect to $\beta_2 f'_{22}$ and setting the equation equal to zero gives

$$\beta_2 f'_{22}|_{\text{opt}}=\beta_1 \tag{8.3.18}$$

which are then both equal to $\eta_{\max}/2$. Then we find max ($\max|\mathbf{X}|$) to be $\eta_{\max}/2\beta_2$ or, equivalently, $\eta_{\max}^2/4\beta_1\beta_2$. (Much smaller $\max|\mathbf{X}|$ values are found for certain other $\beta_2 f'_{22}$ values; for example, it goes to zero for both $\beta_2 f'_{22}$ approaching zero and infinity.) The optimum two measurement points are shown in Fig. 8.6c and are $(\beta_1 X_1, \beta_2 X_2)=(\eta_{\max}/2, 0)$ and $(\eta_{\max}/2, \eta_{\max}/2)$.

8.3.2.2 *Nonlinear Example for* $p=2$

A model studied first by Box and Lucas [1] and later by Atkinson and Hunter [8] is next considered. Preliminary estimates of $\beta_1=0.7$ and $\beta_2=0.2$ yield the model and sensitivities of (see Problem 8.10)

$$\eta=1.4\left[\exp(-0.2t)-\exp(-0.7t)\right] \tag{8.3.19a}$$

$$\beta_1 X_1=0.7\left[(0.8+1.4t)\exp(-0.7t)-0.8\exp(-0.2t)\right] \tag{8.3.19b}$$

$$\beta_2 X_2=0.2\left[(2.8-1.4t)\exp(-0.2t)-2.8\exp(-0.7t)\right] \tag{8.3.19c}$$

which are plotted in Fig. 8.7. The operability range of η is between 0 and 0.6; $X_1(t)$ and $X_2(t)$ are also finite but may be negative as well as positive.

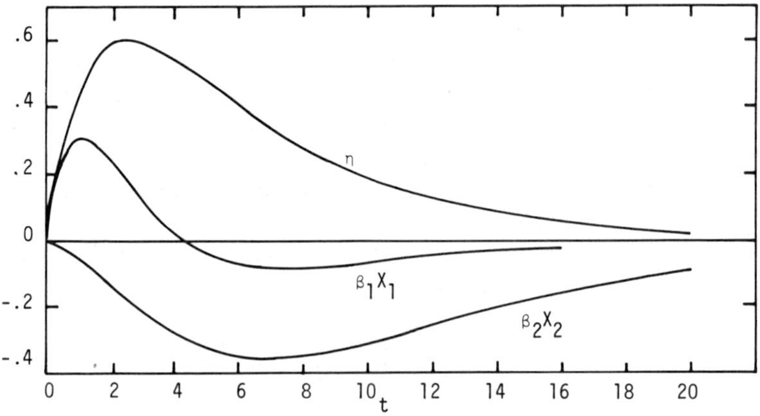

Figure 8.7 η and sensitivities for Box and Lucas [1] example.

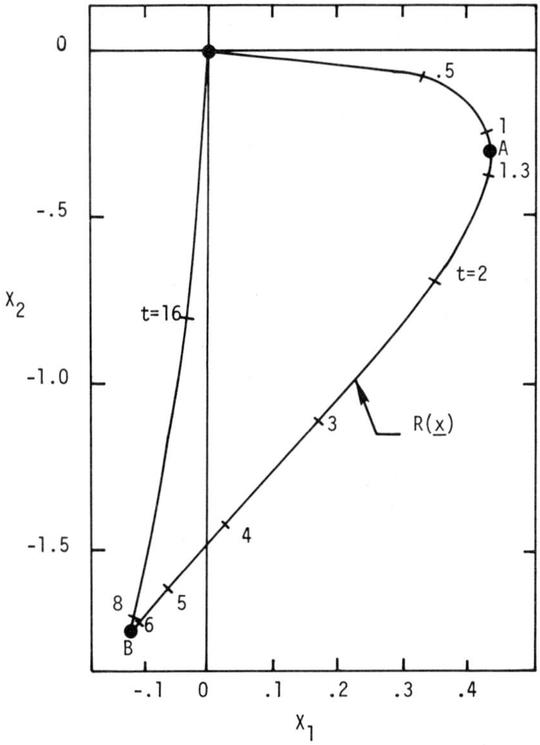

Figure 8.8 $R(\mathbf{x})$ for Box and Lucas [1] example. (Printed by permission of the Biometrika Trustees.)

Note that X_1 and X_2 are uncorrelated and have maximum absolute values at different t values. Plotting X_2 versus X_1 as in Fig. 8.8 provides the attainable region $R(\mathbf{x})$ which is a curved line in this case.

The points of the optimal design, shown in Fig. 8.8 by heavy dots labeled A and B, together with the origin, form the triangle of maximum area within $R(\mathbf{x})$. Associated values of t are 1.23 and 6.86, which values are affected by the choice of the parameters. Since the β's are not precisely known when the experiment is designed, one might wish to relate these values to associated measured η values. For example, at $t_2 = 6.86$, η has reduced to $\frac{1}{2}$ of its maximum value.

Atkinson and Hunter also studied optimal designs for up to 20 measurements. For these cases Δ^n is given by

$$\Delta^n = \frac{\Delta}{n^2} = \frac{\left(\sum_{i=1}^{n} X_{i1}^2\right)\left(\sum_{j=1}^{n} X_{j2}^2\right) - \left(\sum_{i=1}^{n} X_{i1}X_{i2}\right)^2}{n^2} \tag{8.3.20}$$

Their results for maximum values of Δ^n are given in Table 8.1. In each case the optimal design is found to consist of measurements solely at the two times indicated above. When n is even, equal numbers of measurements at each point maximize Δ^n. For odd n an extra measurement at either of the two conditions give the same maximal Δ^n.

Table 8.1 Optimal Designs for up to 20 Measurements for the Box and Lucas Model Given by (8.3.19)*

	Number of measurements at		maximum
n	$t=1.23$	$t=6.86$	Δ^n
2	1	1	0.1642
3	1	2	0.1459
3	2	1	0.1459
4	2	2	0.1642
5	2	3	0.1576
5	3	2	0.1576
6	3	3	0.1642
10	5	5	0.1642
20	10	10	0.1642

*Reprinted by permission from Technometrics [8].

Noted by many is the conclusion that mp optimal conditions for determining p parameters consist of m repeated optimal experiment. However, this conclusion is not always valid, as pointed out in [8].

Note that to obtain the 20 measurements in Table 8.1, 10 different experiments must be run. Because it is wasteful to disregard data at other times when the transient experiment has been performed, the emphasis in this chapter is upon many equally spaced measurements.

8.4 ALGEBRAIC EXAMPLES FOR TWO PARAMETERS AND LARGE n

8.4.1 Linear Model $\eta=\beta_1+\beta_2\sin t$

To illustrate the case of a large number of uniformly spaced measurements, consider the model

$$\eta=\beta_1+\beta_2\sin t, \qquad 0<t<t_n \tag{8.4.1}$$

Assume that the assumptions of additive, zero mean, independent errors apply with the others denoted in 11111-11. No constraints are to be included for η or t_n.

For this case the optimal value of t_n is found by maximizing (8.3.5). The sensitivity coefficients are $X_1=1$ and $X_2=\sin t$ and the C_{ij} values are

$$C_{11}=1, \qquad C_{22}=\frac{1}{2}-\frac{1}{4t_n}\sin 2t_n, \qquad C_{12}=\frac{1}{t_n}(1-\cos t_n) \tag{8.4.2}$$

These expressions are plotted in Fig. 8.9 along with Δ^n. The optimal t_n is 5.5 which is considerably larger than 2.25, the optimal t_n for estimating only β_2.

8.4.2 Exponential Models with One Linear and One Nonlinear Parameter

Exponentially decaying solutions commonly occur in science and engineering. One is

$$\eta=\beta_1\exp(-\beta_2 t) \tag{8.4.3}$$

This could describe the temperature in a fin (Section 7.5.1) or that of a cooling billet (Section 7.5.2) [T_∞ would be assumed known in (7.5.4) and (7.5.10).] For the assumptions denoted 11111-11 and no constraints on η or t the criterion to maximize again is Δ^n, given by (8.3.5). The sensitivities are

$$X_1=\frac{\partial\eta}{\partial\beta_1}=\exp(-t^+), \qquad X_2=\frac{\partial\eta}{\partial\beta_2}=-\left(\frac{\beta_1}{\beta_2}\right)t^+\exp(-t^+) \tag{8.4.4}$$

where $t^+=\beta_2 t$. Functions similar to X_1 and X_2 are shown in Fig. 8.4.

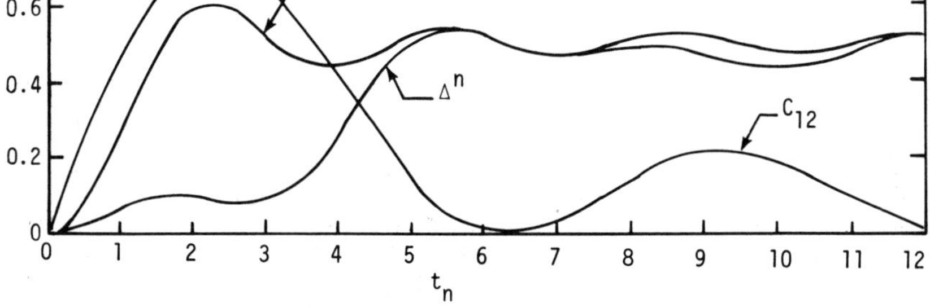

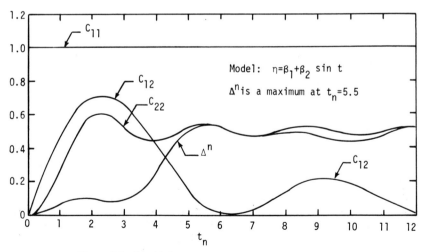

Figure 8.9 Sensitivity curves for the model $\eta=\beta_1+\beta_2\sin t$.

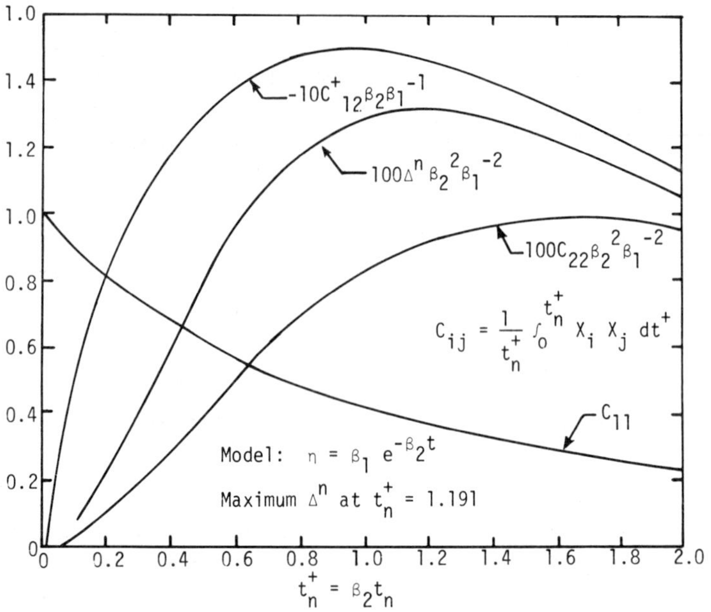

Figure 8.10 Sensitivity curves for the model $\eta=\beta_1\exp(-\beta_2 t)$.

If measurements were desired at only two locations, the optimal locations are at $t^+=0$ and 1, the former being where X_1 is a maximum and the latter where $|X_2|$ is. One can demonstrate this by plotting X_2 versus X_1 as in Fig. 8.8 and then finding the maximum triangle including the origin. If only two measurements are to be taken from each experiment in a series of experiments, the measurements should be made at just these two times in all the experiments.

The integrals C_{ij} associated with X_1 and X_2 are plotted in Fig. 8.10 along with Δ^n. A large number n of equally spaced observations in $0<t<t_n$ is used. The optimal duration of an experiment for determining both β_1 and β_2 is the time at which Δ^n is a maximum, $t_n^+=1.191$. This maximum occurs between the times of the maxima of $t_n^+=0$ for C_{11} and $t_n^+=1.69$ for C_{22}. These latter times are the optimal values if only β_1 and only β_2 were to be estimated.

A model similar to (8.4.3) is

$$\eta=\beta_1\left[1-\exp(-\beta_2 t)\right] \tag{8.4.5}$$

This could represent the same physical cases mentioned above except now T_0 is assumed known. Both models are illustrated on the following page.

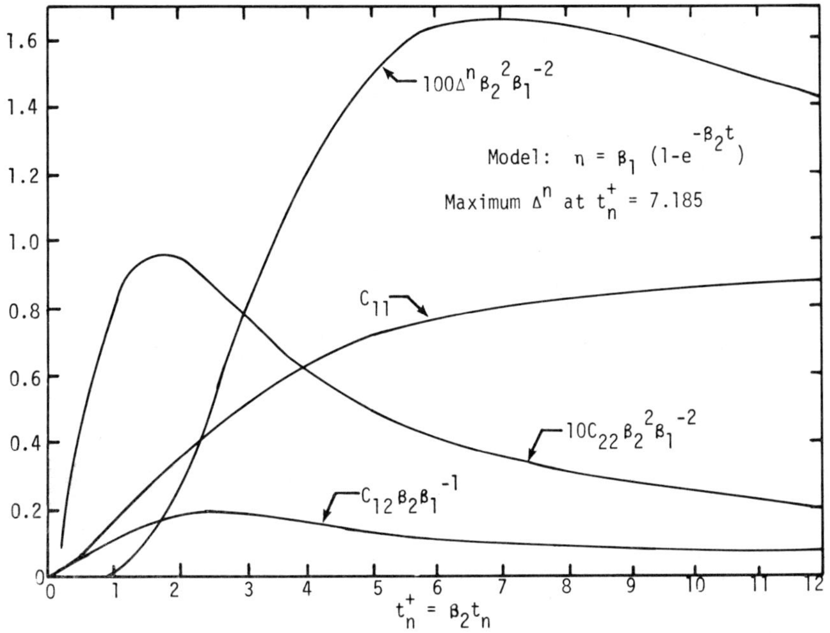

Figure 8.11 Sensitivity curves for the model $\eta = \beta_1(1-e^{-\beta_2 t})$ with no constraint on maximum η.

Sensitivities for (8.4.5) are

$$X_1 = 1 - \exp(-t^+), \qquad X_2 = \left(\frac{\beta_1}{\beta_2}\right) t^+ \exp(-t^+) \tag{8.4.6}$$

where $t^+ = \beta t$. The integrals C_{ij} and Δ^n are depicted in Fig. 8.11. The maximum of Δ^n is at $t_n^+ = 7.184$ which is between the value of $t_n^+ = \infty$ and 1.69, maxima values for C_{11} and C_{22}.

It is significant to note that at time $t_n^+ = 1.191$, the optimal value for model (8.4.3), the Δ^n value shown in Fig. 8.11 is still very small. Hence an experiment design that is optimal for (8.4.3) is very poor for the similar exponential model, (8.4.5).

Example 8.4.1

Consider again the cooling billet example studied in Example 8.2.4 and other sections. The model can be in the following forms

$$T - T_\infty = (T_0 - T_\infty)\exp(-\beta t) \tag{a}$$

$$T - T_0 = (T_\infty - T_0)[1 - \exp(-\beta t)] \tag{b}$$

Consider two cases. For the first case, (a), assume temperature T_∞ is accurately known and $(T_0 - T_\infty)$ and β are the parameters. This describes the billet problem because T_∞ is accurately known. The second case corresponds to (b) for which the initial temperature T_0 is considered to be known and $(T_\infty - T_0)$ and β are now β_1 and β_2, respectively. The optimum durations for both cases for a large number of equally spaced measurements from 0 to t_n are to be found. No constraints on T or t are to be used. Assume that the measurements satisfy the standard conditions denoted 11111-11. An estimate of β in (a) and (b) is 2.7/hr.

Solution

For (a) the dependent variable can be considered to be $T - T_\infty$; this model is similar to (8.4.3) and the optimal duration is $t_n = 1.191/\beta \cong 1.191/2.7 = 0.44$ hr. See Fig. 8.10. For (b) the dependent variable is $T - T_0$ which is analogous to η of (8.4.5); from Fig. 8.11 the optimal duration is $t_n = 7.185/\beta \cong 2.66$ hr. The duration of the optimal experiment is relatively long when T_∞ is unknown; in fact, at $t_n^+ = 7.185$, $T = 199.92$ compared to the value of 200 which is approached as $t \to \infty$.

8.5 OPTIMAL PARAMETER ESTIMATION INVOLVING THE PARTIAL DIFFERENTIAL EQUATION OF HEAT CONDUCTION

To illustrate design of optimal experiments in more complex cases, studied next are cases involving the partial differential equation of heat conduction. Considerations not encountered in the algebraic models given above enter when the model involves this equation. For example, space as well as time dependence is met. Thus in addition to finding optimal duration of experiments, optimal locations of sensors are needed. Furthermore, the response at any location is affected by the prescribed time variation of boundary conditions. Another significant aspect of estimation involving partial differential equations is that the parameters can be present in the equation and/or in the boundary conditions.

The criteria derived in Appendix 8A apply to estimation involving ordinary and partial differential equations. For simplicity, the cases considered in this section were selected because they have solutions in terms of known functions; similar methods of analysis can be used, however, even if the equations must be solved numerically as commonly occurs for nonlinear differential equations.

The criterion utilized is that of maximizing $\Delta = |\mathbf{X}^T\mathbf{X}|$ subject to appropriate constraints. This is the condition to employ when the standard conditions denoted 11111-11 apply. When many transient measurements are obtained using a single sensor, the standard assumption of independent measurement errors may not be valid. If the correlation parameters are not known, however, it is still reasonable to choose the maximization of $|\mathbf{X}^T\mathbf{X}|$ as the criterion.

The transient heat conduction equation for heat flow in a plane wall with constant thermal conductivity k and density–specific heat product c can be written as

$$k\frac{\partial^2 T}{\partial x^2}=c\frac{\partial T}{\partial t} \qquad \text{or} \qquad \alpha\frac{\partial^2 T}{\partial x^2}=\frac{\partial T}{\partial t} \tag{8.5.1}$$

where $\alpha=k/c$ is called the thermal diffusivity. The differential equation can be written in terms of the single parameter α but sometimes there are boundary conditions which involve k. In the following analyses when only α appears, it is the parameter, but when k appears in boundary conditions, k and c are the parameters. The parameters k and c are chosen because of their physical significance although others can be used as indicated in Section 7.10.

For the standard assumptions 11111-11, a fixed large number of equally spaced observations and a constraint on the maximum range of η [which is the increase in T of (8.5.1)], the criterion for one parameter is to maximize Δ^+ given by (8.2.13). If the same conditions are valid for two parameters, Δ^+ is given by (8.3.7,8) where the i and j subscripts could be 1 and 2 with the subscript 1 referring to k and the subscript 2 to c.

Several examples are given in this section. First considered are semi-infinite bodies for which the body starts at $x=0$ and continues indefinitely in the plus x direction. Although such bodies do not exist in nature, many heat-conducting bodies can be so modeled, at least for some period of time. Also considered are finite bodies. Temperature measurements in a finite plate heated on one side and insulated on the other are tabulated in Table 7.14 and illustrated in Fig. 7.17. These measurements also illustrate a semi-infinite body; until time 6 sec, the temperatures in the plate are the same as those that would be measured if the plate were thicker.

8.5.1 Semi-Infinite Body Examples

8.5.1.1 Temperature Boundary Condition (Single Parameter)

Suppose that the temperature in a semi-infinite body is initially uniform at the temperature T_0. Let the temperature at $x=0$ have a step increase to T_∞. The temperature in dimensionless form can be given as [9]

$$T^+\equiv\frac{T-T_0}{T_\infty-T_0}=\operatorname{erfc}\left[(4t^+)^{-1/2}\right]; \qquad t^+=\frac{\alpha t}{x^2} \tag{8.5.2}$$

where erfc(z) is called the complementary error function and is the integral,

$$\operatorname{erfc}(z) = \frac{2}{\pi^{1/2}} \int_z^{\infty} \exp(-u^2)\, du \tag{8.5.3}$$

Note that although T is a function of x and t, the dimensionless temperature can be plotted in terms of the single dimensionless variable t^+. For *temperature* boundary conditions involving the heat conduction equation (8.5.1), the only parameter that enters is α (if temperatures T_0 and T_∞ are not parameters). Note that T is a nonlinear function of α. Thermal diffusivity (α) is also called a "property" and has been estimated for many materials by many different experimentalists, some of whom have used (8.5.2) as their model.

The solution given by (8.5.2) has a natural constraint on the range of temperature T because T must be between T_0 and T_∞. Even though at some interior location x and at some time t the temperature may be much less than T_∞, the temperature near $x=0$ approaches T_∞. Instead of requiring the temperature at x to reach the same maximum value at the end of the experiment, we apply the constraint at the heated surface ($x=0$) where the temperature rise is the greatest. Hence the "nominal" rise in T is taken to be $T_\infty - T_0$.

The dimensionless α sensitivity is

$$X_\alpha^+ \equiv \frac{\alpha}{T_\infty - T_0} \frac{\partial T}{\partial \alpha} = (4\pi t^+)^{-1/2} \exp\left(\frac{-1}{4t^+}\right) \tag{8.5.4}$$

The Δ^+ function for a large number of uniformly spaced measurements starting at $t=0$ and for the maximum T in the body being T_∞ is

$$\Delta^+ = (t_n^+)^{-1} \int_0^{t_n^+} (X_\alpha^+)^2\, dt^+ \tag{8.5.5}$$

Note that Δ^+ is a function of t_n^+, the maximum time in Δ^+.

Plotted in Fig. 8.12 are T^+ and X_α^+ versus t^+ and Δ^+ versus t_n^+. For a given location x for measurement of temperatures, the sensitivity X_α^+ has a maximum at $t^+ = \alpha t/x^2 = \alpha t/x^2 = 0.5$ at which time $T^+ = 0.3173$. Hence if only one measurement is to be taken from those produced by one sensor, it should be selected at a time corresponding to $t^+ = 0.5$; if instead the time of measurement is fixed but any *one* location is to be selected, then the optimum x is $(2\alpha t)^{1/2}$. If many equally spaced-in-time measurements are used, the optimal duration for using data is when Δ^+ is maximized; it is time $t_n^+ = 1.2$ (when $T^+ \cong 0.5$). If a good estimate of α is not initially available, the optimal times can be estimated using the corresponding T^+ values indicated.

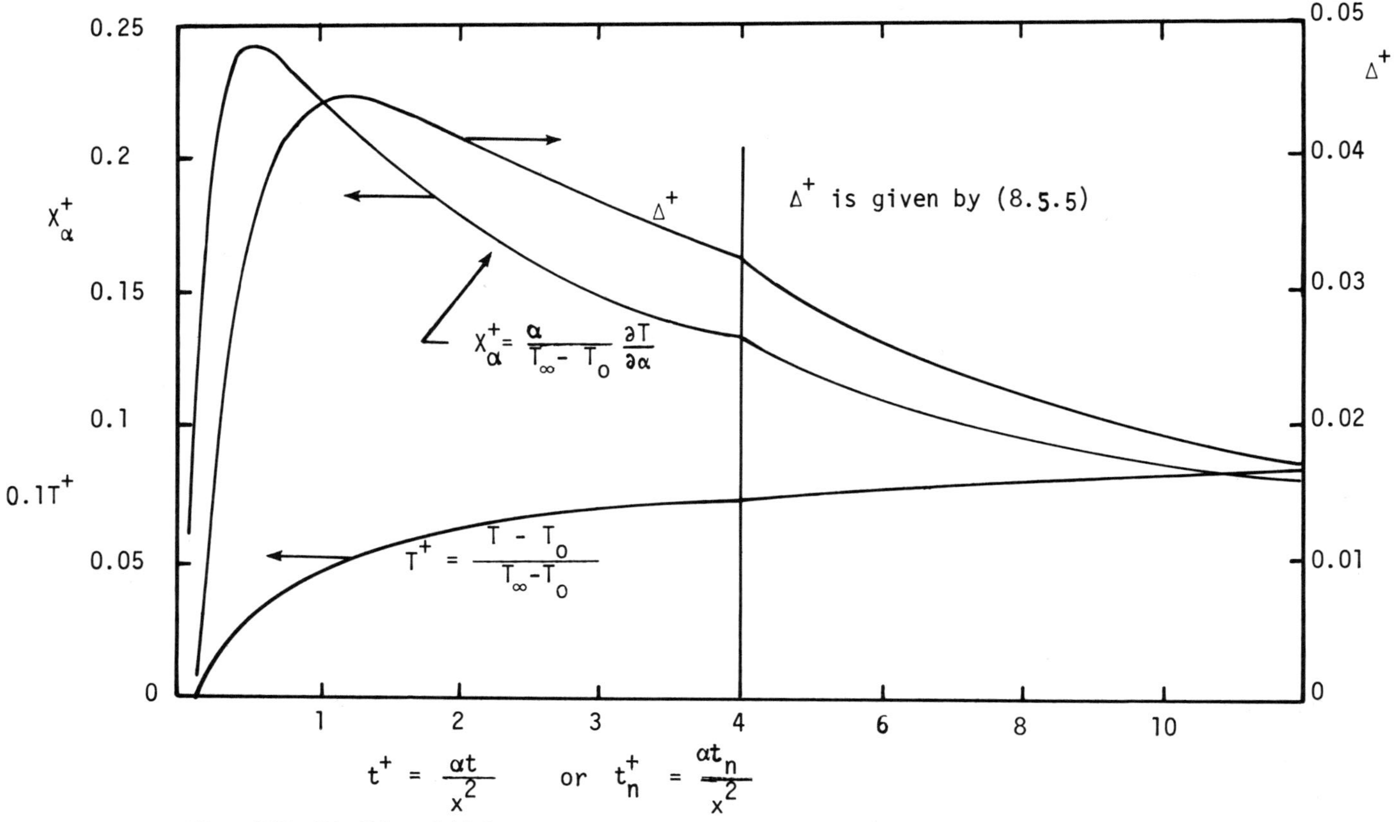

Figure 8.12 T^+, X_α^+, and Δ^+ for a point x in a semi-infinite body with a step change in surface temperature.

8.5.1.2 Constant Heat Flux Boundary Condition (Two Parameters)

If a flat electric heater is affixed to the surface of a large body and a constant current is passed through the heater, the surface heat flux into the body is constant. The surface temperature will respond in a similar manner from 3.3 to 12 sec as that shown in Fig. 7.17. If the body is semi-infinite with an initial temperature T_0 and is subjected to the constant heat flux q, the temperature response can be written as

$$T-T_0=2\left(\frac{qx}{k}\right)(t^+)^{1/2}\,\mathrm{ierfc}\left[(4t^+)^{-1/2}\right] \tag{8.5.6}$$

$$\mathrm{ierfc}(z)=\pi^{-1/2}\exp(-z^2)-z\,\mathrm{erfc}(z) \tag{8.5.7}$$

where t^+ is again $\alpha t/x^2$. In this case T is a nonlinear function of the *two* parameters, k and c (since $\alpha=k/c$). Another combination of parameters is α and k^{-1}; in this case T is nonlinear in α but linear in k^{-1}. [See (7.10.11).]

Dimensionless sensitivities for the parameters k and c are

$$X_1^+\equiv\frac{k}{qx/k}\frac{\partial T}{\partial k}=-\left(\frac{t^+}{\pi}\right)^{1/2}\exp\left(\frac{-1}{4t^+}\right)+\mathrm{erfc}\left[(4t^+)^{-1/2}\right] \tag{8.5.8}$$

$$X_2^+\equiv\frac{c}{qx/k}\frac{\partial T}{\partial c}=-\left(\frac{t^+}{\pi}\right)^{1/2}\exp\left(\frac{-1}{4t^+}\right) \tag{8.5.9}$$

[Verify that the relation given by (7.10.9) is satisfied by (8.5.6), (8.5.8), and (8.5.9).] These two sensitivities are depicted in Fig. 8.13; X_1^+ starts positive and goes negative whereas X_2^+ is always negative and larger in magnitude. At the time that X_1^+ goes to zero, the temperature T is insensitive (i.e., unchanged) by small changes in k. One significance of X_2^+ being larger in magnitude than X_1^+ is that, if only k or c were to be estimated, there would be on the average less relative uncertainty in c than k.

It is also instructive to evaluate T and the sensitivities at the surface $(x=0)$; we get

$$T(0,t)-T_0=2q\left(\frac{t}{kc\pi}\right)^{1/2} \tag{8.5.10}$$

$$\frac{k\partial T(0,t)}{\partial k}=-q\left(\frac{t}{kc\pi}\right)^{1/2}=\frac{c\partial T(0,t)}{\partial c} \tag{8.5.11}$$

Since the two sensitivities at $x=0$ are proportional, Δ is equal to zero and measurements at $x=0$ alone, no matter how accurate, cannot permit the independent estimation of both parameters.

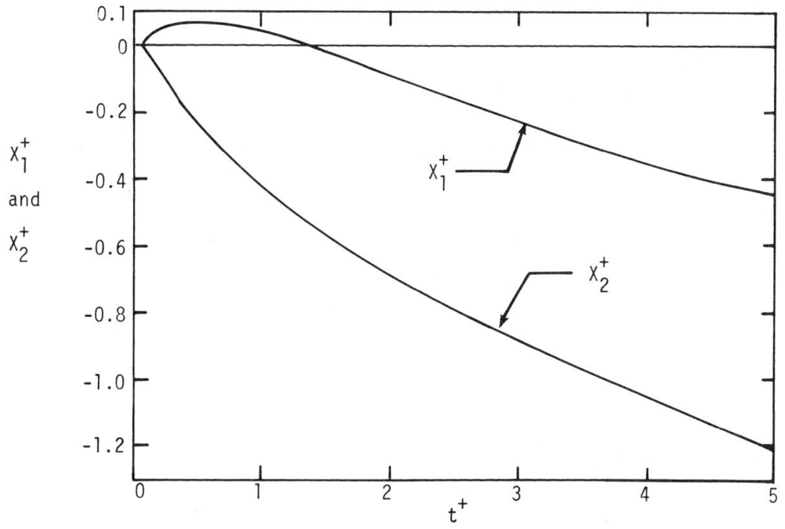

Figure 8.13 Sensitivities X_1^+ and X_2^+ for k and c for semi-infinite body with q=constant.

Because the sensitivities for $x>0$ are not proportional as shown in Fig. 8.13, any interior location can be used to provide data for estimating k and c. Not all locations or durations of the experiments are equally as effective, however. In order to find a meaningful optimal experiment, a constraint for the temperature rise is needed because as shown by (8.5.10) T goes to infinity as t increases without limit. From physical considerations only a finite maximum temperature is possible (materials melt or vaporize).

The constraint of the *same* maximum temperature rise can be introduced using (8.3.6, 7). Let η_{nom} be equal to qx/k; q is analogous to the adjustable constant C in Section 8.2. The quantity $\eta_{\max}$ in (8.3.6b) is the maximum rise of $T_{\max}-T_0$; thus η_m^+, also in (8.3.6b), is $(T_{\max}-T_0)/(qx/k)$, which from (8.5.10) is

$$\eta_m^+=\frac{T_{\max}-T_0}{qx/k}=\frac{T(0,t_n)-T_0}{qx/k}=\frac{2}{\pi^{1/2}}\left(\frac{kt_n}{cx^2}\right)=2(t_n^+/\pi)^{1/2} \quad (8.5.12)$$

The maximum temperature which occurs at $x=0$ and at time t_n is made to be the same in each case by appropriately adjusting q. (The x given explicitly in (8.5.12) refers to the location $x>0$ of a sensor.)

A plot of Δ^+ defined by (8.3.7) versus t_n^+ for one interior measurement yields a maximum Δ^+ value of 0.000167 at $t_n^+=\alpha t_n/x^2=8.5$. Again this results can be interpreted in two ways. First, for a given location of the temperature sensor, say, at $x=0.02m$ in an iron block ($\alpha=2\times10^{-5}$

m^2/sec), the optimal duration is $t_n = 8.5\ x^2/\alpha = 170$ secs. Second, for the same example if the optimal duration were desired to be 170 secs, then the sensor should be located 0.02 m from the heated surface.

It is instructive to study the case of two sensors, each producing equally spaced, independent measurements starting at $t=0$. If two thermocouples are located at the same x, the use of C_{ij}^+ [defined by (8.3.8)] in (8.3.7) would give the same optimal value of Δ^+. If a search is made for the optimal two locations, they are found to be at $x=0$ and at any $x>0$ so that $t^+ = \alpha t_n / x^2 = 1.5$; the associated Δ^+ value is 0.00263, which is almost 16 times the maximal value mentioned above for one sensor. Hence a design involving two sensors positioned as indicated would result in much greater accuracy in the estimates of k and c than if only a single sensor were used or if two were used at the same x.

8.5.1.3 Heat Flux Boundary Condition to Cause a Step Change in Surface Temperature

Temperatures inside the semi-infinite body change most for a given temperature range when the heated surface takes a maximal step increase. Both k and c can be estimated if this change in temperature is caused by a prescribed heat flux. (If the surface temperature is the specified boundary condition, only α can be estimated. See Section 8.5.1.1.) When the temperatures change most, the sensitivity coefficients would also be expected to be greatest in magnitude. [See (7.10.9) for a relation between T, $\partial T/\partial k$, and $\partial T/\partial c$.] We would anticipate for this reason that this case may have the optimal heat flux boundary condition.

A surface heat flux having the time dependence

$$q = a(t\pi)^{-1/2} \tag{8.5.13}$$

produces a step rise in surface temperature of $T_m - T_0$. The constant a is related to $T_m - T_0$ by $a = (kc)^{1/2}(T_m - T_0)$. The temperature distribution [9] and the k and c sensitivities are

$$T^+(t^+) \equiv \frac{T(x,t) - T_0}{T_m - T_0} = \operatorname{erfc}\left[(4t^+)^{-1/2}\right] = A, \quad t^+ = \frac{\alpha t}{x^2} \tag{8.5.14}$$

$$X_1^+ = \frac{k}{T_m - T_0}\frac{\partial T}{\partial k} = -\frac{1}{2}(A - B),$$

$$X_2^+ = \frac{c}{T_m - T_0}\frac{\partial T}{\partial c} = -\frac{1}{2}(A + B) \tag{8.5.15}$$

$$B \equiv (\pi t^+)^{-1/2}\exp\left(\frac{-1}{4t^+}\right) \tag{8.5.16}$$

In Fig. 8.3*a*, X_1^+ is the X_7 curve and C_{11}^+ is the Δ_7^n curve in Fig. 8.3*b*. Because the above case has a limited range of T, a constraint on T is incorporated in the solution.

An optimal location for one sensor again cannot be for $x=0$ as the sensitivities are proportional there. Optimal t_n^+ for one sensor occurs at $t_n^+=10$ at which time Δ^+ is the maximal value of 0.00232. If two sensors are optimally placed, they are at $x=0$ and at the x corresponding to $t_n^+=\alpha t_n/x^2=1.25$ where Δ^+ is the much larger value of 0.0113. Again two sensors located as indicated are much more effective than one.

8.5.1.4 Summary of Optimal Designs for Semi-Infinite Bodies Subjected to Heat Flux Boundary Conditions

A summary of results for the heat flux boundary condition is given in Table 8.2. Cases 1 and 4 are for a single sensor at $x=0$; precise measurements at only that location cannot be used to estimate independently k and c. However, if only k or c is estimated, $x=0$ is the optimal location. Also given are cases 2 and 5 which are for a single sensor at $x>0$. The optimal results are given by cases 3 and 6 for two sensors.

Table 8.2 Summary of Maximum Values of Δ^+ for Semi-Infinite Bodies with Heat Flux Boundary Condition. Δ^+ and the C_{ij}^+ are Normalized to Contain the Same Number of Measurements in Each Case

Case	Boundary Condition	Location of Sensors	Maximum Δ^+	Time of Max. Δ^+, $t_n^+=\alpha t_n/x^2$	Components of Maximum Δ^+ C_{11}^+	C_{22}^+	C_{12}^+
1	$q=$const.	$x=0$	0	—	0.125	0.125	0.125
2	$q=$const.	$x=x>0$	0.000167	8.5	0.0181	0.1119	0.0431
3	$q=$const.	$x=0,x$	0.00263	1.5	0.0631	0.0981	0.0597
4	q for $T=T_m$	$x=0$	0	—	0.25	0.25	0.25
5	q for $T=T_m$	$x=x>0$	0.002317	10.0	0.0585	0.2325	0.1062
6	q for $T=T_m$	$x=0,x$	0.0113	1.25	0.1275	0.2003	0.1192

The covariance matrix of the estimated parameter vector **b** having elements $\hat{k}$ and $\hat{c}$ is given by $(\mathbf{X}^T\mathbf{X})^{-1}\sigma^2$ provided standard assumptions of additive, zero mean, constant variance, independent normal errors apply (more specifically, assumptions denoted 11111-11). Then for n being the total number of measurements, the covariance of **b** is

$$\operatorname{cov}(\mathbf{b}) \cong \begin{bmatrix} C_{11} & C_{12} \\ C_{12} & C_{22} \end{bmatrix}^{-1} \sigma^2 = \begin{bmatrix} C_{22} & -C_{12} \\ -C_{12} & C_{11} \end{bmatrix} \frac{\sigma^2}{\Delta} \tag{8.5.17}$$

$$= \begin{bmatrix} \dfrac{C_{22}^+}{c^2} & -\dfrac{C_{12}^+}{kc} \\ -\dfrac{C_{12}^+}{kc} & \dfrac{C_{11}^+}{k^2} \end{bmatrix} \frac{\sigma^2 k^2 c^2}{n(T_m - T_o)^2 \Delta^+} \tag{8.5.18}$$

Values of C_{ij}^+'s are given in the last three columns of Table 8.2. We can use them, for example, to give the approximate standard deviation of k as

$$\frac{\sigma_k}{k} \cong \left(\frac{C_{22}^+}{n\Delta^+} \right)^{1/2} \left(\frac{\sigma}{T_m - T_o} \right) \tag{8.5.19}$$

The second factor in (8.5.19) can be considered to be relative measurement error in the temperature and the factor with the square root is an amplification factor for the conductivity. The smaller the amplification, the more precise are the k estimates. For $n=25$ the amplification factor is 5.2 for case 2 and 0.84 for case 6. This corroborates that larger values of Δ^+ result in experiments that permit estimating parameters with greater accuracy. Another use for expressions such as (8.5.19) is in determining the number n of measurements needed for specified accuracy.

Conclusions that can be drawn from Table 8.2 for estimating k and c are as follows:

1. A single sensor at $x=0$ is not permitted.
2. When one sensor at $x \neq 0$ is used the optimal time t_n^+ is about 10 for both heat flux boundary conditions.
3. When two sensors are used, one should be at $x=0$ and the other at $x>0$. Note that the optimal conditions for one sensor are *not* repeated.
4. The heat flux condition causing a step change in surface temperature (cases 5, 6) is much superior to the constant flux condition, cases 2 and 3.
5. The optimum of the optimal designs given in Table 8.2 is case 6. Hence, when k and c are estimated in a semi-infinite body, this would be the recommended design. It can be shown [13] that if more than two sensors are to be used, about half should be placed at $x=0$ and the remainder at $x=(\alpha t_n/1.25)^{1/2}$.

6. The number of optimal conditions can be less than, equal to, or more than the number of parameters. For a given heat flux boundary condition and one sensor, Δ^+ is maximized only with respect to t_n^+. Also for given $q(t)$ but with two sensors, Δ^+ is maximized with respect to two parameters relating to the location of the parameters. Finally for arbitrary $q(t)$, Δ^+ can be maximized by varying the function $q(t)$ which involves an infinite set of functions, two of which are illustrated in Table 8.2. Of all these possible functions none can yield larger Δ^+ values for semi-infinite bodies than the heat flux function of cases 5 and 6.

8.5.2 Finite Body Examples

8.5.2.1 Sinusoidal Initial Temperature in a Plate

Consider for the first finite body example the case of a plate which has a sinusoidal initial temperature and zero temperature boundary conditions,

$$T(x,0) = T_m \sin\left(\frac{\pi x}{L}\right), \qquad T(0,t)=0, \qquad T(L,t)=0 \tag{8.5.20}$$

The solution of (8.5.1) with these conditions is

$$T(x,t) = T_m \exp(-\pi^2 t^+)\sin\left(\frac{\pi x}{L}\right), \qquad t^+ \equiv \frac{\alpha t}{L^2} \tag{8.5.21}$$

Again for temperature boundary conditions, only the thermal diffusivity α appears—not k and c independently. The dimensionless α sensitivity is

$$X^+\left(\frac{x}{L}, t^+\right) = \frac{\alpha}{T_m}\frac{\partial T}{\partial \alpha} = -\pi^2 t^+ \exp(-\pi^2 t^+)\sin\left(\pi \frac{x}{L}\right) \tag{8.5.22}$$

This expression has maximal magnitude at $x/L = 0.5$ and $\pi^2 t^+ = 1$ (replace t^+ in Fig. 8.4 by $\pi^2 t^+$ to see the t^+ dependence). Consequently if only one sensor location is chosen, it should be at $x/L = 0.5$. Further, if only one time is selected, it should be at $t = L^2/\pi^2\alpha$.

Since the range of T is constrained to between 0 and T_m, the max Δ^+ criterion is appropriate for n equally spaced measurements starting at $t=0$ (n is "large"). Using (8.2.13) with $\eta_m^+ = T_m/T_m = 1$, an expression for Δ^+ is given. Necessary conditions for a maximum are

$$\frac{\partial \Delta^+}{\partial t_n^+} = 0, \qquad \frac{\partial \Delta^+}{\partial (x/L)} = 0 \tag{8.5.23}$$

Using Fig. 8.4 the optimal duration is $t_n^+ = \alpha t_n/L^2 = 1.691817/\pi^2$; the optimal x/L is 0.5 as for one measurement. Note that though there is only *one* parameter (namely, α), Δ^+ is maximized with respect to two variables.

We can also locate optimal positions for two sensors. In this case of one parameter, Δ^+ is given by C_{ij}^+ as defined by (8.3.8) with $m=2$, $\eta_m^+=1$, and $i=j=1$; Δ^+ is maximized by putting both sensors at $x/L=0.5$.

8.5.2.2 Constant Heat Flux at $x=0$, Insulated at $x=L$

A case permitting the two parameters k and c to be estimated is a plate exposed to a constant heat flux q on one side and insulated on the other. Mathematically this problem is described by (8.5.1) and

$$T(x,0)=T_0, \qquad -\frac{k\partial T(0,t)}{\partial x}=q, \qquad \frac{\partial T(L,t)}{\partial L}=0 \tag{8.5.24}$$

The dimensionless temperature [9] is

$$T^+=t^+ +\frac{1}{3}-x^+ +\frac{1}{2}(x^+)^2-\frac{2}{\pi^2}\sum_{n=1}^{\infty}\frac{1}{n^2}e^{-n^2\pi^2t^+}\cos n\pi x^+ \tag{8.5.25}$$

where $T^+=(T-T_0)/(qL/k)$, $x^+=x/L$, and $t^+=\alpha t/L^2$. In Figures 8.14, 15, and 16 the dimensionless temperature and k and c sensitivities are

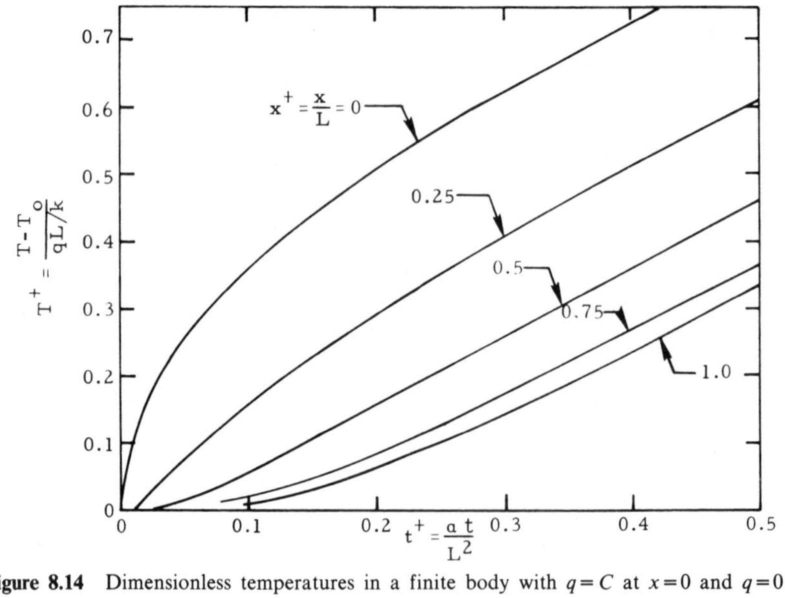

Figure 8.14 Dimensionless temperatures in a finite body with $q=C$ at $x=0$ and $q=0$ at $x=L$.

plotted versus t^+ for various positions in the plate. (X_1^+ and X_2^+ are determined in Problem 8.11). After an initial period, T^+ and $-X_2^+$ increase linearly with time whereas X_1 approaches various constant values including zero. Since X_1 goes to zero near $x/L=0.5$, this is a poor location for a temperature sensor in this case.

Suppose that both k and c are to be estimated using many equally spaced measurements. Assume that the standard assumptions denoted 11111-11 are valid. Since T increases without limit as $t \to \infty$, a constraint is needed. The Δ^+ criterion given by (8.3.7) can be used with C_{ij}^+ defined by (8.3.8) to include this constraint. The term η_m^+ is $(T_m - T_0)/(qL/k)$ which

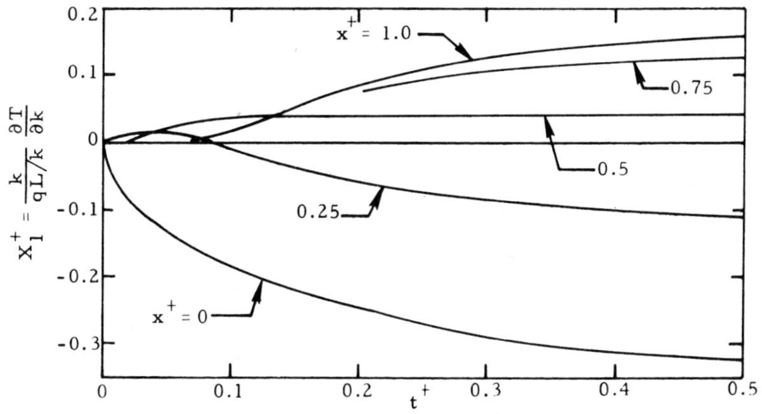

Figure 8.15 Dimensionless sensitivity X_1^+ for $q=C$ at $x=0$ and $q=0$ at $x=L$.

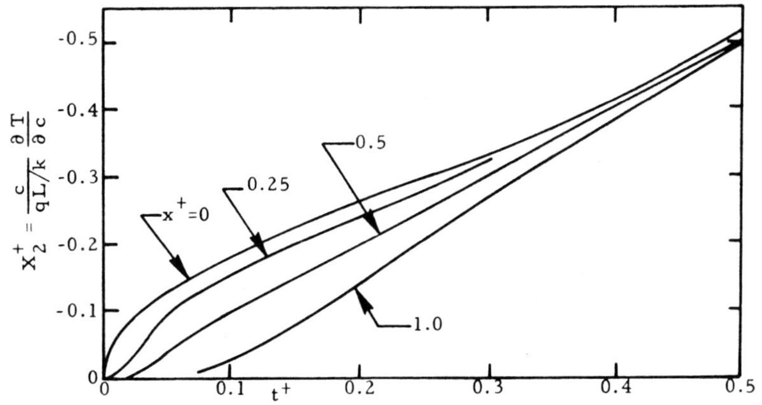

Figure 8.16 Dimensionless sensitivity X_2^+ for $q=C$ at $x=0$ and $q=0$ at $x=L$.

is given by (8.5.25) evaluated at $x^+=0$ and t_n^+; notice that η_m^+ is a function of only t_n^+. By using η_m^+ in this manner the maximum temperature, T_m, is made to be same value for each duration t_n.

Consider first the case of a single sensor. The optimal location is at $x=0$ and the optimal duration for taking uniformly spaced measurements is $t_n^+=1.2$. See case 1 of Table 8.3. This location is suggested from an inspection of Figs. 8.15 and 16 because the magnitude of the k and c sensitivities are largest at $x=0$. Their magnitudes were also largest for the semi-infinite body but we found that a single sensor at $x=0$ for the semi-infinite body would *not* permit both k and c to be estimated. The difference between the two cases is that though the k and c sensitivities are *proportional* [see (8.5.11)] at the heated surface of the semi-infinite body, they are *not* proportional at $x=0$ for the finite body (since X_1^+ approaches a constant and $-X_2^+$ increases with time). It does happen that the k and c sensitivities at $x=0$ for the finite body are nearly proportional until time $t^+=0.3$; clearly Δ^+ must have a maximum at a larger time than that.

Table 8.3 Summary of Maximum Values of Δ^+ for Finite Bodies Insulated on One Side

Case	Boundary conditions at $x=0$	Location of Temperature Sensors	Maximum Δ^+	Time of Maximum Δ^+, $t_n^+=\alpha t_n/L^2$
1	$q=$ constant	$x=0$	0.00098	1.2
2	$q=$ constant	$x=L$	0.00019	1.3
3	$q=$ constant	$x=0$ and L	0.00588	0.65
4	q for $T=T_m$	$x=0$	0.0291	1.8
5	q for $T=T_m$	$x=0$ and L	0.0358	0.76

Two additional optimal cases for T given by (8.5.25) are listed in Table 8.3. Case 2 is for a single sensor at $x=L$. Case 3 is for two sensors optimally located; of all possible two locations the best are at $x=0$ and L. If more than two sensors are used, the optimal design is approximated by having $m/2$ sensors at $x=0$ and $m/2$ at $x=L$. See Problem 8.13.

Recall from the way Δ^+ is defined that having a multiple number of sensors at the same location does not change the Δ^+ values. Notice that Δ^+ of case 1 is about one-sixth of Δ^+ for case 3. Hence the use of one sensor at $x=0$ and another at $x=L$ is much more effective for accurately estimating k and c than placing both at $x=0$.

In addition to optimal experiment durations and optimal sensor locations, optimal boundary conditions could be sought. The optimal heat flux boundary condition at $x=0$ is a heat flux history which causes the surface temperature to take a step increase to the maximum temperature. Cases 4 and 5 in Table 8.3 are for this boundary condition. Notice that Δ^+ of case 5, which is for measurements at $x=0$ and L, is the largest of all those listed in Table 8.3. A still larger value is found if an optimal boundary condition at $x=L$ is used [10].

In Tables 8.2 and 8.3 a number of optimal experiments are given. If we have the freedom to choose (1) the location and number of the temperature sensors, (2) the time variation of the heat flux, and (3) the geometry, an optimal experiment of those listed can be selected. In each case the decision is simply based on the size of Δ^+, with the largest values being best. Notice for comparable heating conditions and locations of sensors that the plate insulated at $x=L$ is always better. One could continue this search by modifying the insulation boundary condition and by investigating other geometries such as cylinders and spheres.

8.5.3 Additional Cases

Applications of the optimal criteria for various ordinary and partial differential equations are unlimited. The purpose of this subsection to provide more references.

Some analyses of optimal experiments involving ordinary equations are given by Heineken et al. [11] and Seinfeld and Lapidus [12, p. 432]. These references relate to optimal design for chemical rate constants. An ordinary differential in connection with the optimal design for heat transfer coefficients is studied by Van Fossen [13].

Further cases involving optimal estimation of parameters in the heat conduction equation or associated boundary conditions are given in references 14–20. Most of these cases involve consideration of linear partial differential equations. The dependent variable is usually a nonlinear function of the parameters even though the differential equation model is linear; nonlinear differential equations introduce further complications in the design of experiments. Two papers studying nonlinear differential equations models are [21], which considers the case of temperature variable k, and [22], which contains a study of optimal experiments for freezing–melting problems. One difficulty is that the sensitivities must be obtained numerically (see Section 7.10); the integrals in the C_{ij}^+'s must then be evaluated using trapezoidal or Simpson's rule. This is not a bothersome difficulty. One more complexity is including the constraint of maximal range of η when η is obtained from a nonlinear equation. In that case η_m^+ in (8.3.ba) is not a simple function of t_n.

8.5.4 Optimal Heat Conduction Experiment

As noted above there are many possible optimal experiments differing in geometry, number of sensors, boundary conditions, and so on. We naturally wish to design "best" experiments but practical aspects frequently mean that the optimum of all the optimal experiments cannot be chosen. Section 7.9.4 describes an experiment that is optimal in many respects for estimating k and c; this section is devoted to a description of the design of that experiment.

From a comparison of optimal results in Tables 8.2 and 8.3 the finite plate heated on one side and insulated on the other is found to be better than the semi-infinite geometry. It is also experimentally practical.

The locations for two or more thermocouples are at the heated ($x=0$) and insulated ($x=L$) surfaces. An equal number should be placed at each surface. Because eight were available, four were at $x=0$ and four at $x=L$. In order to ensure no direct heat losses from the heater, the heater was placed between two identical specimens, both of which had two sensors at $x=0$ and two at L. This placement of multiple sensors at the same location is contrary to intuition—one feels that a better design would be to place each sensor at a different position relative to the heated surface. If the heat conduction model used is correct, then the optimal locations *are* at $x=0$ and $x=L$. Placing them in this manner one maximizes Δ^+ which minimizes the variances of k and c. Furthermore the assumptions of constant variance and independent errors can be checked more readily than if measurements are not replicated.

The insulation boundary condition at $x=L$ can only be approximated since there are no perfect thermal insulators. The validity of this assumption can be investigated by noting if there is a charactersitic "signature" in the residuals.

With an electric heater a step increase in heat flux (i.e., constant flux) of finite duration is easily introduced. The heat flux to cause a step change in temperature at $x=0$ (which is the optimal experiment in Table 8.3) is not as readily applied. For that reason a constant heat flux for a finite duration was used. Figure 8.17 shows the Δ^+ criterion for this geometry for an equal number of sensors at $x=0$ and L. The heat flux is constant between times 0 and t_q. The constraints of a fixed large number of measurements and same maximum temperature rise are used. It is found that a shorter duration of heating than the interval over which data are used results in increased values of Δ^+. This means that there are two optimal times in this experiment: the duration of heating ($t_q^+=0.5$) and the maximum time at which data are used ($t_n^+\cong 0.75$). The experiment was designed to be near these conditions.

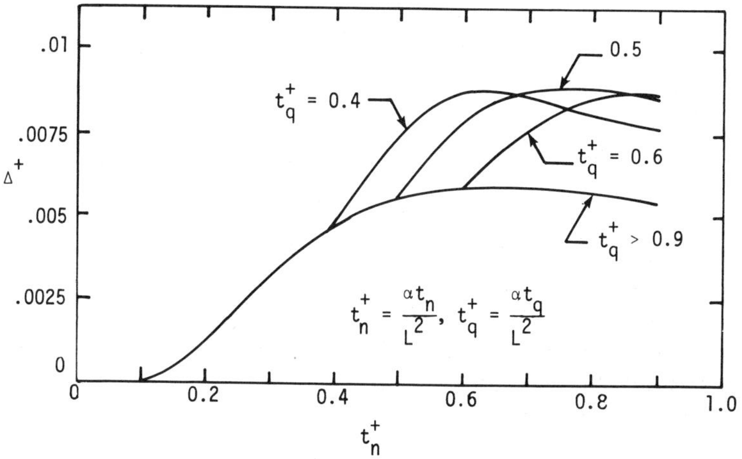

Figure 8.17 The Δ^+ criterion for a finite plate insulated at $x=L$ and heated at $x=0$ with a constant heat flux during times $0<t<t_q$ after which the flux is zero. There are an equal number of temperature sensors at $x=0$ as at $x=L$.

After the experiment is performed and parameters estimated, one should check the validity of the assumptions. Residuals for an actual experiment are shown in Figs. 7.20 and 7.21. Most of the residuals tend to decrease with time for the last third of the experiment. This suggests heat losses at $x=L$ and thus an imperfect model. Moreover, the residuals are highly correlated rather than being uncorrelated. In careful work both conditions would be further considered. It is anticipated, however, that the experiment design would not be greatly altered as the result of such investigation. See the next section for a brief discussion of the treatment of correlated errors.

8.6 NONSTANDARD ASSUMPTIONS

In this section the basic criterion is modified for cases when two standard assumptions are no longer valid. The cases of nonconstant variance measurement errors and correlated errors are considered.

8.6.1 Nonconstant Variance

For all the standard assumptions being valid except that the error variance is not constant [i.e., $E(\varepsilon_i^2)=\sigma_i^2$], the error covariance matrix is given by $\psi=\text{diag}[\sigma_1^2\ldots\sigma_n^2]$. For maximum likelihood estimation the criterion to maximize is (8.3.2). All the equations given above which include various constraints still may be used for nonconstant variance by simply replacing X_{ij} by $X_{ij}\sigma_i^{-1}$.

8.6.2 Correlated Errors

A particular type of correlated errors is the first-order autoregressive error which is described by

$$\varepsilon_i = \rho_i \varepsilon_{i-1} + u_i, \qquad i = 1, 2, \ldots, n \tag{8.6.1}$$

where the u_i are normal and independent with zero mean and variance σ^2. When maximum likelihood estimation is used, this case can also build on the previous results by replacing X_{ij} by

$$Z_{ij} = X_{ij} - \rho_i X_{i-1,j}; \qquad i = 1, \ldots, n; \quad j = 1, \ldots, p \tag{8.6.2}$$

where X_{0j} is defined to be zero for all permissible j values. For many equally spaced measurements in time, Z_{ij} can be approximated by

$$Z_{ij} \cong X_{ij}(1 - \rho_i) + \rho_i \Delta t \frac{\partial X_{ij}}{\partial t} \tag{8.6.3}$$

which indicates that as ρ_i approaches unity (perfect correlation) the time derivatives of the sensitivity coefficients become important.

8.7 SEQUENTIAL OPTIMIZATION

Suppose that a set of experiments have been performed and the associated parameters and parameter covariance matrix have been estimated. These experiments need not have been optimally designed but the next experiment (or set of measurements) is to be optimally designed. Suppose also that (subjective) MAP estimation is being used and that the standard assumptions denoted 11--1113 are valid. The criterion to maximize in this case is (see Appendix 8A)

$$\Delta = \left| \mathbf{X}^T \boldsymbol{\psi}^{-1} \mathbf{X} + \mathbf{V}_\beta^{-1} \right| \tag{8.7.1}$$

where $\mathbf{X}^T \boldsymbol{\psi}^{-1} \mathbf{X}$ is for the proposed experiment and $\mathbf{V}_\beta$ is the covariance matrix of the estimated parameter values based on data of the previous experiments and prior information. The dimensions of $\mathbf{X}^T \boldsymbol{\psi}^{-1} \mathbf{X}$ and $\mathbf{V}_\beta$ must be the same, that is, involve the same number of parameters.

To illustrate the criterion given by (8.7.1) assume that one previous experiment has been performed and that negligible prior information is available so that $\mathbf{V}_\beta^{-1}$ is

$$\mathbf{V}_\beta^{-1} = \left[\mathbf{X}^T \psi^{-1} \mathbf{X}\right]_1 \tag{8.7.2}$$

Then the second experiment would be designed so that

$$\Delta = \left| \left[\mathbf{X}^T \psi^{-1} \mathbf{X}\right]_1 + \left[\mathbf{X}^T \psi^{-1} X\right]_2 \right| \tag{8.7.3}$$

is maximized by the varying, the experiment duration, etc. Only the terms in $[\mathbf{X}^T \psi^{-1} \mathbf{X}]_2$ would be changed. In some cases the second experiment might be similar to the first one while in other cases it would be quite different.

The criterion given by (8.7.1) can also be expressed in a different form. By multiplying (8.7.1) by $|\mathbf{V}_\beta|$ we find

$$\Delta |\mathbf{V}_\beta| = |\mathbf{I} + \mathbf{X}^T \psi^{-1} \mathbf{X} \mathbf{V}_\beta| = |\mathbf{I} + \psi^{-1} \mathbf{X} \mathbf{V}_\beta \mathbf{X}^T| \tag{8.7.4}$$

Now using the identity $|\mathbf{I} + \mathbf{AB}| = |\mathbf{I} + \mathbf{BA}|$, (8.7.4) can be written as

$$\Delta = \frac{|\psi + \mathbf{X} \mathbf{V}_\beta \mathbf{X}^T|}{|\mathbf{V}_\beta| |\psi|} \tag{8.7.5}$$

Since $|\mathbf{V}_\beta||\psi|$ is a positive constant, maximizing Δ is equivalent to maximizing

$$T = |\psi + \mathbf{X} \mathbf{V}_\beta \mathbf{X}^T| \tag{8.7.6}$$

Hence we have a choice between maximizing Δ or T. Our choice should depend on the relative dimensions of the two matrices, which are $p \times p$ and $n \times n$, respectively. The determinant of lower dimension would be chosen. A case favorable to using T is for $p \leqslant 2$ and for $n = 1$, that is, a single measurement of the dependent variable is made.

8.8 NOT ALL PARAMETERS OF INTEREST

There are parameter estimation problems that require the estimation of parameters in addition to those of primary interest. The extra parameters are sometimes termed nuisance parameters. In Example 8.2.4 the parameter β (reciprocal time constant of the billet) might be the one of interest; however, it might also be necessary to estimate simultaneously the fluid temperature T_∞. Another type of problem is when statistical parameters such as the correlation ρ in the autoregression error model (8.6.1) are found. Though the ρ value may be needed to estimate the confidence region, generally its value is not needed as accurately as those of the "physical" parameters. Further examples are given by Hunter and Hill [23, 24].

Appendix 8B gives a derivation of a criterion when out of a total of p estimated parameters only the first q $(p>q)$ are of interest. For the standard assumptions designated 11111-11 the criterion is to maximize

$$\Delta_{qp} = |\mathbf{X}_1^T\mathbf{X}_1 - \mathbf{X}_1^T\mathbf{X}_2(\mathbf{X}_2^T\mathbf{X}_2)^{-1}\mathbf{X}_2\mathbf{X}_1^T| = \frac{\Delta_p}{|\mathbf{X}_2^T\mathbf{X}_2|} \tag{8.8.1}$$

where $\mathbf{X}_1$ is an $n\times q$ matrix and is for the first q parameters and where $\mathbf{X}_2$ is an $n\times r$ matrix which is for the remaining $r=p-q$ parameters that are not of primary interest. The symbol Δ_p means the usual determinant of all the parameters, i.e.,

$$\Delta_p = |\mathbf{X}^T\mathbf{X}| \qquad \text{where} \quad \mathbf{X} = [\mathbf{X}_1 \ \mathbf{X}_2]$$

The minimum q and r values are $q=1$ and $r=1$. In summation form this simple case results in

$$\Delta_{12} = \Sigma X_{i1}^2 - (\Sigma X_{i1}X_{i2})^2(\Sigma X_{j2}^2)^{-1} \tag{8.8.2}$$

Let the condition of a fixed large number of measurements equally spaced in time be valid; by using the notation given by (8.3.5), Δ_{12} can be approximated by

$$\Delta_{12} = n\Delta_{12}^n \cong n\left[C_{11} - C_{12}^2 C_{22}^{-1}\right] = n\Delta_2^n C_{22}^{-1} \tag{8.8.3}$$

A comparison of this expression with (8.5.17) shows that Δ_{12}^n is proportional to the reciprocal of the variance of b_1. Hence maximizing Δ_{12}^n has the beneficial effect of minimizing the variance of b_1.

As an example of the use of the $\max \Delta_{12}^n$ criterion, consider the exponential model

$$\eta = \beta_1 e^{-\beta_2 t}, \qquad t^+ = \beta_2 t \tag{8.8.4}$$

which has one linear and one nonlinear parameter. For β_1 being the parameter of interest, the criterion is to maximize Δ_{12}^n, as implied by (8.8.3); here

$$C_{11}=\frac{1}{t_n}\int_0^{t_n} e^{-2\beta_2 t}\,dt=\frac{1}{2t_n^+}\left[1-\exp(-2t_n^+)\right] \tag{8.8.5}$$

$$C_{12}=\frac{1}{t_n}\int_0^{t_n} \beta_1 t e^{-2\beta_2 t}\,dt=\frac{\beta_1}{4\beta_2 t_n^+}\left[1-e^{-2t_n^+}(1+2t_n^+)\right] \tag{8.8.6}$$

For C_{22}, see Problem 8.1. From these expressions we see that Δ_{12}^n can be plotted as a function of $t_n^+=\beta_2 t_n$ as depicted in Fig. 8.18. For β_1 being the primary parameter of interest, the optimal value of t_n^+ is small as possible. If instead β_2 is the parameter of primary interest, the subscripts 1 and 2 of the C's in (8.8.3) are interchanged. Since the resulting Δ_{12}^n is proportional to $(\beta_1/\beta_2)^2$, plotted also in Fig. 8.18 is $\beta_2^2\Delta_{12}^n/\beta_1^2$ versus t_n^+. The optimal time for this case is $t_n^+=2.0$. This dimensionless time can be compared with the optimal times for estimating β_1 alone of zero, β_2 alone of 1.692, and both β_1 and β_2 of 1.191. Hence the optimal durations of the experiment can be quite different for the various objectives of β_1 only being of interest, and so on.

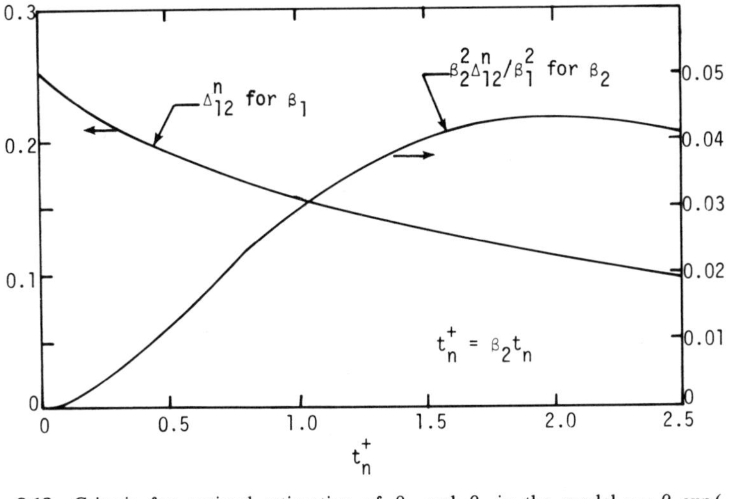

Figure 8.18 Criteria for optimal estimation of β_1 and β_2 in the model $\eta=\beta_1\exp(-\beta_2 t)$ where each may be of primary interest.

8.9 DESIGN CRITERIA FOR MODEL DISCRIMINATION

Sometimes the physical model is not known but several alternate models can be proposed. In such cases the problem is to select the "best" model, that is, the one that best fits the data. A method of model selection, termed model discrimination, involves experimental designs that maximize differences between predicted responses of two or more models.

A chemical engineering example of a case where discrimination is needed occurs when substance A reacts in the presence of a catalyst to form substance B, which in turn forms C. Two possible models are $A \rightarrow B \rightarrow C$ and $A \rightarrow B \rightleftarrows C$. The predicted concentrations of substance B versus time for the two models are shown in Fig. 8.19. If the reaction is observed only until time t_1, no discrimination can be accomplished because the predicted responses are nearly identical until t_1. Measured values of the B concentration are required after time t_1 (and preferable near t_2) to determine the best model.

Many methods of model discrimination have been proposed. Given first is a method that results in a criterion similar to Δ. Next discussed is a method utilizing information theory. The former method is simpler in application but the latter has a more satisfying basis. In each case the analyses start with consideration of two competing mathematical models.

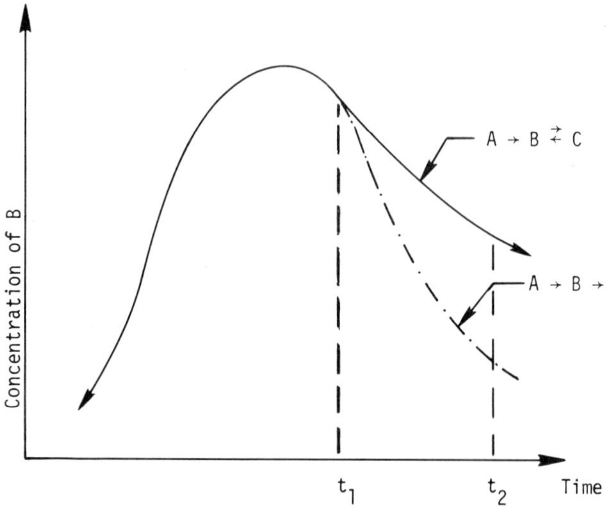

Figure 8.19 Discrimination example involving concentration of substance B for models $A \rightarrow B \rightarrow C$ and $A \rightarrow B \rightleftarrows C$.

8.9.1 Linearization Method

In this method the objective is to seek experiments that cause the minimum values of the sum of squares functions to be quite different for two competing models. Suppose two models are available and the best one is to be determined. Let the standard assumptions 1111--11 be valid and OLS estimation be used. (The analysis can be modified for other cases). The sum of squares function for model i can be written as

$$S^{(i)} = (\mathbf{Y} - \boldsymbol{\eta}^{(i)})^T (\mathbf{Y} - \boldsymbol{\eta}^{(i)}) \tag{8.9.1}$$

Let the model equation be $\boldsymbol{\eta}^{(i)} = \mathbf{f}^{(i)}(\mathbf{x}, \boldsymbol{\beta}_c, \boldsymbol{\beta}^{(i)})$ where $\mathbf{x}$ is the independent variable vector, $\boldsymbol{\beta}_c$ is the vector of parameters common to both models (if there are common ones), and $\boldsymbol{\beta}^{(i)}$ is the q vector of parameters distinctive to model i. Suppose that a nominal set of parameters is chosen and that $\boldsymbol{\eta}^{(i)}$ is expressed in terms of a Taylor series near this nominal set so that

$$\boldsymbol{\eta}^{(i)} \cong \boldsymbol{\eta}^{(0)} + \mathbf{X}^{(i)} \Delta\boldsymbol{\beta}^{(i)} \tag{8.9.2}$$

where $\mathbf{X}^{(i)}$ is the sensitivity matrix for $\boldsymbol{\beta}^{(i)}$. Introducing (8.9.2) into (8.9.1) where the $\Delta\boldsymbol{\beta}^{(i)}$ values are chosen to minimize $S^{(i)}$ yields

$$\min S^{(i)} = (\mathbf{Y} - \boldsymbol{\eta}^{(0)})^T (\mathbf{Y} - \boldsymbol{\eta}^{(0)}) + 2(\mathbf{X}^{(i)} \Delta\boldsymbol{\beta}^{(i)})^T (\boldsymbol{\eta}^{(0)} - \mathbf{Y})$$

$$+ (\mathbf{X}^{(i)} \Delta\boldsymbol{\beta}^{(i)})^T \mathbf{X}^{(i)} \Delta\boldsymbol{\beta}^{(i)} \tag{8.9.3}$$

which implies the $\Delta\boldsymbol{\beta}^{(i)}$ vector of

$$\Delta\boldsymbol{\beta}^{(i)} = (\mathbf{X}^{T(i)} \mathbf{X}^{(i)})^{-1} \mathbf{X}^{T(i)} (\mathbf{Y} - \boldsymbol{\eta}^{(0)}) \tag{8.9.4}$$

Let us now subtract $\min S^{(2)}$ from $\min S^{(1)}$ and attempt to find the maximum of the absolute value of the difference or

$$C = \max\left|\min S^{(1)} - \min S^{(2)}\right|$$

$$= \max\left|(\mathbf{Y} - \boldsymbol{\eta}^{(0)})^T \left[\mathbf{X}^{(2)} (\mathbf{X}^{T(2)} \mathbf{X}^{(2)})^{-1} \mathbf{X}^{T(2)} - \mathbf{X}^{(1)} (\mathbf{X}^{T(1)} \mathbf{X}^{(1)})^{-1} \mathbf{X}^{T(1)}\right] (\mathbf{Y} - \boldsymbol{\eta}^{(0)})\right| \tag{8.9.5}$$

Although we do not know $\mathbf{Y} - \boldsymbol{\eta}^{(0)}$, let us assume temporarily that Model 1 is correct and that the measurement errors are sufficiently small so that

$$\mathbf{Y} - \boldsymbol{\eta}^{(0)} \cong \mathbf{X}^{(1)} \Delta\boldsymbol{\beta}^{(1)} \tag{8.9.6}$$

Then C given by (8.9.5) becomes

$$C = \min|\Delta\boldsymbol{\beta}^{T(1)}\mathbf{M}^{(1)}\Delta\boldsymbol{\beta}^{(1)}| \tag{8.9.7a}$$

$$\mathbf{M}^{(1)} = \mathbf{X}^{T(1)}\mathbf{X}^{(1)} - \mathbf{X}^{T(1)}\mathbf{X}^{(2)}\left(\mathbf{X}^{T(2)}\mathbf{X}^{(2)}\right)^{-1}\mathbf{X}^{T(2)}\mathbf{X}^{(1)} \tag{8.9.7b}$$

The $q \times q$ matrix $\mathbf{M}^{(1)}$ is exactly the same matrix whose determinant is maximized when $\mathbf{X}^{(1)}$ is for q parameters of primary interest and $\mathbf{X}^{(2)}$ is for r parameters of less interest, see (8.8.1).

If instead of Model 1 being correct, Model 2 which involves r parameters is correct, (8.9.7) becomes

$$C = \min|\Delta\boldsymbol{\beta}^{T(2)}\mathbf{M}^{(2)}\Delta\boldsymbol{\beta}^{(2)}| \tag{8.9.8a}$$

$$\mathbf{M}^{(2)} = \mathbf{X}^{T(2)}\mathbf{X}^{(2)} - \mathbf{X}^{T(2)}\mathbf{X}^{(1)}\left(\mathbf{X}^{T(1)}\mathbf{X}^{(1)}\right)^{-1}\mathbf{X}^{T(1)}\mathbf{X}^{(2)} \tag{8.9.8b}$$

The problem now is to select some criterion that has the effect of maximizing C. If C is fixed at some value, (8.9.7a) and (8.9.8a) both describe the surfaces of hyperellipsoids since both are very similar to the confidence region expression given by (6.8.39). The coordinates are the $\Delta\beta$'s. In the case of (8.9.7a), for a given hypervolume C is maximized by maximizing the determinant of $\mathbf{M}^{(1)}$. For Model 2 being correct, the analogous criterion is the maximization of $\mathbf{M}^{(2)}$. But since we do not know which model is correct, we choose a criterion that does not prefer one model over the other. Such a criterion is simply formed from the augmented $\mathbf{X}^T\mathbf{X}$ matrix. That is, we propose that discrimination can be improved by designing experiments so that

$$\Delta = \left|\begin{bmatrix}\mathbf{X}^{T(1)}\\ \mathbf{X}^{T(2)}\end{bmatrix}\begin{bmatrix}\mathbf{X}^{(1)} & \mathbf{X}^{(2)}\end{bmatrix}\right| = \begin{vmatrix}\mathbf{X}^{T(1)}\mathbf{X}^{(1)} & \mathbf{X}^{T(2)}\mathbf{X}^{(2)}\\ \mathbf{X}^{T(2)}\mathbf{X}^{(1)} & \mathbf{X}^{T(2)}\mathbf{X}^{(2)}\end{vmatrix}$$

$$= |\mathbf{X}^{T(2)}\mathbf{X}^{(2)}||\mathbf{M}^{(1)}| = |\mathbf{X}^{T(1)}\mathbf{X}^{(1)}||\mathbf{M}^{(2)}| \tag{8.9.9}$$

is maximized. Note now that the $\mathbf{X}$ matrix is composed of sensitivity matrices from two different models and that $\mathbf{X}^{(1)}$ has dimensions $n \times q$ and $\mathbf{X}^{(2)}$ has dimensions $n \times r$. An advantage of the Δ criterion given by (8.9.9) is that it is simple; its use is similar to the Δ criterion discussed in Sections 8.1–8.7. A further advantage is that no data are needed for the design of experiments using this criterion; one needs only the models and some approximate values of the parameters.

The effect of maximizing Δ given by (8.9.9) is to emphasize the differences between the models. All models fail at some point and it may be that at these points the greatest discrimination power is present. For example, there are certain heat conduction problems in which changes occurring during heating of a material may be due to a change of phase or a chemical reaction. One of these is reversible and the other is not. This suggests that the critical temperature range be covered using a cooling after a heating process. The behavior of the change of phase and reaction models are quite different during the cooling period.

Another example where discrimination might be used in determining if $h=\beta_1+\beta_2 t$ or $h=\beta_1+\beta_2(T-T_\infty)$ is the better model of Section 7.5.2.

Example 8.9.1

Consider the two competing models

$$\eta_1=\beta_1+\beta_2(1-e^{-t}),\qquad \eta_2=\beta_1+\beta_2\sin t$$

The standard error assumptions are valid. The optimal duration of experiments for a large fixed number of uniformly spaced measurements starting at $t=0$ is to be found. No constraints on the ranges of the η's are to be used.

Solution

Since the constant parameter β_1 appears in both models, both models are alike to that extent. Hence the emphasis should be upon the β_2 terms. Using the above notation we have

$$\mathbf{X}^{T(1)}=[1-e^{-t_1}\ \dots\ 1-e^{-t_n}],\qquad \mathbf{X}^{T(2)}=[\sin t_1\ \dots\ \sin t_n]$$

The quantity to maximize is Δ given by (8.9.9). To include the assumption of a fixed large number of uniformly spaced measurements, Δ should be modified to Δ^n as indicated by (8.3.5). In this case C_{11} would be C_{11} of Fig. 8.11 and C_{22} would be C_{22} of Fig. 8.9. The resulting Δ^n is nearly zero until time $t_n=2.5$ at which time Δ^n rises quickly to the first local maximum of about 0.4 at time 5.5. After this time the Δ^n criterion gradually oscillates to larger values with the global maximum being 0.5 at $t_n\to\infty$. These results are reasonable because $\sin t$ and $1-e^{-t}$ are similar until $t=1.5$ but are quite dissimilar for $t\geqslant 3$.

According to the max Δ^n criterion which assumes many equally spaced measurements starting at $t=0$, then, the experiment should be of infinite duration but for practical purposes it could be any time greater than $t_n=5$ to discriminate between the two models.

8.9.2 Information Theory Method

Suppose that two rival models are available, $\boldsymbol{\eta}^{(i)}=\mathbf{f}^{(i)}(\mathbf{x},\boldsymbol{\beta}^{(i)})$ where $i=1$ and 2. Assume that estimates $\mathbf{b}^{(i)}$ for the parameters appearing in the ith model are available and that the associated estimated covariance matrix $\mathbf{V}_0^{(i)}$ is known. Typically these are obtained by fitting each model in turn to data from previously performed experiments. Using the parameter values $\mathbf{b}^{(i)}$ the values of the dependent variable $\boldsymbol{\eta}^{(i)}$ can be predicted for any proposed experiment, assuming the ith model is correct. This prediction is designated

$$\hat{\mathbf{Y}}^{(i)} = \mathbf{f}^{(i)}(\mathbf{x}, \mathbf{b}^{(i)}) \tag{8.9.10}$$

The covariance matrix of the prediction error in (8.9.10), assuming that model i is correct, can be shown to be approximately

$$\mathbf{V}^{(i)} = \boldsymbol{\psi} + \mathbf{X}^{(i)}\mathbf{V}_0^{(i)}\mathbf{X}^{T(i)} \tag{8.9.11}$$

where $\boldsymbol{\psi}$ is the covariance of the measurement errors of $\mathbf{Y}$ and $\mathbf{X}^{(i)}$ is the sensitivity matrix for the ith model and the experiment being considered. The second term on the right side of (8.9.11) is similar to that given by (6.2.12a) or (6.5.6).

The hypothesis that the ith model is correct leads to regarding the outcome of a proposed experiment $\mathbf{x}$ as a random variable $\boldsymbol{\eta}$ with probability density function $p^{(i)}(\boldsymbol{\eta}|\mathbf{x})$ having mean and covariance given by (8.9.10) and (8.9.11), respectively. If Model 1 is correct, then $\boldsymbol{\eta}$ is distributed as $p^{(1)}(\boldsymbol{\eta}|\mathbf{x})$; if Model 2 is correct, $\boldsymbol{\eta}$ is distributed as $p^{(2)}(\boldsymbol{\eta}|\mathbf{x})$. Kullback [28] has suggested that the quantity $\ln[p^{(1)}(\boldsymbol{\eta}|\mathbf{x})/p^{(2)}(\boldsymbol{\eta}|\mathbf{x})]$ is a measure of the favorability of hypothesis 1 over hypothesis 2. The expected information in favor of Model (or hypothesis) 1 is

$$\int_{-\infty}^{\infty} p^{(1)}(\boldsymbol{\eta}|\mathbf{x}) \ln \frac{p^{(1)}(\boldsymbol{\eta}|\mathbf{x})}{p^{(2)}(\boldsymbol{\eta}|\mathbf{x})}\, d\boldsymbol{\eta}$$

But since it is not known whether Model 1 or 2 is correct, Kullback suggested that the measure of total information $J_{1,2}$ be maximized where

$$J_{1,2}(\mathbf{x}) = \int_{-\infty}^{\infty} \left[p^{(1)}(\boldsymbol{\eta}|\mathbf{x}) \ln \frac{p^{(1)}(\boldsymbol{\eta}|\mathbf{x})}{p^{(2)}(\boldsymbol{\eta}|\mathbf{x})} + p^{(2)}(\boldsymbol{\eta}|\mathbf{x}) \ln \frac{p^{(2)}(\boldsymbol{\eta}|\mathbf{x})}{p^{(1)}(\boldsymbol{\eta}|\mathbf{x})} \right] d\boldsymbol{\eta} \tag{8.9.12}$$

The objective is to select an experiment $\mathbf{x}$ that maximizes $J_{1,2}(\mathbf{x})$. A large value of $J_{1,2}$ can be obtained only if $p^{(1)}$ is much larger than $p^{(2)}$, or vice versa. In either case the result is a strong preference for one model over the other. The quantity $J_{1,2}$ is called by Kullback the *information for discrimination* and is similar to (8A.4).

Let the measurement errors be normal (more specifically, 11--1011) and let the model errors have covariance matrices $\mathbf{V}^{(1)}$ and $\mathbf{V}^{(2)}$. Then it can be shown that

$$J_{1,2}(\mathbf{x}) = -m + \tfrac{1}{2}\mathrm{tr}\left[\mathbf{U}^{(1)}\mathbf{V}^{(2)} + \mathbf{U}^{(2)}\mathbf{V}^{(1)}\right]$$

$$+\tfrac{1}{2}(\hat{\mathbf{Y}}^{(2)} - \hat{\mathbf{Y}}^{(1)})^T(\mathbf{U}^{(1)} + \mathbf{U}^{(2)})(\hat{\mathbf{Y}}^{(2)} - \hat{\mathbf{Y}}^{(1)}) \tag{8.9.13}$$

where $\mathbf{U}^{(i)} \equiv (\mathbf{V}^{(i)})^{-1}$. An important special case occurs when one dependent variable is present in the model and only one measurement of it is made. Then for $i = 1,2$, $\mathbf{V}^{(i)} = \sigma_i^2$ and $\mathbf{U}^{(i)} = \sigma_i^{-2}$ where

$$\sigma_i^2 = s_i^2 + \sum_{k=1}^{p}\sum_{l=1}^{p} V_{kl}^{(i)} X_k^{(i)} X_l^{(i)} \tag{8.9.14}$$

and s_i^2 is an estimate of $V(Y)$. Then (8.9.13) becomes

$$J_{1,2}(x) = -1 + \tfrac{1}{2}\left\{\left(\frac{\sigma_1}{\sigma_2}\right)^2 + \left(\frac{\sigma_2}{\sigma_1}\right)^2 + \left[\sigma_1^{-2} + \sigma_2^{-2}\right]\left[\hat{Y}^{(2)} - \hat{Y}^{(1)}\right]^2\right\} \tag{8.9.15}$$

The information regarding the experiment x is contained in σ_1^2, σ_2^2, $\hat{Y}^{(1)}$, and $\hat{Y}^{(2)}$. The objective is now to choose the measurement so that $J_{1,2}(x)$ is maximized.

Box and Hill [29] were the first to derive (8.9.15); they have been pioneers in the application of sequential design of experiments for model discrimination.

Let us briefly consider some implications of (8.9.13)–(8.9.15). Hypothetical plots of the predicted values $\hat{Y}^{(1)}$ and $\hat{Y}^{(2)}$ are shown in Fig. 8.20 as a function of time t (which is $\mathbf{x}$ in this case). If a single time is to be chosen to decide between models 1 and 2, time t_1, where the responses coincide, would not be helpful; time t_2, where $(\hat{Y}^{(2)} - \hat{Y}^{(1)})^2$ is a maximum, would be better. The single best measurement *time* according to (8.9.15) is when $(\hat{Y}^{(2)} - \hat{Y}^{(1)})^2$ is a maximum provided σ_1^2 and σ_2^2 vary only slightly with t. The decision of which *model* to choose depends upon how $\hat{Y}^{(1)}$ and $\hat{Y}^{(2)}$ compare with the measured value Y at the same time. If $\hat{Y}^{(i)}$ is nearer to Y than $\hat{Y}^{(j)}$, then model i would be selected (for $i,j = 1,2$ and $i \neq j$).

Should Y be midway between $\hat{Y}^{(1)}$ and $\hat{Y}^{(2)}$, there is no basis for model discrimination. It is interesting to compare the criterion for this case with the one previously given (Δ). For this latter criterion, $S^{(1)} - S^{(2)}$ would be zero and thus the observation would be sought at some other t where

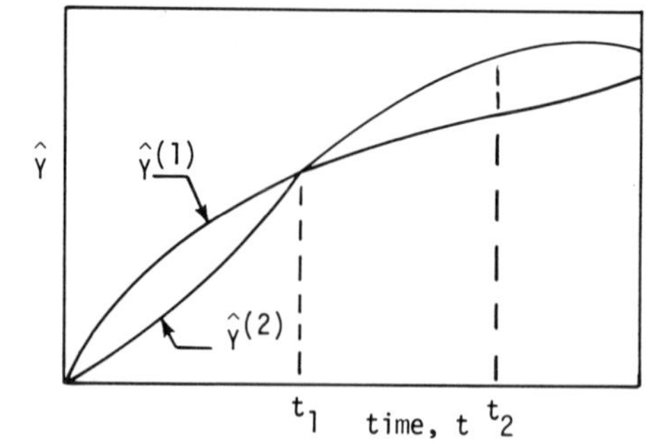

Figure 8.20 Discrimination between two predicted Y's using the information theory method.

$|S^{(1)} - S^{(2)}|$ is a maximum. Hence the two criteria may not yield the same optimal experiments.

There are several ways to treat more than two models. One is the following. After each experiment is performed, the likelihood $L^{(i)}$ associated with each model and its current parameters is computed. We then design the next experiment in such a manner as to discriminate between the two models having the largest likelihood values. Another method for discrimination between more than two models is given by Box and Hill [29].

8.9.2.1 Termination Criteria

A general sequential procedure of mechanistic model building can be visualized as including the steps in Fig. 8.21. (By "mechanistic" we mean a model that can be derived from basic principles.) Note that on the left are tasks performed by the analyst, in the center by the computer, and on the right by the laboratory. After starting one can propose some competing models. G. E. P. Box has made the point that one should not be timid in proposing models. The process itself should lead to discarding unsuitable models. Next comes performing experiments, followed by estimating all the parameters for all the models (block 3). In block 4, optimal experiments are sought to discriminate between the competing models. The method of Box and Hill could be used for this purpose. If desired, the experiment in block 2 could have been designed using the method in Section 8.9.1, which does not directly utilize experimental data.

After the optimal experimental conditions (designated $\mathbf{x}_j$) in block 4 are found, the new experiment is performed (block 5), after which the estimates for all the parameters are found, $b_1^{(1)}, b_2^{(2)}, \ldots$. Then in block 7 a test

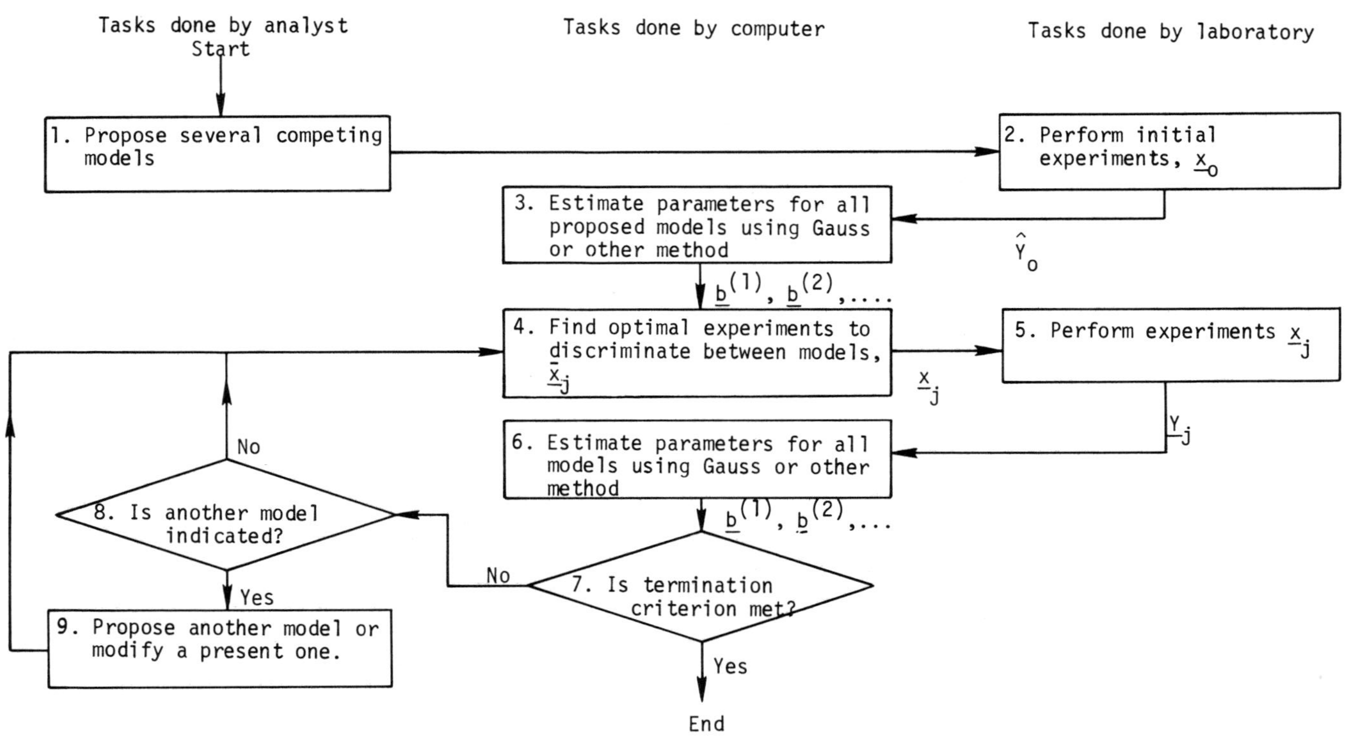

Figure 8.21 A sequential procedure for mechanistic model building including design of experiments to discriminate between competing models.

is made to ascertain if any of the proposed models is satisfactory. At the same time certain of them may be discarded. The rest of this section is a discussion of a termination criterion and suggestions for determining if another model is needed.

The wide applicability of the maximum likelihood method of estimation of parameters and of generalized likelihood ratio tests suggest the consideration of likelihood ratios in selecting the better of two models.

Suppose that the objective is to choose one of two hypotheses, H_1 (Model 1 is correct) or H_2 (Model 2 is correct). Let $L^{(i)}(\mathbf{Y}, b^{(i)})$ be the maximum joint probability density function associated with the data obtained thus far for the ith model and the associated parameters $\mathbf{b}^{(i)}$.

A likelihood ratio test can be constructed as follows:

1. If $L^{(1)}/L^{(2)} \leqslant A$, accept hypothesis 2.
2. If $L^{(1)}/L^{(2)} \geqslant B$, accept hypothesis 1.
3. If $A < L^{(1)}/L^{(2)} < B$, investigate alternate models and perform more experiments.

Methods for choosing A and B are discussed from differing points of view by Ghosh [30], by Fedorov [3], and by Bard [31]. Bard suggests that the relations between A and B and the probabilities of error which Wald [32] gave for testing simple versus simple hypotheses where sample sizes are large will work approximately in this situation. That is, if we let α_1 be the probability that H_1 is accepted when H_2 is true and α_2 the probability that H_2 is accepted when H_1 is true, then for independent observations

$$A \cong \frac{\alpha_2}{1-\alpha_1}, \qquad B \cong \frac{1-\alpha_2}{\alpha_1} \tag{8.9.16a}$$

$$\alpha_1 \cong \frac{1-A}{B-A}, \qquad \alpha_2 \cong \frac{A(B-1)}{B-A} \tag{8.9.16b}$$

These relations mean, for example, that if we wish to be 90% certain that we accept H_1 only if H_1 is true and 80% certain that we accept H_2 only if H_2 is true, then $\alpha_1 = 0.1$ and $\alpha_2 = 0.2$. Then using (8.9.16a), $A \cong 0.2/0.9 = 0.222$ and $B \cong 0.8/0.1 = 8$. If we had started with A and B, then the corresponding probabilities would be found using (8.9.16b).

In addition to continuing experimentation when the likelihood ratio is between A and B, we should also inspect the residuals to see if any insight can be gained for improving any of the models or for proposing another model. This would then lead to blocks 8 and 9 in Fig. 8.21. Regions of large departure in the residuals from random conditions can sometimes imply improvements in the models. Also insight into statistical assumptions can be gained through inspection of the residuals. If, for example, the residuals are highly correlated for a proposed model, then either the model should be improved or the errors must be considered as being correlated. If repeated experiments continue to show high correlation in the residuals, one should examine them to see if there is some characteristic "signature" in the residuals. If there is, one should attempt to improve the model to remove these signatures; if there is no signature one would model the errors as being autoregressive, moving average, etc., processes.

Example 8.9.2

Two models have been proposed for a process in which m different thermocouples have been used to make n measurements each. The assumptions of additive, zero mean, constant variance, independent, normal errors are made. The variance is unknown, there are no errors in the independent variables, and there is no prior information. (These assumptions are designated 11111011.) Find the likelihood ratio.

Solution

The parameters for the ith model are found by maximizing the natural logarithm of the joint probability density function (pdf) of independent normal errors with respect to $\boldsymbol{\beta}^{(i)}$. The maximum value of the pdf is

$$L^{(i)}=(2\pi)^{-mn/2}\sigma^{-mn}\exp\left(-\frac{R^{(i)}}{2\sigma^2}\right)$$

$$R^{(i)}=\sum_{j=1}^{m}\sum_{k=1}^{n}\left[Y_{jk}-\eta_{jk}^{(i)}(\mathbf{b}^{(i)})\right]^2$$

provided $L^{(i)}$ is also maximized with respect to σ^2 which leads to

$$\sigma^{(i)}=\left(\frac{R^{(i)}}{mn}\right)^{1/2}$$

Then $L^{(i)}$ becomes

$$L^{(i)}=(2\pi)^{-mn/2}(\sigma^{(i)})^{-mn}\exp\left(\frac{-mn}{2}\right)$$

and thus the likelihood ratio is

$$\frac{L^{(1)}}{L^{(2)}}=\left(\frac{\sigma^{(2)}}{\sigma^{(1)}}\right)^{mn}=\left(\frac{R^{(2)}}{R^{(1)}}\right)^{mn/2}$$

After obtaining this ratio, we can determine to a given confidence whether Model 1 or Model 2 is to be accepted using the procedure described above. Before accepting a given model, one should investigate if the postulated assumptions are actually reasonable.

REFERENCES

1. Box, G. E. P. and Lucas, H. L., "Design of Experiments in Nonlinear Situations," *Biometrika* **46** (1959), 77–90.
2. Box, G. E. P. and Hunter, W. G., "Non-sequential Designs for the Estimation of Parameters in Nonlinear Models," Tech. Rep. No. 28, University of Wisconsin, Dept. of Statistics, Madison, Wis., 1964.
3. Fedorov, V. V., *Teoriya Optimal'nogo Eksperimenta*, Izdatel'stvo Moskovskogo Universiteta, 1969, translated by W. J. Studden and E. M. Klimko, *Theory of Optimal Experiments*, Academic Press, Inc., New York, 1972.
4. Badavas, P. C. and Saridis, G. W., "Response Identification of Distributed Systems with Noisy Measurements at Finite Points," *Inf. Sci.* **2** (1970), 19–34.
5. McCormack, D. J. and Perlis, H. J., "The Determination of Optimum Measurement Locations in Distributed Parameter Processes," Proceedings of the 3rd Annual Princeton Conference on Information Sciences and Systems, 1969, pp. 510–518.
6. Nahi, N. E., *Estimation Theory and Applications*, John Wiley and Sons, Inc., New York, 1964.
7. Smith, K., "On the Standard Deviations of Adjusted and Interpolated Values of an Observed Polynomial Function and its Constants and the Guidance they give Towards a Proper Choice of the Distribution of Observations," *Biometrika* **12** (1918), 1–85.
8. Atkinson, A. C. and Hunter, W. G., "The Design of Experiments for Parameter Estimation," *Technometrics*, **10** (1968), 271–289.
9. Carslaw, H. S. and Jaeger, J. C., *Conduction of Heat in Solids*, 2nd ed., Oxford University Press, London, 1959.
10. Beck, J. V., "The Optimum Analytical Design of Transient Experiments for Simultaneous Determinations of Thermal Conductivity and Specific Heat," Ph.D. Thesis, Dept. of Mechanical Engineering, Michigan State University, 1964.
11. Heineken, F. G., Tsuchiya, H. M. and Aris, R., "On the Accuracy of Determining Rate Constants in Enzymatic Reactions," *Math. Biosci.* **1** (1967), 115–141.
12. Seinfeld, J. H. and Lapidus, L., *Mathematical Methods in Chemical Engineering Vol. 3 Process Modeling, Estimation, and Identification*, Prentice-Hall, Inc., Englewood Cliffs, N.J., 1974.
13. Van Fossen, G. J., Jr., "Design of Experiments for Measuring Heat-Transfer Coefficients with a Lumped-Parameter Calorimeter," *NASA TN D-7857*, 1975.
14. Beck, J. V., "Analytical Determination of Optimum Transient Experiments for Measurement of Thermal Properties," *Proc. 3rd Int. Heat Transfer Conf.* **44** (1966), 74–80.
15. Beck, J. V., "Transient Sensitivity Coefficients for the Thermal Contact Conductance," *Int. J. Heat Mass Transfer* **10** (1967), 1615–1617.
16. Beck, J. V., "Determination of Optimum Treatment Experiments for Thermal Contact Conductance," *Int. J. Heat Mass Transfer* **12** (1969), 621–633.
17. Bonacina, C. and Comini, G., "Calculation of Convective Heat Transfer Coefficients for Time-Temperature Curves," *Int. Inst. Refrig. Freudenstadt* (1972), 157–167.
18. Comini, G., "Design of Transient Experiments for Measurements of Convective Heat Transfer Coefficients," *Int. Inst. Refrig. Freudenstadt* (1972), 169–178.
19. Cannon, J. R. and Klein, R. E., "Optimal Selection of Measurement Locations in a Conductor for Approximate Determination of Temperature Distributions," *J. Dyn. Sys. Meas. Control*, **93** (1971), 193–199.

20. Seinfeld, J. H., "Optimal Location of Pollutant Monitoring Stations in an Airshed," *Atmos. Environ.* **6** (1972), 847–858.
21. Beck, J. V., "Analytical Determination of High Temperature Thermal Properties of Solids Using Plasma Arcs," *Thermal Conductivity, Proceedings of the Eighth Conference*, 1969.
22. Van Fossen, G. J., Jr., "Model Building Incorporating Discrimination Between Rival Mathematical Models in Heat Transfer," Ph.D. Thesis, Dept. of Mechanical Engineering, Michigan State University, 1973.
23. Hunter, W. G. and Hill, W. J., "Design of Experiments for Subsets of Parameters," Tech. Rep. No. 330, University of Wisconsin, Dept. of Statistics, Madison, Wis., March 1973.
24. Hunter, W. G., Hill, W. J., and Henson, T. L., "Designing Experiments for Precise Estimation of All or Some of the Constants in a Mechanistic Model," *Can. J. Chem. Eng.* **47** (1969), 76–80.
25. Graybill, F. A., *Introduction to Matrices with Applications in Statistics*, Wadsworth Publishing Company, Inc., Belmont, Calif., 1969.
26. Meyers, G. E., *Analytical Methods in Conduction Heat Transfer*, McGraw-Hill Book Company, New York, 1971.
27. Parker, W. J., Jenkins, R. J., Butler, C. P., and Abbott, G. L., "Flash Method of Determining Thermal Diffusivity, Heat Capacity, and Thermal Conductivity," *J. Appl. Phys.* **32** (1961), p. 1679.
28. Kullback, S., *Information Theory and Statistics*, John Wiley and Sons, Inc., New York, 1959.
29. Box, G. E. P. and Hill, W. J., "Discrimination among Mechanistic Models," *Technometrics* **9** (1967), 57–71.
30. Ghosh, B. K. *Sequential Tests of Statistical Hypotheses*. Addison-Wesley, Reading, Mass., 1970.
31. Bard, Y., *Nonlinear Parameter Estimation*, Academic Press, Inc., New York, 1974.
32. Wald, A., *Sequential Analysis*, John Wiley and Sons, Inc., New York, 1947.

APPENDIX 8A OPTIMAL EXPERIMENT CRITERIA FOR ALL PARAMETERS OF INTEREST

For the standard assumptions of additive, zero mean, normal measurement errors in the dependent variable, the joint probability density of the estimated parameter vector **b** is

$$p(\mathbf{b})=(2\pi)^{-p/2}|\mathbf{P}|^{-1/2}\exp\left[-\tfrac{1}{2}(\mathbf{b}-\boldsymbol{\beta})^T\mathbf{P}^{-1}(\mathbf{b}-\boldsymbol{\beta})\right] \qquad (8A.1)$$

where **P** is the covariance matrix of **b**. This expression also assumes errorless independent variables. We also assume that the error covariance matrix ψ is known to within a multiplicative constant σ^2. These assumptions are designated 11--101-. (8A.1) is exact if the dependent variable is linear in the parameters; if η is nonlinear in the parameters, then the expression is approximate.

For the assumptions given above the confidence region can be found from an expression similar to [see (6.8.38)]

$$(\mathbf{b}-\boldsymbol{\beta})^T\mathbf{P}^{-1}(\mathbf{b}-\boldsymbol{\beta})=\text{constant}=C^2 \tag{8A.2}$$

For a given value of C^2 this equation describes a hyperellipsoid which has a hypervolume given by

$$\text{volume}=\pi^{p/2}C(\lambda_1\lambda_2\cdots\lambda_p)^{1/2}\left[\Gamma\left(\frac{p}{2}+1\right)\right]^{-1} \tag{8A.3}$$

where p is the number of parameters, $\Gamma(\cdot)$ is the gamma function, and λ_i is the ith eigenvalue of $\mathbf{P}$. Now the determinant of $\mathbf{P}$ is equal to the product of its eigenvalues. Thus to minimize the hypervolume of a confidence region, the determinant of $\mathbf{P}$ should be minimized. This is equivalent to *maximizing* the determinant of the inverse of $\mathbf{P}$. For the standard assumptions of 11111-11 this leads to the criterion of maximizing $\Delta=|\mathbf{X}^T\mathbf{X}|$, which has been given by Box and Lucas [1]. The criterion of max $|\mathbf{P}^{-1}|$ is more general, however.

Exactly the same criterion can be derived using the Shannon [28] concept of a measure of uncertainty which is related to information theory. He showed that the unique (except for a positive multiplicative factor) suitable measure of uncertainty associated with the probability density function of the random parameter vector $\mathbf{b}$, which is denoted $p(\mathbf{b})$, is given by

$$H(p)\equiv -E(\ln p)=-\int p(\mathbf{b})\ln p(\mathbf{b})\,d\mathbf{b} \tag{8A.4}$$

Information is gained when uncertainty is reduced. Suppose $p_0(\mathbf{b})$ is the prior density of $\mathbf{b}$, that is, resulting from previous experiments. Let $p_1(\mathbf{b})$ be the posterior density after another experiment has been performed. The amount of information I gained by the experiment is [28]

$$I=H(p_0)-H(p_1) \tag{8A.5}$$

Our goal is to select an experiment that maximizes I. Since $H(p_0)$ is unaffected by the new experiment, we simply minimize $H(p_1)$.

Let us evaluate $H(p)$ for the standard assumptions 11--1013. Then $p(\mathbf{b})$ is given by (8A.1) and thus

$$H(p(\mathbf{b}))=-E\left[\ln p(\mathbf{b})\right]=-E\left\{\tfrac{1}{2}\left[-p\ln 2\pi-\ln|\mathbf{P}|-(\mathbf{b}-\boldsymbol{\beta})^T\mathbf{P}^{-1}(\mathbf{b}-\boldsymbol{\beta})\right]\right\}$$

$$=\tfrac{1}{2}\left\{p\ln 2\pi+\ln|\mathbf{P}|+\operatorname{tr}\left[\mathbf{P}^{-1}E(\mathbf{b}-\boldsymbol{\beta})^T(\mathbf{b}-\boldsymbol{\beta})\right]\right\}$$

$$=\tfrac{1}{2}\left\{p\ln 2\pi+\ln|\mathbf{P}|+\operatorname{tr}\left[\mathbf{P}^{-1}\mathbf{P}\right]\right\}=\tfrac{1}{2}\left\{p(1+\ln 2\pi)+\ln|\mathbf{P}|\right\} \tag{8A.6}$$

where p on the right side designates the number of parameters. Discarding irrelevant constants, a measure of uncertainty is

$$H^*[p(\mathbf{b})] \equiv \ln|\mathbf{P}| \tag{8A.7}$$

But minimizing this function is equivalent to maximizing $|\mathbf{P}^{-1}|$ which was given above using the minimum confidence volume approach.

APPENDIX 8B OPTIMAL EXPERIMENT CRITERIA FOR NOT ALL PARAMETERS OF INTEREST

Suppose that of the total number p of the estimated parameters only a subset of them need be estimated accurately. Let the estimated parameter vector $\mathbf{b}$ be partitioned into two vectors $\mathbf{b}_1$ and $\mathbf{b}_2$ so that

$$\mathbf{b}^T = [\mathbf{b}_1^T \; \mathbf{b}_2^T] \tag{8B.1}$$

where $\mathbf{b}$ is $p \times 1$, $\mathbf{b}_1$ is $q \times 1$, and $\mathbf{b}_2$ is an r vector where $r = p - q$. The vector $\mathbf{b}_1$ consists of those b's of primary interest and $\mathbf{b}_2$ contains the others. Let the same statistical assumptions denoted by 11--101- and discussed in the beginning of Appendix 8A be valid. Let the covariance matrix of all the estimated parameters be designated $\mathbf{P}$ and be partitioned as

$$\mathbf{P} = \begin{bmatrix} \mathbf{P}_{11} & \mathbf{P}_{12} \\ \mathbf{P}_{21} & \mathbf{P}_{22} \end{bmatrix} \tag{8B.2}$$

where $\mathbf{P}_{11}$ is $q \times q$ and is for the $\mathbf{b}_1$ vector, etc.

For this case the joint probability density of $\mathbf{b}$ is given by (8A.1). If the experimenter desires precise estimates of only $\mathbf{b}_1$, Hunter and Hill [23, 24] state that the marginal distribution of $\mathbf{b}_1$ is then needed. It is obtained by integrating (8A.1) with respect to $\mathbf{b}_2$. From Theorem 10.6.1 of Graybill [25] the marginal probability density of $\mathbf{b}_1$ is

$$p(\mathbf{b}_1) = (2\pi)^{-q/2}|\mathbf{P}_{11}|^{-1/2}\exp\left[-\tfrac{1}{2}(\mathbf{b}_1 - \boldsymbol{\beta}_1)^T \mathbf{P}_{11}^{-1}(\mathbf{b}_1 - \boldsymbol{\beta}_1)\right] \tag{8B.3}$$

Following the same reasoning as in Appendix 8A, the criterion is to maximize

$$\Delta_{qp} = |\mathbf{P}_{11}^{-1}| \tag{8B.5}$$

The terms in Δ_{qp} should be related to the sensitivity matrix. Let $\mathbf{X}$ be partitioned as

$$\mathbf{X} = [\mathbf{X}_1 \; \mathbf{X}_2] \tag{8B.5}$$

where $\mathbf{X}_1$ is $n \times q$ and $\mathbf{X}_2$ is $n \times r$. Then for maximum likelihood estimation, $\mathbf{P} = (\mathbf{X}^T\psi^{-1}\mathbf{X})^{-1}$ where $\mathbf{X}^T\psi^{-1}\mathbf{X}$ can be written as (see(6.1.17a))

$$\mathbf{X}^T\psi^{-1}\mathbf{X} = \begin{bmatrix} \mathbf{X}_1^T\psi^{-1}\mathbf{X}_1 & \mathbf{X}_1^T\psi^{-1}\mathbf{X}_2 \\ \mathbf{X}_2^T\psi^{-1}\mathbf{X}_1 & \mathbf{X}_2^T\psi^{-1}\mathbf{X}_2 \end{bmatrix} \tag{8B.6}$$

Taking the inverse of (8B.6) and identifying the upper left matrix as $\mathbf{P}_{11}$ results in the criterion being to maximize

$$\Delta_{qp} = |X_1^T\psi^{-1}\mathbf{X}_1 - \mathbf{X}_1^T\psi^{-1}\mathbf{X}_2(\mathbf{X}_2^T\psi^{-1}\mathbf{X}_2)^{-1}\mathbf{X}_2^T\psi^{-1}\mathbf{X}_1| \tag{8B.7}$$

If the errors are independent and have a constant variance (i.e., 11001-11), this expression reduces to the one given by Hunter and Hill [23,24] which is

$$\Delta_{qp} = |\mathbf{X}_1^T\mathbf{X}_1 - \mathbf{X}_1^T\mathbf{X}_2(\mathbf{X}_2^T\mathbf{X}_2)^{-1}\mathbf{X}_2^T\mathbf{X}_1| \tag{8B.8}$$

Using (6.1.17), Δ_{qp} given by (8B.7) can be related to the usual Δ by

$$\Delta_{pq} = \frac{\Delta_p}{|\mathbf{X}_2^T\psi^{-1}\mathbf{X}_2|} \tag{8B.9}$$

where Δ_p is the determinant of the expression given by (8B.6).

PROBLEMS

Unless otherwise stated, assume that the standard conditions designated 11111-11 are valid for the following problems.

8.1 For $X = te^{-t+1}$ show that Δ^n given by (8.2.3) becomes

$$\Delta^n = e^2\left[t_n^{-1} - e^{-2t_n}(2 + t_n^{-1} + 2t_n)\right]/4$$

Verify that at $t_n = 1.691817$, $d\Delta^n/dt_n = 0$. At the same value of t_n show that the sufficient condition for a maximum, $d^2\Delta^n/dt_n^2 < 0$, is also satisfied.

8.2 For $\eta = \beta C \sin t$ show that Δ^+ given by (8.2.13) becomes (for $t_n > \pi/2$)

$$\Delta^+ = \frac{1}{2}\left[1 - \frac{1}{2t_n}\sin 2t_n\right]$$

Also show that Δ^+ has extrema when $\tan\tau = \tau$ is satisfied. Use Myers [26, p. 442] to find the first three nonzero positive roots of $\tan\tau = \tau$.

8.3 Derive (8.2.14).

8.4 Consider the model $\eta_i = \beta f(t_i)$ where $f(t_i)$ can assume only the values indicated below. The optimal conditions for estimating β are needed.

i	$f(t_i)$	i	$f(t_i)$	i	$f(t_i)$
1	0	5	2	9	−2
2	1	6	1	10	−3
3	2	7	0	11	−4
4	2.5	8	−1	12	−3

(*a*) What single i should be chosen if only one measurement could be taken?
(*b*) What i value(s) should be selected if two observations are to be taken? Repeated observations at any t_i are permitted.
(*c*) Same as (*b*) except repeated observations are not permitted.
(*d*) What three i values would be selected if repeated i values are not allowed?

8.5 For the model $\eta_i = \beta_1 f_1(t_i) + \beta_2 f_2(t_i)$ the below discrete values are permitted

i	$f_1(t_i)$	$f_2(t_i)$
1	10	2
2	8	4
3	6	5
4	3	0
5	2	−4

(*a*) What are the two optimum locations to take measurements?
(*b*) What are the best three locations to take observations? Repeated values are not permitted.
(*c*) Same as (*b*) except repeated values are permitted.

8.6 For the model $\eta = \beta_1 \exp(-\beta_2 t)$ verify that the optimal locations for $n=4$ are at $\beta_2 t = 0$ and 1. There are no constraints on η or t. Study the region $0 < t < 1.2$ using the spacing of $\Delta t = 0.1$. Use a programmable calculator or a computer.

8.7 Find the optimal two values of $t^+ = \beta_2 t$ for estimating β_1 and β_2 in the model $\eta = \beta_1 \sin \beta_2 t$. There are no constraints on η or t.

8.8 Find the optimal value of $t_n^+ = \beta_2 t_n$ for a large number of uniformly spaced measurements in $0 < t < t_n$ for the model $\eta = \beta_1 \sin t^+$. Use a computer if necessary. No constraints are to be used on η or t.

8.9 For the model of the cooling billet, $T = T_\infty + (T_0 - T_\infty)\exp(-\beta t)$, find the optimal duration of the experiment for a large number of equally spaced measurements. The parameters are T_0, T_∞, and β.

8.10 For the model $\eta = [\beta_1/\beta_1 - \beta_2)][\exp(-\beta_2 t) - \exp(-\beta_1 t)]$ find expressions for the β_1 and β_2 sensitivity coefficients. See (8.3.19).

8.11 Find general expressions for the sensitivity coefficients plotted in Figs. 8.15 and 8.16.

8.12 A plate which is subjected to a large instantaneous pulse of energy Q at $x=0$ and is insulated at $x=L$ has the solution for the temperature of

$$T(x,t)-T_0=\frac{Q}{cL}\left[1+2\sum_{n=1}^{\infty} e^{-n^2 t^+}\cos\frac{n\pi x}{L}\right]$$

where $t^+=\pi^2\alpha\, t/L^2$, c is the density–specific heat product, and Q has units of energy (Btu or J) per unit area. For $x=0$ the temperature is infinity at time zero and decays to T_0+Q/cL for large time. At $x=L$ the temperature starts at T_0 and increased to T_0+Q/cL.

(*a*) Find an expression for the α sensitivity at $x/L=1$.
(*b*) Evaluate using a computer the expression found in (*a*) for $0<t^+<3$. For a fixed value of Q (and no restriction on the range of T) show that the optimum time to take a single measurement is $t^+=1.38$. Also show that this time corresponds to the time that the temperature at $x=L$ has reached one half of the maximum temperature rise. This "one-half" time is the basis of finding α in pulse or flash experiments. See the paper by Parker, Jenkins, Butler, and Abbott [27].
(*c*) Also using a computer find the optimum experiment duration for many equally spaced measurements at $x/L=1$.

8.13 (*a*) A large number of measurements uniformly spaced in time have been made at $x=0$ and $x=L$ in the heat conducting body discussed in Section 8.5.2.2. For m_0 and m_1 sensors at $x=0$ and L, respectively, show that Δ^+ given by (8.3.7) can be written as

$$\Delta^+=\left[zC_{11,0}^++(1-z)C_{11,1}^+\right]\left[zC_{22,0}^++(1-z)C_{22,1}^+\right]$$
$$-\left[zC_{12,0}^++(1-z)C_{12,1}^+\right]^2$$

where $z=m_0/m$ and $1-z=m_1/m$ and where

$$C_{ij}^+=C_{ij,0}^++C_{ij,1}^+$$

The third subscript in $C_{ij,0}^+$ or $C_{ij,1}^+$ refers to $x=0$ or L, respectively. The standard statistical assumptions are valid.
(*b*) Derive an expression for z at which Δ^+ is a maximum, assuming that z can assume any value in the interval 0 to 1.
(*c*) The following values are for the heat conducting body discussed in Section 8.5.2.2:

$C_{11,0}^+=0.07609$ $\quad C_{12,0}^+=0.1062$ $\quad C_{22,0}^+=0.1552$

$C_{11,1}^+=0.0148$ $\quad C_{12,1}^+=-0.0422$ $\quad C_{22,1}^+=0.126$

The values correspond to the dimensionless time $t_n^+=0.65$. The first two subscripts correspond to k (a 1 subscript) or c (a 2 subscript). Using the expression derived in part (*b*), find a value for z.
(*d*) What conclusions can you draw from the results of this problem?

APPENDIX A

Identifiability condition

A.1 INTRODUCTION

The problem of investigating the conditions under which parameters can be uniquely estimated is called the identifiability problem. A convenient means of anticipating slow convergence or even nonconvergence in estimating parameters can save unnecessary time and expense. Also if easy-to-apply identifiability conditions are known, many times insight can be provided to avoid the problem of nonidentifiability, through either the use of a different experiment or a smaller set of parameters that are identifiable.

The purpose of this appendix is to derive the identifiability criterion that the sensitivity coefficients in the *neighborhood* of the minimum sum of squares function must be linearly independent over the range of the measurements. This criterion applies for linear and *nonlinear* estimation. This criterion is derived only for a weighted sum of squares function which includes least squares, weighted least squares, and ML estimation with normal errors, in each case with no constraints on the parameters. For MAP estimation with prior parameter information it *might* be possible to estimate the parameters even if the sensitivity coefficients are linearly dependent.

This condition of independence of the sensitivity coefficients is particularly convenient if the number of the parameters is not large, say, less than six. Even if the number is larger, linear dependence between two or three of the parameters can sometimes be readily detected from graphs of the sensitivity coefficients. The *plotting* of the coefficients is extremely important and should be done for each new problem before attempting to estimate the parameter.

A.2 THEORY

Consider a general sum of squares function for measurements given by

$$S=\sum_{u=1}^{n}\sum_{v=1}^{n}(Y_u-\eta_u)w_{uv}(Y_v-\eta_v) \tag{A.1}$$

where w_{uv} is an element of $\mathbf{W}$, a square, symmetric, positive-definite matrix. Let the function S possess continuous derivatives in the neighborhood of its minimum in the parameter space which occurs when $\boldsymbol{\eta}$ is evaluated at $\boldsymbol{\beta}^*$,

$$(\boldsymbol{\beta}^*)^T=[\,\beta_1^*\ \ \beta_2^*\ \cdots\ \beta_p^*\,] \tag{A.2}$$

A Taylor series expansion of S in the neighborhood of its minimum is

$$S(\boldsymbol{\beta})=S(\boldsymbol{\beta}^*)+\sum_{i=1}^{p}S_{\beta_i}^*(\beta_i-\beta_2^*)$$

$$+\sum_{i=1}^{p}\sum_{j=1}^{p}S_{\beta_i\beta_j}^*(\beta_i-\beta_i^*)(\beta_j-\beta_j^*)+\cdots \tag{A.3a}$$

where

$$S_{\beta_i}^*\equiv\frac{\partial S(\boldsymbol{\beta}^*)}{\partial\beta_i},\qquad S_{\beta_i\beta_j}^*\equiv\frac{\partial^2 S(\boldsymbol{\beta}^*)}{\partial\beta_i\,\partial\beta_j} \tag{A.3b}$$

Using S defined by (A.1) in (A.3b) gives

$$S_{\beta_i}^*=-2\sum_{u=1}^{n}\sum_{v=1}^{n}w_{uv}(Y_v-\eta_v^*)X_{ui}^*;\qquad X_{ui}\equiv\frac{\partial\eta_u}{\partial\beta_i} \tag{A.4a}$$

$$S_{\beta_i\beta_j}^*=2\sum_{u=1}^{n}\sum_{v=1}^{n}w_{uv}\left[X_{vi}^*X_{uj}^*-(Y_v-\eta_v^*)X_{uij}^*\right] \tag{A.4b}$$

$$X_{uij}\equiv\frac{\partial^2\eta_u}{\partial\beta_i\,\partial\beta_j} \tag{A.4c}$$

The expression X_{ui} in (A.4a) is called a sensitivity coefficient.

For a model linear in the parameters the cross-derivative X_{uij} in (A.4b) is equal to zero as are also the third and higher order derivatives of S. Note that the condition of continuous derivatives of S with respect to $\boldsymbol{\beta}$ will be satisfied if $\boldsymbol{\eta}$ and its derivatives are continuous functions of $\boldsymbol{\beta}$. A necessary condition for S to possess a minimum at $\boldsymbol{\beta}^*$ is that

$$S_{\beta_i}^*=0\qquad\text{for } i=1,2,\ldots,p \tag{A.5}$$

Define the determinant D_r as

$$D_r \equiv \begin{vmatrix} S^*_{\beta_1\beta_1} & S^*_{\beta_1\beta_2} & \cdots & S^*_{\beta_1\beta_r} \\ S^*_{\beta_2\beta_1} & S^*_{\beta_2\beta_2} & \cdots & S^*_{\beta_2\beta_r} \\ \vdots & \vdots & & \vdots \\ S^*_{\beta_r\beta_1} & S^*_{\beta_r\beta_2} & \cdots & S^*_{\beta_r\beta_r} \end{vmatrix} \tag{A.6}$$

Then S, approximated by the terms explicitly given by (A.3a), has a unique local minimum if in addition to (A.5) being true it is also true that

$$D_r > 0 \qquad \text{for } r = 1, 2, \ldots, p \tag{A.7}$$

which is the condition that D_r be positive-definite; see reference 1. A minimum can exist with weaker conditions, however. For example, if $D_p = 0$ a minimum may exist but it may not be unique; that is, the minimum could be along a line rather than at a point. The conditions given by (A.5) and (A.7) are necessary and sufficient conditions for a unique local minimum.

We wish to relate the conditions of $D_r > 0$ and $D_r = 0$ to the sensitivity coefficients.

Let us define

$$\Delta\beta_i^+ \equiv (\beta_i - \beta_i^*)/\beta_i^*, \qquad S_{ij}^+ \equiv \beta_i^* \beta_j^* S^*_{\beta_i\beta_j} \tag{A.8a}$$

$$X_{ui}^+ \equiv \beta_i^* X_{ui}^*, \qquad X_{uij}^+ \equiv \beta_i^* \beta_j^* X_{uij}^* \tag{A.8b}$$

Then using $S^*_{\beta_i} = 0$ for all i, (A.3a) can be written

$$S(\boldsymbol{\beta}) - S(\boldsymbol{\beta}^*) \cong \tfrac{1}{2} \sum_{i=1}^{p} \sum_{j=1}^{p} S_{ij}^+ \Delta\beta_i^+ \Delta\beta_j^+ \tag{A.9}$$

where (A.8a) and (A.8b) are employed in S_{ij}^+,

$$S_{ij}^+ = 2 \sum_{u=1}^{n} \sum_{v=1}^{n} w_{uv} \left[X_{ui}^+ X_{vj}^+ - (Y_v - \eta_v^*) X_{uij}^+ \right] \tag{A.10}$$

Now (A.9) is a quadratic form and if a unique minimum is to exist it is necessary that all the determinants D_r^+

$$D_r^+ = \begin{vmatrix} S_{11}^+ & \cdots & S_{1r}^+ \\ \vdots & & \vdots \\ S_{r1}^+ & \cdots & S_{rr}^+ \end{vmatrix}, \qquad r = 1, 2, \ldots, p \tag{A.11}$$

be greater than zero. Suppose that a minimum exists at $\boldsymbol{\beta}^*$ but that the minimum is not unique, for example, it exists along a line or in a plane. This results in $D_r^+=0$ for some r.

Suppose first that the term in (A.10) involving X_{uij}^+ is negligible in its contribution to S_{ij}^+. Notice that this term becomes negligible as the residuals, $Y_v-\eta_v^*$, become small but this is not true for the $X_{ui}^+X_{vj}^+$ terms. (For linear-in-the-parameters cases, X_{uij}^+ is always zero.) Furthermore, assume that

$$\begin{aligned} w_{uv}&=\sigma_u^{-2} \qquad u=v \text{ and } \sigma_u^2\neq 0 \\ w_{uv}&=0 \qquad u\neq v \end{aligned} \tag{A.12}$$

and then S_{ij}^+ given by (A.10) becomes

$$S_{ij}^+=\sum_{u=1}^{n}\left(\sigma_u^{-1}X_{ui}^+\right)\left(\sigma_u^{-1}X_{uj}^+\right) \tag{A.13}$$

The summation in (A.13) can be considered to form an inner product involving vectors $\mathbf{a}_i$, $i=1,2,\ldots,r$:

$$\mathbf{a}_i=\begin{bmatrix}\sigma_1^{-1}X_{1i}^+\\ \vdots \\ \sigma_n^{-1}X_{ni}^+\end{bmatrix} \tag{A.14}$$

Use this interpretation in (A.13) and introduce (A.13) into (A.11) to get

$$D_r^+=\begin{vmatrix}\mathbf{a}_1^T\mathbf{a}_1 & \cdots & \mathbf{a}_1^T\mathbf{a}_r\\ \vdots & & \vdots\\ \mathbf{a}_r^T\mathbf{a}_1 & \cdots & \mathbf{a}_r^T\mathbf{a}_r\end{vmatrix} \tag{A.15}$$

which can be considered a Gram determinant of $\mathbf{a}_1,\mathbf{a}_2,\ldots,\mathbf{a}_r$. It is known that D_r^+ is equal to zero if and only if the vectors $\mathbf{a}_i$ are linearly dependent, which means

$$C_1'\sigma_k^{-1}X_{k1}+C_2'\sigma_k^{-1}X_{k2}^+ +\cdots+C_r'\sigma_k^{-1}X_{kr}^+=0 \tag{A.16a}$$

or

$$C_1X_{k1}+C_2X_{k2}+\cdots+C_rX_{kr}=0 \tag{A.16b}$$

for $k=1,2,\ldots,n$ and for not all C_i being equal to zero. In other words, *if the (continuous) sensitivity coefficients are linearly dependent in the neighborhood of the minimum there is no unique minimum and all the* r *parameters cannot be simultaneously and uniquely estimated*. This is the desired relation. Note, however, that this result assumes that the term involving X_{uij}^+ in (A.10) can be dropped; w_{uv} is given by (A.12); there is no prior information; and there are no parameter constraints.

Suppose (A.16) is written in the form

$$\sum_{j=1}^{r} C_j X_{uj}^{+} = 0, \qquad u = 1,2,\ldots,n; \quad r = 1,2,\ldots,p \tag{A.17}$$

where at least one C_j is not equal to zero. Also form the summation involving a row in (A.11)

$$\sum_{j=1}^{r} S_{ij}^{+} C_j = \sum_{j=1}^{r} \sum_{u=1}^{n} \sum_{v=1}^{n} w_{uv}\left[X_{ui}^{+} X_{vj}^{+} - (Y_v - \eta_v^*) X_{uij}^{+}\right] C_j$$

$$= \sum_{u=1}^{n} \sum_{v=1}^{n} w_{uv}\left[X_{ui}^{+} \sum_{j=1}^{r} C_j X_{uj}^{+} - (Y_v - \eta_v^*) \sum_{j=1}^{r} C_j X_{uij}^{+}\right] \tag{A.18}$$

Differentiating (A.17) with respect to β_i yields

$$\sum_{j=1}^{r} C_j X_{uij}^{+} = 0, \qquad u = 1,\ldots,n; \quad i = 1,\ldots,p; \quad r = 1,\ldots,p \tag{A.19}$$

Using (A.17) and (A.19) in (A.18) then produces

$$\sum_{j=1}^{r} S_{ij}^{+} C_j = 0, \qquad r = 1,\ldots,p; \quad i = 1,\ldots,p \tag{A.20}$$

We have shown for linear dependence of the sensitivity coefficients, (A.17), that a given column of the square matrix D_r^{+} given by (A.11) can be considered to be a linear combination of the other columns. But if any column of a square matrix is a linear combination of the other columns, then the determinant of that matrix is zero. Consequently, the sum of squares function S does not have a unique minimum in the $\beta_1, \beta_2, \ldots, \beta_r$ space and thus not all of these parameters can be uniquely determined.

The results given above apply for r equal to 1 to p parameters. Note, however, if the linear dependence condition of the sensitivities given by (A.17) is satisfied for $r < p$, then it is also satisfied for $r+1, r+2, \ldots, p$ because $C_{r+1}, C_{r+2}, \ldots, C_p$ can be set equal to zero in (A.17).

A.3 COMMENTS

(*a*) Parameters cannot all be uniquely estimated for η being linear or nonlinear in the parameters if the sensitivity coefficients are linearly dependent over the range of the measurements. This is true if (i) the S function is formed by some weighted least square function, (ii) the sensitivities are continuous functions of the parameters, (iii) there is no prior information regarding the parameters, and (iv) there are no constraints on the parameters.

(*b*) If the X_{uij}^{+} term in (A.10) is negligible or is dropped, the determinant of D_p^{+} is proportional to $|\mathbf{X}^T\mathbf{W}\mathbf{X}|$ which must not be zero when attempting to estimate parameters using the Gauss method discussed in Section 7.4. Hence, using the Gauss method, *none* of the parameters can be estimated if the sensitivities are linearly dependent. Other methods might permit one to obtain certain parameters, but not all since there would be no unique minimum of S if the conditions in (*a*) above are true.

(*c*) Regardless of the form of W if $|\mathbf{X}^T\mathbf{X}|=0$ it is also true that $|\mathbf{X}^T\mathbf{W}\mathbf{X}|=0$. Hence if the sensitivities are linearly dependent, there is no choice of $\mathbf{W}$ possible that will cause $|\mathbf{X}^T\mathbf{W}\mathbf{X}|$ to be not zero.

(*d*) If $|\mathbf{X}^T\mathbf{X}|\neq 0$, then $|\mathbf{X}^T\mathbf{W}\mathbf{X}|$ may or may not be equal to zero. But if ML estimation is used as mentioned in (*b*), $|\mathbf{X}^T\mathbf{W}\mathbf{X}|$ would not be zero if $|\mathbf{X}^T\mathbf{X}|\neq 0$.

A.4 RELATION TO EIGENVALUES

The determinant D_r^+ is numerically equal to the product of its eigenvalues $\lambda_1, \lambda_2, \ldots, \lambda_r$. Now the matrix on the right side of (A.11) is real and symmetric which results in all the eigenvalues being real. Also since

$$D_r^+ = \lambda_1 \lambda_2 \cdots \lambda_r \tag{A.21}$$

D_r^+ will be then equal zero if and only if at least one of the λ_i values is equal to zero.

Since S_{ij}^+ is normalized so that the scale of the λ_i values (or the choice of their units) is unimportant, the relative magnitudes of the λ_i values is significant. If one value is much smaller than the others (but not zero), the $\mathbf{X}^T\mathbf{W}\mathbf{X}$ matrix is probably ill-conditioned and the minimum of S is not well-defined. This would occur when there is "almost" linear dependence of the sensitivity coefficients. In such cases there will be relatively large inaccuracy (large variances) in the parameters.

Consider the case of two parameters and then the eigenvalues λ_1 and λ_2 in

$$\begin{vmatrix} S_{11}^+ - \lambda & S_{12}^+ \\ S_{12}^+ & S_{22}^+ - \lambda \end{vmatrix} = 0 \tag{A.22}$$

are

$$\lambda_1, \lambda_2 = \tfrac{1}{2}\left\{ S_{11}^+ + S_{22}^+ \pm \left[(S_{11}^+ + S_{22}^+)^2 - 4\Delta \right]^{1/2} \right\}; \qquad \Delta \equiv S_{11}^+ S_{22}^+ - S_{12}^{+2} \tag{A.23}$$

where Δ is the determinant in (A.22) with λ set equal to zero. Let λ_1 be the smaller eigenvalue. The the ratio of the eigenvalues, λ_1/λ_2, is always between zero and one; λ_1/λ_2 is given by

$$\frac{\lambda_1}{\lambda_2} = \frac{1-(1-\xi)^{1/2}}{1+(1-\xi)^{1/2}} \tag{A.24}$$

where

$$\xi \equiv \frac{4\Delta}{(S_{11}^{+} + S_{22}^{+})^2} \tag{A.25}$$

which is also limited to between zero and one. For small ξ it can be demonstrated that

$$\frac{\lambda_1}{\lambda_2} \approx \frac{\xi}{4} \tag{A.26}$$

Only for $\xi = 1$ does λ_1/λ_2 equal unity; ξ equals one only when $S_{12}^{+} = S_{21}^{+} = 0$ and $S_{11}^{+} = S_{22}^{+}$.

The above analysis suggests for more than two parameters that the criterion of small

$$\xi_p = \frac{|\mathbf{X}^{+T}\mathbf{W}\mathbf{X}^{+}|}{\left[p^{-1}\operatorname{tr}(\mathbf{X}^{+T}\mathbf{W}\mathbf{X}^{+})\right]^p} \tag{A.27}$$

could be used to see if there is near linear dependence of the sensitivity coefficients. (Note that if the X_{uij}^{+} term in (A.10) can be dropped, the components of $(\mathbf{X}^{+})^T\mathbf{W}\mathbf{X}^{+}$ are given by S_{ij}^{+}.) When ξ_p goes to zero, at least one eigenvalue is equal to zero; the maximum value of ξ_p is unity.

Thus in addition to plotting the sensitivity coefficients, one could examine ξ_p to see if it is near zero. If it is, the experiment is poorly designed and one or more of the parameters should not be estimated, but rather certain groups of parameters. If possible, the experiment should be redesigned so that ξ_p is not so small. However, the recommended criterion for accomplishing this is not ξ_p, but rather the numerator of (A.27), subject to certain constraints. See Chapter 8.

REFERENCES

1. G. S. G. Beveridge and R. S. Schechter, *Optimization: Theory and Practice*, McGraw-Hill Book Company, New York, 1970, p. 217.

APPENDIX B

Appendix B

Estimators and Covariances[a] for Various Estimation Methods for the Linear Model $\eta = X\beta$

Name of Estimator	Assumptions Used	Estimator	Covariance Matrix of **b**	$\text{cov}(\hat{\mathbf{Y}})$ for $\hat{\mathbf{Y}} = \mathbf{X}_1\mathbf{b}$	Usual Estimator for Unknown σ^2 For $\psi = \sigma^2\Omega$, Ω Known
Ordinary	--------	$(\mathbf{X}^T\mathbf{X})^{-1}\mathbf{X}^T\mathbf{Y}$			
Least Squares	11----11	Same as above	$(\mathbf{X}^T\mathbf{X})^{-1}\mathbf{X}^T\psi\mathbf{X}(\mathbf{X}^T\mathbf{X})^{-1}$	$\mathbf{X}_1(\mathbf{X}^T\mathbf{X})^{-1}\mathbf{X}^T\psi\mathbf{X}(\mathbf{X}^T\mathbf{X})^{-1}\mathbf{X}_1^T$	
(OLS)	1111--11	Same as above	$\sigma^2(\mathbf{X}^T\mathbf{X})^{-1}$	$\sigma^2\mathbf{X}_1(\mathbf{X}^T\mathbf{X})^{-1}\mathbf{X}_1^T$	$s^2 = (\mathbf{Y}-\hat{\mathbf{Y}})^T(\mathbf{Y}-\hat{\mathbf{Y}})/(n-p)$ where $\hat{\mathbf{Y}} = \mathbf{X}\mathbf{b}_{LS}$
Maximum	11--1111	$(\mathbf{X}^T\psi^{-1}\mathbf{X})\mathbf{X}^T\psi^{-1}\mathbf{Y}$	$(\mathbf{X}^T\psi^{-1}\mathbf{X})^{-1}$	$\mathbf{X}_1(\mathbf{X}^T\psi^{-1}\mathbf{X})^{-1}\mathbf{X}_1^T$	(σ^2 assumed known)
Likelihood	11--1011	$(\mathbf{X}^T\Omega^{-1}\mathbf{X})^{-1}\mathbf{X}^T\Omega^{-1}\mathbf{Y}$	$\sigma^2(\mathbf{X}^T\Omega^{-1}\mathbf{X})^{-1}$	$\sigma^2\mathbf{X}_1^T(\mathbf{X}^T\Omega^{-1}\mathbf{X})^{-1}\mathbf{X}_1^T$	$s^2 = (\mathbf{Y}-\hat{\mathbf{Y}})^T\Omega^{-1}(\mathbf{Y}-\hat{\mathbf{Y}})/(n-p)$
(ML)					where $\hat{\mathbf{Y}} = \mathbf{X}\mathbf{b}_{ML}$
Gauss–	11---011	$(\mathbf{X}^T\Omega^{-1}\mathbf{X})^{-1}\mathbf{X}^T\Omega^{-1}\mathbf{Y}$	$\sigma^2(\mathbf{X}^T\Omega^{-1}\mathbf{X})^{-1}$	$\sigma^2\mathbf{X}_1(\mathbf{X}^T\Omega^{-1})\mathbf{X})^{-1}\mathbf{X}_1^T$	$s^2 = (\mathbf{Y}-\hat{\mathbf{Y}})^T\Omega^{-1}(\mathbf{Y}-\hat{\mathbf{Y}})/(n-p)$
Markov					where $\hat{\mathbf{Y}} = \mathbf{X}\mathbf{b}_{G\text{-}M}$
Maximum a posteriori (MAP)	11--1112	$\boldsymbol{\mu} + \mathbf{P}_{MAP}\mathbf{X}^T\psi^{-1}(\mathbf{Y}-\mathbf{X}\boldsymbol{\mu})$ where $\mathbf{P}_{MAP} = [\mathbf{X}^T\psi^{-1}\mathbf{X}+\mathbf{V}^{-1}]^{-1}$	$\text{cov}(\mathbf{b}-\boldsymbol{\beta}) = \mathbf{P}_{MAP}$	$\mathbf{X}_1\mathbf{P}_{MAP}\mathbf{X}_1^T$	(σ^2 assumed known)
	11--1012	$\boldsymbol{\mu} + \mathbf{P}_{MAP}^{\Omega}\mathbf{X}^T\Omega^{-1}(\mathbf{Y}-\mathbf{X}\boldsymbol{\mu})$ where $\mathbf{P}_{MAP}^{\Omega} = [\mathbf{X}^T\Omega^{-1}\mathbf{X}+\hat{\sigma}^2\mathbf{V}^{-1}]^{-1}$	$\text{cov}(\mathbf{b}-\boldsymbol{\beta}) = \sigma^2\mathbf{P}_{MAP}^{\Omega}$	$\sigma^2\mathbf{X}_1\mathbf{P}_{MAP}^{\Omega}\mathbf{X}_1^T$	$\hat{\sigma}^2 = (\mathbf{Y}-\hat{\mathbf{Y}})^T\Omega^{-1}(\mathbf{Y}-\hat{\mathbf{Y}})/n$ where $\hat{\mathbf{Y}} = \mathbf{X}\mathbf{b}_{MAP}$ (Note iteration is required for $\hat{\sigma}^2$).

[a] $\psi = \text{cov}(\boldsymbol{\varepsilon})$ for assumptions denoted 11------. (See Section 6.1.5 or inside rear cover for list of standard assumptions).

APPENDIX **C**

LIST OF SYMBOLS

ENGLISH SYMBOLS

a_i	Coefficients in S; see (7.6.3)
$a1$	First-order AR errors
$A_j(i+1)$	Term used in sequential methods, (6.7.8a) and (7.8.23a)
AR	Autoregressive
$\mathbf{b}$	Parameter vector; estimated from estimation equation; $[p \times 1]$
C_{ij}	Component of $\mathbf{X}^T\mathbf{W}\mathbf{X}$; $C_{ij} = \Sigma_k \Sigma_l w_{kl} X_{ki} X_{lj}$, (7.4.13)
$\mathbf{C}$	$\equiv \mathbf{X}^T\mathbf{W}\mathbf{X}$, (7.4.12); $[p \times p]$
$\operatorname{cov}(\cdot,\cdot)$	Covariance, $\operatorname{cov}(A,B) = E\{[A - E(A)][B - E(B)]\}$
$\mathbf{D}$	Matrix used for dependent observations, $\boldsymbol{\varepsilon} = \mathbf{D}\mathbf{u}$; $[n \times n]$
$\mathbf{e}$	Residual vector $= \mathbf{Y} - \hat{\mathbf{Y}}$
$\mathbf{e}$	Eigenvector; Section 6.8
$E(\ )$	Expectation operator
$f(\cdot)$	Probability density
$f(\mathbf{Y}\mid\boldsymbol{\beta})$	Probability density of $\mathbf{Y}$ given $\boldsymbol{\beta}$
$f(\boldsymbol{\beta}\mid\mathbf{Y})$	Probability density of $\boldsymbol{\beta}$ given $\mathbf{Y}$
$\mathbf{F}$	Modified observation vector, $\mathbf{F} = \mathbf{D}^{-1}\mathbf{Y}$ where $\mathbf{D}$ comes from $\boldsymbol{\varepsilon} = \mathbf{D}\mathbf{u}$
$F_{1-\alpha}(p, n-p)$	F statistic associated with $(1-\alpha)100\%$ confidence region and p and $n-p$ degrees of freedom
G	Related to the slope of S, (7.6.8a)
h	Acceleration factor (7.6.1)
$\mathbf{H}$	$\equiv \mathbf{X}^T\mathbf{W}(\mathbf{Y} - \boldsymbol{\eta})$, (7.4.15); $[p \times 1]$
H_i	$= \Sigma_{l=1}^{n} \Sigma_{m=1}^{n} w_{lm} X_{li} (Y_m - \eta_m)$, (7.4.16)
i	Subscript or superscript

j	Subscript
k	Subscript or superscript
$l_{1-\alpha}(p)$	Coefficient for confidence region, Section 6.8
L	Likelihood function, Sections 3.2.5 and 6.1.6
LS	Least squares
m	Number of observations at a given time
MA	Moving average
MAP	Maximum a posteriori
ML	Maximum likelihood
n	Number of observations or number of observation times
p	Number of parameters
$P(\cdot)$	Probability
$\mathbf{P}$	Covariance matrix of estimators; $[p\times p]$
$\mathbf{P}_{LS}$	$=(\mathbf{X}^T\mathbf{X})^{-1}\mathbf{X}^T\boldsymbol{\psi}^{-1}\mathbf{X}(\mathbf{X}^T\mathbf{X})^{-1}$
$\mathbf{P}_{ML}$	$=(\mathbf{X}^T\boldsymbol{\psi}^{-1}\mathbf{X})^{-1}$
$\mathbf{P}_{MAP}$	$=\left(\mathbf{X}^T\boldsymbol{\psi}^{-1}\mathbf{X}+\mathbf{V}_\beta^{-1}\right)^{-1}$
Q	Quadratic form, $\mathbf{A}^T\boldsymbol{\Phi}\mathbf{A}$, (6.1.27)
R	Minimum S; for LS, $R=(\mathbf{Y}-\hat{\mathbf{Y}})^T(\mathbf{Y}-\hat{\mathbf{Y}})$
s	Estimated standard deviation of observation errors, $s=(s^2)^{1/2}$ where $s^2=\hat{\sigma}^2$; for independent constant variance errors, $s^2=R/(n-p)$
S	Sum of squares function, scalar
S_{LS}	Least squares sum of squares, $(\mathbf{Y}-\boldsymbol{\eta})^T(\mathbf{Y}-\boldsymbol{\eta})$
S_{ML}	Maximum likelihood sum of squares; for standard assumptions, $(\mathbf{Y}-\boldsymbol{\eta})^T\boldsymbol{\psi}^{-1}(\mathbf{Y}-\boldsymbol{\eta})$
S_{MAP}	MAP loss function; for standard MAP assumptions, $S_{MAP}=(\mathbf{Y}-\boldsymbol{\eta})^T\boldsymbol{\psi}^{-1}(\mathbf{Y}-\boldsymbol{\eta})+(\boldsymbol{\mu}_\beta-\boldsymbol{\beta})^T\mathbf{V}_\beta^{-1}(\boldsymbol{\mu}_\beta-\boldsymbol{\beta})$
t	Time
$t_{1-\alpha/2}(n-p)$	t statistic associated with $(1-\alpha)$ 100% confidence region and $n-p$ degrees of freedom
u_i	Random component for correlated errors, Appendix 6A. $\boldsymbol{\varepsilon}=\mathbf{D}\mathbf{u}$
$u_{\alpha/2}$	$100(1-\alpha/2)$ percentage point of the normal distribution
$V(\cdot)$	Variance operator; $V(A)=E\left\{\left[A-E(A)\right]^2\right\}$
$\mathbf{V}_b$	Covariance matrix of $\mathbf{b}$, $[p\times p]$
$\mathbf{V}_\beta$	Covariance matrix of $\boldsymbol{\mu}_\beta$, $[p\times p]$ ($\boldsymbol{\mu}_\beta$ is prior vector of $\boldsymbol{\beta}$)
w_{ij}	Component of weighting matrix $\mathbf{W}$
$\mathbf{W}$	Weighting matrix; for ML, $\mathbf{W}=\boldsymbol{\psi}^{-1}$, $[n\times n]$
x	Coordinate or independent variable
$\mathbf{X}$	Sensitivity matrix; $\mathbf{X}=\left(\nabla_\beta\boldsymbol{\eta}^T\right)^T$. If $\boldsymbol{\eta}$ is linear in the parameter as in $\boldsymbol{\eta}=\mathbf{X}\boldsymbol{\beta}, \left(\nabla_\beta\boldsymbol{\eta}^T\right)^T$ reduces to $\mathbf{X}$.

$\mathbf{Y}$	Observation vector, $[n \times 1]$
$\hat{\mathbf{Y}}$	Predicted vector of observations, $[n \times 1]$; for linear case $\hat{\mathbf{Y}} = \mathbf{X}\mathbf{b}$
$\mathbf{Z}$	Modified sensitivity matrix, $\mathbf{Z} = \mathbf{D}^{-1}\mathbf{X}$; $[n \times p]$

GREEK SYMBOLS

α	Associated with confidence interval or region percent confidence; see Section 6.8
α	See Section 7.6 for parameter related to reducing the interval for calculating $S^{(k)}$
$\boldsymbol{\beta}$	Parameter vector, $[p \times 1]$
$\Gamma(\cdot)$	Gamma function; see Section 6.8
Δ	Optimum experiment criterion (see Chapter 8); for standard assumptions of $\mathbf{Y} = \boldsymbol{\eta} + \boldsymbol{\varepsilon}$, $E(\boldsymbol{\varepsilon}) = 0$, $\boldsymbol{\varepsilon}$ with normal density, known independent variable values, $\boldsymbol{\psi}$ known within a multiplicative constant, we have $\Delta = \lvert\mathbf{X}^T\boldsymbol{\psi}^{-1}\mathbf{X}\rvert$
$\boldsymbol{\varepsilon}$	Error vector, $[n \times 1]$; usually $\mathbf{Y} = \boldsymbol{\eta} + \boldsymbol{\varepsilon}$
$\boldsymbol{\eta}$	Expected value vector, regression vector, model vector, $[n \times 1]$
θ	Moving average parameter
λ	Eigenvalue, Section 6.8
$\boldsymbol{\mu}_\beta$	Parameter vector known from prior information $[p \times 1]$
ρ	Autoregressive parameter
ρ	Correlation coefficient; (2.6.17)
σ	Standard deviation of constant variance observation errors
σ^2	Constant variance of observation errors
σ_i^2	Variance of ε_i; $V(\varepsilon_i) = \sigma_i^2$
$\sigma_{u_i}^2$	Variance of u_i; $V(u_i) = \sigma_{u_i}^2$
σ_ε^2	Variance of ε_i, used for the AR case designated $a1$; $\sigma_\varepsilon^2 = \sigma_u^2(1-\rho^2)^{-1}$. See below (6.9.9).
$\boldsymbol{\phi}$	Diagonal matrix; usually $\boldsymbol{\phi} = E(\mathbf{u}\mathbf{u}^T)$ for $E(\mathbf{u}) = \mathbf{0}$ and where $E(u_iu_j) = 0$ for $i \neq j$; $[n \times n]$
χ^2	Chi-squared statistic
$\boldsymbol{\psi}$	Covariance matrix of the observation errors; for $E(\boldsymbol{\varepsilon}) = \mathbf{0}$, $\boldsymbol{\psi} = E(\boldsymbol{\varepsilon}\boldsymbol{\varepsilon}^T)$
$\boldsymbol{\Omega}$	Known part of $\boldsymbol{\psi}$, as in $\boldsymbol{\psi} = \sigma^2\boldsymbol{\Omega}$ where σ^2 is unknown; $[n \times n]$

OTHER SYMBOLS

$\nabla_\beta(\ )$	Matrix derivative operator, $\nabla_\beta = [\partial/\partial\beta_1 \cdots \partial/\partial\beta_p]$

APPENDIX D

SOME ESTIMATION PROGRAMS

In this appendix a few computer programs are referenced. Many others are available. For additional references see Himmelblau [D1, pp. 170, 171, 203], Bard [D2, pp. 323, 324], and Kuester and Mize [D3].

LINEAR ESTIMATION PROGRAMS

LINFIT A linear least squares program with optional constraints to make the parameters nonnegative, add to a constant, etc. This is one of eighteen statistical routines written by J. R. Miller [D4].

LINREG A linear least squares program that is described in reference D3 where an example and the listing are given.

OMNITAB A general purpose computer program for statistical and numerical analysis [D5].

NONLINEAR ESTIMATION PROGRAMS

BARD A nonlinear least squares program that uses the Gauss method [D3, p. 218].

BSOLVE A nonlinear least squares program that uses Marquardt's method [D3].

NLIN IBM Share Program SD 3094 written by Marquardt and others. Written in FORTRAN IV for IBM 7040. Uses Marquardt's method with derivatives or finite difference approximations to solve weighted least squares problems.

NLINA This is a program written at Michigan State University by J. V. Beck and available from him. It uses the sequential and Box–Kanemasu modifications of the Gauss method.

SSQMIN This program uses the Powell procedure and is discussed in reference D3.

REFERENCES

D1. Himmelblau, D. M., *Process Analysis by Statistical Methods*, John Wiley & Sons, Inc., New York, 1970.

D2. Bard, Y., *Nonlinear Parameter Estimation*, Academic Press, Inc., New York, 1974.

D3. Keuster, J. L. and Mize, J. H., *Optimization Techniques With Fortran*, McGraw-Hill Book Co., New York, 1973.

D4. Miller, J. R., *On-Line Analysis for Social Scientists*, MAC-TR-40, Project MAC, Massachusetts Institute of Technology, Cambridge, Mass., 1967.

D5. Hilsenrath, J., Ziegler, G., Messina, C. G., Walsh, P. J., and Herbold, R., *OMNITAB, A Computer Program for Statistical and Numerical Analysis*, Nat. Bur. of Std. Handbook 101, U. S. Government Printing Office, Washington, D. C., 1966. Reissued Jan. 1968, with corrections.

Index

Applied Probability and Statistics (Continued)

GUTTMAN, WILKS and HUNTER · Introductory Engineering Statistics, *Second Edition*

HAHN and SHAPIRO · Statistical Models in Engineering
HALD · Statistical Tables and Formulas
HALD · Statistical Theory with Engineering Applications
HARTIGAN · Clustering Algorithms
HILDEBRAND, LAING and ROSENTHAL · Prediction Analysis of Cross Classifications
HOEL · Elementary Statistics, *Fourth Edition*
HOLLANDER and WOLFE · Nonparametric Statistical Methods
HUANG · Regression and Econometric Methods
JAGERS · Branching Processes with Biological Applications
JOHNSON and KOTZ · Distributions in Statistics
Discrete Distributions
Continuous Univariate Distributions-1
Continuous Univariate Distributions-2
Continuous Multivariate Distributions
JOHNSON and KOTZ · Urn Models and Their Application: An Approach to Modern Discrete Probability Theory
JOHNSON and LEONE · Statistics and Experimental Design: In Engineering and the Physical Sciences, Volumes I and II, *Second Edition*
KEENEY and RAIFFA · Decisions with Multiple Objectives
LANCASTER · The Chi Squared Distribution
LANCASTER · An Introduction to Medical Statistics
LEWIS · Stochastic Point Processes
McNEIL · Interactive Data Analysis
MANN, SCHAFER and SINGPURWALLA · Methods for Statistical Analysis of Reliability and Life Data
MEYER · Data Analysis for Scientists and Engineers
OTNES and ENOCHSON · Digital Time Series Analysis
PRENTER · Splines and Variational Methods
RAO and MITRA · Generalized Inverse of Matrices and Its Applications
SARD and WEINTRAUB · A Book of Splines
SEARLE · Linear Models
THOMAS · An Introduction to Applied Probability and Random Processes
WHITTLE · Optimization under Constraints
WILLIAMS · A Sampler on Sampling
WONNACOTT and WONNACOTT · Econometrics
WONNACOTT and WONNACOTT · Introductory Statistics, *Third Edition*
WONNACOTT and WONNACOTT · Introductory Statistics for Business and Economics, *Second Edition*
YOUDEN · Statistical Methods for Chemists
ZELLNER · An Introduction to Bayesian Inference in Econometrics

Tracts on Probability and Statistics

BHATTACHARYA and RAO · Normal Approximation and Asymptotic Expansions
BILLINGSLEY · Convergence of Probability Measures
JARDINE and SIBSON · Mathematical Taxonomy
RIORDAN · Combinatorial Identities